Developing circular agricultural production systems

It is widely recognised that agriculture is a significant contributor to global warming and climate change. Agriculture needs to reduce its environmental impact and adapt to current climate change whilst still feeding a growing population, i.e. become more 'climate-smart'. Burleigh Dodds Science Publishing is playing its part in achieving this by bringing together key research on making the production of the world's most important crops and livestock products more sustainable.

Based on extensive research, our publications specifically target the challenge of climate-smart agriculture. In this way we are using 'smart publishing' to help achieve climate-smart agriculture.

Burleigh Dodds Science Publishing is an independent and innovative publisher delivering high quality customer-focused agricultural science content in both print and online formats for the academic and research communities. Our aim is to build a foundation of knowledge on which researchers can build to meet the challenge of climate-smart agriculture.

For more information about Burleigh Dodds Science Publishing simply call us on +44 (0) 1223 839365, email info@bdspublishing.com or alternatively please visit our website at www.bdspublishing.com.

Related titles:

Assessing the environmental impact of agriculture
Print (ISBN 978-1-78676-228-3); Online (ISBN 978-1-78676-230-6, 978-1-78676-231-3)

Achieving carbon-negative bioenergy systems from plant materials
Print (ISBN 978-1-78676-252-8); Online (ISBN 978-1-78676-254-2, 978-1-78676-255-9)

Climate change and agriculture
Print (ISBN 978-1-78676-320-4); Online (ISBN 978-1-78676-322-8; 978-1-78676-323-5)

Reducing greenhouse gas emissions from livestock production
Print (ISBN 978-1-78676-439-3); Online (ISBN 978-1-78676-441-6, 978-1-78676-442-3)

Energy-smart farming: Efficiency, renewable energy and sustainability
Print (ISBN 978-1-78676-835-3); Online (ISBN 978-1-78676-837-7, 978-1-78676-838-4)

Chapters are available individually from our online bookshop:
https://shop.bdspublishing.com

climate•SMART•publishing
IN AGRICULTURAL SCIENCE

BURLEIGH DODDS SERIES IN AGRICULTURAL SCIENCE
NUMBER 135

Developing circular agricultural production systems

Edited by Professor (UZ) Dr Barbara Amon, University of Zielona Góra, Poland and Leibniz Institute for Agricultural Engineering and Bioeconomy (ATB), Germany

burleigh dodds
SCIENCE PUBLISHING

Published by Burleigh Dodds Science Publishing Limited
82 High Street, Sawston, Cambridge CB22 3HJ, UK
www.bdspublishing.com

Burleigh Dodds Science Publishing, 1518 Walnut Street, Suite 900, Philadelphia, PA 19102-3406, USA

First published 2023 by Burleigh Dodds Science Publishing Limited

Library of Congress Control Number: 2023940886

British Library Cataloguing in Publication Data
A catalogue record for this book is available from the British Library

ISBN 978-1-80146-256-3 (print)
ISBN 978-1-80146-258-7 (PDF)
ISBN 978-1-80146-257-0 (ePub)
ISSN 2059-6936 (print)
ISSN 2059-6944 (online)

DOI: 10.19103/AS.2023.0120

Typeset by Deanta Global Publishing Services, Dublin, Ireland

Contents

Part 3 Co-products

Series list

Title	Series number
Achieving sustainable cultivation of maize - Vol 1 From improved varieties to local applications Edited by: Dr Dave Watson, CGIAR Maize Research Program Manager, CIMMYT, Mexico	001
Achieving sustainable cultivation of maize - Vol 2 Cultivation techniques, pest and disease control Edited by: Dr Dave Watson, CGIAR Maize Research Program Manager, CIMMYT, Mexico	002
Achieving sustainable cultivation of rice - Vol 1 Breeding for higher yield and quality Edited by: Prof. Takuji Sasaki, Tokyo University of Agriculture, Japan	003
Achieving sustainable cultivation of rice - Vol 2 Cultivation, pest and disease management Edited by: Prof. Takuji Sasaki, Tokyo University of Agriculture, Japan	004
Achieving sustainable cultivation of wheat - Vol 1 Breeding, quality traits, pests and diseases Edited by: Prof. Peter Langridge, The University of Adelaide, Australia	005
Achieving sustainable cultivation of wheat - Vol 2 Cultivation techniques Edited by: Prof. Peter Langridge, The University of Adelaide, Australia	006
Achieving sustainable cultivation of tomatoes Edited by: Dr Autar Mattoo, USDA-ARS, USA and Prof. Avtar Handa, Purdue University, USA	007
Achieving sustainable production of milk - Vol 1 Milk composition, genetics and breeding Edited by: Dr Nico van Belzen, International Dairy Federation (IDF), Belgium	008
Achieving sustainable production of milk - Vol 2 Safety, quality and sustainability Edited by: Dr Nico van Belzen, International Dairy Federation (IDF), Belgium	009
Achieving sustainable production of milk - Vol 3 Dairy herd management and welfare Edited by: Prof. John Webster, University of Bristol, UK	010
Ensuring safety and quality in the production of beef - Vol 1 Safety Edited by: Prof. Gary Acuff, Texas A&M University, USA and Prof. James Dickson, Iowa State University, USA	011
Ensuring safety and quality in the production of beef - Vol 2 Quality Edited by: Prof. Michael Dikeman, Kansas State University, USA	012
Achieving sustainable production of poultry meat - Vol 1 Safety, quality and sustainability Edited by: Prof. Steven C. Ricke, University of Arkansas, USA	013
Achieving sustainable production of poultry meat - Vol 2 Breeding and nutrition Edited by: Prof. Todd Applegate, University of Georgia, USA	014

Introduction

This volume reviews the emergence of circular agriculture as an approach to improving the sustainability of the agricultural sector. It addresses recent advances in understanding and developing closed-loop systems to optimise crop-livestock nutrient cycles and resource use, as well as ways agricultural wastes can be recycled back into agricultural production or used as feedstock to produce a range of bio-based materials.

The chapters are split into four parts: Part 1 chapters focus on general issues associated with circular agricultural production. Chapters in Part 2 draw attention to re-using agricultural and other types of waste and finally, in Part 3, chapters cover the importance of co-products within circular agricultural production systems. Part 4 illustrates circular agricultural practices through case studies.

Part 1 General issues

The first chapter of the book focuses on the key principles and challenges of circular, closed-loop agricultural systems. Chapter 1 begins by providing an overview of the principles and practice of circular agriculture, then moves on to review circular biomass use. The chapter also examines circular nutrition provision, which is then followed by a review of circular energy provision as well. A section on wood cascading a special form of circularity is also provided, before the chapter concludes with an overview of why developing circular agriculture research is important in improving agri-food system sustainability.

Chapter 2 draws attention to developing closed-loop nutrient cycles in crop production. It starts by first discussing the importance of understanding nutrient cycling in agri-food systems, then goes on to highlight current options for achieving closed-loop nutrient cycles. A section on manure management in different types of region is also included, focusing specifically on region types such as Delta, Metropole, Hinterland and Intensive Livestock. The chapter also includes an overview of the role of livestock and mixed farming systems in closed-loop nutrient cycling.

Moving on from Chapter 2, Chapter 3 discusses closed-loop crop-livestock farming systems. The chapter begins by providing an overview of the various solutions that have been suggested in terms of improving the environmental impact livestock production has. It then goes on to review the concept of ecological intensification, something that is considered essential for future farming systems. The chapter also examines the potential of mixed cropping and integrated crop-livestock systems. A section on the role of grasslands in cropping and livestock systems is also provided, focusing specifically on the case of grass-clover leys. This is followed by a case study that compares specialised and integrated systems to improve the eco-efficiency of dairy farming in Germany.

The final chapter of Part 1 focuses on closed-loop precision farming technologies and how they can be used to optimise resource use. Chapter 4 first highlights the emerging technologies enhancing precision agriculture, such as the prediction of soil fertility, frost events and pest attacks as well as the use of sensors to improving production quality and quantity. Several case studies on closed-loop agriculture are also included, focusing specifically on the application of Internet of Things technology, big data and artificial intelligence. The chapter concludes by emphasising the importance of closed-loop agriculture and how it can be enhanced further through the use of precision farming technologies.

Part 2 Re-using agricultural and other wastes

Part 2 opens with a chapter that examines using crop residues/byproducts as livestock feed in a circular economy. Chapter 5 begins by providing an overview of current agro-industrial activities and main byproduct feeds used in animal agriculture, focusing specifically on by-product feeds generated by the biofuel industry, the human food production industry and the brewing industry. This is then followed by an analysis of the chemical composition and nutritional value of selected byproducts. The chapter also discusses the use of dried distillers' grain with solubles in animal agriculture, highlighting how this grain can be fed to swine, poultry, beef cattle and dairy cattle species.

The next chapter discusses optimising slurry management in agricultural practices. Chapter 6 begins by first highlighting current decision tools for optimising manure management, then goes on to review modifying animal slurry pH to enhance its value as a biobased fertilizer through methods such as bioacidification and alkalinisation. A section on improving manure management systems to minimise trade-offs is also provided, followed by an overview of combining manure management with anaerobic digestion. The chapter also reviews pre- and post-treatment for anaerobic digestion as well as the optimisation of anaerobic digestion operations to optimise digestate quality.

The subject of Chapter 7 is optimising livestock manure as a biofertiliser and bioenergy source. The chapter begins by first providing an overview of anaerobic digestion and how this can be implemented in order to treat livestock manure. This is then followed by an overview of current mechanical separation technologies that are commonly used as a first step in manure treatment. The chapter also examines the use of composting, followed by an overview of how struvite precipitation can be used within livestock manure treatment. A section on ammonia stripping/scrubbing is also included. The chapter also discusses the use of a physical process known as membrane filtration then moves on to review the use of bioelectrochemical systems. A case study known as 'farm for the future' is also included.

Part 2 of this volume closes with a chapter that draws attention to the safe and sustainable use of bio-based fertilisers in agricultural production systems. Chapter 8 begins by discussing some common issues regarding the use of bio-based fertilisers, such as the concentration of nutrients leading to losses of reactive nitrogen and phosphorus into the environment. Potential issues with heavy metals, pathogens, antibiotics and other contaminants that can pose a health risk to humans, animals, crops and the ecosystem are also discussed. Finally, a case study investigating the effects of intensive livestock production on both reactive nitrogen losses and antibiotic resistance in the environment is presented.

Part 3 Co-products

The first chapter of Part 3 discusses producing biogas from livestock manure and agricultural biomass. Chapter 9 begins by addressing the different aspects of manure-based anaerobic digestion. The chapter then goes on to review recent developments in co-digestion research. This is then followed by a section that addresses the availability of biomass (manure, energy crops and agricultural residues) and its biomethane potential. A section on the types of anaerobic digestion plants and their economics is also included, focusing specifically on on-farm plants and centralised plants. To support this research, the chapter includes case studies of different types of agriculture-based biogas plants, such as small-scale plants using mono or co-digestion of manure and centralised plants, which co-digest manures collected from nearby farms with agro-industrial waste.

The next chapter draws attention to multi-feedstock biorefineries for converting agricultural wastes and microalgae into co-products. Chapter 10 begins by assessing availability of biomass feedstock from agricultural sources. It then reviews the range of biological and thermochemical biomass conversion processes together with the growing number of co-products from these sources. The chapter then focuses on microalgae as an emerging biomass feedstock. It reviews the key steps and technologies in growing and processing microalgae as well as characterisation techniques to optimise process control.

The subject of Chapter 11 is developing bioplastics from agro-industrial wastes for applications in food packaging. The chapter begins by first discussing the potential use of bioplastic as an alternative to petroleum-based plastics. It then goes on to highlight the types of bio-based polymers currently available, emphasising that, depending on their origin, these polymers can be classified into two macro-groups: synthetic and natural. The chapter also reviews how bioplastics can be created from renewable feedstocks, using the example of polyhydroxyalkanoates from agricultural waste as a basis for discussion. A section on the application of bioplastics in food packaging is also included, which is followed by a review of innovative food packaging and the various packaging production techniques. A case study on the Horizon 2020 Usable Packaging Project is also provided.

Expanding on areas previously touched upon in Chapter 11, Chapter 12 focuses on developing polyphenols from agricultural wastes. It starts by first providing an overview of two types of phenolic compounds and their chemical structures: flavonoids and non-flavonoids. The chapter focuses specifically on non-flavonoid compounds such as phenolic acids, stilbenes and tannins. It also highlights the types of organic waste containing phenolic compounds. The then goes on to discuss the nutraceutical properties of polyphenols and potential applications, such as their antioxidant properties, anti-inflammatory properties, anti-diabetic and anti-obesity properties as well as several others which are summarised in a single section. It also provides an overview of current extraction methods available for obtaining phenolic compounds, such as conventional solid–liquid extraction technologies and various green technologies. The chapter then provides a section comparing these extraction techniques and how successful they are in terms of phenolic compound extraction.

Part 4 Case studies

The first chapter of the case studies section reviews developing closed-loop dairy value chains and tools to support decision makers. Chapter 13 starts by providing an overview of the frameworks used for identifying and managing food loss and waste in developed dairy chains. It then goes on to review the key stages and management practices in place, such as the production stage, processing, the retail and wholesale stage and consumption and service. It also highlights how soil nutrients, energy and water not reused in human and animal feed can be recovered through aerobic and anaerobic digestion. A section on improving decision-making in dairy value chains to manage food loss and waste is also included, focusing specifically on decision support tools and the factors that need to be considered in order to move towards closed-looped chains.

The final chapter of the book discusses the challenges and opportunities of aquaponics in terms of its commercial application. In Chapter 14, along with a brief description of hydroponic technologies and recirculating aquaculture systems, several issues concerning the application of aquaponics on a commercial scale are discussed, such as: 1) the selection of fish and crop species; 2) the development of salt-water aquaponics systems for the diversification of water sources, fish, and crops; 3) the system monitoring; 4) fish and crop protection; 5) food safety; 6) animal welfare; 7) how to increase economic and environmental sustainability.

Acknowledgements

We wish to acknowledge the following for their help in reviewing a particular chapter:

- Chapter 11: Dr Manuela Facchin, Institute for Polymers, Composites and Biomaterials-CNR IPCB, Italy

Part 1

General issues

Chapter 1

Circular, closed-loop agricultural systems: key principles and challenges

Rolf Meyer, Institute for Technology Assessment and Systems Analysis (ITAS), Karlsruhe Institute of Technology (KIT), Germany

1 Introduction

Ideas about circular agriculture have come about due to growing scientific attention to and increasing political promotion of the circular economy (CE). CE concepts are designed to overcome the linear economic model - commonly described as 'take-make-use-dispose' (Lazarevic and Valve, 2017) - by closing resource loops (Murray et al., 2017). Slowing, narrowing and closing material and energy loops can be achieved through design for long use, easy maintenance, repair, reuse, remanufacturing, refurbishing and recycling (Geissdoerfer et al., 2017). Instead of the current production and consumption model based on continuous growth and increasing resource throughput, CE aims to increase the efficiency of resource use, with particular focus on urban and industrial waste (Ghisellini et al., 2016). It is also often described as regenerative and restorative (Ellen MacArthur Foundation, 2015; Jarre et al., 2020). Several reviews have highlighted that a wide variety of CE definitions, principles, interpretations and origins are used in scientific discussions (e.g. Blomsma and Brennan, 2017; Ghisellini et al., 2016; Kirchherr et al., 2017; Murray et al., 2017).

At the political level, the European Commission published in 2015 a Circular Economy Action Plan (EC, 2015). The Action Plan is targeted at achieving a transition, in which the value of products, materials and resources is maintained

http://dx.doi.org/10.19103/AS.2023.0120.02

in the economy for as long as possible, and the generation of waste minimized, as an essential contribution to the EU's efforts to develop a sustainable, low carbon, resource-efficient and competitive economy. Reuse of food waste and production of bio-based products from renewable sources are among the priority areas (EC, 2015). National CE strategies backed up by legislation have been adopted in many EU Member States and elsewhere (e.g. China and Japan) (POST, 2016; Weber and Stuchtey, 2019).

Enthusiasm for CE by different stakeholders has been tempered by the harsh reality of current low levels of circularity. For example, a rough estimate of exemplary energy-intensive materials (e.g. aluminium, steel, concrete, plastic, paper) shows that the gaps between theoretical circularity and the reality are very high, making the transition to a CE at any significant scale challenging. Issues such as material loss and the additional energy input to sustain circularity are often not considered sufficiently in CE concepts (Cullen, 2017).

In parallel to CE debates, concepts and official strategies for developing a bioeconomy (BE) have developed since the 2000s (Meyer, 2017). A distinction between CE and BE can be seen in the McArthur Foundation's popular butterfly graph (2015) - used by many scholars and stakeholders - showing a clear separation between the 'technosphere' and the 'biosphere', with CE research focused for a long time purely on the technosphere (Leipold and Petit-Boix, 2018). Differentiation between technical and biological cycles has been criticized as misleading, as the majority of bio-based (industrial) products enter a technical cycle (Carus and Dammer, 2018).

In recent years, the terms 'circular bioeconomy' and 'bio-based circular economy' have been introduced (e.g. Borrello et al., 2016; Carus and Dammer, 2018; Kardung et al., 2021). The connection between CE and BE was recognized based on the shared focus on resource use efficiency (D'Amato et al., 2017). Applying a systems-based approach and reducing the impact on the environment are common factors to both approaches. Since the mid-2010s, there has been an exponential growth in scientific publications on the joint discussion and application of CE and BE (Abad-Segura et al., 2021).

The Updated Bioeconomy Strategy of the European Commission from 2018 points out that 'to be successful, the European BE needs to have sustainability and circularity at its heart' (EC, 2018, p. 4), thereby emphasizing the importance of circularity. In national and regional strategies, bio-based CE is rarely included as a term but circularity aspects are recognized through objectives concerning waste management and bioenergy (Vanhamaki et al., 2019). To describe the complex relationship between the CE and bio-based businesses or the BE, stakeholders use a number of relationship archetypes (Leipold and Petit-Boix, 2018):

- the CE and the BE as separate but reinforcing;
- both concepts as completely integrated;

- both concepts as partially antagonistic;
- the BE as a precondition to a CE; and
- the CE as a tool to move from a fossil fuel economy to a BE.

This chapter discusses the relationship between CE and BE as they relate to agriculture.

2 Circular agriculture: principles and practice

In industrial processes and value chains, industrial products and their incorporated resources still exist - in principle - at the end of their life cycle after the use phase. In contrast, the life cycle of agricultural products for food, feed and fuel are inherently linear due to the consumption of final food and feed products or energy use of biofuels. Therefore, some key approaches of CE - such as reuse, repair, remanufacturing and refurbishing - are not applicable to agriculture and the food system.

Nonetheless, circular agriculture (CA) concepts have emerged recently, particularly in the Netherlands. CA is placed in opposition to conventional agriculture and linear food chains (Dagevos and de Lauwere, 2021). CA and circular food systems are not clearly distinguished and are partly used as synonyms. The concept of CA is generally described by principles since a broadly accepted definition does not exist. It is seen as nearly impossible to define a unique CE model for the agro-food sector due to the complexity of agri-food supply chains (Esposito et al., 2020). The published principles for CA have a number of common points but also some variations.

In one of the first explicit proposals for a CA, de Boer and van Ittersum (2018) outlined three principles for circular food production:

- Plant biomass is the basic building block of food and should be used by humans first.
- By-products from food production, processing and consumption should be recycled back into the food system.
- Animals should mainly be fed on biomass that is not directly edible by humans.

The key focus is on increasing resource use efficiency (e.g. raising yield potential and narrowing yield gaps, better soil management with soil-organic carbon sequestration and processing residual flows), with a broader implementation of production-ecological concepts. The last principle diverges the most from the current animal production system. Instead of using feed crops from arable land, a CA requires that animal production should be based on biomass inedible to humans (e.g. from grassland) and leftovers from the production of plant-based food (de Boer and van Ittersum, 2018, pp. 30–36). More fundamentally, rethinking economic growth is also discussed (de Boer and van Ittersum, 2018, pp. 44–47).

In 2019, the Dutch government published its 'Vision on Circular Agriculture' (Ministry of Agriculture, 2019), with high ambitions for structural change in agriculture, stating: 'The vision entails a paradigm shift from growth in production volumes and cost price reductions towards optimization in resource use and food production in harmony with nature' (Ministry of Agriculture, 2019, p. 2). CA is defined as agriculture with the lowest amount of harmful emissions to the environment and the highest possible resource efficiency. The aim is to maintain a leading position in agriculture globally by taking the lead in the transition towards more circular farming systems (Vrolijk et al., 2020). An essential part of circular production is seen in putting an end to waste that occurs in various parts of the food chain, from losses in primary production to the wasted food by consumers. In a circular system, arable and livestock farming and horticulture should primarily use raw materials from each other's supply chains and waste flows from the food industry. The focus is more on the use of waste rather than avoidance. It should be noted that, to date, concrete policy measures are missing, the policy for CA is mainly aimed at adaptability to normal business operations, and CA is primarily regarded as a technological challenge requiring smart innovations that improve the efficiency of production processes (Dagevos and de Lauwere, 2021).

Based on new research about development policies, a discussion paper from the Dutch Research Council has established three key principles of CA (Bianchi et al., 2020):

- Regenerative agriculture: preserving and enhancing natural resources.
- Efficient use of resources: closing nutrient loops.
- Multi-purpose use and recovering value: minimizing food losses by using waste streams and turning these into valuable inputs for the food production chain.

The incorporation of the concept of 'regeneration' is the main difference from other principles of CA. Regeneration forms part of many definitions of the circular economy but is rarely defined or explained in the CE literature. To date, regeneration has been a symbolic term with little practical application in the context of CE approaches, except in the case of some agricultural practices (Morseletto, 2020).

Regenerative agriculture has recently received significant attention from producers, retailers, researchers and consumers, as well as politicians and the mainstream media. The claim by advocates is that regenerative agriculture as an alternative means of producing food could lead to lower - or even net positive - environmental and/or social impacts. No widely accepted definition has emerged for regenerative agriculture, and no legal or regulatory definition - as for organic agriculture - exists. The different definitions and descriptions of regenerative agriculture are based on processes (e.g. use of cover crops, the integration of livestock and reducing or eliminating tillage), outcomes (e.g. to improve soil health, to sequester carbon and to increase biodiversity) or combinations of the two (Newton et al., 2020). A literature-based analysis identified the following key thematic themes characterizing regenerative agricultural discourses (Gordon et al., 2022):

- Agricultural production is conducted within nested, complex living systems.
- Farms are relational and characterized by co-evolution between humans and landscape biota.
- The innate potential of living systems is place-sourced.
- Openness to alternative thinking and practice is transformative.
- Multiple regenerative cultures are necessary for deeply regenerative agriculture.
- Regenerative approaches depart from conventional agriculture to varying degrees.

Overall, the concepts and principles of CA build on the general discussion of CE and on the recognition of increased demand for food, inefficient resource use, environmental impacts and high rates of food waste at all stages of the food system, calling for a transition towards a more sustainable food system (Jurgilevich et al., 2016). CA can include more resource use efficiency, new pathways in food chains and bio-based value chains, and/or closed-loop applications. CA remains mainly programmatic and practical applications are very restricted. Instead of a 'blueprint' for adopting a CA approach, a collective effort by farmers, citizens, businesses, researchers and policymakers is proposed to find optimum combinations of ecological principles with modern technology (Bianchi et al., 2020).

Looking closer at circularity, and closed loops in a narrower sense, some differentiation is important for better understanding. Circularity can be achieved at different spatial levels:

- on the farm;
- between farms, from the local level up to regional;
- cross-regional and nationwide; and
- across national borders.

Generally, it is postulated that circular chains should be built up as locally as possible. Circularity can encompass different stages in the food supply chain as regards sectors of the food system:

- Primary circulation – circularity inside the agricultural production system.

- Secondary circulation - circularity along the food supply chain, from agriculture via food processing to retail.
- Tertiary circulation - closed loops from the end of consumption such as from food consumption and municipal waste.

Closing of circles can be applied to biomass, plant nutrients, energy or a combination of these. Some key points are discussed in the following sections.

3 Circular biomass use

Building circles of biomass in agriculture relates closely to the circulation of nutrients. At farm level, crop residues and organic fertilizers form the backbone of biomass circulation in the field and are essential for preservation and enhancement of soil organic matter (SOC) and soil fertility. The amount of crop residues and biomass below the ground for restoring SOC is influenced by the total biomass of the cultivated crop, the crop rotation, intercropping ('green manure') and mixed cultivations.

In discussions on CA, little attention is given to concepts and practices of organic farming. This is surprising in the light of the fact that the circulation of biomass and nutrients in the farm system and an understanding of soil, plants and animals as an integrated system represent key principles of organic farming. Learning from organic agriculture could lead to huge potential for improving biomass circulation on the farm, especially in terms of enhancing soil fertility.

Traditionally, a key element of circularity in agriculture was mixed farming. However, specialization and regional concentration of livestock farming have in the last few decades separated plant and livestock production to a significant extent, abolishing the resulting recirculation of manure to plant production. A revival of the integrated crop-livestock systems is proposed as a key form of ecological intensification for achieving future food security and environmental sustainability (Lemaire et al., 2014). Innovative reintegration of animals, cropping and pasture systems could take place on various territorial scales, from individual farms to cooperation between different farms at a local level and to collaboration at a regional level or between neighbouring regions (Peyraud et al., 2014). However, the obstacles facing reintegration are high. Beyond the investments required to revert to diversified and integrated crop-livestock systems - such as purchasing specific machinery, fencing plots to hold animals and building or renovating animal buildings, which are often impossible for individual farmers - the skills and knowledge required to manage crop-livestock systems have generally been lost since specialization (Martin et al., 2016).

Beyond circularity on the farm and within agriculture, biomass from the food chain is going back or should increasingly go back into the agricultural system. These secondary circulations from the food chain include by-products, co-products, residues and wastes. The recirculation of by-products from food processing is already well established. From the first stage of food processing, by-products from sugar refineries, cereal and oil mills return to agriculture as feedstuff. The potential of by-products has been explored in industrialized countries. In addition to by-products, different types of residues and wastes accrue at different stages of the food chain. A specific and detailed assessment for Germany shows that established uses already exist for biogenic by-products, residues and wastes from food chains, and unexplored potential lies mainly in primary circulations (e.g. cereal straw and livestock manure) (Brosowski et al., 2016; FNR, 2015). Consequently, increasing efficiency rather than creating new cycles is a priority.

In the context of tertiary circulation from the end of consumption, the reduction of food losses at the food consumption level is high on the political agenda and broadly discussed in scientific circles and society. In its 'Farm to Fork Strategy' (EC, 2020), the European Commission has set the objective of halving per capita food waste at retail and consumer levels by 2030. In 2022, the Commission will set a baseline and propose legally binding targets to reduce food waste across the EU (EC, 2020).

Besides reduction, the use of unavoidable food waste as animal feed is discussed. Current legal regulations in many countries, and the EU as a whole, ban the use of food waste as animal feed due to the potential risks to human health, such as foot-and-mouth disease, classical swine fever and bovine spongiform encephalopathy. However, it is argued that feeding food waste to animals (especially monogastric animals) can be safe when appropriately heat-treated (de Boer and van Ittersum, 2018).

Overall, a couple of trade-offs exist in the context of biomass circularity. Energy and material uses of crop residues in new value chains (e.g. in biorefineries for biofuels and/or bio-based products) are in conflict with the biomass circulation of crop residues in the field, which is particularly important for maintenance and enhancement of SOC and soil fertility. In the case of residues and waste in the food chain, the production of energy (e.g. in biogas plants) competes with direct recycling to agriculture. In addition, a conflict of objectives can accrue between recycling biomass and reduction of waste generation along the food supply chain. Reducing food loss and waste is seen as a major action for a circular food system, but a successful reduction implies greater linearity in food chains.

4 Circular nutrition provision

Historically, crop production has relied on and was restricted by natural levels of soil phosphorus, nitrogen-fixing bacteria and the addition of locally available organic matter such as manure and human excreta (Cordell et al., 2009; Gorman, 2013). Thereby, nutrient cycling existed for relevant parts of the limited nutrient budget. The technological innovation of fixing nitrogen industrially and the industrial mining of phosphate rock have been key preconditions for intensification and productivity growth in crop production. At the same time, nutrient supply became a more put-through system. Significantly more chemically active nitrogen and phosphorus were placed in circulation and lost to water bodies, with resulting environmental impacts such as eutrophication (Ashley et al., 2011; Gorman, 2013). Moreover, due to specialization and globalization, nutrients accumulate in areas with high animal (due to manure) and human (food waste or human excreta) density and get depleted in other areas (de Boer and van Ittersum, 2018).

At the farm level, the circularity of nutrients is dependent on organic fertilizers. Facilitating closed loops with organic fertilizers brings a number of challenges (de Boer and van Ittersum, 2018). First, nutrients in organic fertilizers are available in ratios that may not necessarily match crop requirements. Second, the mineralization of at least partly organically bound nutrients may be poorly synchronized with crop demand and is hard to calculate and impossible to influence, increasing the risk of losses. Third, the composition of organic fertilizers is highly variable across farms and seasons, which can also impair the efficiency of organic fertilization. Finally, the spatial distribution of farm manure generation and crop production do not often match. In summary, the trade offs between nutrient circularity and environmental impacts in the management of organic waste can be challenging (Cobo et al., 2018).

New digital technologies - ranging from precision fertilization to farm management systems - present new possibilities for better coordination of nutrient supply with nutrient demand in animal and crop production. The most progress in technological development and agricultural application has been achieved for precision nitrogen fertilization (Kehl et al., 2021a,b). Digital applications aim to better control agricultural production processes, thereby enabling more efficient resource use and a reduction in external inputs. The central premise of digitization is to achieve environmental benefits and operational savings. However, the magnitude of these savings is not easy to determine, as local production conditions have a strong influence on the reduction effects that can be achieved in practice (Kehl et al., 2021a,b).

At the end of the life cycle, digested and excreted food products can end up in organic recycling (e.g. as composts), energy production or landfill (Carus and Dammer, 2018). Organic waste and sewage from urban origins have proven to be a source of nutrients that can be recycled and used in agriculture (Velasco-Munoz et al., 2021). In 2005, the EU-27 phosphorus recycling rates stood at 73% for animal production, 29% for food processing and 21% for human consumption (Van Dijk et al., 2016).

Wastewater has traditionally been viewed as a waste requiring treatment in order to reduce negative environmental impacts before it is released back into the receiving water system. Reuse, historically, has not been a design objective in wastewater treatment. Phosphorus in wastewater is currently removed using flocculating agents such as aluminium or iron sulphate and iron chloride, and the resulting floc is not generally used as fertilizer. Nitrogen in wastewater is traditionally reduced to volatile N gas by exploiting the biological process of denitrification which occurs under anaerobic conditions and is therefore lost (Rosemarin et al., 2020). The use of sewage sludge in agriculture has in the past been reduced due to contamination by chemicals (e.g. heavy metals and persistent organic pollutants) and other contaminants (e.g. pharmaceutical residues) in the wastewater.

New approaches to recover nitrogen and phosphorus and to close nutrient cycles are under development. Nitrogen recovery can be achieved through ammonia stripping and adsorbed to an acid, producing ammonium sulphate fertilizer (Rosemarin et al., 2020). Phosphate rock reserves are limited, geopolitically concentrated in a few countries, and therefore regarded as a critical raw material. Against this background, different technologies for phosphorus recovery from sewage water and/or sludge are currently being developed and tested (Egle et al., 2015; Ohtake and Tsuneda, 2019). Phosphorus recovery and recycling have received more and more political and legislative support in the last few years. Following Switzerland's example, Germany implemented an amendment to its sewage sludge ordinance in 2017 that - after a transition phase of 12-15 years - means any large German sewage treatment plant will have to recover phosphorus from sewage water (Scholz and Wellmer, 2019; Kind, 2020). New recovery technologies involve the addition of a further technological processing step in wastewater management and the reuse of single nutrients, moving further away from a combined closed loop on carbon, nitrogen and phosphorus with sewage use than in the past.

5 Circular energy provision

Historically, a great deal of the energy needed in agricultural production was supplied by animal traction and human work on the farm and was based on on-farm biomass production. With the industrialization of agriculture production, bio-based energy supply has been replaced by external fossil fuel inputs (Harchaoui and Chatzimpiros, 2018).

In the last two decades, with the cultivation of energy crops, agriculture has once again become a producer of renewable energy resources and renewable energy itself. In 2016, Europe's production of renewable energy from agriculture played an important role in supplying renewable energy, totalling

11.7% of renewable energy production in the EU-28. The Netherlands had the highest share of national renewable energy production from agriculture with 35%, followed by Belgium with 27.2% and Germany with 26.6% (EC, 2018).

Besides the use of renewable energy on the farm itself, agricultural renewable energy production is mainly used in overall energy supply systems, from local heat networks to national electricity grids to international biofuel markets. Circularity within bioenergy provision takes place not so much at farm level but in the context of renewable energy systems. As broadly discussed in science and society, the bioenergy production of agriculture faces important trade-offs with land use change, preservation of biodiversity and food security (WBGU, 2009).

The energy neutrality of agriculture - equivalent to circularity at the sector level - is designed to produce as much energy from internal sources as is consumed from external sources for food production. Only a phasing out of feed from cropland, combined with very high energy recovery from agricultural residues, could enable energy neutrality and transform agriculture into a net energy source for society (Harchaoui and Chatzimpiros, 2018).

6 Cascade use as a special form of circularity

Wood cascading is a concept with a long history of debate and analysis in EU bio-based industries (Olsson et al., 2016). Conceptual considerations and practical applications of cascading are very popular in the wood sector (Jarre et al., 2020). Cascade use (CU) strongly overlaps with CE as both concepts promise to increase resource efficiency by supporting multiple uses of a resource. Although the two concepts have many similarities, research has mostly been conducted independently, and the number of connections between the two concepts is surprisingly low (Mair and Stern, 2017).

The cascading analogy to 'a river flowing over a sequence of plateaus' was first introduced by Sirkin and ten Houten (1994) as a general tool to achieve sustainability in resource use. The concept of cascading aims to increase the efficiency of biomass utilization by reusing, recycling and ultimately generating energy, but there is no common definition and understanding of it to date (Jarre et al., 2020).

Based on Odegard et al. (2012), three approaches of CU can be distinguished. The first approach, cascading in time, is conceptualized as a sequential use of biomass. This implies reusing or recycling a bio-based product, with energy production at the end of the life cycle. Paper recycling and particleboards are conventional examples, but more innovative solutions including bioplastics are also possible. In the second approach, cascading in value, the time steps of the cascade can be optimized by prioritizing the highest possible value over the whole life cycle. The third approach, cascading in function, optimizes co-production. In these two cases, more efficient biomass use is intended by successive processing of the total biomass into different products for varying uses. This understanding is primarily applied in the context of biorefineries, which involve both conventional waste-to-energy strategies and new pathways for waste wood, such as chemicals or bioplastics (Jarre et al., 2020).

The concept of CU is not really discussed in the context of CA and food production systems. Closed-loop approaches of organic matter and nutrients inside agricultural production systems do not fit well into the concept of cascading. However, the proposed and implemented recycling of by-products and residues from food processing back into the food system represents a step in cascading. Nonetheless, it remains the fact that food products as primary products are consumed in the end and cannot be recycled in different steps as can wood products in CU.

7 Conclusion and future trends

Having reviewed the most significant EC research and innovation programmes from 2007 to 2020, it seems that the will to increase the sustainability of the agri-food system and to foster the socio-technical transition towards circularity is evident but not very relevant (Muscio and Sisto, 2020). The scientific debate regarding the pros and cons of the giant leap transition versus small-wins approach remains. Against this background, the vision for CA takes a small-wins approach as its point of departure (Vrolijk et al., 2020).

CA is, to a large extent, a programme or promise. Overarching challenges are to move forward from principles to practice and to step up from specific examples to overall transformation of agriculture. Substantially more knowledge regarding barriers and enablers, including their determining political and socioeconomic factors and their interdependencies, is needed to handle the upcoming challenges successfully.

8 Where to look for further information

A first basic paper of the Wageningen University gives a good introduction on the concept of Circular Agricultural:

- de Boer, I. J. M. and van Ittersum, M. K. (2018). *Circularity in Agricultural Production*. Wageningen University & Research.

An example for international cooperation is the 'Circular Food Systems Network' whose objective is to develop an active international network of researchers in the field of circular food systems, where knowledge can be shared and collaboration between research groups can increase the development and implementation of circularity:

- https://www.wur.nl/en/Research-Results/Research-Institutes/livestock-research/show-wlr/Circular-Food-Systems-Network.htm.

The evolution and status of research on CE and agriculture, based on bibliometric analyses, is discussed in three papers:

- Batlles-delaFuente, A., Abad-Segura, E., González-Zamar, M.-D. and Cortés-García, F. J. (2022). An evolutionary approach on the framework of circular economy applied to agriculture, *Agronomy* 12(3), 620. DOI: 10.3390/agronomy12030620.
- Duque-Acevedo, M., Belmonte-Ureña, L. J., Yakovleva, N. and Camacho-Ferre, F. (2020). Analysis of the circular economic production models and their approach in agriculture and agricultural waste biomass management, *International Journal of Environmental Research and Public Health* 17(24), 9549. DOI: 10.3390/ijerph17249549.
- Esposito, B., Sessa, M. R., Sica, D. and Malandrino, O. (2020). Towards circular economy in the Agri-food sector, A systematic literature review, *Sustainability* 12(18), 7401. DOI: 10.3390/su12187401.

A policy brief of the United Nations Department of Economic and Social Affairs gives a first insight on the international dimension of CA, with a focus on the Global South:

- United Nations Department of Economic and Social Affairs. (2021). Circular agriculture for sustainable rural development. Policy Brief No 105.

9 References

Abad-Segura, E., Batlles-delaFuente, A., González-Zamar, M.-D. and Belmonte-Ureña, L. J. (2021). Implications for sustainability of the joint application of bioeconomy and circular economy: A worldwide trend study, *Sustainability* 13(13), 7182. DOI: 10.3390/su13137182.

Ashley, K., Cordell, D. and Mavinic, D. (2011). A brief history of phosphorus: From the philosopher's stone to nutrient recovery and reuse, *Chemosphere* 84(6), 737–746. DOI: 10.1016/j.chemosphere.2011.03.001.

Bianchi, F., van Beek, C., de Winter, D. and Lammers, E. (2020). Opportunities and barriers of circular agriculture, Insights From a Synthesis Study of the Food & Business Research Programme, NOW, Dutch Research Council, Den Haag, The Netherlands.

Blomsma, F. and Brennan, G. (2017). The emergence of circular economy, A new framing around prolonging resource productivity, *Journal of Industrial Ecology* 21(3), 603–614. DOI: 10.1111/jiec.12603.

Borrello, M., Lombardi, A., Pascucci, S. and Cembalo, L. (2016). The seven challenges for transitioning into a bio-based circular economy in the agri-food sector, *Recent Patents on Food, Nutrition and Agriculture* 8(1), 39–47.

Brosowski, A., Thrän, D., Mantau, U., Mahro, B., Erdmann, G., Adler, P., Stinner, W., Reinhold, G., Hering, T. and Blanke, C. (2016). A review of biomass potential and current utilisation – Status quo for 93 biogenic wastes and residues in Germany, *Biomass and Bioenergy* 95, 257–272. DOI: 10.1016/j.biombioe.2016.10.017.

Carus, M. and Dammer, L. (2018). The "Circular Bioeconomy" – Concepts, opportunities and limitations. Nova Paper 9. Available at: https://bioplasticfeedstockalliance.org/wp-content/uploads/2020/01/External_Resource_Nova_Paper_9_The_Circular_Bioeconomy.pdf, accessed January 17, 2022.

Cobo, S., Dominguez-Ramos, A. and Irabien, A. (2018). Trade-offs between nutrient circularity and environmental impacts in the management of organic waste, *Environmental Science and Technology* 52(19), 10923–10933. DOI: 10.1021/acs.est.8b01590.

Cordell, D., Drangert, J. O. and White, S. (2009). The story of phosphorus: Global food security and food for thought, *Global Environmental Change* 19(2), 292–305. DOI: 10.1016/j.gloenvcha.2008.10.009.

Cullen, J. M. (2017). Circular economy: Theoretical benchmark or perpetual motion machine?, *Journal of Industrial Ecology* 21(3), 483–486. DOI: 10.1111/jiec.12599.

Dagevos, H. and de Lauwere, C. (2021). Circular business models and circular agriculture: Perceptions and practices of Dutch farmers, *Sustainability* 13(3), 1282. DOI: 10.3390/su13031282.

D'Amato, D., Droste, N., Allen, B., Kettunen, M., Lähtinen, K., Korhonen, J., Leskinen, P., Matthies, B. D. and Toppinen, A. (2017). Green, circular, bio economy: A comparative analysis of sustainability avenues, circular, *Journal of Cleaner Production* 168, 716-734. DOI: 10.1016/j.jclepro.2017.09.053.

de Boer, I. J. M. and van Ittersum, M. K. (2018). *Circularity in Agricultural Production*, Wageningen University & Research.

EC (European Commission) (2015). Closing the loop - An action plan for the circular economy. Communication from the Commission to the European Parliament, the Council, the European Economic and Social Committee and the Committee of the Regions. COM(2015) 614, Brussels.

EC (2018a). *A Sustainable Bioeconomy for Europe: Strengthening the Connection Between Economy, Society and the Environment*, Updated Bioeconomy Strategy, Brussels. DOI: 10.2777/792130.

EC (2018b). Production of renewable energy from agriculture and forestry, *Cap Context Indicators, 2014-2020, 2018 Update.*

EC (2020). *A Farm to Fork Strategy for a Fair, Healthy and Environmentally-Friendly Food System*, Brussels.

Egle, L., Rechberger, H. and Zessner, M. (2015). Overview and description of technologies for recovering phosphorus from municipal wastewater, *Resources, Conservation and Recycling* 105, 325–346. DOI: 10.1016/j.resconrec.2015.09.016.

Ellen MacArthur Foundation (2015). *Towards a Circular Economy: Business Rationale for an Accelerated Transition.* Available at: https://ellenmacarthurfoundation.org/towards-a-circular-economy-business-rationale-for-an-accelerated-transition, accessed January 17, 2022.

Esposito, B., Sessa, M. R., Sica, D. and Malandrino, O. (2020). Towards circular economy in the agri-food sector: A systematic literature review, *Sustainability* 12(18), 7401. DOI: 10.3390/su12187401.

FNR (Fachagentur Nachwachsende Rohstoffe) (2015). Biomassepotentiaziale von Rest- und Abfallstoffen, *Status Quo in Deutschland*, Schriftenreihe Nachwachsende Rohstoffe Band 36, FNR, Gülzow, Germany. Available at: https://mediathek.fnr.de/band-36-biomassepotenziale-von-rest-und-abfallstoffen.html, accessed February 22, 2022.

Geissdoerfer, M., Savaget, P., Bocken, N. M. P. and Hultink, E. J. (2017). The circular economy – A new sustainability paradigm?, *Journal of Cleaner Production* 143, 757-768. DOI: 10.1016/j.jclepro.2016.12.048.

Ghisellini, P., Cialani, C. and Ulgiati, S. (2016). A review on circular economy: The expected transition to a balanced interplay of environmental and economic systems, *Journal of Cleaner Production* 114, 11–32. DOI: 10.1016/j.jclepro.2015.09.007.

Gordon, E., Davila, F. and Riedy, C. (2022). Transforming landscapes and mindscapes through regenerative agriculture, *Agriculture and Human Values* 39(2), 809–826. DOI: 10.1007/s10460-021-10276-0.

Gorman, H. S. (2013). Learning from 100 years of ammonia synthesis: Establishing human-defined limits through adaptive systems of governance, *GAIA - Ecological Perspectives for Science and Society* 22(4), 263–270.

Harchaoui, S. and Chatzimpiros, P. (2018). Can agriculture balance its energy consumption and continue to produce food? A framework for assessing energy neutrality applied to French agriculture, *Sustainability* 10(12), 4624. DOI: 10.3390/su10124624.

Jarre, M., Petit-Boix, A., Priefer, C., Meyer, R. and Leipold, S. (2020). Transforming the bio-based sector towards a circular economy – what can we learn from wood cascading?, *Forest Policy and Economics* 110, 101872. DOI: 10.1016/j.forpol.2019.01.017.

Jurgilevich, A., Birge, T., Kentala-Lehtonen, J., Korhonen-Kurki, K., Pietikäinen, J., Saikku, L. and Schösler, H. (2016). Transition towards circular economy in the food system, *Sustainability* 8(1), 69. DOI: 10.3390/su8010069.

Kardung, M., Cingiz, K., Costenoble, O., Delahaye, R., Heijman, W., Lovrić, M., van Leeuwen, M., M'Barek, R., van Meijl, H., Piotrowski, S., Ronzon, T., Sauer, J., Verhoog, D., Verkerk, P. J., Vrachioli, M., Wesseler, J. H. H. and Zhu, B. X. (2021). Development of the circular bioeconomy: Drivers and indicators, *Sustainability* 13(1), 413. DOI: 10.3390/su13010413.

Kehl, C., Meyer, R. and Steiger, S. (2021a). Digitalisierung der Landwirtschaft: Technologischer Stand und Perspektiven. TAB-Arbeitsbericht Nr. 193, Büro für Technikfolgen-Abschätzung Beim Deutschen Bundestag, Berlin, Deutschland. DOI: 10.5445/IR/1000142950.

Kehl, C., Meyer, R. and Steiger, S. (2021b). Digitalisierung der Landwirtschaft: Gesellschaftliche Voraussetzungen, Rahmenbedingungen und Effekte. TAB-Arbeitsbericht Nr. 194, Büro für Technikfolgen-Abschätzung Beim Deutschen Bundestag, Berlin, Deutschland. DOI: 10.5445/IR/1000142951.

Kind, S. (2020). Nachhaltige Phosphorversorgung, TAB Themenkurzprofil Nr. 39, April 2020, Büro für Technikfolgen-Abschätzung Beim Deutschen Bundestag, Berlin, Deutschland. DOI: 10.5445/IR/1000133950.

Kirchherr, J., Reike, D. and Hekkert, M. (2017). Conceptualizing the circular economy: An analysis of 114 definitions, *Resources, Conservation and Recycling* 127, 221–232. DOI: 10.1016/j.resconrec.2017.09.005.

Lazarevic, D. and Valve, H. (2017). Narrating expectations for the circular economy: Towards a common and contested European transition, *Energy Research and Social Science* 31, 60–69. DOI: 10.1016/j.erss.2017.05.006.

Leipold, S. and Petit-Boix, A. (2018). The circular economy and the bio-based sector – Perspectives of European and German stakeholders, *Journal of Cleaner Production* 201, 1125–1137. DOI: 10.1016/j.jclepro.2018.08.019.

Lemaire, G., Franzluebbers, A., de Faccio Carvalho, P. C. and Dedieu, B. (2014). Integrated crop-livestock systems: Strategies to achieve synergy between agricultural production and environmental quality, *Agriculture, Ecosystems and Environment* 190, 4–8. DOI: 10.1016/j.agee.2013.08.009.

Mair, C. and Stern, T. (2017). Cascading utilization of wood: A matter of circular economy?, *Current Forestry Reports* 3(4), 281–295. DOI: 10.1007/s40725-017-0067-y.

Martin, G., Moraine, M., Ryschawy, J., Magne, M.-A., Asai, M., Sarthou, J.-P., Duru, M. and Therond, O. (2016). Crop-livestock integration beyond the farm level: A review, *Agronomy for Sustainable Development* 36(3), 53. DOI: 10.1007/s13593-016-0390-x.

Meyer, R. (2017). Bioeconomy strategies: Contexts, visions, guiding implementation principles and resulting debates, *Sustainability* 9(6), 1031. DOI: 10.3390/su9061031.

Ministry of Agriculture, Nature and Food Quality of the Netherlands (2019). *The Dutch Government's Plan to Support the Transition to Circular Agriculture*, Den Haag, The Netherlands.

Morseletto, P. (2020). Restorative and regenerative, exploring the concepts in the circular economy, *Journal of Industrial Ecology* 24(4), 763–773. DOI: 10.1111/jiec.12987.

Murray, A., Skene, K. and Haynes, K. (2017). The circular economy: An interdisciplinary exploration of the concept and application in a global context, *Journal of Business Ethics* 140(3), 369–380. DOI: 10.1007/s10551-015-2693-2.

Muscio, A. and Sisto, R. (2020). Are agri-food systems really switching to a circular economy model? Implications for European research and innovation policy, *Sustainability* 12(14), 5554. DOI: 10.3390/su12145554.

Newton, P., Civita, N., Frankel-Goldwater, L., Bartel, K. and Johns, C. (2020). What is regenerative agriculture? A review of scholar and practitioner definitions based on processes and outcomes, *Frontiers in Sustainable Food Systems* 4, 577723. DOI: 10.3389/fsufs.2020.577723.

Odegard, I., Croezen, H. and Bergsma, G. (2012). Cascading of biomass, 13 Solutions for a Sustainable Bio-based Economy, Making Better Choices for Use of Biomass Residues, By-Products and Wastes, council of Europe, Delft, Delft, The Netherlands.

Ohtake, H. and Tsuneda, S. (Eds.) (2019). *Phosphorus Recovery and Recycling*. Springer Nature, Singapore. DOI: 10.1007/978-981-10-8031-9.

Olsson, O., Bruce, L., Roos, A., Hektor, B., Guisson, R., Lamers, P., Hartley, D., Ponitka, J., Hildebrandt, J. and Thrän, D. (2016). Cascading of Woody biomass: Definitions, policies and effects on international trade, *IEA Bioenergy Task* 40.

Peyraud, J.-L., Taboada, M. and Delaby, L. (2014). Integrated crop and livestock systems in Western Europe and South America: A review, *European Journal of Agronomy* 57, 31–42. DOI: 10.1016/j.eja.2014.02.005.

POST (The Parliamentary Office of Science and Technology, UK Parliament) (2016). Designing a circular economy, POSTNOTE Number 536.

Rosemarin, A., Macura, B., Carolus, J., Barquet, K., Ek, F., Järnberg, L., Lorick, D., Johannesdottir, S., Pedersen, S. M., Koskiaho, J., Haddaway, N. R. and Okruszko,

T. (2020). Circular nutrient solutions for agriculture and wastewater – A review of technologies and practices, *Current Opinion in Environmental Sustainability* 45, 78–91. DOI: 10.1016/j.cosust.2020.09.007.

Scholz, R. W. and Wellmer, F.-W. (2019). Although there is no physical short-term scarcity of phosphorus, its resource efficiency should be improved, *Journal of Industrial Ecology* 23(2), 313–318. DOI: 10.1111/jiec.12750.

Sirkin, T. and ten Houten, Mt. (1994). The cascade chain, *Resources, Conservation and Recycling* 10(3), 213–276. DOI: 10.1016/0921-3449(94)90016-7.

Vanhamaki, S., Medkova, K., Malamakis, A., Kontogianni, S., Marisova, E., Huisman Dellago, D. and Moussiopoulos, N. (2019). Bio-based circular economy in European national and regional strategies, *International Journal of Sustainable Development and Planning* 14(1), 31. DOI: 10.2495/SDP-V14-N1-31-43.

Van Dijk, K. C., Lesschen, J. P. and Oenema, O. (2016). Phosphorus flows and balances of the European Union Member States, *Science of the Total Environment* 542(B), 1078–1093. DOI: 10.1016/j.scitotenv.2015.08.048.

Velasco-Muñoz, J. F., Mendoza, J. M. F., Aznar-Sánchez, J. A. and Gallego-Schmid, A. (2021). Circular economy implementation in the agricultural sector: definition, strategies and indicators, *Resources, Conservation and Recycling* 170, 105618. DOI: 10.1016/j.resconrec.2021.105618.

Vrolijk, H., Reijs, J. and Dijkshoorn-Dekker, M. (2020). Towards sustainable and circular farming in the Netherlands, lessons from the socio-economic perspective, Wageningen Economic Research, Den Haag, The Netherlands.

WBGU (German Advisory Council on Global Change) (2009). *Future Bioenergy and Sustainable Land Use*, Earthscan Publications, London, UK.

Weber, T. and Stuchtey, M. (Eds.) (2019). Deutschland auf dem weg zur circular Economy, Erkenntnisse aus Europäischen Strategien, Circular Economy Initiative Deutschland, acatech – Deutsche Akademie der Technikwissenschaften, München, Germany.

Chapter 2

Understanding and developing closed-loop nutrient cycles in crop production

Theun Vellinga, Wageningen University and Research, The Netherlands

1 Introduction

Crop production is part of agri-food systems (AFS). Many studies have defined food systems, for example Van Berkum et al. (2018): 'Food systems comprise all the processes associated with food production and utilisation: growing, harvesting, packing, processing, transporting, marketing, consuming and disposing of food remains'. AFS also produce feed, fibre, fuels, building materials and other non-food products for consumption by people and animals either directly after harvest or combined with (industrial) processing steps. As Van Berkum et al. (2018) point out, food systems also operate in, and are influenced by social, political, technological, economic and natural environments. There is a strong interaction e.g. between the technical aspects of plant and animal production and economic aspects such as farmers' costs of production and the incomes they receive, which may be influenced by consumer purchasing patterns and the role of market intermediates such as retail. Social aspects such as food habits and religious beliefs may also play an important role. AFS need therefore to be assessed not just as physical systems (e.g. with material flows) but as complex economic and social systems.

This chapter will focus on the (re)cycling of nutrients in AFS. Nutrient flows and cycles in AFS are complex. Crop products are e.g. often transported to, processed and utilised in other places than where the crop is produced, as part of long, complex, globalised agri-food supply chains (Trienekens et al., 2014).

http://dx.doi.org/10.19103/AS.2023.0120.05

Analysing nutrient flows has to be carried out at different scales, from field and individual farm to regional, national and global scales. Economic, social, cultural and political aspects play an important role with many causes and effects interacting with each other (Van Bueren et al., 2003). AFS may involve a wide range of stakeholders with different and sometimes conflicting economic interests and priorities.

The complexity of AFS, not just as biophysical but as economic and social systems, as well as the different needs and priorities of stakeholders within AFS, make attempts to improve them particularly challenging. This type of problem has been identified by Churchman (1967) and Rittel and Webber (1973) as a 'wicked problem'. Solving wicked problems requires understanding of the mechanisms of systems and a dialogue between stakeholders to find solutions (Bremmer et al., 2020). This chapter will discuss a number of aspects of closed-loop nutrient cycles including

- ways of representing cycling of biomass and nutrients in AFS;
- the role of human and livestock densities; and
- the particular role of crop-livestock systems in closed loop cycles.

The aim of the chapter is to increase understanding of the mechanisms involved as a way to support stakeholders in jointly finding solutions.

2 Understanding nutrient cycling in agri-food systems

Figure 1 provides a simplified picture of nutrient cycling in AFS. The figure has been developed and tested with the help of stakeholders from sectors such as farming, the compound feed industry, policy makers and NGOs (Bremmer et al., 2020). The scheme is based on a simple farm nutrient cycling framework defined by Schröder et al. (2003), extended to include human consumption. Three primary processes are identified:

- Plant production: starting with soil, manure and fertilisers as key nutrient sources nourishing crops which can be used for various purposes;
- Animal production: starting with plant and some animal products used as feed for animals, producing animal-based food products such as milk, meat and eggs, with manure as a by-product; and
- Human consumption: plant and animal-based products are utilised by consumers both as food and as non-food products; in the case of food, by-products include faeces and urine.

These primary processes are shown by boxes (see Fig. 1).

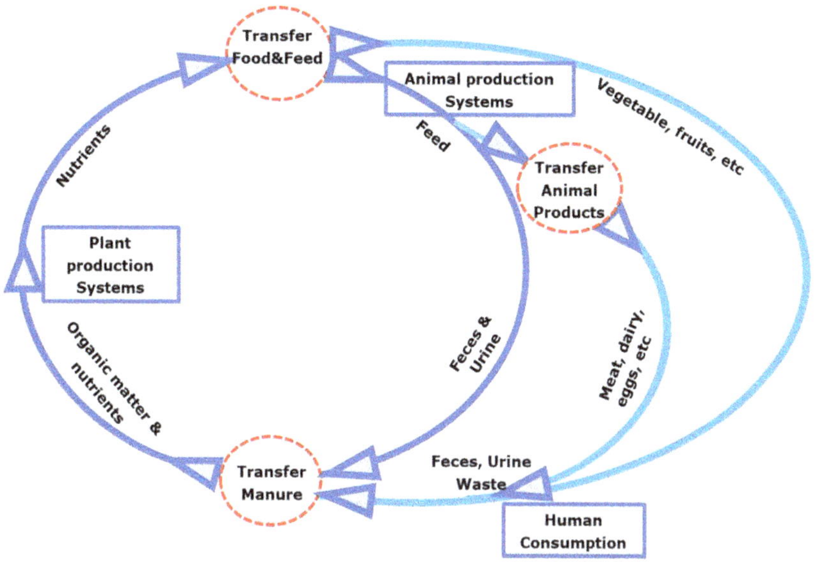

Figure 1 Schematic picture of the nutrient cycling in agri-food systems, with distinction between primary and secondary processes. Source: Bremmer et al. (2020).

These primary processes have a number of characteristics in common:

- Each process in itself can be considered as linear with inputs and outputs. Efficiency can be calculated as output divided by input (expressed as percentages), loss is the complementary fraction (1 minus efficiency). Many studies have been performed to analyse these losses. The use of mineral balance sheets and other nutrient budgeting tools in plant and animal production systems can be an effective tool for understanding and calculating losses as well as overall efficiency (with output divided by input) (Zhang et al., 2020). Each step in the production process provides a material output that can be used for the next step in the AFS. Since, in many industrialised countries, human waste/excreta are not utilised as a fertiliser in agriculture, the point of human consumption breaks this chain.
- Primary processes are bound to one location: a crop is grown in a field and, although more mobile, livestock in most specialised farming systems are also typically raised in one location. Although humans are relatively mobile, consumption can also be considered as location specific.

These primary processes provide the primary building blocks for nutrient cycles.

The three primary linear processes are connected and transformed to a circular agri-food system via three secondary processes, called transfer points (Fig. 2):

- The connection point between plant production as input and animal production and human consumption as outputs, called Transfer Food and Feed (TFF);
- The connection point between animal production as input and human consumption as output, called Transfer Animal Products (TAP); and
- The third connection point between animal production and human consumption as inputs, where excreta and waste are transferred and converted to fertiliser, called the Transfer Manure (TM). Fertiliser for Plant production is the output.

The transfer points process the outputs of the primary processes in various ways. They transport and process input materials and convert them into products that are suitable for the next primary process. The transfer points are the key steps in the circularity: they connect the linear processes and different locations where production and consumption occur (Fig. 2).

On livestock farms TFF and TM are often hypothetical. Although, in many intensive and (partly) footloose livestock farming systems e.g. feed is often from an external source and manure has to be exported out of the system. The food and beverage processing industries are part of the TFF; dairy companies and slaughterhouses producing meat are part of the TAP. Trading and transport companies can be found both in TFF and TAP. Transfer points can be globally connected, involving import and export flows between countries.

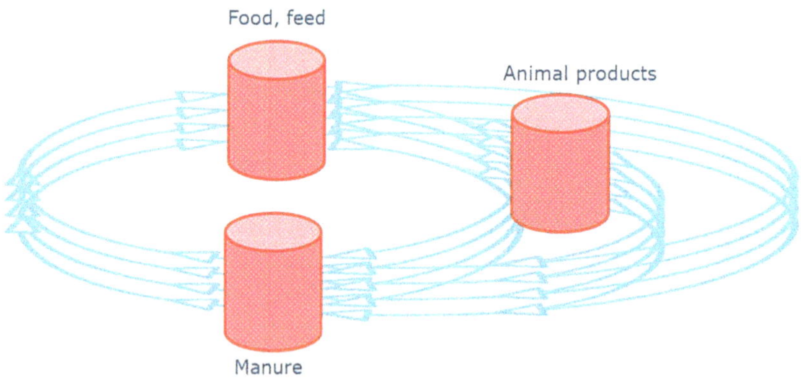

Figure 2 At the three transfer points exchange of products between locations takes place. Source: Bremmer et al. (2020).

The TM does not work well in many systems. In many industrialised countries, human waste is processed into sewage sludge which is often incinerated or dumped. The use of sewage sludge as a fertiliser has been decreased in Europe in the last decades due to food safety and public health concerns. Even in countries with relatively sophisticated water management systems, human waste can still end up in surface waters or is lost in other ways. Animal manure is often a serious pollution problem in livestock-dense areas (particularly as run-off into watercourses), with legislation and enforcement agencies required to control losses.

As an illustration, losses and efficiencies in the EU agri-food system for nitrogen are shown in Fig. 3 (Leip et al., 2011). Although the data are not very recent, they still give a good illustration of efficiencies and losses in the AFS. Efficiency in the plant production systems is 53% with losses at 47%. Efficiencies are higher and losses are lower in the animal production system: 23% output in animal products and 62% in animal manure. The loss is 15% to air, mainly as ammonia. However, as long as manure acts as a throughput from animal production to plant production, inefficiencies related to organic manure are transferred to plant production. Nutrient use efficiency in human consumption is 0%, as there is no output. Losses in human consumption are considered to be nil and efficiency is considered to be 100% as all waste and excreta go to the TM.

Losses also occur at transfer points, including processing losses in TFF and TAP. Animal based feed (slaughter by products) can be an important loss in

Plant Production Systems			
IN	21.20	OUT	-21.20
From TM	85%	products to TFF	53%
Deposition	10%	loss	47%
Fixation	5%		

Transfer Food & Feed			
IN	14.28	OUT	-14.28
total imported	19%	loss	8%
EU produced	81%	exports	1%
		to animal production	81%
		to human consumption	10%

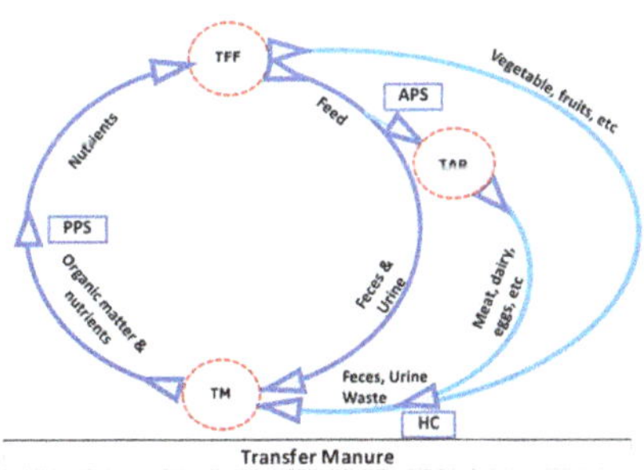

Animal Production Systems			
IN	11.60	OUT	-11.60
From TFF	100%	products to TAP	23%
		manure	62%
		loss	15%

Transfer Animal Products			
IN	4.00	OUT	-4.00
total imported	26%	loss	28%
EU produced	74%	exports	16%
		to human consumption	56%

Transfer Manure			
IN	21.72	OUT	-21.72
total imported	50%	loss (human waste)	17%
EU produced	50%	exports	0%
animal manure	*33%*	to plant production	83%
human manure	*13%*		
waste/swill	*3%*		

Human Consumption			
IN	3.62	OUT	-3.62
From TFF	38%	waste/swill to TM	20%
From TAP	62%	manure to TM	80%
		products	0%

Figure 3 The nitrogen balance for the EU27. Source: Data from Leip et al. (2011).

TAP. The most important material flow in TFF is animal feed. Losses in the TM are linked to the loss of human excreta and food waste/swill. All animal manure is considered to pass this transfer point without losses. Synthetic fertiliser N is imported from outside the AFS (both within and outside Europe) and is part of the balance sheet in TM.

This structure of primary and secondary processes means that options for contributing to closed-loop nutrient cycles are different. In primary processes, the key is to minimise losses. This can be done in various ways, e.g. by reducing inputs, by optimising growth conditions, optimising fertiliser application rates and animal rations, controlling plant and animal diseases (to improve nutrient use efficiency), reducing harvest waste, etc. These issues are the domain of plant and animal sciences. Although food waste is not considered as a loss in human consumption, it is clearly an important factor with assumed losses of 20% in this balance sheet.

Secondary processes are about the organisation of material flows. Movement of plant and animal-based materials and manure is organised at transfer points, supporting linear processes at different locations. Secondary processes convert inputs to utilisable outputs whilst minimising losses. As noted, secondary processes for manure and waste are far from perfect. Where the role of the transfer point manure is to get the manure to the right place at the right time in the right form, the task of the primary process is to apply the right amount with the right application method.

3 Options for achieving closed-loop nutrient cycles

The previous analysis identifies different aspects of closed-loop or circular agricultural systems. A key theme is recirculation of nutrients, with nutrients from plant production going via human consumption or via animal production and human consumption to the transfer point of manure and potential re-use as fertiliser. This recirculation can potentially occur on an international scale, although this is currently not the case. Over many decades, organic nutrient sources have been replaced and replenished in many cases by synthetic fertilisers with a tailored and uniform composition that are often cheaper, easier to apply and with demonstrated benefits on yields. Successful recirculation requires effective systems for manure collection, processing, transport and application. Given the challenges many people opt for small-scale loops at the level of groups of farms within one region which are easier to manage. As long as human waste is not included in recirculation, the system is always leaking nutrients and replenishment from (potentially environmentally damaging and unsustainable) synthetic sources is necessary. The problem has been exacerbated by growing specialisation in agriculture with arable and livestock farming carried out separately, and intensive production concentrated in a smaller number of large farms and in regions specialising in livestock production. These concentrations of livestock production lead to regional imbalances (MacDonald et al., 2011), large flows of plant material used for feed and, consequently, to accumulation of nutrients (Wang et al., 2018).

Another important aspect of closed loops, rather than of circular food systems, is the utilisation by livestock of co-products from processing crops for food, beverages, fuel and fibre (e.g. Schader et al., 2015; van Kernebeek et al., 2015; van Zanten, 2016). The definition of co products is very wide. Extensive livestock production systems in Africa, Asia and Latin America rely heavily on crop residues such as straw and stover which are often low in energy and protein content (e.g. Blümmel et al., 2017; Samireddypalle et al., 2019). More intensive livestock systems in many industrialised countries do make more use of plant-based industrial co-products (e.g. distillers grains) although utilisation of co-products is variable across animal species (Schader et al., 2015; Van Zanten, 2016). From this perspective, the shift from red meat (e.g. cattle and sheep) to white meat (e.g. poultry) consumption is problematic since co-products do not currently play a significant role in poultry nutrition (van Selm et al., 2022).

4 Manure management in different types of region

There are large flows of biomass around the world with clear imbalances demonstrated e.g. for phosphorus by MacDonald et al. (2011) and for nitrogen as well by Wang et al. (2018). Important drivers of imbalances are concentrations of people and livestock. The process of urbanisation is ongoing with more than 50% of the world population now living in urban areas. In industrialised countries this proportion can be 70% or more. An analysis of human and livestock concentrations by Verstand et al. (2021) defines four types of regions, classified by population and livestock densities. The regions are shown in Fig. 4.

The four regions are as follows:

- Delta: this type of region has high concentrations of livestock and humans, often found in river deltas where urban settlement has a long history. The competition for land is high, leading to high prices, high technical efficiency and little room for growing feed crops, resulting in feed being imported. In delta regions in industrialised countries, livestock production (especially for monogastric animals) can be intensive and large-scale. Livestock production exceeds local consumption needs and a significant percentage of livestock products are exported, mostly as processed products but also raw (e.g. fresh meat) or as live animals. High livestock concentrations mean that high densities of emissions to air, ground and surface water, accumulation of manure is problematic and the risk

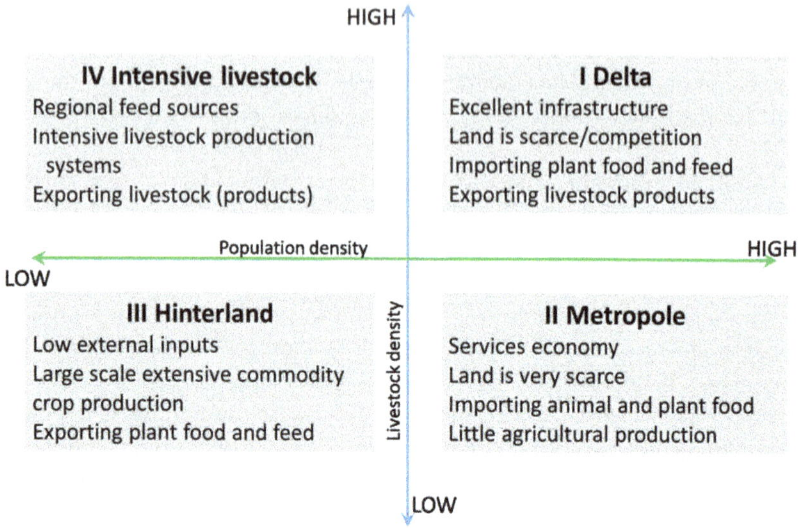

Figure 4 The agrifood system framework and the four types of regional population and stocking and their characteristics. Source: Verstand et al. (2021).

of zoonoses is relatively high. Due to the high population density and the urban character of the population livestock density is often heavily debated.

- Metropole: in this region population density and the degree of urbanisation are relatively high, but most of the land is used for crop production.
- Hinterland: these are regions with low population and livestock densities. Almost all agricultural land is used for crop production. When agriculture is highly developed with relatively high levels of inputs and mechanisation, these regions act as exporters of food and feed.
- Intensive livestock: These regions are characterised by low population densities and high livestock densities. Livestock can be fed from regional feed production and animal products are exported.

These regions are distributed across the world.

Figure 5 shows the distribution of the four types of regions across countries. Different types can be found in individual countries. This is shown for Europe (Fig. 6). Livestock production is often found in deltas (which are historic areas for urban development). The livestock industry here is embedded in a 'hard' institutional infrastructure (e.g. feed industry, slaughterhouses), and a 'soft' infrastructure of extension services such as veterinarian support and research support. Concentration can lead to a high level of efficiency and profitability but can cause significant environmental harm.

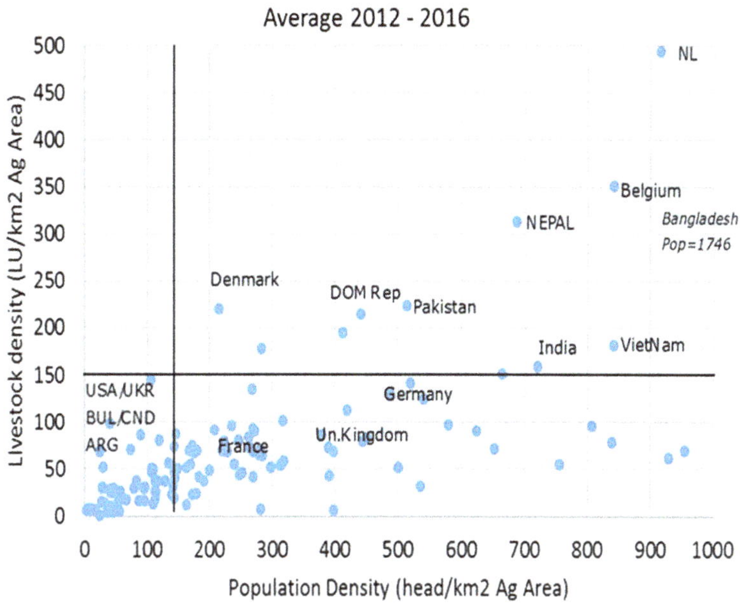

Figure 5 The average population and livestock densities per country (in persons and LU per km² of agricultural area), averaged over the period 2012-16. Source: Verstand et al. (2021), combined with own calculations.

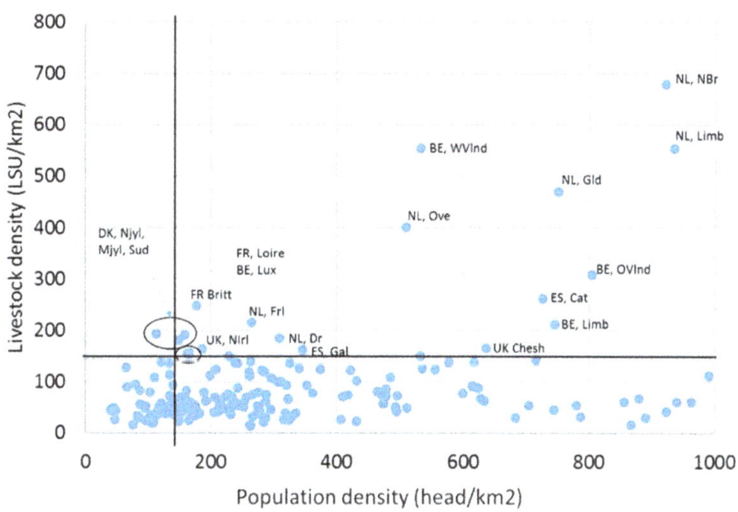

Figure 6 The average population and livestock densities per administrative unit within a country of the EU (in persons and LU per km² of agricultural area), averaged over the period 2012-16. Source: Verstand et al. (2021), combined with own calculations.

Different nutrient flows can be identified within and between regions:

- Feed transport from Hinterland regions to Deltas;
- Livestock products from Delta regions to Metropoles; and
- Hinterland and Intensive Livestock regions are net exporters while Deltas and Metropoles are net importers, due to their differences in population densities. The export of livestock products from Deltas is smaller than the related import of feed materials.

It appears that livestock concentrations play a very big role in nutrient flows and nutrient accumulation. However, nutrient accumulation would occur without livestock, due to the concentration of humans in urban areas. This will increase with ongoing urbanisation.

Verstand et al. (2021) consider two options for solving the problem of nutrient accumulation. The first one is to organise return flows for manure and excreta. Return flows require large investments in manure processing and in governance to get the nutrients in the right form at the right place. Application in the right amount at the right time is the responsibility of the primary process of plant production. The volume of the return flow will be affected by the size of the livestock sector in a region.

The other option is to reorganise livestock systems and the related biomass flows. Livestock systems should be self-supporting in feed, which will reduce the size of the livestock sector in Deltas. This provides large opportunities for Hinterland regions to increase livestock numbers. This pathway also requires large investments in both a hard and soft infrastructure in new livestock production locations, particularly as newer, smaller livestock productions may have fewer economies of scale and agglomeration advantages at sector level are small. Even in this pathway nutrients in the form of animal products will go to human population concentrations and organisation of the return flow is required. The second option can significantly reduce global flows of biomass and nutrients. The European Union currently imports about 51 megatons (FEFAC, 2020) of biomass with about 17-18 megatons of protein to feed the region's livestock (Leip et al., 2011). With feed conversion rates of about 2 kg feed/kg LWG for poultry and 3 for pigs, nutrient flows could be reduced by 50-65%. Nutrients from consumption need to be returned to plant production systems in order to maintain soil fertility.

5 The role of livestock and mixed farming systems

Although the chapter deals with crop production, livestock plays an important role in the issues regarding closing nutrient cycling loops. A solution that is often mentioned in closing nutrient cycles is the development of mixed crop-livestock farms, either via co-operation between crop and livestock farms or by combining crop and livestock production on the same farm. Many different types of mixed crop-livestock systems exist as shown by e.g. Stark et al. (2018) and Hou et al. (2021). In many countries of the world, fully integrated crop-livestock systems can be found. Highly specialised livestock and arable farms also cooperate for mutual benefits; in this case interactions are limited to the exchange of land, manure and feed crops. The advantages of mixed crop-livestock systems have been discussed extensively (e.g. Herrero et al., 2010; Stark et al., 2018; Hou et al., 2021). Analyses focus on increased resilience, productivity, efficiency and reduced dependency on external sources. Despite the agronomic advantages of mixed systems, a process of development of specialised, separated crop and livestock operations has taken place in recent decades, especially in industrialised countries, mainly driven by economic factors (McMichael, 2009; Garrett et al., 2020). This separation has also been facilitated by the development of international production and distribution of cheap feed (Wang et al., 2018).

The rationale behind the advantage of mixed crop-livestock farms is as follows:

- Improved efficiency and productivity of the system as a whole: the utilisation of crops, co-products and crop residues as feed for livestock, with more efficient use compared to just organic matter addition to the soil and removing reliance on purchased compound feed;
- In return for feed products from the arable farm, the use of manure from the livestock farm as provider of nutrients and organic matter for crops;
- The opportunity for improved crop rotation, especially when temporary pasture is part of the crop rotation, both in increasing soil organic matter content and potentially disrupting insect pest and disease cycles affecting crops; and
- Increased resilience by spreading ecological and economic risks over a range of agricultural products and by reducing dependency on external inputs.

Low et al. (2023) also identified potentially better economic returns for mixed farming systems.

In practice the first three items take place to some degree in existing supply chains without mixed systems on farms or cooperation between crop and livestock farms. A significant percentage of feed concentrates for all livestock types is now often based on cereals and protein rich co-products from vegetable oil and biofuel production, a combination of primary crops and co-products from processing crops. Farms with relatively little land and high stocking densities sell manure to intermediaries who can then make it available to farms growing crops. This implies that also highly specialised livestock farms cannot function without crop-livestock interactions. This 'on-distance' interaction is more formalised and only can function properly with a strong institutional and governmental infrastructure, providing a legal framework and enforcement. This is especially the case when regional high livestock densities occur. In the European Union environmental regulations regarding the use of manure (e.g. the Nitrates Directive), play in important role in preventing environmental harm. This institutional and governmental infrastructure, however, does not organise a return flow of nutrients from the 'Delta' regions to 'Hinterland' regions (Verstand et al., 2021). This leads to nutrient imbalances at an international, even intercontinental scale (Wang et al., 2018). More broadly, the concentration of humans in urban regions and the shift to a higher fraction of 'exotic' food products (Muilwijk et al., 2019) also leads to accumulation of nutrients, which need to be recycled and returned to production locations. This is similar to the situations where livestock concentrations occur. An analysis of population and livestock densities shows that there are a number of regions in the world where both population and livestock densities are high.

Although there are many advantages of crop–livestock interactions, as mentioned before, the direct cooperation between livestock and arable farms may not close loops completely, but for some part, shorten them via direct supply of feed and manure. The net export of biomass (meat, milk, eggs, crops) to consumers is still the main function of the crop-livestock system and, thanks to improved efficiency and productivity, the net export of agricultural products to consumers might even increase. Closing loops still requires the return of nutrients from consumers. Whilst nutrient flows from crops (primary crops, co-products and crop residues) to livestock and vice versa (as manure) are mainly internal flows in mixed crop–livestock systems, in specialised crop and livestock systems is the flow of feed and manure are respectively export and import flows.

6 Conclusion

Crop production is the basis of all agri-food systems, directly via processed plant products as food, or indirectly via feed for animals in producing animal-based food. Given the complexity of supply chains and the number of stakeholders involved, improving nutrient cycling in crop production and agri-food systems can be considered a 'wicked problem'. An important source of leakage from the system is the loss of nutrients via sewage systems since, in many cases, sewage sludge is incinerated and not used as a fertiliser.

A distinction between primary processes (production and consumption) and secondary processes is useful to define different options to improve closing nutrient cycles. In fact there is limited opportunity for primary producers to close nutrient cycles, because the return of nutrients from human consumption is not under their control. The contribution of crop producers (including grass production at dairy farms) is by optimising crop, fertiliser and pest management and taking care of good soil health. As discussed, the interaction between crop and livestock production can contribute to this optimisation.

Completely closing nutrient loops consists of two main components: the return flow of nutrients from 'Delta' and 'Metropole' regions to 'Hinterland' regions (Verstand et al., 2021) and the utilisation of human excreta as a source of nutrients. Both will require substantial investment in technical and institutional infrastructure. Organising the return flow requires manure treatment technologies, transport of treated manure of nutrients, in combination with a system of quality and public health control. This will lead to increased costs of production, costs that are currently part of the externalities. Utilising human excreta requires on the one hand large investments in redesign of the existing sewage systems, modifying technology from water purification to nutrient recovery. On the other hand, in many emerging economies, urbanisation takes place and often proper sanitary systems are lacking. In that case, the development of sanitary systems can serve multiple goals as public health, biodiversity and nutrient recovery.

Crop-livestock interactions already occur throughout the whole agri-food system, primarily via specialist intermediaries such as feed manufacturers using crops or their co-products in feed formulations. The advantage of mixed farming systems and re-connecting crop and livestock production lies in creating smaller nutrient loops with reduced nutrient losses, even if nutrient cycles may still not be fully closed. Additional advantages are the improvement of crop rotation and distribution of soil organic matter.

7 References

Blümmel, M., Muller, A., Schader, C., Herrero, M. and Garg, M. R. (2017). The use and abuse of cereals, legumes and crop residues in rations for dairy cattle. In: Webster, J. (Ed.). *Achieving sustainable production of milk (vol 3). Dairy herd management and welfare*, Burleigh Dodds Science Publishing Limited, Cambridge, UK, 273–294. http://hdl.handle.net/10568/78403.

Bremmer, B., Leenstra, F. R. and Vellinga, T. (2020). Nutrient cycle assessment tool: a tool for dialogue and ex ante evaluation of policy interventions aiming at closing nutrient cycles in agriculture. *NJAS: Wageningen Journal of Life Sciences* 92(1): 1–10.

Churchman, C. W. (1967). Guest editorial: wicked problems. *Management Sciences* 14(4): 141–142.

FEFAC (2020). *Feed & Food 2020*, Annual Report FEFAC, Brussels, 19 pp.

Garrett, R. D., Ryschawy, J., Bell, L. W., Cortner, O., Ferreira, J., Garik, A. V. N., Gil, J. D. B., Klerkx, L., Moraine, M., Peterson, C. A., dos Reis, J. C. and Valentim, J. F. (2020). Drivers of decoupling and recoupling of crop and livestock systems at farm and territorial scales. *Ecology and Society* 25(1): 24. https://doi.org/10.5751/ES-11412-250124.

Herrero, M., Thornton, P. K., Notenbaert, A. M., Wood, S., Msangi, S., Freeman, H. A., Bossio, D., Dixon, J., Peters, M., van de Steeg, J., Lynam, J., Parthasarathy Rao, P., Macmillan, S., Gerard, B., McDermott, J., Seré, C. and Rosegrant, M. (2010). Smart investments in sustainable food production: revisiting mixed crop-livestock systems. *Science* 327(5967): 822–825.

Hou, Y., Oenema, O. and Zhang, F. (2021). Integrating crop and livestock production systems—towards agricultural green development. *Frontiers of Agricultural Science and Engineering* 8(1): 1–14. https://doi.org/10.15302/J-FASE-2021384.

Leip, A., Britz, W., Weiss, F. and de Vries, W. (2011). Farm, land and soil nitrogen budgets for agriculture in Europe calculated with Capri. *Environmental Pollution* 159(11): 3243–3253.

Low, G., Dalhaus, T. and Meuwisen, M. P. M. (2023). Mixed farming and agroforestry systems: a systematic review on value chain implications. *Agricultural Systems* 206: 103606. https://doi.org/10.1016/j.agsy.2023.103606.

MacDonald, G. K., Bennetta, E. M., Potter, P. A. and Ramankutty, N. (2011). Agronomic phosphorus imbalances across the world's croplands. *Proceedings of the National Academy of Sciences of the United States of America* 108(7): 3086–3091.

McMichael, P. (2009). A food regime genealogy. *Journal of Peasant Studies* 36(1): 139–169. https://doi.org/10.1080/03066150902820354.

Muilwijk, H., Huitzing, H., de Krom, M., Prins, A. G., Vonk, M. and Westhoek, H. (2019). Dagelijkse kost. Hoe overheden, bedrijven en consumenten kunnen bijdragen aan een duurzaam voedselsysteem. © PBL Planbureau voor de Leefomgeving, Den Haag 2019. PBL-publicatienummer: 2638.

Rittel, H. W. J. and Webber, M. M. (1973). Dilemmas in a general theory of planning. *Policy Sciences* 4(2): 155–169.

Samireddypalle, A., Prasad, K. V. S. V., Ravi, D., Khan, A. A., Reddy, R., Angadi, U. B. and Blümmel, M. (2019). Embracing whole plant optimization of rice and wheat to meet the growing demand for food and feed. *Field Crops Research* 244: 107634. https://doi.org/10.1016/j.fcr.2019.107634.

Schader, C., Muller, A., Scialabba, N. E. H., Hecht, J., Isensee, A., Erb, K. H., Smith, P., Makkar, H. P. S., Klocke, P., Leiber, F., Schwegler, P., Stolze, M. and Niggli, U. (2015). Impacts of feeding less food-competing feedstuffs to livestock on global food system sustainability. *Journal of the Royal Society. Interface/the Royal Society* 12: 20150891.

Schröder, J. J., Aarts, H. F. M., Ten Berge, H. F. M., Van Keulen, H. and Neeteson, J. J. (2003). An evaluation of whole-farm nitrogen balances and related indices for efficient nitrogen use. *European Journal of Agronomy* 20(1–2): 33–44.

Stark, F., González-García, E., Navegantes, L., Miranda, T., Poccard-Chapuis, R., Archimède, H. and Moulin, C. H. (2018). Crop-livestock integration determines the agroecological performance of mixed farming systems in Latino-Caribbean farms. *Agronomy for Sustainable Development* 38(1): 4. https://doi.org/10.1007/s13593-017-0479-x.

Trienekens, J. H., van der Vorst, J. G. A. J. and Verdouw, C. N. (2014). Global food supply chains. In: van Alfen, N. K. (Ed.). *Encyclopaedia of Agriculture and Food Systems*, Academic Press, Cambridge, MA. 499–517.

Van Berkum, S., Dengerink, J. and Ruben, R. (2018). The food systems approach: sustainable solutions for a sufficient supply of healthy food. Wageningen,

Wageningen economic research, memorandum 2018-064. 32 pp.; 9 fig. 39 ref.: 0 tab. https://doi.org/10.18174/451505.

Van Bueren, E. M., Klijn, E. and Koppenjan, J. F. M. (2003). Dealing with wicked problems in networks: analyzing an environmental debate from a network perspective. *Journal of Public Administration Research and Theory* 13(2): 193–212.

Van Kernebeek, H. R., Oosting, S. J., Van Ittersum, M. K. and De Boer, I. J. M. (2015). Saving land to feed a growing population: consequences for consumption of crop and livestock products. *The International Journal of Life Cycle Assessment*. Special issue LCA of nutrition and food consumption:1-11.

Van Selm, B., Frehner, A., de Boer, I. J. M., van Hal, O., Hijbeek, R., van Ittersum, M. K., Talsma, E. F., Lesschen, J. P., Hendriks, C. M. J., Herrero, M. and van Zanten, H. H. E. (2022). Circularity in animal production requires a change in the EAT-Lancet diet in Europe. *Nature Food* 3(1): 66–73. https://doi.org/10.1038/s43016-021-00425-3.

Van Zanten, H. E. (2016). Feed sources for livestock: recycling towards a green planet. PhD thesis, Wageningen University, Wageningen, NL. ISBN: 978-94-6257-805-0.

Verstand, D., Vellinga, T. V., Poppe, K. and de Wolf, Pd. (2021). Manure and metropoles: a global framework for nutrient cycling in food systems. *Frontiers of Agricultural Science and Engineering* 8(1): 182–187. https://doi.org/10.15302/J-FASE-2020364.

Wang, J. M., Liu, Q., Hou, Y., Qin, W., Lesschen, J. P., Zhang, F. S. and Oenema, O. (2018). International trade of animal feed: its relationships with livestock density and N and P balances at country level. *Nutrient Cycling in Agroecosystems* 110(1): 197–211. https://doi.org/10.1007/s10705-017-9885-3.

Zhang, X., Davidson, E. A., Zou, T., Lassaletta, L., Quan, Z., Li, T. and Zhang, W. (2020). Quantifying nutrient budgets for sustainable nutrient management. *Global Biogeochemical Cycles* 34(3): e2018GB006060.

Chapter 3

Closed-loop combined crop–livestock farming systems

Friedhelm Taube, John Kormla Nyameasem and Friederike Fenger, University of Kiel, Germany

1 Introduction

Many national, regional, and international policies aimed at enhancing agricultural productivity focus on high productivity per unit input without considering the sustainability of the practices involved. These policies have led to the specialization and intensification of agriculture, characterized by simplified and standardized production techniques. They have also led to the decoupling of crop and livestock systems and homogenized agricultural landscapes. There is also strong specialization of production on a regional scale, in which only one kind of farm business is practised, such as growing food or feed crops or rearing sheep or dairy cattle.

Specialization and intensification encourage reliance on fertilizers and pesticides to ensure crop health which, if not correctly applied, can result in water, air, and natural ecosystem pollution (Tilman, 1999) and the contamination of food products (Rather et al., 2017), increasing human health risks. Extensive use of farm equipment and fossil fuels to apply these inputs has replaced more traditional measures to maintain soil and crop health such as rotations, fallow systems, animal manure, and legumes in mixed-farm situations. Specialization

http://dx.doi.org/10.19103/AS.2023.0120.08

and intensification have been accelerated by the industrial revolution which resulted in cheap fertilizers and feed concentrates to increase yields. The effect has been increased farm profit margins but led to low nutrient use efficiency, whether in animal nutrition or nitrogen use on crop fields.

Increasing demand for animal-sourced foods is leading to expansion and intensification of livestock production in many regions (Godfray et al., 2010). Specialization has resulted in changes in land use and reduced agro-diversity within landscapes as crop–livestock systems, forage crops, and grassland areas have diminished. More than 80% of the world's arable land is devoted to monocultures for grains and animals, and the amount of crop diversity per unit of arable land continues to decrease (Alteieri et al., 2015). In high-income countries, for instance, specialized dairy farms are typical in the livestock sector, in which highly productive dairy breeds are housed in confined conditions. These animals require diets high in energy and protein. To meet their nutritional needs, they are fed high levels of concentrates produced from arable crops, such as soy and maize, usually imported from external sources. In addition, the cultivation of cereal crops and maize silage on grasslands for ruminant feeding has decreased permanent grassland areas for pure forage crops and land for food crops. Consequently, farmed animals compete with humans for land, depending on the type of feed and feeding regimes adopted.

Specialization is associated with larger farm sizes, resulting in loss of grazing areas and shrinking farm sizes (Udo et al., 2011), threatening the profitability and survival of smallholder farms. Around half of the undernourished population globally are smallholder subsistence farmers, 20% of whom are landless families, mostly agricultural labourers in low- and medium-income countries (Sekaran et al., 2021). However, evidence suggests that smallholder and family farms are crucial to feeding the planet and that policies aimed at poverty alleviation, food security, and protection of biodiversity and natural resources are more successful when smallholder farmers are involved (FAO, 2014; Lowder et al., 2016). In addition, the diversity of crops and animal species on small farm holdings can help regulate ecosystem processes, mitigate the spread of pests and diseases across the landscape, and thus increase system resilience (Walia and Walia, 2015). It has also been shown that small farms in many systems have greater crop productivity per unit area than large farms (Tilman, 1999) and can contribute to creating more sustainable food systems.

Agriculture is responsible for about a quarter of all anthropogenic greenhouse gases (GHGs) (Poore and Nemecek, 2018). Livestock production is responsible for 60% of total agricultural GHG emissions, contributing significantly to climate change. Ruminant livestock excretes 70–80% of ingested N, resulting in surpluses that can increase the risk of N leaching (Ledgard et al., 2009) and N_2O emissions (Flechard et al., 2005), particularly as stocking rates and fertilizer N inputs increase. Besides N losses, ruminants release approximately 70% of the C they ingest into the atmosphere as CO_2 via respiration and as methane (CH_4) via rumen fermentation (Parsons et al., 2011). In some areas, grasslands on peat soils are drained and cultivated to produce feed for livestock, leading to huge losses of CO_2 (e.g. more than 15 tons ha^{-1} $year^{-1}$, Poyda et al., 2016). Climate change characterized by elevated CO_2, higher temperature, altered precipitation, and transpiration regimes and increased frequency of extreme events have impacted food and fibre production (Lobell et al., 2011). There is growing evidence that crop phenology and crop yields have already been impacted by increasing temperatures and days with extreme temperatures (Webber et al., 2014).

In addition to GHG emissions, intensive livestock farms can reduce biosphere integrity (Liebman and Schulte, 2015). This can result from overgrazing or intensive grazing, soil degradation through livestock trampling (Fleischner, 1994), improper disposal of animal waste, and the use of antibiotics and hormones. Low biosphere integrity and simplification of regional landscapes can lead to biodiversity loss, lack of pollinator habitat, and rapid spread of pests and diseases with dire implications for food security. The low nutrient-use efficiencies in livestock systems result in deep groundwater pollution, lake and surface water eutrophication, forest acidification and eutrophication, and nitrous oxide emissions. In addition, livestock production encourages deforestation and takes up 40% of the world's ice- and desert-free land (Poore and Nemecek, 2018). About 70% of all agricultural land (arable land and grassland) is used for livestock production, 40% of which is used for feed production.

Expanding the livestock sector in tandem with increasing global food demand using business-as-usual practices will also further damage an already fragile agro-ecological environment. Global agriculture must ensure higher production levels to support the ever-increasing world population while ensuring its impact on the environment does not undermine already precarious ecological systems. An ever-increasing demand for food due to a growing population combined with yield stagnation in specialized monocropping crop systems requires new paradigms to increase productivity. Looming social costs due to poor production practices, high food prices, and malnutrition, which have resulted in increased social unrest, must be addressed. Relevant questions include whether there are alternatives to high-input specialized systems, whether ecological intensification is possible, and how to utilize livestock to ensure sustainability. This chapter shows that integrating crops and livestock can yield multifunctional benefits, including ecosystem services. It will also demonstrate that ley systems can contribute to resilient, circular agricultural systems and provide examples of how this can be implemented in practice.

2 Solutions to improve the environmental impact of livestock farming

Various solutions have been proposed to reduce the negative impact of livestock production on the environment (which can be understood in terms of production and consumption narratives) (van Zanten et al., 2018). The production narrative focuses on reducing the environmental impact per unit of animal-source foods (ASF) by increasing yields per unit input or per emission produced. Measures to achieve this will usually include precision fertilization to minimize nutrient losses from the field; increasing lifetime productivity of the herd by improving feed digestibility and use efficiency, reproductive rate, and yields; and enhancing disease control and manure management. In this sense, animal breeders will always prefer the highest-producing cows. On the other hand, high genetic merit cows will require high concentrate use, promoting feed-food competition and discouraging grass-based production (Herrero et al., 2016).

As shown by Flachowsky (2008), if we assume that the human population will consume ASF from dairy in a ratio of 1000 tons of milk to 30 tons of meat, which is also the nutritional expectation, then a cow yielding 12 000 kg of milk per year will require additional resources to fulfil this demand with dire implications for GHG emissions and land use efficiency (LUE). However, an intermediate breed producing about 8000 kg of milk has been shown to more efficiently meet this demand with a lower environmental footprint. Thus, maximum milk performance is not always in line with the lowest C footprint.

The circular agriculture concept requires that food production and consumption patterns are adapted and matched with local productive capacity (Schut et al., 2021). Accordingly, the consumption narrative proposes eating less or no ASF to minimize the environmental footprint of food and feed-food competition. A radical shift from animal-based to plant-based foods will reduce the environmental impact on the agricultural sector and make land available for biodiversity conservation, for energy generation via photovoltaic cells, and to produce more and potentially cheaper crops to feed less affluent populations.

However, animal-based foods supply protein, energy, and highly bioavailable micronutrients, such as calcium, iron, and vitamin B1. Farmed animals also represent more than just food in many societies, especially in Asia and sub-Saharan Africa, where herd size is synonymous with wealth and status. Animals represent a significant component of the diet for many smallholders, serve as a form of insurance, and play other cultural roles, such as providing a dowry for marriages. The consumption narrative ignores the interconnection between feed and food systems. It fails to acknowledge that agricultural products come as a package and that non-eatable residues from crops can be used to produce high-value eatable animal products, thus reducing feed-food competition and highlighting animals as recyclers of by-products in crop production systems (van Zanten et al., 2018).

The supply and consumption of animal-based foods in Asia and Africa are within the land-use boundary (van Zanten et al., 2018); however, over-consumption in many high-income regions, food surpluses, and waste must be addressed. This suggests that the current eating habits and land use patterns in wealthier regions are incompatible with climate goals. Aligning eating habits along the lines of the Planetary Health Diet (EAT-Lancet, 2019) would reduce GHG emissions drastically, freeing land to create an additional carbon sink or to feed people across the globe. Many dietary recommendations are characterized by an increasing share of legumes, nuts/seeds, and vegetables instead of meat, butter, egg, and fish products, as well as fruits (Meier and Christen, 2013).

Consuming a vegan diet has been estimated to reduce land requirement per person by 50% in Germany, using 2006 food intake levels as the baseline (Meier and Christen, 2013). For instance, the German Nutrition Society (Deutsche Gesellschaft für Ernährung (DGE)) and the Federation for Independent Health Consultation (Unabhängige Gesundheitsberatung (UGB)) suggest a reduction of 40–61% of eggs and meat relative to 2006 intake levels. This change, if implemented, could release 2.6-3.0 million ha of land and reduce GHG emissions by 25 million tons year^{-1}, reduce NH_3 emissions by 22–28%, and save about 142–182 million kg of N year^{-1}.

Considering the concept of agricultural circularity and the role that ASF can play in feeding the world, a third pathway, which takes advantage of the inherent trade-offs in the food system, has been proposed (van Zanten et al., 2018). This narrative requires that consumption is capped at levels that avoid food–feed competition, reducing the need for arable land. Accordingly, some have suggested that livestock should only exist in our future food system when feed–food competition is minimal (Van Zanten et al., 2018). In this regard, farmed animals are considered converters of by-products that humans cannot or do not want to eat into valuable products (meat, milk, eggs, and manure) while providing various ecosystem services. In their study, van Zanten et al. (2018) showed that eating food from recyclers and using left-over ecological areas for animal production could meet 14–23% of protein demand per capita per day from ASF. This offers the most efficient way of using arable land for food production. This approach is estimated to reduce N and P losses by 40% and 46%, respectively, and GHG emissions by 19–50%, compared with a business-as-usual scenario (Röös et al., 2017).

However, the production of low-cost animals may be limited by the availability and quality of diets and the productivity level of the animals, which might favour less productive breeds. In any case, the debate must go beyond whether white meat is better than red meat but be guided by the ability of the animal species/

breed to take advantage of the inherent trade-offs in food systems and the ability of the animal to convert low-cost diets into valuable products with much higher efficiency. For instance, whilst ruminant livestock production results in relatively higher GHG emissions compared with non-ruminants, the production of the latter results in competition with humans for land.

3 Ecological intensification: a critical concept for future farming systems

In addition to the measures discussed previously, agriculture must deal with the current puzzle of producing more food to meet increasing global food demand without harming the environment and by contributing positively to natural and social capital (Pretty and Bharucha, 2014). Kurth et al. (2020) argue that carbon storage through undersown and catch crops, more complex crop rotations, pasture-based livestock farming on extensive grasslands, input optimization, and nature conservation are potential measures to reduce the external costs associated with agriculture. Agricultural practices must aim to reconcile productivity with long-term environmental sustainability, supporting soil functionality and improved nutrient recycling to reduce the negative externalities of agriculture. This might involve the utilization of crop varieties and animal breeds that have high productive efficiency or whose production does not require the unnecessary use of external inputs. Furthermore, systems for producing food must take advantage of agro-ecological processes (such as nutrient cycling, biological N fixation, etc.), adopting technologies or practices that positively impact environmental and human health, including GHG mitigation, clean water, biodiversity, C sequestration, and minimizing the spread of pests and diseases.

It is natural to assume that such a system might not be able to produce enough food for the ever-growing population. It is noticeable that yield stagnation has been reported in specialized crop systems. For example, yields of cash crops (cereals, oilseed rape, maize, and potatoes) in Germany have shown no significant increases in 20 years (Fig. 1). The current impacts of climate change and major yield gaps in smallholder systems have resulted in insufficient food supply and low income, reducing food security for small farm holdings, particularly in the developing world (Global yield gap atlas, 2022). However, a study (Pretty and Bharucha, 2014) has shown that implementing sustainable measures increased crop yield by 22–146% within 3–10 years, providing about 557 kg of aggregate food per farming household.

Implementing sustainable agricultural schemes does not require a single global solution. Particular solutions must be developed in a local context, taking into consideration production and other ecosystem functions (i.e. ecological intensification) (Garnett and Godfray, 2012; Taube et al., 2014).

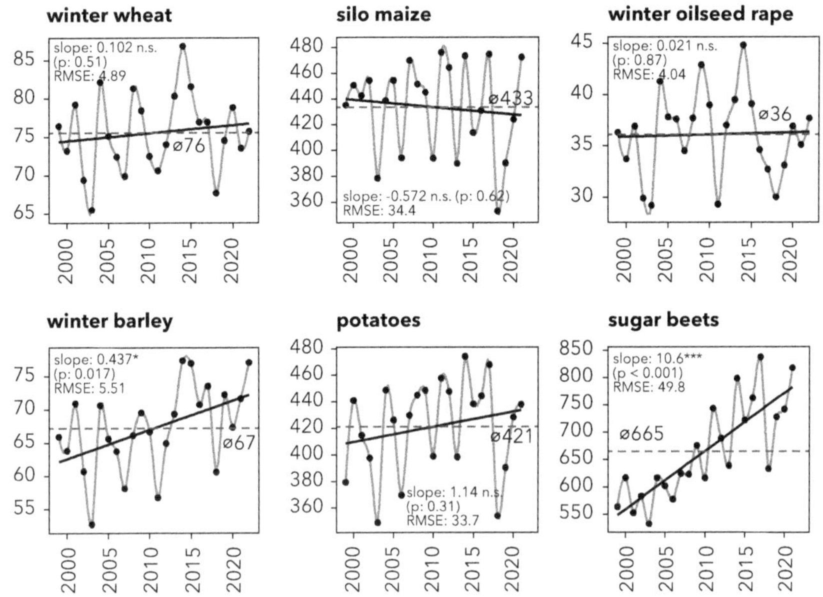

Figure 1 Yield trends of cash crops in Germany: no significant yield increment for 20 years for cereals, oilseed rape, maize, and potatoes (Federal and State Statistical Offices, 2022).

The EU Green Deal, for instance, envisages that crop systems will solve the nitrogen problem if fewer agro-chemicals, fewer nutrient losses, less fertilizer input, more carbon sequestration, and more agro-biodiversity are encouraged. However, this will not work in specialized arable systems. In Europe, e.g. regulations have been implemented to control animal density or the amount of manure to apply on the field. Animal manures may be transported, relocated, or disposed of at a cost to the livestock farmer. Other regulations are aimed at enhancing nutrient use efficiency by regulating times for spreading manures, the amount of N to apply, and using low-emission spreading techniques to increase efficiency and reduce ammonia emissions. Nonetheless, farm-to-gate N balances remain high for specialized ruminant farms; for example, farm-to-gate N surpluses on dairy farms in northern Germany are around 200 kg N ha^{-1} (Taube, 2016). High N-balances mean that nutrient losses to the environment in the form of nitrogen leaching are very likely, especially in sandy soils. Accordingly, around 75% of the N-surplus can be directly linked to environmental pollution. At an N surplus of only 100 kg ha^{-1}, the 'social cost' of environmental pollution in the form of N leaching (ca. 37 kg N ha^{-1}), ammonia volatilization (ca. 30 kg N ha^{-1}), and gaseous emissions in the form of N_2O and NOx (ca. 8 kg N ha^{-1}) is estimated to be around €989 ha^{-1} (Taube, 2016). This is the story in many developed economies and the

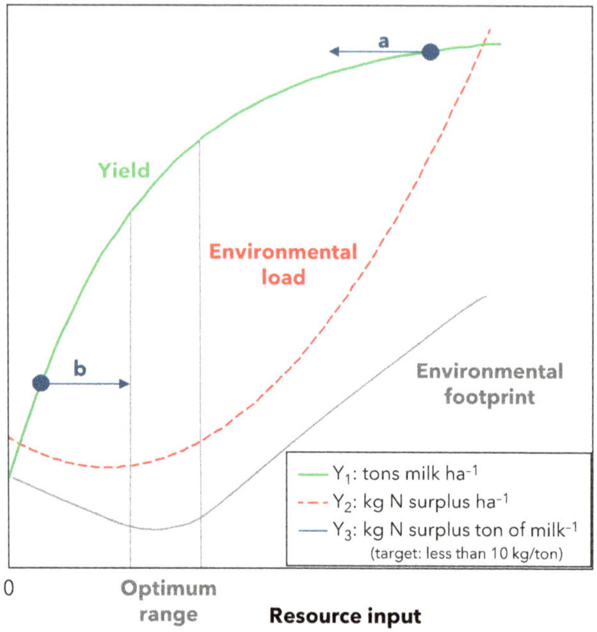

Figure 2 Eco-efficiency (environmental footprint) as a relationship between yield (e.g. milk production) and environmental load [e.g. N input (x) > N surplus > N losses (Y_2)].

most critical environmental issue affecting intensive farming systems in the EU (Taube et al., 2015).

It is imperative that high-input farming systems, characterized by high environmental load, de-intensify the production of agricultural commodities to a point on the production function curve where yield and environmental load are optimal, as illustrated in Fig. 2 (farms employing high input levels at point *a* on the production function curve). On the other hand, many smallholder farms in developing countries are located on the lowest portion of the production function curve (point b). Such farms are characterized by low productivity, efficiency, and environmental load, primarily due to low input levels. For such farms, some level of intensification will be necessary to increase feed/food production and land use efficiency, focusing on enhancing ecological benefits (ecological intensification). In this context, ecological intensification should aim at complementing or replacing artificial inputs with nature-based resources to reduce environmental costs without negatively impacting productivity.

Intensifying agriculture to enhance productivity and increase land use efficiency in livestock production in poorer economies requires specific measures, including increasing the levels of inputs, developing and adopting improved breeds, improving feed quality and supply, especially for dry season feeding, effective grazing management, and disease and parasite control. In a study to

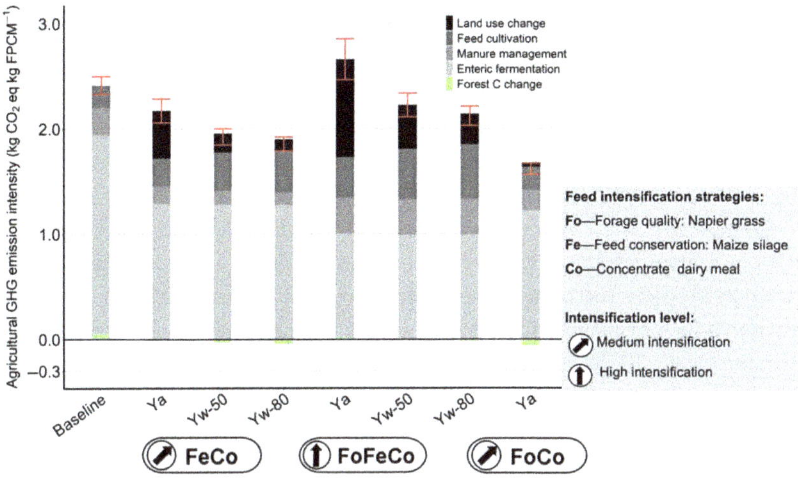

Figure 3 GHG emission intensity per kilogram fat and protein corrected milk (FPCM). Bars indicate the baseline and three scenarios at medium (FoCo, FeCo) and high (FoFeCo) intensification levels. FoCo and FoFeCo included maize silage, where we show actual maize yields (Ya) and yields at 50% (Yw-50) and 80% (Yw-80) of water-limited yield potentials (Yw). Error bars indicate relative SDs (Brandt et al., 2020).

assess the potential of sustainable intensification of dairy production in sub-Saharan Africa, Brandt et al. (2020) showed that increasing concentrate feeding from an average of 0.06 to about 0.74 kg day^{-1} cow^{-1}, combined with either maize silage or Napier grass (medium intensification), increased milk yield by 44–51% and reduced GHG emission intensity from 2.4 to 1.6 kg CO_2 eq kg FPCM^{-1} for Napier and 2.2 kg CO_2 eq kg FPCM^{-1} for maize silage (Fig. 3). However, a high intensification approach, which involved feeding about 1.5 kg of concentrate day^{-1} cow^{-1} combined with Napier and maize silage, increased GHG emission intensity by 13%, attributable to land use change emissions. The authors also showed that whilst intensification increased GHG emissions above the baseline, closing the yield gap for maize via N fertilizer application reduced GHG emission intensity, and intermediate intensification with high-quality grass (Napier) resulted in the lowest GHG emissions and improved land use efficiency.

For cropping systems, the most frequently implemented management practices to increase the sustainability of farming in delivering multiple ecosystem services have included conservation tillage, cover cropping, crop rotation, and mixed cropping. Using cover crops, reduced or no tillage in a conservation agriculture approach, combined with moderate grazing, could further strengthen the coupling of C, N, and P. However, farmers' adoption of these biodiversity and ecosystem service-enhancing practices generally favour practices that interfere the least with normal farming operations which may limit their effectiveness (Kleijn et al., 2019).

4 The potential of mixed cropping and integrated crop–livestock systems

The lack of biogeochemical and ecological controls due to the loss of diversity within cropping systems is partly why synthetic fertilizers and pesticides use are on the rise. Agricultural land management should maintain or improve crop yields and soil quality while limiting GHG emissions. Resilient plant-based production systems can be developed using 'plant teams'. The benefits of such plant associations depend on their ability to better acquire and manage resources in space and time (niche complementarity), the beneficial interactions between species (facilitation), and the selection of the best-performing species (Brooker et al., 2015; Cardinale et al., 2007; Ghani et al., 2022). These plant teams may include grain–legume combinations, harvested simultaneously for both human consumption and livestock feed, and grain–green manure combinations, where the green manure can serve as an alternative to chemical inputs, or undersowing of single- or multi-year crops onto ground-bearing grain crops to maximize the use of both time and space.

Such mixed-crop scenarios or rotation arrangements contribute to soil fertility, the productivity of the 'main' crop, and the suppression of weeds, diseases, and infestations. Other benefits might include an opportunity to enhance biodiversity both above and below ground by providing food and habitat for wildlife. In addition, biological activity is stimulated by plant root exudates and earthworms to promote the formation of microaggregates in soil. Finally, crop rotation can positively affect water dynamics to increase water use efficiency and reduce the risk of yield reduction (Yu et al., 2022).

Historically, integrated crop–livestock systems (ICLS) have been essential for recycling nutrients in the farming system, but the advent of synthetic fertilizers reduced their importance (Schut et al., 2021). ICLS maximize the interaction between system components, i.e. soils, plants, and animals, and, therefore, have the potential to decrease environmental impacts whilst maintaining production levels (Ryschawy et al., 2017).

Coupling crop and livestock systems can be an economically viable strategy for farmers (Wilkins, 2008; Ryschawy et al., 2012), providing flexibility regarding the allocation of land and forage resources in response to climate and price. For instance, cropping dual-purpose cereals and canola (*Brassica napus*) for forage during the vegetative stage – a common practice in southern Australia – has been shown to provide risk management benefits through diversified crop rotations and pressure reduction on other feed resources (Bell et al., 2014). This practice significantly increased livestock and farm crop productivity by 25–75% with little input increase. Further, sacrificial crop grazing by livestock, when expected grain yield was low, and/or livestock prices were attractive relative to grain, provided flexibility in managing crop and livestock during variable climatic conditions. Other examples of crop-livestock integration innovations that provide a range of benefits include replacing temporary pastures with perennial pasture phases in rotation with crops and integrating perennial forages in mixtures with croppings, such as alley cropping and inter-cropping. In addition, crop-livestock integration can reduce herbicide use and mowing in orchards and vineyards (Jeffries, 2015), improve weed management, and sustain yields even at reduced input levels (MacLaren et al. 2018).

Integrating crop and livestock systems at the farm, landscape, or regional level will optimize food production and contribute to the diversification of a cropping system while simultaneously creating the opportunity to exchange feed and manure and mitigate external inputs, such as synthetic fertilizers and feed. This way, plant-bound nutrients are converted into animal-source food, and readily mineralizable substrates are returned to the soil (Karlsson and Röös, 2019; Franzluebbers et al., 2014). This can help close the nutrient loop and contribute to a circular food system (Boer and Ittersum, 2018), particularly in the face of legislative restrictions on the use of artificial N fertilizers in some regions (Vellinga, 2006). Furthermore, the nesting of fields on farms in a regional economy provides opportunities for efficiency gaps to be reduced by internal recycling of waste (van Noordwijk and Brussaard, 2014). This may also overcome constraints, such as increased labour, extra skills, and infrastructure required for a mixed farm (Regan et al., 2017). For this to yield the expected benefits, however, effective management of the agro-environment, with particular consideration for topography, hydrology, and ecology at farm, regional, and territorial level, is needed. However, the reintegration of specialized farms may also lead to further land use intensification, suggesting the development of new approaches within a food system context to offer alternatives to reintegration (Schut et al., 2021).

5 The role of grasslands in cropping and livestock systems: the case of grass-legume leys

Production grasslands, consisting of both permanent and temporary grasslands, not only provide feed for herbivores and ruminants but can also offer additional benefits. Grasslands have a long growing season, a dense, fibrous root system, and a substantially larger below-ground biomass production than annual crops, with positive effects on soil biota and C sequestration (Loges et al., 2018; van Eekeren et al., 2008). The initially high C:N and lignin:N ratios (Rasse et al., 2005) from predominantly dead leaf and root material increase the mean residence time of C in grassland soils. Grasslands promote a strong relationship between C, N, and phosphorous (P) in the soil (Nyameasem et al., 2020), preventing losses of nutrients due to leaching and denitrification (Lemaire et al., 2014). In addition, grasslands may enhance soil biological diversity, controlling weed communities, insects, and disease populations (Franzluebbers et al., 2011). The disappearance of forage crops and grassland areas in crop rotations reduces the potential of an ecosystem to offer services traditionally served by grassland systems. However, the level of intensification (i.e. the inputs of manure and fertilizer, grazing pressure, mowing frequency, and grassland renewal) affects grassland productivity and ecological value.

One effective way to integrate crop and livestock farming might be through sod-based rotations or ley-arable systems (Katsvairo et al., 2006; Allen et al., 2007; Franzluebbers, 2007). Leys are temporary grasses or grass-forb mixtures grown during arable crop rotations. The system can capture the benefits of leys in combination with periods of intensive cropping to minimize environmental impacts. The integration of leys into rotations can take different forms. This may include the integration of various types of livestock grazing with cropland, legumes, and forages rotated with pastured dairy, cotton rotated with pastured beef, and small livestock grazing in fruit and nut orchards and viticulture systems (Garrett et al., 2017). Grass/legume leys are commonly included in the crop rotation of mixed-farming systems to maintain the supply of nutrients and maintain soil fertility. In addition, combining cash-crop production with temporary grasslands, especially those containing legumes, can ensure protein self-sufficiency of the livestock farm and nutrient cycling via biological N fixation (Peyraud et al., 2014).

5.1 Grass-legume leys increase the yield of arable crops in rotations

A mixed-farming system with legume leys can serve a dual purpose, providing forage for livestock, the manure from which can be applied to arable crops. Grass-legume leys have positive above- and below-ground effects, including improving soil structure and moisture content and minimizing soil microbial or fungal populations and plant disease pressure. Furthermore, the termination of grasslands results in a large quantity of mineral N being available to subsequent crops, thus reducing N fertilizer requirements in the arable phase of the rotation with greater yields of the subsequent arable crop due to the additional rotation effect (Lemaire et al., 2015; ten Berge et al., 2016). Grass-legume mixtures can increase dry herbage matter and N yields and improve soil structure more than grass monocultures (de Haas et al., 2019). Moreover, in a related study, Younie and Hermansen (2000) demonstrated that high clover presence in ley farming can control perennial weeds on organic farms.

An experiment (Nevens and Reheul, 2001, 2002, 2003) comparing continuous maize cultivation and a 3-year grass and 3-year maize rotation reported similar maize yield for both systems; however, N yield, ear fractions, and feed energy yield were higher for the rotation system, saving about 231 kg ha^{-1} of mineral N fertilizer. This shows that a crop rotation of grass and maize can be sustainable in terms of efficient nutrient use. Although the long-term effects of rotations and agronomic inputs are evident in both high-yielding and low-yielding years, yield gains of crop rotations over monoculture are more pronounced in low-yielding years (Borrelli et al., 2014).

5.2 Leys can improve nutrient cycling to reduce nitrate pollution

The leaching of nitrates into groundwater, and their presence in surface runoff, depends on excess mineral nitrogen, the hydrological regime, land use, soil type, and climatic conditions. In an arable cropping system, plant-soil interactions are frequently interrupted, decoupling C-N dynamics, and increasing environmental risks of N losses. Leys within ICLS are an option to reduce nutrient surpluses observed in rotations with annual crops (Kunrath et al., 2015).

The presence of multiple functional groups in grasslands can effectively deal with several environmental issues. Grasses spread their roots at soil depths, allowing plants to capture nitrate before it can leach and contaminate water. Moreover, the permanency of carbon flow through perennial vegetation to the soil under grassland enables a rapid N immobilization-mineralization turnover by soil microbes, preventing the accumulation of N as nitrate in the soil to reduce the risk of leaching. By comparing grassland/ley systems to pure cropping systems on silty-loamy soil in France, Kunrath et al. (2015) reported high NO_3 concentrations above acceptable levels in drinking water for the pure cropping system.

Good management practices following grassland cultivation may be required to reduce nitrate losses. For example, N mineralization may exceed the requirements of the subsequent crop or be released asynchronously with crop demand. Good practices might include delaying ploughing until late winter or spring, using efficient catch crops, and reducing fertilizer N application. Mowing, which ensures the continuous growth of grassland, ensures continuous N uptake from the soil to reduce surpluses. Thus, increasing the proportion of grassland in cereal rotations can lead to marked exhaustion of nitrate concentration in drainage water.

5.3 Leys can facilitate greenhouse gas mitigation and carbon storage in soils

Ley-crop rotations (temporary grass of 2–5 years) can provide indirect ecosystem services, such as improved soil fertility and biodiversity, controlled weed and pest populations, and saved energy due to the economy of tillage and N fertilizers (Lemaire et al., 2014). Leys can facilitate soil organic carbon (SOC) sequestration as grassland ecosystems are a general sink for carbon dioxide (Goudriaan, 1992). Multispecies swards in leys, including tannin-rich species, may have additional benefits. For example, tannin-rich species could limit N_2O emission from soils via biological nitrification inhibition (de Klein et al., 2020) and mitigate enteric methane emissions when fed to livestock (Beauchemin et al., 2020) to further reduce its environmental footprint.

In addition, plant secondary metabolites in both above- and below-ground biomass have been shown to impact C and N cycles (Kagiya et al., 2019) and can slow down the decomposition rate of soil organic matter (SOM) (Dong et al., 2016) to promote SOC accumulation in the long term (Tamura and Tharayil, 2014). The equilibrium dynamics of C in soils under different land use is a function of the amount of plant residues returned to the soil, their litter C content, the amount of soluble/nonsoluble C, mineralization rates of C components, their placement, and degree of soil aggregation. Including perennial leys in crop rotations can increase soil fertility due to their positive effect on SOM and the quantity and quality of organic matter inputs from subsequent crops. The organic matter input by grass and forage crops provides the soil mineral matrix with a high degree of physical protection, improving soil structure, water infiltration, and nutrient cycling. Soil aggregation leads to the physical protection of C from mineralization by microorganisms and results in long-term soil C sequestration.

Nevertheless, the breakup of permanent grassland for conversion to arable fields typically leads to increased mineralization from SOM and plant residues. This increases the risk of nitrogen and carbon emissions to the atmosphere and loss of N to groundwater, especially when farmers apply additional organic and/or mineral fertilizers. Accordingly, the beneficial effects of leys on C sequestration are attributed to a reduction in tillage and a change in the amount and distribution of carbon input into soils. When organic crop management is combined with reduced tillage, a grass–clover ley can increase soil aggregate stability, suggesting short-term ley as a strategy to improve soil structure for cropping systems under intensive tillage (Puerta et al., 2018). When vegetation develops a dense root system, it may improve soil aggregate stability and favour the protection and stabilization of SOM and associated nutrients; however, this process requires time. An increasing proportion of grass–clover ley in the crop rotation results in significantly larger topsoil organic carbon content, smaller bulk densities, and consequent increases in total porosity (Jarvis et al. (2017). By comparing a 2-year ley period with a 4-year crop rotation, Blanco-Canqui et al. (2017) concluded that the time under ley (i.e. 2 years) was insufficient for the plants to develop an extensive and deep root system to build soil C concentration in the subsoil.

5.4 Leys can improve biodiversity

Multispecies leys might potentially conserve and improve biodiversity by supporting a range of insect species (Badenhausser et al., 2009), thus contributing to the food chain for birds and other species (Lemaire et al., 2015). Furthermore, grassland leys in crop rotation may improve soil health and microbial activity. Grasslands were reported to increase bacterial biomass compared with arable fields (van Eekeren et al., 2008; Bloem et al., 2006). The ability of grasslands to enhance soil biology might be due to the constant high food or C supply to the soil and the more stable environment compared with arable fields. In addition, high C and N returns from grazing animals increase microbial population due to the positive effect on the accumulation of organic residues in the soil.

Compared to continuous arable cropping, leys are reported to improve soil structure through several biological processes; e.g. larger litter inputs, reduced tillage disturbance, and support of larger populations of earthworms. For example, when land use changed from arable cropping to perennial pasture, Yeates et al. (1998) observed increased earthworm populations. Two years later, Bouwman and Arts (2000) also found a higher number and percentage of herbivorous nematodes in a treatment with a higher grassroot density in the upper soil layers. Regarding the functioning of soil biota, ley-arable crop rotations were found to be intermediate to permanent grassland and continuous arable land (van Eekeren et al., 2008), making permanent grassland preferable.

Additional benefits may accrue to ensure ecological stability and resilience if agriculture is integrated with habitat areas within landscapes. Setting aside areas to serve as habitats can encourage biodiversity across the landscape. Such areas provide adequate food, water, and sheltered areas (brush piles, wooded areas, and rock piles) suitable for wildlife to raise their young. Buffers, such as a band or strip of perennial vegetation established on the edge of a cropland field or areas of grass or other permanent vegetation (filter strips), can also be created to reduce runoff and seepage of agro-chemicals to hydrological systems. These buffers can also serve as wildlife habitats and food for terrestrial and aquatic life, including microorganisms. In addition, vegetative barriers, including deep-rooted trees and permanent grasses, can trap nitrates from polluting water bodies.

6 Case study: comparing specialized and integrated systems to improve the eco-efficiency of dairy farming in Germany

Potential solutions to the challenges discussed earlier have been evaluated in the 'eco-efficient pasture-based milk production' project on the organic experimental farm 'Lindhof' of the University of Kiel in Schleswig-Holstein in northern Germany (www.lindhof.uni-kiel.de). The objective for setting up an ICLS with dairy cows grazing grass-clover leys was to demonstrate the eco-efficiency of pasture-based milk production with multiple ecosystem services (food production, clean water, climate change mitigation, biodiversity, animal welfare, and attractive landscapes) and maximized milk production per hectare from temporary grasslands. It was hypothesized that the combination of low N_2O emissions and carbon sequestration of ley systems, together with the benefits of low-input pasture-based milk production, results in the lowest product carbon footprint (PCF) of milk from an ICLS.

Grass-clover leys form part of a crop rotation in which the cash crops are able to benefit from carry-over effects of nutrients and carbon from the 2-year ley phase. Grazing Jersey cows are managed in a spring calving intensive rotational grazing system with low input of supplemental concentrates. The system has shown very low N leaching losses (Smit et al., 2021), low N_2O emissions (Nyameasem et al., 2020), and low methane emissions per unit of milk output from the grazing jersey cows (Loza et al., 2021) measured with the SF_6 Tracer technique (Fig. 4).

A comparison between our ICLS system and specialized dairy farming typical of the region has recently been conducted in a case study by Reinsch et al. (2021). The objective of the study was to compare four prevailing dairy

Figure 4 Grazing Jersey cow on Lindhof wearing measuring equipment for assessing methane emissions per cow with the SF6 Tracer technique.

systems to assess productivity and the environmental impact of specialized compared to integrated systems. The study was undertaken in a region (Schleswig Holstein) in which farming systems are typically highly specialized in either crop or livestock production, predominantly cereals and dairy. The humid climate and fertile soils of the region allow high yields of both cereals and forage crops. The specialization resulted in high farm N surpluses and associated nitrate leaching losses to groundwater, specifically in areas with a high density of dairy cows where Taube et al. (2015) estimated average farm N surpluses to exceed 150 kg N ha^{-1}.

The study compared four farm types:

- Intensive confinement (IC);
- Semi-confinement (SC);
- Full-grazed (FG); and
- Integrated full-grazed (IFG).

These reflected a gradient in the proportion of feed intake from grazed grass compared to feed rations (Fig. 5). The IC system represented highly specialized intensive dairy farming (>10 000 kg ECM per cow per year). Cows are fed entirely indoors with a total mixed ration of grass and maize silage and concentrates. SC represented an average milk producer of the region with performance indicators comparable to the average found in regional farm surveys (LKSH, 2022). Cows were fed grass silage, maize silage, and concentrates indoors and additionally had access to some pasture during the day from May to September on permanent grasslands. The FG and IFG systems represented farms untypical

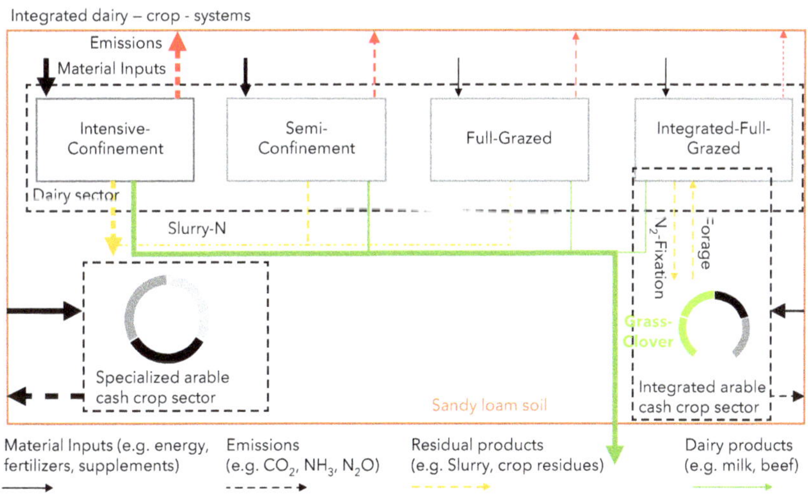

Figure 5 Assumption of input and emission flows within the prevailing dairy systems of the case study by Reinsch et al. (2021). The system boundaries indicate the scope of the conducted farm evaluation.

for the region and were therefore tested as an alternative farming strategy to the 'business as usual' scenario outlined earlier. Cows were managed in a rotational grazing system with full-day access to pastures with diverse grass-clover swards. No mineral N fertilization was applied in either FG or IFG. IC, SC, and FG were all specialized dairy farms whilst IFG was part of the ICLS on Lindhof. N_2O emissions and N leaching were measured directly on the farms. The measured data were used in the calculation of the PCF, which was calculated based on a cradle-to-gate consequential life cycle assessment approach.

The ICLS showed the lowest GHG emissions per hectare and per kilogram of ECM and the lowest nitrogen footprint between the four farm types (Table 1). The low imports of N (no mineral fertilizer and low concentrate input) and the export of manure and biologically fixed N in the form of plant residual matter in the ICLS resulted in a low N surplus in IFG. A clear relationship between N balance and GHG emissions showed that farms with higher N surplus also had higher GHG emissions per unit of milk and per unit area (Fig. 6). The farm type representing an average milk producer in the region (SC) showed the highest PCF due to inadequate feeding management and high N losses. In high-input indoor systems, GHG emissions per product can be negatively correlated with milk yield per cow as the high resource input is compensated by greater output (Lesschen et al., 2011). Hence, IC showed a lower PCF than SC. Nonetheless, the very low PCF of the ICLS shows that the relationship between PCF and milk yield per cow is dependent on the level of resource input as described by Lorenz et al. (2019).

Table 1 Results of the four farm types compared in the case study by Reinsch et al. (2021)

	Intensive confinement (IC)	Semi-confinement (SC)	Full-grazed (FC)	Integrated-full-grazed (IFC)
Milk yield per cow (kg ECM cow^{-1} $year^{-1}$)	11 152	9484	6060	6867
Milk per area (kg ECM ha^{-1})	15 817	11 512	7420	10 394
Farm N balance	229	200	94	50
Nitrogen footprint (g N surplus ECM^{-1})	19	22	15	7
Land use efficiency (kg ECM ha^{-1a})	1.2	1.2	1.4	1.3
Product carbon footprint[b] (kg CO_2 eq kg ECM^{-1})	1.1	1.2	0.9	0.6
GHG emissions[b] (kg CO_2 eq ha^{-1})	17 131	13 800	6647	6130

[a]Land required for feed production on- and off-farm.
[b]Including soil carbon sequestration and credits for N exports.

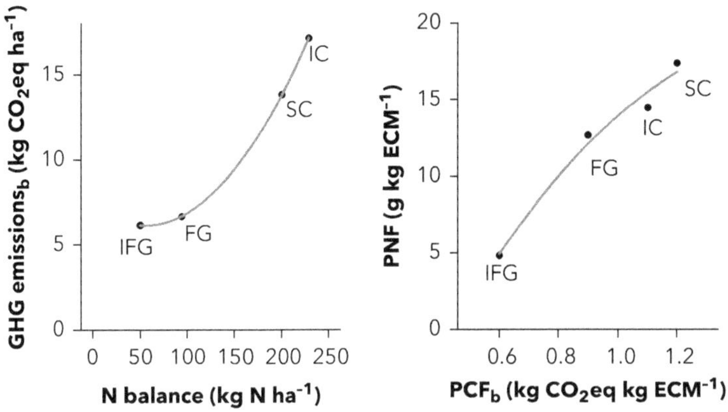

Figure 6 Relationship between farm N balance, GHG emissions, product carbon footprint (PCF), and product nitrogen footprint (PNF), among the different systems (IC, intensive-confinement; SC, semi-confinement; FG, full-grazing; IFG, integrated full-grazing) according to Reinsch et al. (2021).

The study once again highlighted that it is very difficult for farm systems exclusively specializing in dairy production to achieve farm N surpluses below 100 kg N ha^{-1}. A decrease from an average farm N balance of 150 kg N ha^{-1} (Taube et al., 2015) to 50 kg N ha^{-1}, as realized in our ICLS, would therefore result in substantial mitigation potential for social costs caused by environmental N pollution. Nonetheless, milk output per hectare in specialized IC systems was the highest. In combination with the advantages of economies of scale, specialized ICs have greater potential to generate high incomes and compete in the milk market as long as social costs are not accounted for. This depends on milk prices and the extent of the competitive advantage of low feed costs generated from grazed grass. This further highlights that a 'Resilient ICLS' Narrative (ley systems + pasture-based dairy) that provides a significant contribution to more functional diversity in agriculture (Beye et al., 2022) considerably reduces social costs and works towards fulfilling the Farm to Fork Strategy (F2F) and warrants more political support. Only a rapid transformation of the agricultural sector can achieve sustainable food production goals and protect human and environmental health.

7 Conclusion and future trends

The case study presented clearly demonstrates the benefits of a grass-clover ley-based ICLS compared to specialized systems for a range of ecosystem services in terms of water and air quality protection, climate change mitigation, and agro-biodiversity while operating at an overall high level of land use efficiency. The study confirmed that ongoing specialization and the intensification of dairy production do not adhere to the need for mitigating GHG and nitrogen emissions at a regional level and thus are not following the principles of ecological intensification. Coupled with a change in human food consumption patterns lowering the requirement to maximize production per unit of land, ICLS provides a blueprint for more sustainable future food production systems.

In practice, a move from specialized systems to an ICLS within one farm requires a full reorganization of farming infrastructure while potentially sacrificing some of the benefits of specialization. Management complexities and required skill sets are higher (Garrett et al., 2020). In 2016, just 2.5% of farms in the EU 10 states were mixed crop-livestock farms (Schut et al., 2021). Until recently, research on ICLS has mainly focused on the farm level (Garrett et al., 2020). Another way of exploiting both the economic benefits of specialization and the environmental benefits of ICLS would be a redevelopment of crop-livestock integration at the local level with intersectoral cooperation between specialized farms (Ryschawy et al., 2017; Taube, 2021).

The F2F strategy aims to increase the proportion of organic farming at EU level to 25%. Organic production is typically accompanied by lower yields, especially in the field crops sector, and requires a larger land area (Seufert et al., 2012). Hence, there are concerns about whether a food production system, producing solely within the boundaries of organic farming principles will be productive enough to feed the world's population (Muller et al., 2017). We argue that 'Hybrid Agriculture' – a system combining organic and conventional production, together with intersectoral cooperation between specialized farms – is a further essential step towards ecological intensification (Taube, 2021). Organic – or close to organic – forage production from grass-clover leys can be as productive as conventional forage production while reducing negative environmental impacts (Loges et al., 2018). A 2-year ley phase that is followed by an organically managed summer field crop, e.g. oats, can efficiently utilize the N mineralized from plant residues. The following conventional 3-year cash crop phase benefits from pre-cropping effects and broader crop rotations (Taube, 2020). This provides a model that facilitates cooperation between specialized crop and livestock farms and helps to develop regional circularity in agricultural production systems.

8 Where to look for further information

In this section, we are directing readers to sources of further information regarding the issues discussed in this chapter. These links will provide you with comprehensive information on each subject.

Sustainable agriculture:

1. Sustainable agriculture. https://commission.europa.eu/food-farming-fisheries/sustainable-agriculture_en.
2. Ecological intensification of agriculture - sustainable by nature. https://www.wur.nl/upload_mm/6/d/4/1d1499e5-b489-465d-bda7-e6716c711049_TittonellCurrentOpinionInEnvironmentalSustainability2014.pdf.
3. Sustainable intensification in agriculture. Navigating a course through competing food system priorities. https://cgspace.cgiar.org/handle/10568/52201.

Climate change mitigation in agriculture:

1. Ambitious greenhouse gas mitigation in agriculture. https://www.umweltbundesamt.de/en/topics/climate-energy/ambitious-greenhouse-gas-mitigation-in-agriculture#background.
2. Enhancing the mitigation of climate change through agriculture. https://www.oecd.org/publications/enhancing-the-mitigation-of-climate-change-though-agriculture-e9a79226-en.htm.
3. Agriculture, climate change mitigation and climate resilience https://www.bmel.de/EN/topics/farming/climate-stewardship/climate-stewardship_node.html.

Soil health and agriculture:

1. European Commission - soil health. https://agriculture.ec.europa.eu/sustainability/environmental-sustainability/natural-resources/soil_en.
2. Soil health in agricultural systems. https://www.ncbi.nlm.nih.gov/pmc/articles/PMC2610104/.

Plant-based diets and environmental sustainability:

1. Harvard School of Public Health - plant based diets. https://www.hsph.harvard.edu/news/press-releases/healthy-plant-based-diets-better-for-the-environment-than-less-healthy-plant-based-diets/.
2. Plant-based diets and their impact on health, sustainability, and the environment. https://apps.who.int/iris/bitstream/handle/10665/349086/WHO-EURO-2021-4007-43766-61591-eng.pdf?sequence=1&isAllowed=y.
3. Promoting sustainability in food consumption. https://www.bmel.de/SharedDocs/Downloads/EN/_Ministry/promoting-sustainability-in-food-consumption.pdf?__blob=publicationFile&v=2.

Precision agriculture and environmental efficiency:

1. Association of equipment manufacturers – environmental benefits of precision agriculture. https://www.aem.org/news/the-environmental-benefits-of-precision-agriculture-quantified.
2. Environmental benefits of precision agriculture. https://www.atlas-h2020.eu/the-environmental-benefits-of-precision-agriculture/.
3. Handbook digital farming: digital transformation for sustainable agriculture. Chapter 6.3. https://link.springer.com/book/10.1007/978-3-662-64378-5.

Livestock production and environmental impact:

1. Livestock production. https://www.econstor.eu/obitstream/10419/66085/1/332311899.pdf.
2. Livestock and environment. https://www.fao.org/livestock-environment/en.

Crop-livestock integration:

1. Crop-livestock integration. https://rodaleinstitute.org/science/crop-livestock-integration/.
2. Action against hunger – crop and livestock integration. https://www.actioncontrelafaim.org/wp-content/uploads/2022/01/1.-Crop-and-livestock-integration-VF.pdf.
3. Integrated crop-livestock systems. https://www.fao.org/agriculture/crops/thematic-sitemap/theme/spi/scpi-home/managing-ecosystems/integrated-crop-livestock-systems/icls-how/en/.
4. Weidemilch (German). https://www.lindhof.uni-kiel.de/de/forschung/weidemilch.
5. Yields and nitrogen dynamics in Ley-Arable systems – comparing different approaches in the APSIM model. https://doi.org/10.3390/agronomy12030738.

Land use and Agro-diversity:

1. Agrobiodiversity and land use. https://www.genres.de/fileadmin/SITE_MASTER/content/SITE_GENRES/downloads/docs/Beirat-GR/Gutachten_Stellungnahmen/agrobiodiv_and_landuse.pdf.
2. Agriculture & food security – agrobiodiversity. https://agricultureandfoodsecurity.biomedcentral.com/articles/10.1186/2048-7010-2-2.
3. Land use and biodiversity. https://www.pnas.org/doi/10.1073/pnas.1910023117.

Biodiversity conservation in agriculture:

1. Biodiversity conservation. https://royalsocietypublishing.org/doi/10.1098/rspb.2014.1358.
2. Biodiversity in agriculture. https://agriculture.ec.europa.eu/sustainability/environmental-sustainability/biodiversity_en.
3. Biodiversity conservation. https://icem.com.au/documents/biodiversity/pad/tlp-10.pdf.
4. Species-enriched grass-clover mixtures can promote bumblebee abundance compared with intensively managed conventional pastures. https://doi.org/10.3390/agronomy12051080.

Production grasslands:

1. Grassland report. https://ec.europa.eu/eurostat/documents/2393397/8259002/Grassland_2014_Final+report.pdf/58aca1dd-de6f-4880-a48e-1331cafae297.
2. Production grasslands. https://academic.oup.com/aob/article/110/6/1263/112127.
3. How much complexity is required for modelling grassland production at regional scales? https://www.mdpi.com/2073-445X/12/2/327.

Multi-species leys and grassland biodiversity:

1. Multi-species grassland: Is it time to consider your roots? https://businesswales.gov.wales/farmingconnect/news-and-events/technical-articles/multi-species-grassland-it-time-consider-your-roots.
2. Multi-species for multifunctions: Combining four complementary species enhances the multifunctionality of sown grassland. https://www.nature.com/articles/s41598-021-82162-y.
3. Species diversity effects on productivity, persistence and quality of multispecies swards in a four-year experiment. The future role of ley-farming in cropping systems. Grassland Science in Europe 28:42-53. https://www.europeangrassland.org/fileadmin/documents/Infos/Printed_Matter/Proceedings/EGF2023.pdf.

9 References

Allen, V. G., Baker, M. T., Segarra, E. and Brown, C. P. 2007. Integrated irrigated crop-livestock systems in dry climates. *Agron. J.* 99(2), 346–360. doi: 10.2134/agronj2006.0148.

Altieri, M. A., Nicholls, C. I., Henao, A. and Lana, M. A. 2015. Agroecology and the design of climate change-resilient farming systems. *Agron. Sustain. Dev.* 35(3), 869–890. doi: 10.1007/s13593-015-0285-2.

Badenhausser, I., Amouroux, P., Lerin, J. and Bretagnolle, V. 2009. Acridid (Orthoptera: Acrididae) abundance in Western European Grasslands: Sampling methodology and temporal fluctuations. *J. Appl. Entomol.* 133(9-10), 720-732. doi: 10.1111/j.1439-0418.2009.01437.x.

Beauchemin, K. A., Ungerfeld, E. M., Eckard, R. J. and Wang, M. 2020. Review: fifty years of research on rumen methanogenesis: Lessons learned and future challenges for mitigation. *Animal* 14(S1), s2-s16. doi: 10.1017/S1751731119003100.

Bell, L. W., Moore, A. D. and Kirkegaard, J. A. 2014. Evolution in crop-livestock integration systems that improve farm productivity and environmental performance in Australia. *Eur. J. Agron.* 57, 10-20. doi: 10.1016/j.eja.2013.04.007.

Beye, H., Taube, F., Lange, K., Hasler, M., Kluß, C., Loges, R. and Diekötter, T. 2022. Species-enriched grass-clover mixtures can promote bumblebee abundance compared with intensively managed conventional pastures. *Agronomy* 12(5) doi: 10.3390/agronomy12051080.

Blanco-Canqui, H., Wienhold, B. J., Jin, V. L., Schmer, M. R. and Kibet, L. C. 2017. Long-term tillage impact on soil hydraulic properties. *Soil Till. Res.* 170, c. doi: 10.1016/j.still.2017.03.001.

Bloem, J., Schouten, A. J., Sorensen, S. J., Rutgers, M., Van der Werf, A. and Breure, A. M. 2006. Monitoring and evaluating soil quality. In: Bloem, J., Hopkins, D. W. and Benedetti, A. (Eds). *Microbiological Methods for Assessing Soil Quality*. CABI Publishing, Wallingford, UK.

Boer, M. D. and Ittersum, M. K. V. n.d. *Imke J Circularity in Agricultural Production*.

Borrelli, L., Castelli, F., Ceotto, E., Cabassi, G. and Tomasoni, C. 2014. Maize grain and silage yield and yield stability in a long-term cropping system experiment in Northern Italy. *Eur. J. Agron.* 55, 12-19. doi: 10.1016/j.eja.2013.12.006.

Bouwman, L. A. and Arts, W. B. M. 2000. Effects of soil compaction on the relationships between nematodes, grass production and soil physical properties. *Appl. Soil Ecol.* 14(3), 213-222. doi: 10.1016/S0929-1393(00)00055-X.

Brandt, P., Yesuf, G., Herold, M. and Rufino, M. C. 2020. Intensification of dairy production can increase the GHG mitigation potential of the land use sector in East Africa. *Glob. Change Biol.* 26(2), 568-585. doi: 10.1111/gcb.14870.

Brooker, R. W., Bennett, A. E., Cong, W. F., Daniell, T. J., George, T. S., Hallett, P. D., Hawes, C., Iannetta, P. P. M., Jones, H. G., Karley, A. J., Li, L., McKenzie, B. M., Pakeman, R. J., Paterson, E., Schöb, C., Shen, J., Squire, G., Watson, C. A., Zhang, C., Zhang, F., Zhang, J. and White, P. J. 2015. Improving intercropping: A synthesis of research in agronomy, plant physiology and ecology. *New Phytol.* 206(1), 107-117. doi: 10.1111/nph.13132.

Cardinale, B. J., Wright, J. P., Cadotte, M. W., Carroll, I. T., Hector, A., Srivastava, D. S., Loreau, M. and Weis, J. J. 2007. Impacts of plant diversity on biomass production increase through time because of species complementarity. *Proc. Natl. Acad. Sci. U. S. A.* 104(46), 18123-18128. doi: 10.1073/pnas.0709069104.

de Haas, B. R., Hoekstra, N. J., van der Schoot, J. R., Visser, E. J. W., de Kroon, H., van Eekeren, N., de Haas, B. R., Hoekstra, N. J., van der Schoot, J. R., Visser, E. J. W., de Kroon, H. and van Eekeren, N. 2019. Combining agroecological functions in grass-clover mixtures. *AIMS Agric. Food* 3, 547-567. doi: 10.3934/agrfood.2019.3.547.

de Klein, C. A. M., van der Weerden, T. J., Luo, J., Cameron, K. C. and Di, H. J. 2020. A review of plant options for mitigating nitrous oxide emissions from pasture-based systems. *N. Z. J. Agric. Res.* 63(1), 29-43. doi: 10.1080/00288233.2019.1614073.

Dong, L., Mao, Z. and Sun, T. 2016. Condensed tannin effects on decomposition of very fine roots among temperate tree species. *Soil Biol. Biochem.* 103, 489–492. doi: 10.1016/j.soilbio.2016.doi: 10.003.

EAT-Lancet 2019. *Summary Report of the EAT-Lancet Commission*. Available at: https://eatforum.org/content/uploads/2019/07/EAT-Lancet_Commission_Summary_Report.pdf.

FAO 2014. The state of food and agriculture. Innovation in family farming. *Food and Agriculture Organization of the United Nations, Rome*. Available at: https://www.fao.org/3/i4040e/i4040e.pdf.

Flachowsky, G. 2008. Treibhausgase und Ressourceneffizienz. Aspekte der Erzeugung von Lebensmitteln tierischer Herkunft[Greenhouse gases and resource efficiency: Aspects of production of food of animal origin]. *Ernährungsumschau* 55, 414419. Available at: http://www.ernaehrungs-umschau.de/media/pdf/pdf_2008/07_08/EU07_414_419.qxd.pdf.

Flechard, C. R., Neftel, A., Jocher, M., Ammann, C. and Fuhrer, J. 2005. Bi-directional soil/atmosphere N2O exchange over two mown grassland systems with contrasting management practices. *Glob. Change Biol.* 11(12), 2114–2127 doi: 10.1111/j.1365-2486.2005.01056.x.

Fleischner, T. L. 1994. Ecological costs of livestock grazing in western north America. *Conserv. Biol.* 8(3), 629–644. doi: 10.1046/J.1523-1739.1994.08030629.X.

Franzluebbers, A. J. 2007. Integrated crop-livestock systems in the Southeastern USA. *Agron. J.* 99(2), 361–372. doi: 10.2134/AGRONJ2006.0076.

Franzluebbers, A. J., Sawchik, J. and Taboada, M. A. 2014. Agronomic and environmental impacts of pasture-crop rotations in temperate North and South America. *Agric. Ecosyst. Environ.* 190, 18–26. doi: 10.1016/j.agee.2013.09.017.

Franzluebbers, A. J., Sulc, R. M. and Russelle, M. P. 2011. Opportunities and challenges for integrating North-American crop and livestock systems. In: *Grassland Productivity and Ecosystems Services*. CABI Publishing, 208–218.

Garnett, T. and Godfray, C. 2012. Sustainable Intensification in Agriculture: Navigating a Course through Competing Food System Priorities. *Food Climate Research Network* Oxford Martin Programme on the Future of Food: Oxford, UK.

Garrett, R. D., Niles, M., Gil, J., Dy, P., Reis, J. and Valentim, J. 2017. Policies for reintegrating crop and livestock systems: A comparative analysis. *Sustainability* 9(3). doi: 10.3390/su9030473.

Garrett, R. D., Ryschawy, J., Bell, L. W., Cortner, O., Ferreira, J., Garik, A. V. N., Gil, J. D. B., Klerkx, L., Moraine, M., Peterson, C. A., Dos Reis, J. C. and Valentim, J. F. 2020. Drivers of decoupling and recoupling of crop and livestock systems at farm and territorial scales. *Ecol. Soc.* 25(1). doi: 10.5751/ES-11412-250124.

Ghani, M. U., Kamran, M., Ahmad, I., Arshad, A., Zhang, C., Zhu, W., Lou, S. and Hou, F. 2022. Alfalfa-grass mixtures reduce greenhouse gas emissions and net global warming potential while maintaining yield advantages over monocultures. *Sci. Total Environ.* 849, 157765. doi: 10.1016/j.scitotenv.2022.157765.

Global Yield Gap Atlas. 2022. Available at: https://www.yieldgap.org/gygaviewer/index.html.

Godfray, H. C. J., Beddington, J. R., Crute, I. R., Haddad, L., Lawrence, D., Muir, J. F., Pretty, J., Robinson, S., Thomas, S. M. and Toulmin, C. 2010. Food security: the challenge of feeding 9 billion people. *Science* 327(5967), 812–818. doi: 10.1126/science.1185383.

Goudriaan, J. 1992. Biosphere structure, carbon sequestering potential and the atmospheric 14C carbon record. *J. Exp. Bot.* 43(8), 1111–1119. doi: 10.1093/jxb/43.8.1111.

Herrero, M., Henderson, B., Havlík, P., Thornton, P. K., Conant, R. T., Smith, P., Wirsenius, S., Hristov, A. N., Gerber, P., Gill, M., Butterbach-Bahl, K., Valin, H., Garnett, T. and Stehfest, E. 2016. Greenhouse gas mitigation potentials in the livestock sector. *Nat. Clim. Change* 6(5), 452–461. doi: 10.1038/NCLIMATE2925.

Jarvis, N., Forkman, J., Koestel, J., Kätterer, T., Larsbo, M. and Taylor, A. 2017. Long-term effects of grass-clover leys on the structure of a silt loam soil in a cold climate. *Agric. Ecosyst. Environ.* 247, 319–328. doi: 10.1016/J.AGEE.2017.06.042.

Jeffries, A.-M. 2015. Employing sheep for weed control. Growing produce. Available at: https://www.growingproduce.com/fruits/grapes/employing-sheep-for-weed-control/.

Kagiya, N., Reinsch, T., Taube, F., Salminen, J. P., Kluß, C., Hasler, M. and Malisch, C. S. 2019. Turnover rates of roots vary considerably across temperate forage species. *Soil Biol. Biochem.* 139. doi: 10.1016/j.soilbio.2019.107614.

Karlsson, J. O. and Röös, E. 2019. Resource-efficient use of land and animals—environmental impacts of food systems based on organic cropping and avoided food-feed competition. *Land Use Policy* 85, 63–72. doi: 10.1016/j.landusepol.2019.03.035.

Katsvairo, T. W., Wright, D. L., Marois, J. J., Hartzog, D. L., Rich, J. R. and Wiatrak, P. J. 2006. Sod-Livestock integration into the Peanut-Cotton rotation: A systems farming approach. *Agron. J.* 98(4), 1156–1171. doi: 10.2134/AGRONJ2005.0088.

Kleijn, D., Bommarco, R., Fijen, T. P. M., Garibaldi, L. A., Potts, S. G. and van der Putten, W. H. 2019. Ecological intensification: bridging the gap between science and practice. *Trends Ecol. Evol.* 34(2), 154–166. doi: 10.1016/j.tree.2018.11.002.

Kunrath, T. R., de Berranger, C., Charrier, X., Gastal, F., de Faccio Carvalho, P. C., Lemaire, G., Emile, J. C. and Durand, J. L. 2015. How much do sod-based rotations reduce nitrate leaching in a cereal cropping system? *Agric. Water Manag.* 150, 46–56. doi: 10.1016/J.AGWAT.2014.11.015.

Kurth, T., zum Felde, A. M., Rubel, H. and Zielcke, S. 2020. The True Cost of Food. Accessed on October 5, 2022 from bcg-the-true-cost-of-food-oct-2020.pdf.

Ledgard, S., Schils, R., Eriksen, J. and Luo, J. 2009. Environmental impacts of grazed clover/grass pastures. *Ir. J. Agric. Food Res.* 48, 209–226.

Lemaire, G., Franzluebbers, A., de Faccio Carvalho, P. C. and Dedieu, B. 2014. Integrated crop-livestock systems: strategies to achieve synergy between agricultural production and environmental quality. *Agric. Ecosyst. Environ.* 190, 4–8. doi: 10.1016/j.agee.2013.08.009.

Lemaire, G., Gastal, F., Franzluebbers, A. and Chabbi, A. 2015. Grassland-Cropping rotations: an avenue for agricultural diversification to reconcile high production with environmental quality. *Environ. Manag.* 56(5), 1065–1077. doi: 10.1007/s00267-015-0561-6.

Lesschen, J. P., van den Berg, M., Westhoek, H. J., Witzke, H. P. and Oenema, O. 2011. Greenhouse gas emission profiles of European livestock sectors. *Anim. Feed Sci. Technol.* 166–167, 16–28. doi: 10.1016/j.anifeedsci.2011.04.058.

Liebman, M. and Schulte, L. A. 2015. Enhancing agroecosystem performance and resilience through increased diversification of landscapes and cropping systems. *Elementa* 3, 3. doi: 10.12952/journal.elementa.000041.

LKSH 2022. Ergebnisse der Vollkostenauswertung der Rinderspezialberatungsringe in Schleswig-Holstein. *Auswertungsjahr 2020/1. Landwirtschaftskammer Schleswig*

-Holstein (LKSH). Available at: https://www.lksh.de/fileadmin/PDFs/Landwirtschaft/Tier/Rinder-Report_2021_gesamt.pdf.

Loaiza Puerta, V., Pujol Pereira, E. I., Wittwer, R., van der Heijden, M. and Six, J. 2018. Improvement of soil structure through organic crop management, conservation tillage and grass-clover ley. *Soil Till. Res.* 180, 1-9. doi: 10.1016/j.still.2018.02.007.

Lobell, D. B., Schlenker, W. and Costa-Roberts, J. 2011. Climate trends and global crop production since 1980. *Science* 333(6042), 616-620. doi: 10.1126/science.1204531.

Loges, R., Bunne, I., Reinsch, T., Malisch, C., Kluß, C., Herrmann, A. and Taube, F. 2018. Forage production in rotational systems generates similar yields compared to maize monocultures but improves soil carbon stocks. *Eur. J. Agron.* 97, 11-19. doi: 10.1016/j.eja.2018.04.010.

Lorenz, H., Reinsch, T., Hess, S. and Taube, F. 2019. Is low-input dairy farming more climate friendly? A meta-analysis of the carbon footprints of different production systems. *J. Cleaner Prod.* 211, 161-170. doi: 10.1016/j.jclepro.2018.11.113.

Lowder, S. K., Skoet, J. and Raney, T. 2016. The number, size, and distribution of farms, smallholder farms, and family farms worldwide. *World Dev.* 87, 16-29 doi: 10.1016/j.worlddev.2015.doi: 10.041.

Loza, C., Reinsch, T., Loges, R., Taube, F., Gere, J. I., Kluß, C., Hasler, M. and Malisch, C. S. 2021. Methane emission and milk production from jersey cows grazing perennial ryegrass-white clover and multispecies forage mixtures. *Agriculture* 11(2), 175. doi: 10.3390/agriculture11020175.

MacLaren, C., Storkey, J., Strauss, J., Swanepoel, P. and Dehnen-Schmutz, K. 2018. Livestock in diverse cropping systems improve weed management and sustain yields whilst reducing inputs. *J. Appl. Ecol.* doi: 10.5281/ZENODO.1308220.

Meier, T. and Christen, O. 2013. Environmental impacts of dietary recommendations and dietary styles: Germany as an example. *Environ. Sci. Technol.* 47(2), 877-888. doi: 10.1021/ES302152V.

Muller, A., Schader, C., El-Hage Scialabba, N., Brüggemann, J., Isensee, A., Erb, K. H., Smith, P., Klocke, P., Leiber, F., Stolze, M. and Niggli, U. 2017. Strategies for feeding the world more sustainably with organic agriculture. *Nat. Commun.* 8(1), 1290. doi: 10.1038/s41467-017-01410-w.

Nevens, F. and Reheul, D. 2001. Crop rotation versus monoculture; yield, N yield and ear fraction of silage maize at different levels of mineral N fertilization. *Neth. J. Agric. Sci.* 49(4), 405-425. Available at: https://library.wur.nl/ojs/index.php/njas/article/view/660/361.

Nevens, F. and Reheul, D. 2002. The nitrogen- and non-nitrogen contribution effect of ploughed grass leys on the following arable forage crops: Determination and optimum use. *Eur. J. Agron.* 16(1), 57-74. doi: 10.1016/S1161-0301(01)00115-0.

Nevens, F. and Reheul, D. 2003. Permanent grassland and 3-years leys alternating with 3 years of arable land: 31 years of comparison. *Eur. J. Agron.* 19(1), 77-90. doi: 10.1016/S1161-0301(02)00021-7.

Nyameasem, J. K., Reinsch, T., Taube, F., Domozoro, C. Y. F., Marfo-Ahenkora, E., Emadodin, I. and Malisch, C. S. 2020. Nitrogen availability determines the long-term impact of land use change on soil carbon stocks in grasslands of southern Ghana. *Soil* 6(2), 523-539. doi: 10.5194/soil-6-523-2020.

Parsons, A. J., Rowarth, J. S. and Rasmussen, S. 2011. High-sugar grasses. *CAB Rev.* 2011, 1-12. doi: 10.1079/PAVSNNR20116046.

Peyraud, J. L., Taboada, M. and Delaby, L. 2014. Integrated crop and livestock systems in Western Europe and South America: a review. *Eur. J. Agron.* 57, 31–42. doi: 10.1016/j.eja.2014.02.005.

Poore, J. and Nemecek, T. 2018. Reducing food's environmental impacts through producers and consumers. *Science* 360(6392), 987–992. doi: 10.1126/science.aaq0216.

Poyda, A., Reinsch, T., Kluß, C., Loges, R. and Taube, F. 2016. Greenhouse gas emissions from fen soils used for forage production in northern Germany. *Biogeosciences* 13(18), 5221–5244. doi: 10.5194/bg-13-5221-2016.

Pretty, J. and Bharucha, Z. P. 2014. Sustainable intensification in agricultural systems. *Ann. Bot.* 114(8), 1571–1596. doi: 10.1093/aob/mcu205.

Rasse, D. P., Rumpel, C. and Dignac, M. F. 2005. Is soil carbon mostly root carbon? Mechanisms for a specific stabilization. *Plant Soil* 269(1–2), 341–356. doi: 10.1007/S11104-004-0907-Y.

Rather, I. A., Koh, W. Y., Paek, W. K. and Lim, J. 2017. The sources of chemical contaminants in food and their health implications. *Front. Pharmacol.* 8, 830. doi: 10.3389/fphar.2017.00830.

Regan, J. T., Marton, S., Barrantes, O., Ruane, E., Hanegraaf, M., Berland, J., Korevaar, H., Pellerin, S. and Nesme, T. 2017. Does the recoupling of dairy and crop production via cooperation between farms generate environmental benefits? A case-study approach in Europe. *Eur. J. Agron.* 82, 342–356. doi: 10.1016/j.eja.2016.08.005.

Reinsch, T., Loza, C., Malisch, C. S., Vogeler, I., Kluß, C., Loges, R. and Taube, F. 2021. Toward specialized or integrated systems in northwest Europe: On-farm eco-efficiency of dairy farming in Germany. *Front. Sustain. Food Syst.* 5. doi: 10.3389/fsufs.2021.614348.

Röös, E., Bajželj, B., Smith, P., Patel, M., Little, D. and Garnett, T. 2017. Greedy or needy? Land use and climate impacts of food in 2050 under different livestock futures. *Glob. Environ. Change* 47, 1–12. doi: 10.1016/j.gloenvcha.2017.09.001.

Ryschawy, J., Choisis, N., Choisis, J. P., Joannon, A. and Gibon, A. 2012. Mixed crop-livestock systems: an economic and environmental-friendly way of farming? *Animal* 6(10), 1722–1730. doi: 10.1017/S1751731112000675.

Ryschawy, J., Martin, G., Moraine, M., Duru, M. and Therond, O. 2017. Designing crop-livestock integration at different levels: Toward new agroecological models? *Nutr. Cycl. Agroecosyst.* 108(1), 5–20. doi: 10.1007/s10705-016-9815-9.

Schut, A. G. T., Cooledge, E. C., Moraine, M., Van de Ven, G. W. J., Jones, D. L. and Chadwick, D. R. 2021. Reintegration of crop-livestock systems in Europe : an overview. *Front. Agric. Sci. Eng.* 8(1), 111–129. doi: 10.15302/J-FASE-2020373.

Seufert, V., Ramankutty, N. and Foley, J. A. 2012. Comparing the yields of organic and conventional agriculture. *Nature* 485(7397), 229–232. doi: 10.1038/nature11069.

Sekaran, U., Lai, L., Ussiri, D. A. N., Kumar, S. and Clay, S. 2021. Role of integrated crop-livestock systems in improving agriculture production and addressing food security – a review. *J. Agric. Food Sys.* 5. doi: 10.1016/j.jafr.2021.100190.

Smit, H. P. J., Reinsch, T., Kluß, C., Loges, R. and Taube, F. 2021. Very low nitrogen leaching in grazed ley-arable-systems in northwest Europe. *Agronomy* 11(11), 2155. doi: 10.3390/agronomy11112155.

Tamura, M. and Tharayil, N. 2014. Plant litter chemistry and microbial priming regulate the accrual, composition and stability of soil carbon in invaded ecosystems. *New Phytol.* 203(1), 110–124. doi: 10.1111/nph.12795.

Taube, F., Gierus, M., Herrmann, A., Loges, R. and Schönbach, P. 2014. Grassland and globalization - challenges for northwest European grass and forage research. *Grass Forage Sci.* 69(1), 2-16. doi: 10.1111/gfs.12043.

Taube, F. 2021. Vorschläge der Zukunftskommission Landwirtschaft mit DüV umsetzen. *Agrar. Eur. Sonderbeilage* 36/21(6. Sept.) AgE-SB 36-21 Taube.pdf.

Taube, F. 2020. Klimaschutz über Ökologische Intensivierung und Hybridlandwirtschaft umsetzen ASG-Herbsttagung und Schwerpunkt Klimakrise! Landwirtschaft als Täter - Opfer - Retter? H 20781, 35-39. Available at: https://www.asg-goe.de/pdf/LR0420.pdf.

Taube, F., Henning, C., Albrecht, E., Reinsch, T. and Kluß, C. 2015. Nährstoffbericht des landes Schleswig-Holstein 2016. *Im Auftrag des Ministeriums für Energiewende, Landwirtschaft, Umwelt und Ländliche Räume des Landes S-H (MELUND).* Available at: https://www.schleswig-holstein.de/DE/fachinhalte/G/grundwasser/naehrstoffberichte.html.

ten Berge, H. F. M., Pikula, D., Goedhart, P. W. and Schröder, J. J. 2016. Apparent nitrogen fertilizer replacement value of grass-clover leys and of farmyard manure in an arable rotation. Part I: grass-clover leys. *Soil Use Manag.* 32, 9-19. doi: 10.1111/sum.12246.

Tilman, D. 1999. Global environmental impacts of agricultural expansion: the need for sustainable and efficient practices. *Proc. Natl. Acad. Sci. U. S. A.* 96(11), 5995-6000. doi: 10.1073/pnas.96.11.5995.

Udo, H. M. J., Aklilu, H. A., Phong, L. T., Bosma, R. H., Budisatria, I. G. S., Patil, B. R., Samdup, T. and Bebe, B. O. 2011. Impact of intensification of different types of livestock production in smallholder crop-livestock systems. *Livest. Sci.* 139(1-2), 22-29. doi: 10.1016/j.livsci.2011.03.020.

van Eekeren, N., Bommelé, L., Bloem, J., Schouten, T., Rutgers, M., de Goede, R., Reheul, D. and Brussaard, L. 2008. Soil biological quality after 36 years of ley-arable cropping, permanent grassland and permanent arable cropping. *Appl. Soil Ecol.* 40(3), 432-446. doi: 10.1016/j.apsoil.2008.06.010.

van Noordwijk, M. and Brussaard, L. 2014. Minimizing the ecological footprint of food: closing yield and efficiency gaps simultaneously? *Curr. Opin. Environ. Sustain.* 8, 62-70 doi: 10.1016/j.cosust.2014.08.008.

Vellinga, T. 2006. Management and nitrogen utilisation of grassland on intensive dairy farms. PhD Thesis. Wageningen University. ISBN 90-8504-373-5. Available at: https://edepot.wur.nl/121768.

Walia, U. S. and Walia, S. S. 2015. *Crop Management*. Scientific Publishers, ISBN9788172339319.

Webber, H., Kahiluoto, H., Rötter, R. and Ewert, F. 2014. Enhancing climate resilience of cropping systems. In: *Clim. Change Impact Adapt. Agric. Syst.*, 167-185. doi: 10.1079/9781780642895.0167.

Wilkins, R. J. 2008. Eco-efficient approaches to land management: A case for increased integration of crop and animal production systems. *Philos. Trans. R. Soc. Lond., B Biol. Sci.* 363(1491), 517-525. doi: 10.1098/rstb.2007.2167.

Yeates, G. W., Shepherd, T. G. and Francis, G. S. 1998. Contrasting response to cropping of populations of earthworms and predacious nematodes in four soils. *Soil Till. Res.* 48(4), 255-264. doi: 10.1016/S0167-1987(98)00134-2.

Younie, D. and Hermansen, J. E. 2000. The role of grassland in organic livestock farming. *Grassl. Sci. Eur.* 5, 493-509.

Yu, T., Mahe, L., Li, Y., Wei, X., Deng, X. and Zhang, D. 2022. Benefits of crop rotation on climate resilience and its prospects in china. *Agronomy* 12(2), 436. doi: 10.3390/agronomy12020436.

Van Zanten, H. H. E., Herrero, M., Van Hal, O., Röös, E., Muller, A., Garnett, T., Gerber, P. J., Schader, C. and De Boer, I. J. M. 2018. Defining a land boundary for sustainable livestock consumption. *Glob. Change Biol.* 24(9), 4185-4194. doi: 10.1111/gcb.14321.

Chapter 4

Closed-loop precision farming technologies to optimize resource use

C. Maraveas and T. Bartzanas, Agricultural University of Athens, Greece

1 Introduction

Agriculture plays an integral role as a source of food and livelihood for the world's population, in both developing and developed countries. Different studies further indicate that small-scale farmers are mainly dependent on producing the largest share of the world's food supply (70–80%) (Ricciardi et al., 2018; Lowder et al., 2021). However, this aspect leads to further implications which challenge the efficiency of agricultural production; for instance, since most farmers operate on a small scale, they are constrained by the lack of necessary resources and knowledge to improve their production capacities. Stringer et al. (2008) support this assertion by revealing that small-scale farmers across the world face similar socioeconomic and biophysical challenges ranging from the degradation of land and disease outbreaks to climatic variability and changes in market dynamics. In another research article, Mpandeli and Maponya (2014) reveal that small-scale farmers are further constrained by the lack of productive resource inputs such as fertilizer and herbicides, irrigation, and market access. As such, there is a need for strategies that can enable small-scale farmers to increase their yields while reducing the costs incurred in acquiring different types of production inputs.

However, insights reveal that even where farmers have access to required production resources such as fertilizer and herbicides, they are still inefficient in using them to optimize the production capacities of their farms (Awunyo-Vitor et al., 2016). In the research, Awunyo-Vitor et al. (2016) reported that maize

http://dx.doi.org/10.19103/AS.2023.0120.07

farmers in Ghana inefficiently used their available sources, whereby labor and capital were overutilized, whereas pesticides, herbicides, seed, and land were underutilized. A similar insight was reported by Konja et al. (2019), who examined technical and resource-use efficiency among rice farmers in Ghana and reported that most farmers were underutilizing or overutilizing the available production resources, including nutrients, fertilizer, pesticides, land, and water. Pereira Filho and Ferreira Filho (2003) also revealed that the inefficiency of 79.1% of small farms in the Recôncavo region in Brazil emerged from technical inefficiency (9.3%) and scaled inefficiency (11.7%). In this perspective, the productivity debate among small-scale farmers shifts from whether they have adopted and are using technologies to enhance productivity to whether the farmers are maximizing the available inputs and resources available to them.

Kumar et al. (2020) describe the concept of eco-efficiency, whereby there is a concern about maximizing outputs from agriculture in terms of quality and quantity by using fewer resources such as land, water, energy, labor, nutrients, and capital. Through eco-efficiency, agricultural production practices become more sustainable as natural resources are used efficiently to meet current and future needs in both large and smallholder farms. This chapter elaborates on closed-loop precision farming (PF) technologies, which process data from sensors to optimize resource usages such as water and energy. The significance of PF in modern agriculture is further underscored by the fact that smallholder farms contribute the largest share of available food in the world (Ricciardi et al., 2018). As a result, equipping small-scale farmers with knowledge on how to optimize resource usage is a fundamental strategy that can boost sustainable agricultural practices. Drawing from the current body of knowledge, it was deduced that PF focus has mainly been on large farms and concepts for small farms are under-researched.

2 Precision and closed-loop farming improving resource usage

According to the International Society of Precision Agriculture, precision agriculture is identified as a management strategy that gathers, processes, and analyzes temporal, spatial, and individual data and combines it with other information to provide support for management decisions according to the estimated variability to improve resource usage efficiency, profitability, quality, productivity, and sustainability of agricultural production (ISPA, 2019). Close inspection of the definition underscores two core aspects that encompass precision agriculture: (i) collection and processing of different kinds of data using information technologies, and using the processed data to make decisions that improve resource usage efficiency, thereby enhancing the sustainability of agricultural production; (ii) technologies to support the implementation of the decisions (such as site-specific fertilization). A recent research study noted that the adoption of site-specific fertilization guidelines in developing countries were grounded on traditional field experimentation, which was inadequate compared to crop modeling, analysis of satellite imagery, and mapping of soil fertility to determine levels of electrical conductivity (EC), organic matter, and pH (McKenzie et al., 2017; Tsibart et al., 2022). Beluhova-Uzunova and Dunchev (2019) reiterate this argument and reveal that in the contemporary world, PF is associated with diverse data collected via GPS, satellite, unmanned planes, GIS, and drones and is analyzed using complicated software processes.

Closed-loop agriculture is defined as a farming practice, whereby all nutrients and organic matter are recycled back to the soil that it grew in (FEASTA, 2016). In this perspective, the farming practice preserves the nutrients and levels of carbon in the soil and, in turn, promotes sustainable agricultural practices. Figure 1 distinguishes conventional and closed-loop agriculture.

As detailed in Fig. 1, conventional farming involves significant wastage of nutrients as organic matter is removed from crops and livestock is treated before it can be returned to the soil. However, with closed-loop agriculture, the extracted organic matter is returned directly to the soil, thereby minimizing wastage. The side-by-side comparison of closed-loop agriculture and PF reveals a similarity in that both strategies underscore the need to minimize resource wastage. However, a difference between them arises from the fact that PF leverages information technologies to improve resource usage efficiency in agriculture, whereas closed-loop farming does not necessarily emphasize this aspect.

2.1 Emerging technologies enhancing precision agriculture

A characteristic feature of agriculture in the contemporary world is that advanced technologies (big data, internet of things (IoT) and artificial intelligence (AI))

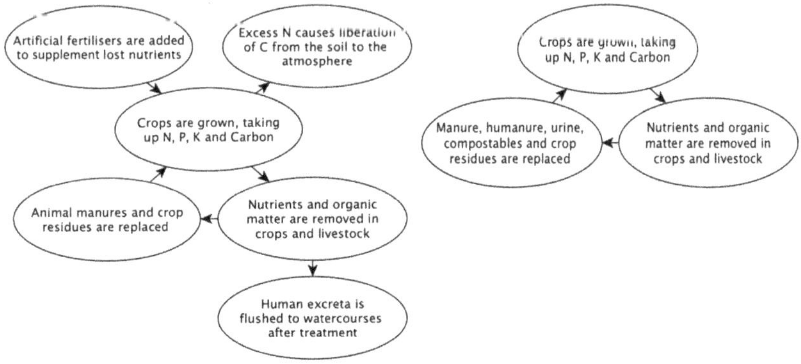

Figure 1 Conventional versus closed-loop agriculture (FEASTA, 2016).

are widely adopted to empower farmers to maximize the use of resources such as fertilizers and herbicides, monitor soil conditions, predict pests, and minimize costs with a view to enhancing the sustainability of agriculture. As such, the central question addressed in this section regards investigating how these modern technologies are enhancing precision agriculture and closed-loop farming. The reviewed studies in this section illuminate how the advanced technologies are enabling farmers to better utilize resources used in production such as fertilizers as well as minimize waste by excessive use of pesticides or herbicides.

2.1.1 Prediction of soil fertility

Soil fertility is predicted by nutritional composition, particularly the ratios of nitrogen (N), phosphorous (P), and potassium (K). The ratios are moderated by organic matter, pH, EC, soil bacteria, and fungi (European Commission, 2019; Sofo et al., 2020). Any imbalance in the EC and pH has a detrimental effect on soil fertility. Considering that it was not practical to monitor N, P, and K levels, organic matter, pH, and EC in large farms, farmers had integrated precision agriculture technologies, including IoT, to facilitate monitoring of soil fertility and address any imbalances through automated application of fertilizers using farm robots (Madushanki et al., 2019; Ruan et al., 2019; Verma et al., 2020). The adoption of IoT in agriculture led to significant cost savings and improvements in productivity.

A study by Roy and De (2022) reported various benefits of IoT-based precision agriculture, whereby data was collected from sensors to facilitate decision-making regarding a wide range of farming activities such as when to spray crops using drones, weather prediction, and soil predictions for different crops. The research further proposed the use of genetic algorithms to predict rainfall and identify whether manual water supply was required based on the moisture content of the soil. Roy and De (2022) further used unmanned aerial vehicles (UAVs) to water the plants in situations where the moisture level did not meet a predefined threshold. The analysis of the study showed that IoT was improving the sustainability of agriculture by enabling farmers to not only manually supply water to the soil when moisture levels were lower than expected, but also eliminate human intervention with regard to watering plants as UAVs facilitated the activity. Rehman et al. (2022) supported Roy and De (2022) by further revealing that IoT sensors facilitated the collection of useful data such as soil fertility, amount of water required by the plants, and pest detection among others which directly impacted the productivity of smart agriculture. As a result, increased productivity was reported with minimal human intervention as automation of monitoring activities was already achieved via sensors.

The further review also highlights the benefit of IoT sensors in providing intelligence regarding soil fertility, the likelihood of pest attacks, and the soil water requirements for the different plants (Nuchhi et al., 2020). In their study, Nuchhi et al. (2020) revealed that the use of IoT sensors with wireless sensor networks (WSNs) allowed farmers to monitor a variety of information pertaining to the soil such as moisture content, humidity, temperature, and levels of major nutrients, including N, P, and K in real time with the analyzed results being transmitted from the cloud platform to mobile applications. Madhumathi et al. (2020) also proposed a similar solution that monitored soil macronutrients, pH, temperature, and moisture content and recommended farmers on the fertilizer and water requirements required to improve soil quality as well as optimal growth of the crops. In the two studies, (Madhumathi et al., 2020; Nuchhi et al., 2020) different types of IoT sensors that measured humidity, temperature, pH, and moisture were embedded in the soil. The sensors transmitted the data in a digital form to the cloud platform where further analysis was undertaken to ascertain the soil conditions and provide recommendations regarding how to improve soil fertility and quality. Shylaja and Veena (2017) advanced the discussion further, revealing that using IoT and WSNs also enabled farmers to remotely ascertain the fertility level of the soil in real time and receive recommendations on the kind of fertilizers and crops that would perform best in the particular soil conditions. The synthesis of these studies indicates that IoT sensors and WSNs are pivotal in precision agriculture as they provide important soil information regarding the nutrient levels, temperature, and moisture content and further enable farmers to overcome the challenges associated with manual processes of soil testing that are often inaccurate, time-consuming, and unreliable. An added benefit is that the intelligence is provided in a convenient manner as updates are sent directly to a mobile application.

2.1.2 Prediction of frost events

Second, IoT sensors have also been adopted to gather intelligence on microclimates affecting the production of crops such as predicting pest attacks and frost to prevent adverse effects. For instance, Maraveas and Bartzanas (2021a) reported that state-of-the-art IoT sensors were facilitating the development of optimized microclimates for crop production, whereby the sensors enabled applications such as frost prevention, remote monitoring of crops, and precise control of nutrients in the soilless greenhouse cultivation with a view to minimize costs while increasing yields. In another study, Diedrichs et al. (2018) described the use of IoT sensors to predict frost events within peach orchards in Argentina, where a loss of 85% was reported in 2013 when more than 35 000 ha of vineyards were affected. The study revealed that frost was disastrous to orchards as water icing on the leaves and fruits significantly damaged the plants. However, manual methods of eliminating frost on the plants, such as using air heaters and large fans, were considered too expensive (Diedrichs et al., 2018). To overcome the issue, Diedrichs et al. (2018) deployed IoT sensors in the orchards to collect data on temperature and humidity conditions. The data was thereafter fed to machine learning algorithms which were trained on past readings of temperature and humidity which were associated with frost events. By relying on machine learning algorithms, the farmers were able to make more informed decisions and better prepare for frost events by implementing countermeasures at the right time (Diedrichs et al., 2018). Brun-Laguna et al. (2016) also deployed IoT sensors to predict the likelihood of frost events where findings showed that the combination of WSN and IoT technologies facilitated their accurate prediction, thereby avoiding adverse impacts on crops. Close inspection of these studies reveals that IoT sensors are effective in the prediction of climatic conditions such as the likelihood of frost events, thereby allowing farmers to prepare countermeasures in advance and prevent adverse effects on crops. As a result, the farmers were able to minimize costs of losses from frost events through better preparation as they implemented different countermeasures.

2.1.3 Prediction of pest attacks

A third application area of the IoT sensors regards the prediction of pest attacks to allow farmers to mitigate losses from purchasing inappropriate pesticides that would otherwise remain unused. A study by Sharma et al. (2021) demonstrated that WSNs and IoT monitoring infrastructure coupled with fuzzy logic algorithms were effective in predicting the likelihood of pest attacks in millet and rice. In the research, it was argued that since it was difficult to predict individual causes of the diseases, a more effective approach regarded deploying IoT sensors to predict the ambient breeding conditions of the pests (Sharma et al., 2021). As a result, correlation results between the temperature, rainfall, and relative humidity conditions were studied to predict the likelihood of pests breeding and attacking the crops.

Almeyda et al. (2020) also deployed IoT sensors in situ in organic banana crops to generate data on the climatic conditions that facilitated the growth of different pests. The conditions were thereafter input into a logistic regression and support vector machine model, and output results were generated with an accuracy of 79% (Almeyda et al., 2020). Materne and Inoue (2018) also identified similar results to those of Almeyda, Paiva, and Ipanaque's study (2020), where they deployed the IoT sensors in a farm to collect data on eight important parameters which have a high correlation to pests and disease outbreaks. The data were thereafter evaluated via machine learning algorithms in order to accurately predict the outbreak of diseases with high accuracy levels. As such, a confluence was observed from the synthesis of the studies (Materne and Inoue, 2018; Almeyda et al., 2020; Sharma et al., 2021) where it was reported that predicting the ambient conditions that pests bred in via machine learning algorithms were an accurate indicator of the likelihood of pest attacks.

2.1.4 Improving quality and quantity of production

Fourth, IoT sensors are also important in promoting the quality and quantity of crop production by providing large datasets. For instance, Akhter and Sofi (2022) reported on the combined use of machine learning, IoT data analytics, and WSNs to increase the quality and quantity of crop production from the crop fields to ensure the growing food demand was met. In the research, Akhter and Sofi (2022) revealed that the large datasets produced by IoT sensors such as plant sensors, soil sensors, and environmental sensors could be stored in the cloud and, thereafter, analyzed using machine learning and big data techniques. As such, big data analysis and machine learning were argued to make precision agriculture more practical.

Roukh et al. (2020) identified similar insights as they revealed that in modern farms, IoT sensors were capable of generating huge data amounts every second, totaling up to half-a-million data points per day. However, to gain in-depth insights from the voluminous data, big data analytics techniques enabled stakeholders in agriculture to analyze and visualize different types of insights. Hajjaj et al. (2021a) similarly reported that the integration of big data and IoT was leading to exciting opportunities for real-world applications involving monitoring, protection, and facilitating the improvement of natural resources. The synthesis of these studies (Roukh et al., 2020; Hajjaji et al., 2021b; Akhter and Sofi, 2022) underscores two important findings related to precision agriculture. The first one is that only by combining IoT sensors and big data analytics, there is a possibility to make precision agriculture practices. The view arises from the fact that over time, the number of IoT sensors embedded within the smart farms is increasing and, as a result, voluminous datasets are generated. Roukh et al. (2020), for instance, report that by 2023, up to 12 million agricultural sensors are anticipated to be installed globally. Subsequently, a high number of sensors are expected to generate large datasets which can only be analyzed using big data analytics techniques.

Finally, with the use of IoT sensors, a wide variety of data can be collected within smart farms that also impact overall productivity, for instance, by monitoring the smart farm structures. Maraveas and Bartzanas (2021b) reiterate this argument by revealing that sensors are still effective in the monitoring of the health of agricultural structures such as the durability of concrete and metallic structures which are used in the construction of smart greenhouses. Mishra et al. (2022) further revealed that IoT sensors could also be deployed to monitor the structural health of structures such as forecasting the deterioration of masonry construction, predicting the early stages of the compressive strength of concrete, corrosion diagnosis, and crack detection in buildings. In this perspective, the sensors allowed the farmers to ensure the smart farms are secure enough to withstand different types of environmental effects and also prepare countermeasures in advance to handle any unprecedented disasters.

3 Case studies on closed-loop agriculture

Several case studies have revealed the benefits of adopting emerging technologies including big data and IoT sensors to enhance closed-loop agriculture. The reviewed case studies in this section illuminate the application of IoT and emerging technologies such as big data and AI in facilitating closed-loop agriculture.

3.1 Case study one

The first case study regarded the development of a closed-loop irrigation system for northeastern Australian sugarcane farms using IoT technologies (Wang et al., 2020). The premise of the pilot study was that although farmers were increasingly adopting automated irrigation systems for their farms in order to save the time required to open and close the valves, the traditional approach was still limited by two core issues. First, the farmers were unable to identify insights on the amount and timing of the irrigation required for crops, and, second, farmers were also unaware of how to change the irrigation based on the conditions of the climate and type of soils (Wang et al., 2020). To overcome the problem and improve irrigation management, the automated irrigation system was linked to the decision support tool via IoT sensors. An uplink system uploaded data on irrigation and rainfall to the irrigation decision support tool, whereby the downlink system calculated and automatically applied irrigation schedules by considering constraints such as the capability of the pump, priorities of irrigation, energy, and farmer irrigation preferences (Wang et al., 2020). Figure 1 displays the schematic structure of the closed-loop irrigation system.

As revealed in Fig. 1, the uplink collected data from the farm pertaining to moisture content from rainfall and irrigation. The downlink thereafter provided water after processing the data generated and, as a result, controlled the irrigation schedule. Based on the fact that the irrigation was only initiated after soil moisture content was examined, this was indicative that the system was closed-loop as water resources were used effectively.

Outcomes from the case study showed that by using IoT technologies, a closed-loop solution could effectively manage irrigation schedules and improve the efficiency of water use in sugarcane farms, thereby delivering profitable, practical, and economic benefits to the farmers (Wang et al., 2020). The cost of the PF equipment was included in the economic assessment. Further analysis of the case study also revealed that IoT technologies were improving the reuse of rainwater within the sugarcane farms as the automated systems were only turned on based on rainfall patterns and soil conditions. As a result, where the soil had sufficient water or moisture levels, the automatic irrigation systems were not activated by downlink instructions, thereby saving water resources.

3.2 Case study two

The findings from Wang et al. (2020) were reiterated in a different case study by Anguraj et al. (2021), whereby the researchers also deployed IoT and WSNs in precision agriculture to monitor the use of irrigation water. In the case study, it was argued that although irrigation water was crucial in semiarid and arid lands, a lot of water evaporated and was lost through leaks and different types of losses (Anguraj et al., 2021). Additionally, relying on weekly irrigation to water the plants was ineffective as the soil moisture conditions were not ascertained beforehand. To tackle the problem and ensure that the irrigation water was appropriately used for plant growth, IoT sensors were deployed on the farm where they transmitted raw data on the real-time moisture content of the farm. AI algorithm-deep learning Convolutional Neural Support Vector Machines Hybrid Classifier (CNSVMHC) was thereafter introduced to process the data and facilitate decision-making on when the automated irrigation system would be turned on. Similar to Wang et al. (2020), Anguraj et al. (2021) also revealed that relying on IoT sensors in the farm optimized water usage and reduced wastage by ensuring that real-time conditions of the soil determined the frequency of irrigation as opposed to relying on predictions.

3.3 Case study three

A third case study by Rajkumar et al. (2017) also proposed a closed-loop irrigation system to conserve irrigation water. However, unlike Wang et al. (2020) and Anguraj et al. (2021), where the automatic irrigation system was initiated without the control of the farmer (based on soil moisture conditions), the system by Rajkumar et al. (2017) was used to control the pumps using on and off switches via Arduino microcontrollers. Additionally, the case study also collected more data on the soil conditions, including temperature, humidity, and soil moisture, via sensors in the farm. The decision engine was also developed using partial least-squares regression (PLSR) and Adaptive Neuro-Fuzzy Inference System (ANFIS) machine learning algorithms where real-time data on soil conditions influenced the pumps to be switched off or on. The decisions to control the pump were transmitted via a GSM line to the farmer where a mobile phone allowed them to control the pumps. Figure 2 displays the system architecture adopted for the closed-loop irrigation system.

As detailed in Fig. 2, the IoT sensors collected data on diverse aspects, including humidity, temperature, and soil moisture via the Arduino hardware. The data was processed via machine learning algorithms and decisions were transmitted to the motor driver where the pump could be controlled by the farmer. As such, the farmer was in a position to make informed decisions on how water was used in the irrigation system.

The uniqueness of Rajkumar et al. (2017) was that it allowed the farmer to also have control over the automated closed-loop irrigation system as they could determine whether to turn the water pumping systems off or on based on the provided details. On the contrary, Wang et al. (2020) and Anguraj et al. (2021) relied on automatic irrigation systems, whereby the pumps turned on from downlinks sent by the machine learning algorithms. In another case study, Kashyap et al. (2021) adopted a system similar to Rajkumar et al. (2017), whereby their proposed closed-loop irrigation system also provided control to the farmer, thereby allowing to determine whether to follow the decisions provided by the system or to alter them. In the particular case study, the system was subdivided into four parts as shown in Fig. 3.

From Fig. 3, four parts of the system were identified. The first regarded the IoT sensors in the farm, while the second comprised the internet servers which processed data from the farm. The third component regarded the LSTM RNN model which predicted the volumes of soil moisture for the next day based on details from the previous day. The output from the LSTM RNN model which involved water prediction levels was sent to the farmer via their end user interface such as a smartphone. The farmer would agree with the predictions of the irrigation planner or alter them before they were sent back to the farm via

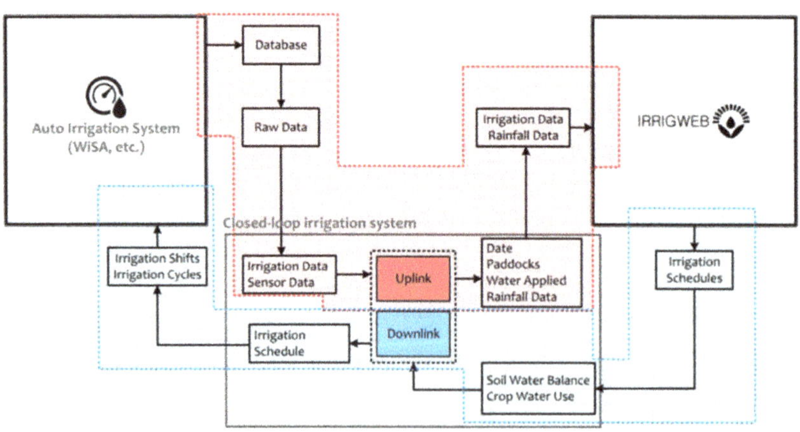

Figure 2 System architecture closed-loop irrigation (Rajkumar et al., 2017).

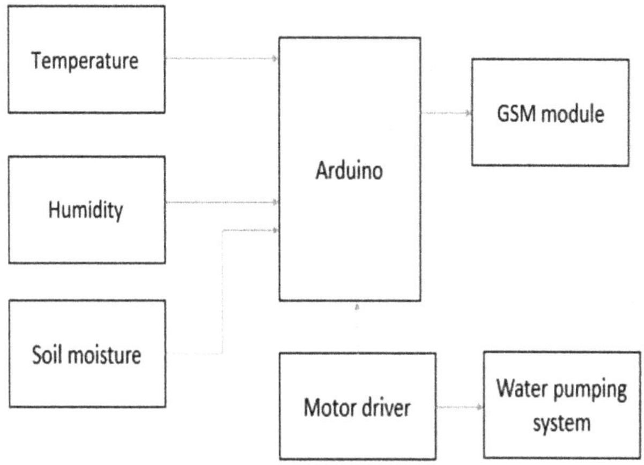

Figure 3 Closed-loop irrigation system (Kashyap et al., 2021).

the gateway. Once sent, the water sprinklers and gate actuators were opened and water flowed to the farm.

Soil nutrient is a critical issue in agriculture because N, P, and K deficiency has a deleterious impact on yields (Achour et al., 2021). Advances in technology have led to the adoption of IoT-enabled systems for balancing soil nutrients. For example, smart greenhouses feature sensors for monitoring temperature, humidity, nutrient, light intensity, and CO_2 concentration (Achour et al., 2021). If the threshold is exceeded, the IoT systems regulate the microclimate and trigger automated fertilizer application to balance soil nutrients. In brief, smart technology systems were indispensable in agriculture.

3.4 Case study four

The fourth case study on closed-loop agriculture concerns the interoperability of IoT technologies and openness in communication protocols in farms with large automated operations. According to Katamreddy et al. (2019), closed-loop farming in agriculture 4.0 is limited to specific stages of crop production, particularly where the continuous monitoring of nutrient content, pH, and moisture is vital. The closed-loop system depicted in Fig. 4 shows that the crop database contains information drawn from the national food census.

The information is incorporated into the crop data analytics dashboard and AI predictive models, which guide farm equipment initialization, controller network, and the on-farm gateway and control unit that is in charge of precision fertilizer application, pesticide spraying, soil preparation, and harvesting as noted by Katamreddy et al. (2019). Even though full automation of the system affords unique benefits, the monitoring of moisture, pH, and soil nutrients is not entirely automated with sensors.

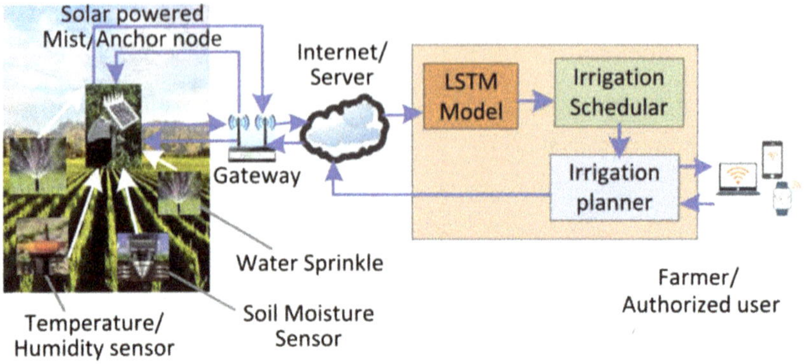

Figure 4 Unified IoT architecture (Katamreddy et al., 2019).

Selected operations are conducted manually, including the collection of soil samples (Katamreddy et al., 2019). The problem is anticipated to persist owing to the inadequacy of existing systems, interoperable skills, and information about farms and nutrient levels. These challenges might be resolved with advances in research and engineering research (Katamreddy et al., 2019). Once the challenges are addressed, it would be possible for farmers to seamlessly control farm operations; this would enhance the scope of PF beyond the limited stages of crop production.

4 Conclusion

The focus of this chapter was to examine how emerging technologies such as big data analytics, IoT, AI, machine learning, and deep learning were transforming closed-loop agriculture in modern society through enhancing precision agriculture. From the analysis of the definitions of precision agriculture (ISPA, 2019; Beluhova-Uzunova and Dunchev, 2019) and closed-loop agriculture (FEASTA, 2016), a confluence emerged in that the two concepts focused on minimizing resource wastage and maximizing available inputs such as nutrients, water, and pesticides. However, with precision agriculture, the definition emphasized the reliance on technology to collect data, analyze it, and utilize it in facilitating complex decision-making in agriculture. Due to this difference, the research underscored the role of precision agriculture in improving closed-loop agriculture where resources were reused without necessarily relying on technological input. Therefore, the reviewed literature and case studies focused on precision agriculture applications that could be ported to closed-loop agriculture in order to achieve higher benefits.

The evaluation of the different literature studies also revealed that coupling IoT sensors and WSNs with emerging technologies such as big data analytics and AI enhanced precision agriculture, particularly through decision-making processes. For instance, the AI provided farmers with intelligent recommendations for fertilizers and water requirements from diversified information regarding the conditions of the soil such as temperature, soil macronutrients, pH, and moisture content (Madhumathi et al., 2020). As a result, the technologies were enabling the farmers to minimize reliance on manual guesswork or inaccurate predictions when purchasing fertilizers or determining the frequency of watering the crops to ensure optimal growth. A similar insight was also reported with regard to frost prevention, whereby different studies revealed that big data analytics and IoT technologies allowed farmers to accurately predict frost events and prevent them beforehand by implementing countermeasures such as air heaters (Diedrichs et al., 2018; Maraveas and Bartzanas, 2021b). Synthesis of the studies, therefore, underscored the role of predictive analytics in closed-loop agriculture as farmers were able to implement countermeasures beforehand, thereby preventing losses from frost events, pest activity, and diseases.

The analysis of the case studies also showed that closed-loop irrigation systems which provided direct information to the farmers achieved similar benefits as farmers were able to work alongside technology to make more informed decisions regarding resource usage in the farms (Kashyap et al., 2021; Kashyap et al., 2021). In the examined closed-loop irrigation systems case studies, farmers obtained insights on the soil moisture and temperature conditions from IoT sensors and were able to control automated pumps to ensure the plants received the necessary amount of water to support their growth. As a result, water from rainfall was effectively reused without any waste being reported. The similarity between previous literature and case studies emphasized the value of modern technologies in enhancing closed-loop agriculture by allowing farmers to access information that would otherwise be too complicated or tedious to obtain. For example, from the literature studies, manual methods of measuring soil nutrient levels such as potassium and phosphorous require trained laboratory specialists and significant time investment before soil conditions can be deciphered.

Similarly, in the case studies, trained specialists would be required to assess the soil conditions before recommending irrigation schedules for the farmers. However, with the adoption of IoT sensors and actuators, raw data on the soil conditions were collected and transmitted to decision support systems which analyzed it and send actionable information insights to the farmers. In this perspective, technology was identified to replace human expertise as well as manual labor and tedious processes of evaluating soil conditions which were difficult to access for traditional farmers. On the same note, the IoT technologies provided convenience as farmers in remote locations could access information from their farms and make informed and accurate decisions regarding fertilizer procurement and its application in farms.

A second finding from the evaluation of the case studies was that the farmers were also able to influence the technology or make their own decisions as opposed to relinquishing all control to the automated systems. In this case, the farmers were able to view the recommendations by the decision support systems and determine whether they could adhere to the generated instructions before the automated irrigation pumps were turned on. A case in point was the case study by Kashyap et al. (2021), whereby insights showed that farmers were able to assess the irrigation planner predictions provided by deep learning LSTM RNN models before providing approval for downlink instructions that were sent to the actuators and water sprinklers to the farm. Likewise, the case study by Rajkumar et al. (2017) revealed that the farmers were able to control the automatic irrigation pumps via on and off switches which were provided via mobile GSM connection. The findings resonated with previous literature studies where the farmers relied on prediction results from the analysis by machine learning and AI systems to prepare countermeasures against frost attacks (Brun-Laguna et al., 2016; Diedrichs et al., 2018; Maraveas and Bartzanas, 2021a). The findings also aligned with similar studies which relied on IoT and machine learning to predict pest and disease attacks against crops (Almeyda et al., 2020; Sharma et al., 2021). Close analysis of these findings revealed that although the technologies provided predictions on different agricultural aspects, the farmer was however provided with the final control regarding whether to implement the decisions or forfeit them. A further implication was that by working alongside the farmers, the IoT and big data technologies were not replacing the role of the farmer in influencing crop production.

The conclusion from the examination of the literature and various case studies in this chapter is that closed-loop agriculture in the contemporary world can be enhanced further by integrating concepts of precision agriculture, especially by coupling IoT technologies with big data analytics and AI. The findings from this research revealed that farmers who only partially adopted advanced technologies such as automatic irrigation systems to reduce manual labor and improve productivity were still limited with regard to knowledge of soil conditions and the efficiency of such systems in delivering necessary water requirements to the crops. For instance, they were unaware of the soil moisture conditions when the pumps were switched off and on as well as whether the crops received the required amount of water for growth. In other studies, similar conclusions were also reached, whereby the farmers were unaware of soil fertility conditions in terms of the concentration of potassium and phosphorous levels at the time when more fertilizers were added to the soil.

Therefore, IoT and big data analytics technologies were underscored in this research as a solution that provided insights to the farmers to improve their decision-making. The findings discussed showed that farmers were able to receive insights regarding soil conditions such as temperature, pH, moisture levels, and required minerals for crop growth. However, an important finding was that the IoT sensors were only effective where they worked alongside big data analytics and AI algorithms including deep and machine learning. The algorithms were observed to process the raw data received from the sensors in the farm and derive actionable insights regarding decisions such as the right kind of fertilizer to be used in the form and frequency of irrigation. In this perspective, farmers adopting the technologies required to adopt both IoT sensors and advanced decision support systems which include machine learning and deep learning algorithms.

5 Future trends in research

A further conclusion from the research is that the case studies were mainly limited to small-scale farms and minor projects due to the cost implications associated with the implementation of the technology. As a result, there is a need for future work in the research area in order to identify the feasibility of implementing IoT and AI on a larger scale. Such projects are recommended in order to identify the effectiveness of the technology in improving closed-loop agriculture on a wider scale. A further recommendation regards the need for researchers to undertake more empirical studies in the area as the research revealed that most studies focused on closed-loop irrigation systems. In this regard, there is a need for research on closed-loop agriculture applications with regard to soil fertility, pest and disease prediction, and frost prevention areas among others. The dearth of available literature for the different case studies indicates that the research area is not widely explored and that there is a need for research focus.

Future trends in the research area also reveal that aside from coupling IoT sensors with AI and big data, other technologies, including blockchain, UAVs, and agricultural robots, promote precision agriculture and, as a result, can improve closed-loop agriculture. For instance, Pincheira et al. (2021) report that IoT sensors are an important source of trustworthy data used in blockchain-based water management for precision agriculture. The research reveals that farmers are increasing their interest in blockchain technology to manage water usage in farms where multiple stakeholders are involved. Similarly, further research also indicates that unmanned aerial vehicles (UAVs) are also being coupled with IoT sensors and networks to enhance activities such as fertilization, crop-disease management, application of pesticides, and weed management among others (Boursianis et al., 2022). In other studies, agricultural robots are also emerging as positive solutions to facilitate farming practices such as seeding, crop weeding, and harvesting as they replace manual laborers. The different technologies are therefore minimizing reliance on manual labor to improve crop productivity as UAVs and robots outperform human laborers in undertaking activities such as fertilizer application, seeding, and irrigation. Therefore, as sensors transmit information from the soil, they can be leveraged by these technologies to improve agricultural practices on the farm.

6 Where to look for further information

For further information on precision agriculture and advances in the field, several journals are recommended:

- An International Journal on Advances in Precision Agriculture - the journal has 23 volumes on different studies regarding the application of precision agriculture. https://link.springer.com/journal/11119/volumes-and-issues.
- Special issue of precision agriculture - MDPI journal. https://www.mdpi.com/journal/agriculture/special_issues/Precision_Agriculture#published.

Information available on the internet:

- https://lowerblackwood.com.au/new-events/2020/5/14/farming-in-a-closed-loop-system.
- https://www.resilience.org/stories/2016-04-26/what-is-closed-loop-agriculture/.
- https://www.threemilecanyonfarms.com/sustainability/closed-loop-system.
- https://www.soilassociation.org/blogs/2018/june/closed-loop-agriculture-for-the-21st-century/.
- https://www.spartagroup.ca/closed-loop-urban-agriculture/.
- https://asi.ucdavis.edu/programs/rr/research/closing-loop.

7 References

Achour, Y., Ouammi, A. and Zejli, D. 2021. Technological progresses in modern sustainable greenhouses cultivation as the path towards precision agriculture. *Renewable and Sustainable Energy Reviews* 147, 111251. doi: 10.1016/j.rser.2021.111251.

Akhter, R. and Sofi, S. A. 2022. Precision agriculture using IoT data analytics and machine learning. *Journal of King Saud University - Computer and Information Sciences* 34(8), 5602–5618. Available at: https://www.sciencedirect.com/science/article/pii/S1319157821001282 (Accessed 4 February 2022).

Almeyda, E., Paiva, J. and Ipanaque, W. 2020. Pest incidence prediction in organic banana crops with machine learning techniques. 21–23 Oct. 2020, Lima, Peru. In:

IEEE Engineering International Research Conference (EIRCON). Available at: https://ieeexplore.ieee.org/abstract/document/9254034. (Accessed 4 February 2022).

Anguraj, D. K., Mandhala, V. N., Bhattacharyya, D. and Kim, T. 2021. Hybrid neural network classification for irrigation control in WSN based precision agriculture. *Journal of Ambient Intelligence and Humanized Computing*.

Awunyo-Vitor, D., Wongnaa, C. A. and Aidoo, R. 2016. Resource use efficiency among maize farmers in Ghana. *Agriculture and Food Security* 5(1).

Beluhova-Uzunova, R. P. and Dunchev, D. M. 2019. Precision farming – Concepts and perspectives. *Problems of Agricultural Economics* 360(3), 142–155. Available at: https://ssrn.com/abstract=3716802 (Accessed 17 January 2022).

Boursianis, A. D., Papadopoulou, M. S., Diamantoulakis, P., Liopa-Tsakalidi, A., Barouchas, P., Salahas, G., Karagiannidis, G., Wan, S. and Goudos, S. K. 2022. Internet of things (IoT) and agricultural unmanned aerial vehicles (UAVs) in smart farming: A comprehensive review. *Internet of Things* 18, 100187.

Brun-Laguna, K., Diedrichs, A. L., Chaar, J. E., Dujovne, D., Taffernaberry, J. C., Mercado, G. and Watteyne, T. 2016. A demo of the PEACH IoT-based frost event prediction system for precision agriculture. In: *13th Annual IEEE International Conference on Sensing, Communication, and Networking (SECON)*, London, UK.

Diedrichs, A. L., Bromberg, F., Dujovne, D., Brun-Laguna, K. and Watteyne, T. 2018. Prediction of frost events using machine learning and IoT sensing devices. *IEEE Internet of Things Journal* 5(6), 4589–4597.

European Commission. 2019. EIP-AGRI focus group circular horticulture, pp. 1-36. Available at: https://ec.europa.eu/eip/agriculture/sites/agri-eip/files/eip-agri_fg_circular_horticulture_final_report_2019_en.pdf.

FEASTA. 2016. Closed loop agriculture for environmental assessment. Available at: http://www.feasta.org/wp-content/uploads/2016/04/Feasta-Closed-Loop-Agriculture-Report-2016.pdf (Accessed 4 February 2022).

Hajjaj, S. S. H., Azmi, H. N., Gsangaya, K. R., Sultan, M. T. H., Mail, M. F. and Hua, L. S. 2021a. Design and fabrication of an agricultural robot for crop seeding. *Materials Today: Proceedings*.

Hajjaji, Y., Boulila, W., Farah, I. R., Romdhani, I. and Hussain, A. 2021b. Big data and IoT-based applications in smart environments: A systematic review. *Computer Science Review* 39, 100318. Available at: https://www.sciencedirect.com/science/article/pii/S1574013720304184 (Accessed 4 February 2022).

ISPA. 2019. Precision agriculture. Available at: https://www.ispag.org/ (Accessed 4 February 2022).

Kashyap, P. K., Kumar, S., Jaiswal, A., Prasad, M. and Gandomi, A. H. 2021. Towards precision agriculture: IoT-enabled intelligent irrigation systems using deep learning neural network. *IEEE Sensors Journal* 21(16), 17479–17491.

Katamreddy, S., Walsh, J., Ward, S. and Riordan, D. 2019. Closed loop process control for precision farming: An Agriculture 4.0 perspective. In *2019 30th Irish Signals and Systems Conference (ISSC)*, Maynooth, Ireland, 1–6, doi: 10.1109/ISSC.2019.8904938.

Konja, D. T., Mabe, F. N. and Alhassan, H. 2019. Technical and resource-use-efficiency among smallholder rice farmers in Northern Ghana. *Cogent Food and Agriculture* 5(1).

Kumar, S., Meena, R. S. and Meena, R. S. 2020. *Resources Use Efficiency in Agriculture*. Singapore: Springer.

Lowder, S. K., Sánchez, M. V. and Bertini, R. 2021. Which farms feed the world and has farmland become more concentrated? *World Development* 142, 105455.

Madhumathi, R., Arumuganathan, T. and Shruthi, R. 2020. Soil NPK and moisture analysis using wireless sensor networks. In: *11th International Conference on Computing, Communication and Networking Technologies (ICCCNT)*, Kharagpur, India.

Madushanki, A. A. R., N, M., A., W. and Syed, A. 2019. Adoption of the Internet of Things (IoT) in agriculture and smart farming towards urban greening: A review. *International Journal of Advanced Computer Science and Applications* 10(4), 11-28. doi: 10.14569/IJACSA.2019.0100402.

Maraveas, C. and Bartzanas, T. 2021a. Application of Internet of Things (IoT) for optimized greenhouse environments. *AgriEngineering* 3(4), 954-970.

Maraveas, C. and Bartzanas, T. 2021b. Sensors for structural health monitoring of agricultural structures. *Sensors* 21(1), 314.

Materne, N. and Inoue, M. 2018. IoT monitoring system for early detection of agricultural pests and diseases. In: *12th South East Asian Technical University Consortium (SEATUC)*, Yogyakarta, Indonesia.

McKenzie, R. M., Barraclough, T. W. and Stokes, A. A. 2017. Integrating soft robotics with the robot operating system: A hybrid pick and place arm. *Frontiers in Robotics and AI* 4(39), 1-7. doi: 10.3389/frobt.2017.00039.

Mishra, M., Lourenço, P. B. and Ramana, G. V. 2022. Structural health monitoring of civil engineering structures by using the internet of things: A review. *Journal of Building Engineering* 48, 103954.

Mpandeli, S. and Maponya, P. 2014. Constraints and challenges facing the small scale farmers in Limpopo Province, South Africa. *Journal of Agricultural Science* 6(4), 135.

Nuchhi, S., Bagali, V. and Annigeri, S. 2020. *IoT Based Soil Testing Instrument for Agriculture Purpose*. Vijiyapur, India: IEEE Bangalore Humanitarian Technology Conference (B-HTC).

Pereira Filho, C. A. and Ferreira Filho, J. BdS. 2003. Sources of small family farm production inefficiency, Recôncavo region, Bahia, Brazil. *Revista de Economia e Sociologia Rural* 41(1), 169-182.

Pincheira, M., Vecchio, M., Giaffreda, R. and Kanhere, S. S. 2021. Cost-effective IoT devices as trustworthy data sources for a blockchain-based water management system in precision agriculture. *Computers and Electronics in Agriculture* 180, 105889.

Rajkumar, M. N., Abinaya, S. and Kumar, V. V. 2017. Intelligent irrigation system - An IOT based approach. *IEEE Xplore*. Available at: https://ieeexplore.ieee.org/abstract/document/8094057 (Accessed 29 May 2021).

Rehman, A., Saba, T., Kashif, M., Fati, S. M., Bahaj, S. A. and Chaudhry, H. 2022. A revisit of Internet of things technologies for monitoring and control strategies in smart agriculture. *Agronomy* 12(1), 127.

Ricciardi, V., Ramankutty, N., Mehrabi, Z., Jarvis, L. and Chookolingo, B. 2018. How much of the world's food do smallholders produce? *Global Food Security* 17, 64-72.

Roukh, A., Fote, F. N., Mahmoudi, S. A. and Mahmoudi, S. 2020. Big data processing architecture for smart farming. *Procedia Computer Science* 177, 78-85.

Roy, S. K. and De, D. 2022. Genetic algorithm based internet of precision agricultural things (IopaT) for agriculture 4.0. *Internet of Things* 18, 100201.

Ruan, J., Jiang, H., Zhu, C., Hu, X., Shi, Y., Liu, T., Rao, W. and Chan, F. T. S. 2019. Agriculture IoT: Emerging trends, cooperation networks, and outlook. *IEEE Wireless Communications* 26(6), 56-63. doi: 10.1109/MWC.001.1900096.

Sharma, R. P., Ramesh, D., Pal, P., Tripathi, S. and Kumar, C. 2021. IoT enabled IEEE 802.15.4 WSN monitoring infrastructure driven Fuzzy-logic based crop pest prediction. *IEEE Internet of Things Journal* 9(4), 3037–3045.

Shylaja, S. N. and Veena, M. B. 2017. Real-time monitoring of soil nutrient analysis using WSN. In: *International Conference on Energy, Communication, Data Analytics and Soft Computing (ICECDS)*, Chennai, India.

Sofo, A., Mininni, A. N. and Ricciuti, P. 2020. Soil macrofauna: A key factor for increasing soil fertility and promoting sustainable soil use in fruit orchard agrosystems. *Agronomy* 10(4), 1–20. doi: 10.3390/agronomy10040456.

Stringer, L. C., Twyman, C. and Gibbs, L. M. 2008. Learning from the South: Common challenges and solutions for small-scale farming. *Geographical Journal* 174(3), 235–250.

Tsibart, A., Haumont, J. and Saeys, W. 2022. Site-specific fertilization. *Future Learn*. Available at: https://www.futurelearn.com/info/courses/revolutionising-the-food-chain/0/steps/170928.

Verma, S., Bhatia, A., Chug, A. and Singh, A. P. 2020. *Recent advancements in multimedia big data computing for IoT applications in precision agriculture: Opportunities, issues, and challenges*. Singapore: Springer, pp. 391–416. doi: 10.1007/978-981-13-8759-3_15.

Wang, E., Attard, S., Linton, A., McGlinchey, M., Xiang, W., Philippa, B. and Everingham, Y. 2020. Development of a closed-loop irrigation system for sugarcane farms using the internet of things. *Computers and Electronics in Agriculture* 172, 105376. Available at: https://www.sciencedirect.com/science/article/pii/S0168169919322859#b0115 (Accessed 1 November 2020).

Part 2

Re-using agricultural and other wastes

Chapter 5

Using crop residues/by-products as livestock feed in a circular economy

Andre F. Brito and Kleves V. Almeida, University of New Hampshire, USA; and Luiz H. P. Silva, Western Kentucky University, USA

1 Introduction

The human population is projected to reach 9.8 billion people by 2050[1], indicating that demand for food will also increase to meet the needs of additional calories required to nourish children and adults worldwide. Specifically, the global demand for meat and milk is expected to increase by approximately 50% between 2005 and 2050 due to rising incomes, population growth, and urbanization (Alexandratos and Bruinsma, 2012; Mottet et al., 2017). To keep up with a growing demand for animal-based protein, the animal agriculture sector must adapt to supply the need for more calories while improving its production efficiency to reduce the environmental impact from livestock production systems.

Animal sources contribute with 18% and 25% of calories and protein consumed by humans, respectively, on a global scale (FAOSTAT, 2016). In addition to protein, livestock supply essential macro- and micronutrients, increase crop production through manure and draught power, and generate income, thus improving livelihoods (Mottet et al., 2017). However, there are potential negative tradeoffs associated with the livestock industry including

[1] www.un.org.

http://dx.doi.org/10.19103/AS.2023.0120.10

rations containing ingredients that can be used as food for humans (i.e. feed/food competition), land used for feed production that could otherwise grow crops for producing foods, and low conversion efficiency of feed into human-edible products by livestock species (Mottet et al., 2017). Nevertheless, the livestock sector is expected to expand in line with the increased demand for animal-based food sources in response to population growth, so the use of non-edible by-product feeds such as brans, dried distillers' grains with solubles (DDGS), pulps (e.g. beet and citrus), and molasses need to be further optimized in animal rations. With the rapid growth of the ethanol industry in the last decades in the United States, DDGS became widely available and, therefore, fed to different animal species from cattle to hogs to chickens.

By-product feeds are key players in a circular economy as they are converted into animal protein at expense of traditional feedstuffs such as corn that can be also used for humans' consumption. According to the Association of American Feed Control Officials (2016), by-product feeds are defined as '… secondary products produced in addition to the principal product', and they can be originated from multiple sources including the food, fiber, beverage, and bioenergy industries (NASEM, 2021). It should be noted that by repurposing by-products into animal feeds less waste is disposed in the environment and less grains are used in animal rations resulting in the dual benefit of decreasing feed costs and improving the economic sustainability of farms. However, complete replacement of human-edible products such as ground corn with non-forage fiber sources has been shown to reduce milk production in high-producing dairy cows. Specifically, milk production decreased by 5.5% when ground corn, fed at 30% of the diet dry matter (DM), was replaced by an equal amount of soyhulls and beet pulp in rations of dairy cows (Zang et al., 2021).

Although livestock consume about 6 billion tons of feed DM annually, approximately 86% of the ingredients used in animal rations are not eaten by people (Mottet et al., 2017), including humans' industry-derived by-product feeds (NASEM, 2021). Globally, 2.5 billion hectares of land are required for feed production totaling about 50% of the total agricultural area available on the planet (Mottet et al., 2017). However, 2 billion hectares are grasslands of which 1.3 billion hectares are not suitable for crops, indicating that more than half of the land required to produce feed cannot be used for food production (Mottet et al., 2017). Regardless, the livestock industry is under increasing scrutiny, particularly the dairy and beef cattle sectors that are known to generate more greenhouse gas and are perceived to be less efficient than the non-ruminant sector. Interestingly, Mottet et al. (2017) estimated that to produce 1 kg of boneless meat requires 2.8 kg and 3.2 kg of human-edible feed in ruminant and non-ruminant (chicken layers excluded) production systems, respectively. Another aspect that is often overlooked when making comparisons across different livestock species is the environmental and economic evaluation of the whole-farm system through life-cycle analysis (LCA). de Vries and de Boer (2010) concluded after reviewing 16 LCA studies done with livestock that the differences in the environmental impact among swine, poultry, and beef production systems could be explained primarily by differences in utilization of nutrients and energy in feed, enteric methane emissions, and reproduction efficiency.

In this chapter, we will provide an overview of the nutrient profile of commonly used by-product feeds, how they are generated, and the limitations associated with their use in animal rations. We will focus on DDGS because of its use across different livestock production systems and will specifically assess the impact of DDGS on the production performance of swine, poultry, beef, and dairy cattle, and on the environment through published LCA.

2 Agro-industrial activities and main by-product feeds used in animal agriculture

Most by-product feeds included in livestock diets originate from the processing of crops used for producing biofuels (e.g. bioethanol from corn), humans' food (e.g. sugar from beet pulp or sugarcane), and beverages (e.g. beer from barley, wheat, rice, or corn). These agricultural by-products are classified as secondary products because they originated from industrial processes where most nutrients were removed or extracted to produce the main products (NASEM, 2021).

By-product feeds are not consumed by humans, which has prompted their inclusion in livestock diets. There are many advantages of feeding by-products to livestock, with one of the primary benefits being the reduction in feed costs and the potential increase in farm profitability. It should also be noted that nutrients from by-product feeds are converted to animal protein to be used as food for humans, with less nutrients leaching into the environment. Therefore, feeding by-products can improve food security while positively impacting the environment by diverting industrial waste into livestock diets, thereby improving the sustainability of animal production systems.

2.1 By-product feeds generated by the biofuel industry

Biofuels can be produced from different sources of oilseed crops (e.g. soybean and canola) or corn (Fig. 1), and they have been considered as alternative and sustainable fuels compared to petro-diesel. They are also known as 'green diesel' because their sources can be renewable and biodegradable. Bioethanol from grains is generally manufactured in conjunction with many distilled grains by-products such as wet distiller's grains with solubles (WDGS) and DDGS (Taheripour et al., 2010).

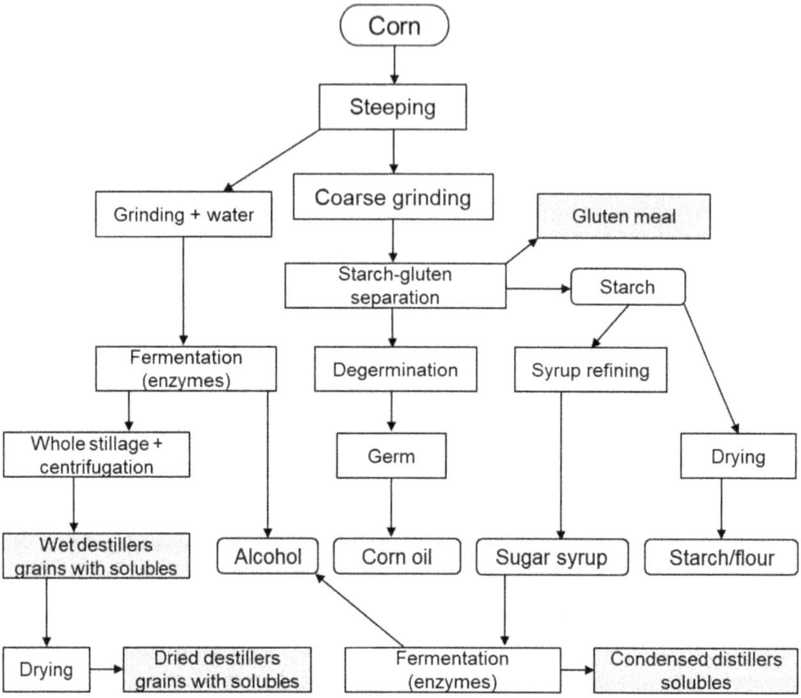

Figure 1 Flowchart representation of corn processing for ethanol, vegetal oil, high-fructose syrup, and dry flour (in green) and their respective by-products (in gray). Source: Adapted from NASEM (2021).

Over the past years, biofuel production has rapidly increased worldwide resulting in the generation of many associated by-products (Taheripour et al., 2010). In 2021, the United States was the largest ethanol (e.g. biofuel usually made from corn) producer in the world, with 56 836 million liters, accounting for nearly 55% of the total global production. Brazil and European Union ranked second and third place, producing roughly 27 706 and 5110 million liters, respectively.[2]

Biofuel by-products are generated based on the crop source (e.g. corn or oilseed), industrial processing, and technical specification of the fuel and its destination. Overall, biofuels are divided into three main categories: (1) first-generation biofuels, produced from food or animal feed crops; (2) second-generation biofuels, produced from non-food feedstocks, such as agricultural residues, forest residues, and other waste materials; and (3) third-generation biofuels, produced from microalgae or algal oil (Jeswani et al., 2020).

In terms of a circular economy, a positive aspect of using biofuel by-product feeds is that they may help to mitigate the environmental impact of the biofuel

[2] www.ethanolrfa.org.

industry through incorporation of by-products into livestock diets. In fact, the use of by-products as animal feed has been extensively studied, particularly DDGS that can replace both corn and soybean meal in livestock rations.

2.2 By-product feeds generated by the humans' food production industry

Many by-products from food production have the potential to be used in livestock rations. Therefore, they are no longer considered waste (Mirzaei-Aghsaghali and Maheri-Sis, 2008). Numerous experiments have demonstrated that by-products such as citrus pulp and beet pulp can be safely fed to ruminants. In fact, the *Nutrient Requirements of Dairy Cattle* (NASEM, 2021) book dedicated an entire chapter describing these by-products along with others that originated from the processing of food and beverages for humans.

Citrus fruits are consumed by humans as fresh fruits or processed juices, either fresh or concentrated. After the juice is extracted, the remaining residue is comprised of peel, pulp, and seeds. These components, individually or in various combinations, are the source materials from which citrus pulp is produced (Bampidis and Robinson, 2006). Citrus by-products are high in fiber, so that they are mostly fed to ruminants because of their ability to ferment high fiber feeds in the rumen (Grasser et al., 1995). In general, citrus pulp can replace corn or barley without negatively affecting intake and production of ruminants (Bampidis and Robinson, 2006).

Corn is considered one of the most valuable commodities worldwide as it plays an important role as food for humans while originating several by-products during its processing. The remarkable growth of the corn refining industry has generated large amounts of feedstuffs available for producers (Erickson, 2006). Figure 1 summarizes the main products obtained from corn processing along with the major by-products.

2.3 By-product feeds generated by the brewing industry

The beer industry is ranked as one of the most important sectors in global food production. Breweries produce large amounts of residues, which may lead to pollution if they are disposed of in the environment. Thus, diverting these wastes to livestock as feedstuffs could at the same time mitigate environmental pollution while contributing to a sound circular economy. Brewer spent grains is the major by-product generated from the beer industry. In modern breweries, there are four main ingredients used for beer production: water, malt, yeasts, and hops, but it can also include barley, wheat, rice, or corn (Jackowski et al., 2020). These grains are usually mashed and heated during the beer processing. Thus, the chemical composition of brewer grains would likely vary across and within

Table 1 Characteristics of selected by-product feeds used in animal agriculture

Crop	By-products	Main characteristics	Animal species
Corn	DDGS[a]	By-product of solvent extraction through yeast fermentation	Ruminants, poultry, swine, horses
	Germ	Produced after dehulling and degerming the corn grain	Ruminants, poultry, swine
	Cobs	It comprises the structural makeup of corn residue	Ruminants
	Gluten meal	One of the by-products of wet milling process, it is largely used as a source of protein and energy	Ruminants
	Gluten feed	By-product of starch or ethanol production – originated from the wet milling process	Ruminants
Soybean	Meal	By-product of the extraction of soybean oil	Ruminants, poultry, swine, horses
	Hulls	Produced during the oil recovery process and it accounts for about 8% of total DM (bean)	Ruminants, poultry, swine, horses
Beet	Pulp, dry	Dried after sugar extraction from beet	Ruminants, poultry, swine, horses
	Pulp, wet	Original form produced after sugar extraction from beet	
Citrus	Pulp, dry	Dried after the extraction of juice, and it is comprised of peel (flavedo and albedo), pulp (juice sac residue), rag (membranes and cores) and seeds	Ruminants, poultry, swine, horses
	Pulp, wet	It results from the extraction of juice, and it is comprised of peel (flavedo and albedo), pulp (juice sac residue), rag (membranes and cores) and seeds	
Cottonseed	Hulls	Removed before oil extraction and it is comprised of seed coats from cottonseed.	Ruminants
	Meal	Originated from the mechanical extraction of oil from cottonseed	Ruminants, poultry, swine, horses
Barley	Brewer grains, dry	Dried after beer production, and mostly comprised of husk-pericarp seed coat layers of barley. It can also contain hop residues	Ruminants, poultry, swine, horses
	Brewer grains, wet	Original wet form after beer is produced. Comprised of husk-pericarp seed coat layers of barley. It can also contain hop residues	
Wheat	Wheat middlings	Produced during the wheat milling, and contains wheat bran, shorts, germ, and flour	Ruminants, poultry, swine, horses
	Bran	Originated from the dry milling of wheat for flour production	

(*Continued*)

Table 1 (*Continued*)

Crop	By-products	Main characteristics	Animal species
Sugarcane or beets	Molasses	The residue remaining after sugar extraction. Characterized as a thick, brown-colored liquid and considered an energy feedstuff	Ruminants, swine, horses
Canola	Meal	By-product originated from of the oil extraction from canola (*Brassica* spp.) and is widely used as a source of protein	Ruminants, poultry, swine

[a] Dried distillers' grains with solubles.

a brewery due to differences in grain variety, harvesting time, and mashing and malting conditions (NASEM, 2021). Dried brewer grains are rarely fed to swine because of their low energy concentration; however, in some cases, they can be a substitute for oats in rations for horses as a source of protein (Westendorf and Wohlt, 2002).

Feeding dry brewers grains to crossbred steers (11.7% of the diet DM) did not change ruminal fermentation parameters (i.e. pH, volatile fatty acids, and ammonia concentration) or digestibility of nutrients (Faccenda et al., 2019), and therefore seemed not to affect ruminal function. However, if not dried after beer production, the use of brewer grains may be difficult as its low DM concentration (~22%) can lead to undesirable microbial growth and alteration of its nutritional composition. Thus, drying technologies are needed to optimize the inclusion of low DM by-product feeds such as brewer grains in ruminant diets. Faccenda et al. (2021) reported that wet brewer grains can be effectively dried within 16 h of direct sun exposure, which allowed it to be stored for up to 180 days without negatively affecting its nutritional value. Table 1 summarizes the characteristics of selected agro-industrial by-product feeds according to crop type and their utilization for livestock animals.

3 Chemical composition and nutritional value of selected by-product feeds

The nutritional composition of by-product feeds can vary according to numerous factors such as industrial processes, storage conditions, and physical form (i.e. dry and wet). Therefore, depending on the stages and types of processing including heating, steaming, and level of pressure, the nutritional quality and physical characteristics of plant by-product feeds can be changed. The variation in the nutritional value of by-product feeds associated with processing methods within the same batch or across plants can limit their utilization in livestock diets. However, if used properly, by-product feeds such as soybean meal are widely recommended for livestock feeding, especially for high-producing dairy cows due to its high nutritional value [53% of crude protein (CP)] as shown in Table 2.

Table 2 Chemical composition of selected by-product feeds used in animal agriculture

Item	Chemical composition[a]						
	DM	CP	ADF	NDF	Lignin	Starch	Crude fat
DDGS[b]	91.1 ± 1.40	39.0 ± 2.84	17.7 ± 3.20	37.6 ± 4.80	5.83 ± 1.96	6.20 ± 2.87	7.56 ± 2.00
Corn germ	90.2 ± 2.04	15.4 ± 1.57	10.1 ± 4.82	27.0 ± 6.62	2.69 ± 2.01	27.6 ± 10.4	17.8 ± 3.40
Corn cobs	89.8 ± 8.06	3.00 ± 0.77	46.3 ± 6.32	83.8 ± 5.28	4.20 ± 1.79	1.10 ± 0.80	0.62 ± 0.29
Corn gluten meal	90.5 ± 1.64	68.5 ± 3.52	3.70 ± 2.02	6.80 ± 3.58	1.79 ± 1.24	16.4 ± 4.19	2.44 ± 1.05
Corn gluten feed	89.2 ± 1.77	23.2 ± 2.48	11.5 ± 2.11	35.7 ± 4.62	2.31 ± 1.21	15.5 ± 4.63	3.91 ± 1.11
Soybean meal	89.3 ± 1.12	52.6 ± 1.69	7.20 ± 1.58	11.1 ± 3.07	1.08 ± 0.52	1.90 ± 1.05	1.82 ± 0.87
Soybean hulls	90.4 ± 1.90	11.9 ± 1.43	47.9 ± 3.07	66.7 ± 3.63	2.57 ± 0.75	1.00 ± 0.70	1.89 ± 0.80
Beet pulp, dry	92.3 ± 1.24	9.90 ± 1.36	28.2 ± 2.37	46.9 ± 4.27	3.88 ± 2.10	0.60 ± 0.44	1.06 ± 0.32
Beet pulp, wet	22.7 ± 3.36	9.10 ± 1.08	27.2 ± 2.92	44.1 ± 6.49	3.18 ± 2.33	0.90 ± 0.60	0.99 ± 0.30
Citrus pulp, dry	88.2 ± 1.35	7.20 ± 1.11	20.1 ± 3.31	24.1 ± 2.90	2.49 ± 1.36	1.30 ± 1.20	2.55 ± 0.70
Citrus pulp, wet	18.3 ± 5.49	8.70 ± 1.56	23.2 ± 3.86	25.9 ± 4.16	3.61 ± 3.05	0.90 ± 0.95	3.42 ± 2.12
Cottonseed hulls	91.0 ± 1.75	7.00 ± 2.18	66.0 ± 4.81	80.9 ± 6.13	19.9 ± 2.71	1.20 ± 1.16	3.20 ± 1.60
Cottonseed meal	89.9 ± 2.03	46.7 ± 3.84	19.2 ± 4.01	28.1 ± 5.89	7.01 ± 2.07	1.10 ± 0.76	3.60 ± 2.21
Brewer grains, dry	93.1 ± 2.34	25.3 ± 4.56	24.7 ± 5.61	51.8 ± 10.6	6.66 ± 2.13	6.50 ± 2.76	9.02 ± 2.27
Brewer grains, wet	22.5 ± 3.56	28.1 ± 4.00	23.8 ± 2.66	49.3 ± 4.94	6.64 ± 1.11	5.20 ± 3.88	9.52 ± 1.24
Wheat middlings	88.3 ± 1.38	19.1 ± 1.25	35.1 ± 3.28	51.1 ± 4.25	4.99 ± 0.70	13.0 ± 5.85	3.06 ± 0.44
Wheat bran	90.1 ± 1.16	17.4 ± 1.60	13.8 ± 2.45	40.1 ± 6.22	4.15 ± 1.02	20.8 ± 6.20	4.39 ± 0.74
Molasses	65.4 ± 14.1	9.30 ± 4.36	0.20 ± 0.19	0.60 ± 0.41	-	0.80 ± 0.52	0.61 ± 0.42
Canola meal	89.1 ± 1.15	41.5 ± 1.55	20.3 ± 2.17	29.0 ± 2.78	8.51 ± 1.96	1.60 ± 0.85	3.51 ± 0.85

Adapted from NASEM (2021).

[a] Mean ± standard deviation; % of DM unless otherwise stated; CP, crude protein; ADF, acid detergent fiber; NDF, neutral detergent fiber.

[b] Dried distillers' grains with solubles.

3.1 Nutritional value of by-product feeds

Many by-products have been used to feed ruminant and non-ruminant animals. Their availability mostly depends on crop production and logistics factors. As many of these by-products are considered non-edible for humans, their price is generally lower compared with traditional feedstuffs (e.g. corn and soybean meal). In addition, some by-product feeds have high nutritional value. For instance, soyhulls are rich in highly digestible fiber, differing from the original soybean grain, which makes it a valuable by-product feed for ruminant's diets because it can be used to replace forage or grain sources. Another example of a high value by-product feed is beet pulp (Table 2), which has high fiber concentration and is commonly fed to lactating dairy cows, used in backgrounding diets or as a roughage source in finishing diets for beef cattle (Lardy and Anderson, 2009).

Numerous by-products are good sources of energy, with some of them having high concentration of digestible fiber (e.g. beet pulp, citrus pulp, and soyhulls) and the amount included in the rations should be based on the dietary concentrations of starch and fat (Bernard, 2017). By-product feeds with low fiber concentration can also be used as a source of energy in ruminant diets. For instance, molasses (Table 2), a rich source of sucrose, can be fed to ruminants because ruminal microbes can use sugars for growth and production of microbial protein.

By-product feeds with low fiber digestibility are commonly fed to ruminants due to their ability to digest fiber in the rumen. By-product feeds with low fiber digestibility usually contain peels and hulls that are concentrated during the industrial processing. For instance, cottonseed hulls (Table 2) is a by-product originated from the industrial processing of cotton known to have high neutral detergent fiber (NDF), but low CP concentration and poor fiber digestibility. However, cottonseed hulls can still be included in ruminant diets as a source of fiber. Table 2 summarizes the chemical composition of selected by-product feeds commonly used in animal agriculture.

3.2 Limitations of inclusion of by-product feed in livestock diets

There are several reasons for limiting the inclusion of a certain by-product feed in rations, including factors such as cost, palatability, moisture content of the diet, essential amino acid requirements, and fiber and fat concentrations (Lardy and Anderson, 2009; Bernard, 2017).

In dairy cow diets, moisture levels in the total diet should not exceed 50% under normal circumstances, which may limit the dietary inclusion of wet by-product feeds such as WDGS and brewers spent grains (Bernard, 2017). As mentioned earlier, previous research (Faccenda et al., 2021) has looked at drying and storing strategies for brewer spent grains with high moisture levels. However, further studies aimed at finding different storage and handling technologies at low cost to maximize the utilization of wet by-products are needed.

By-products with high concentrations of fat such as DDGS and corn germ (Table 2) could interfere with normal fiber digestion in the rumen if excessive amounts are included in the diet, especially if oilseeds are fed as well. Rice hulls have high concentrations of silica, which may injure the digestive tract of animals and should be limited if fed. Whole cottonseed contains gossypol, which is toxic to young ruminants and non-ruminants livestock. Although mature dairy cattle can detoxify gossypol, no more than 4.5 kg of cottonseed products should be included in rations to prevent toxicity (Bernard, 2017). According to Stewart (2022), whole cottonseed and cottonseed meal are the primary sources of gossypol in beef cattle diets, while cottonseed hulls contain low concentrations of gossypol.

4 Use of dried distillers' grain with solubles in animal agriculture

Dried distillers' grain with solubles is one of the by-products of ethanol production from cereals, such as corn and sorghum. This by-product is generated after the starch fraction of cereals is fermented with selected yeasts and enzymes producing ethanol. After fermentation, the alcohol is removed by distillation and the remaining residue produced is WDGS, which when dried produces DDGS (Fig. 1). The ethanol production in the United States increased rapidly from 2000 to 2010, but only a modest annual growth (i.e. 3.8%) has been registered in the last 10 years.[3] Currently, the United States produces 56.8 billion liters of ethanol annually, making the country the world's largest ethanol producer.[4] This increase in ethanol production generated several by-products from the distillery industry. Therefore, an appropriate and profitable use of these by-products from the ethanol biorefinery helps to improve the viability of the industry (Gerrior et al., 2022). In the case of DDGS, the market value as animal feed is greater than its fertilizer value (Mahbub et al., 2019).

Considering the current American operational efficiency of the ethanol industry, for each 100 kg of corn used, approximately 43.3 L of ethanol is produced along with 28 kg of DDGS, 1.5 kg of oil, and 1.8 kg of CO_2 (Irwin, 2022). It is worth noting that the distillers corn oil is a human non-edible product, but it can be used as animal feedstuff or biofuel. According to the United States Bioenergy statistics report, 36.3 million metric tons of DDGS was produced from the 2021 to 2022 crop,[5] from which 11.6 million metric tons were exported while 24.7 million metric tons were used domestically. However,

[3] www.ers.usda.gov.
[4] www.ethanolrfa.org.
[5] www.ers.usda.gov.

according to an assessment done by Hoffman and Baker (2011), the current American livestock and poultry inventory is able to utilize 63 million metric tons, considering the recommended inclusion levels for each species.

4.1 Feeding dried distillers' grain with solubles to swine

When using by-products as an alternative feed, it is important to understand the impact of the by-product inclusion on animal production, health, and product quality. A common approach is to compare alternative products with traditional ones, such as corn and soybean meal. Although DDGS can provide a similar level of digestible energy and greater CP concentration than corn to growing hogs, this by-product feed has a few limitations due to its relatively high fiber and fat concentrations (Pedersen et al., 2007). Elevated fiber concentration increases the bulkiness of the diet, which may limit appetite as gut capacity is reached (Thacker, 2006). Increased bulkiness along with a potential decrease in palatability may reduce feed intake in growing hogs (Stein and Shurson, 2009). In addition, it was been shown that DDGS can decrease carcass dressing percentage when fed to growing hogs (Stein and Shurson, 2009). This effect seems to be explained by an increase in non-carcass components than a negative effect on carcass weight (Boler et al., 2015). The higher fiber concentration of DDGS compared with ground corn increases gut fill and stimulates the gastrointestinal tract growth, thus increasing the mass of both large and small intestines (Boler et al., 2015). Therefore, a heavier gastrointestinal tract and increased gut fill are likely responsible for decreased carcass dressing percentage in hogs consuming DDGS.

The fatty acid profile found in DDGS can also be a limiting factor for feeding DDGS to market hogs. Corn oil is rich in monounsaturated fatty acids and polyunsaturated fatty acids, which decrease the fat firmness in carcasses of hogs fed DDGS (Boler et al., 2015; Stein and Shurson, 2009). This is an issue particularly with the belly cut because a soft belly fat produces a slice of bacon that easily breaks apart. However, this adverse effect was observed only when more than 20% of DDGS was included in the diets of growing and finishing hogs (Stein, 2007). Therefore, it has been recommended to include up to 20% of DDGS in the diet DM for growing and finishing hogs, but up to 40% can be included in diets used for gestating sows (Stein, 2007). It is important to note that this is not an issue when feeding DDGS for ruminants because unsaturated fatty acids are extensively biohydrogenated in the rumen. Regardless of the possible negative effect on belly thickness and firmness, apparently, there is no adverse effect on meat quality (Boler et al., 2015).

The CP concentration of DDGS is greater than that of corn, so that when DDGS is included in livestock diets, it can also replace a portion of soybean meal (Hoffman and Baker, 2011). However, the amino acid profile of DDGS may limit swine performance due to the low concentrations of digestible amino acids, specially lysine and tryptophan (Stein, 2007). In addition, drying may negatively affect the amino acid digestibility of DDGS because when high temperatures are applied, Maillard reactions can occur making amino acids less digestible. These non-enzymatic reactions are also known as browning reactions, and studies have demonstrated that DDGS color can be used as a predictor for lysine digestibility, where darker DDGS have been associated with lower amino acids digestibility (Cromwell et al., 1993; Ergul et al., 2003).

4.2 Feeding dried distillers' grain with solubles to poultry

Some of the abovementioned limitations of the use of DDGS in swine's diet are also limitations in poultry's diet as both are non-ruminant species and have limited capacity to use the energy stored in non-starch polysaccharides. The concentration of starch reduces while that of non-starch polysaccharides increases due to the fermentation process in DDGS (Świątkiewicz and Koreleski, 2008). This negative effect can be alleviated by treating DDGS with enzymes that are able to hydrolyze the non-starch polysaccharides fraction (Swiatkiewicz et al., 2016). However, the fermentation process used that generates DDGS can improve nutrient availability because of microbial fermentation. For instance, phosphorus (P) bioavailability is increased by the fermentation process, which is probably an effect of microbial enzymes on phytate. In a study done with growing boiler chickens, Lumpkins and Batal (2005) reported that the P bioavailability of corn DDGS varied from 54% to 68%, which was greater than that from corn. Another benefit of feeding DDGS is its greater concentration of xanthophylls compared with corn as these fat-soluble compounds can intensify the yellowness of the yolk and skin of broiler carcasses, making these products more attractive to consumers (Salim et al., 2010; Hoffman and Baker, 2011).

It has been suggested that DDGS can be included in growing broilers, growing turkeys, and layer chickens diets at up to 15% (DM basis) without adverse results (Świątkiewicz and Koreleski, 2008; Hoffman and Baker, 2011). However, due to the lower amino acid digestibility of DDGS compared to corn, it has been recommended a maximum of 8% inclusion in starter broiler diets (Świątkiewicz and Koreleski, 2008). According to the Renewable Fuels Association report, the amount of DDGS used by the poultry industry accounts for only 7% of the United States domestic consumption.[6]

[6] www.ethanolrfa.org.

4.3 Feeding dried distillers' grain with solubles to beef cattle

Although the fiber concentration is a limitation when feeding DDGS for non-ruminant species, this is not the case for cattle. Some studies have shown a decreased NDF digestibility in response to DDGS inclusion (Hünerberg et al., 2013). However, a greater feeding value has been observed for DDGS than corn grain in finishing diets for steers, indicating that the high fiber concentration seems to be compensated by improved feed efficiency (Klopfenstein et al., 2008). In a meta-analysis conducted with data from growing beef cattle, Klopfenstein et al. (2008) showed that DM intake (DMI) decreased linearly while average daily weight gain increased quadratically as DDGS level in the diet increased from 0% to 40%. Interestingly, they also observed that WDGS had a better feeding value than DDGS, suggesting that the drying process also decreases the nutrient availability of DDGS for cattle (Klopfenstein et al., 2008). Note that cattle's diets are usually comprised of a mixture of forages and concentrates, with water often added to obtain a more homogeneous ration to minimize sorting. Therefore, the use of WDGS can be also advantageous due to the moisture and stickiness of this by-product feed (Klopfenstein et al., 2008). In addition, less energy is required to produce WDGS than DDGS because drying is not needed, thus reducing the environmental impact of the ethanol industry.

The decrease in DMI seen with DDGS inclusion levels above 40% of the diet DM may be partially explained by the greater concentration of fat and sulfur (S) compared to corn (Klopfenstein et al., 2008; Leupp et al., 2009). Due to the fermentation process, the concentrations of both P and S increase in DDGS. Therefore, diets must be adjusted to reduce the concentrations of these macrominerals in the mineral premix supplemented to the animals. This change is economically beneficial considering that P is one of the most expensive minerals in an animal's diet. While overfeeding P does not usually impair animal health, it may increase environmental pollution leading to eutrophication as more P is excreted through feces. On the other hand, overfeeding S may cause polioencephalomalacia, reduce hepatic cupper stores, and negatively affect cattle growth performance (Klopfenstein et al., 2008). Although the high fat concentration of DDGS may reduce DMI, studies have shown that high-fat DDGS can mitigate enteric methane emissions in growing beef cattle (McGinn et al., 2009; Hünerberg et al., 2013). However, increased dietary inclusion levels of DDGS may also increase urinary N excretion thereby generating other greenhouse gases such as nitrous oxide, which is more potent to trap atmospheric heat than methane (McGinn et al., 2009; Hünerberg et al., 2013, 2014)

Overall, DDGS is a good source of energy and protein for beef cattle, and it appears that DDGS is more suitable be used for cattle than other livestock species (Hoffman and Baker, 2011). Experimental data have demonstrated that up to 40% to 50% of diet DM may be replaced with DDGS in feedlot rations resulting in improved production performance and decreased costs. (Klopfenstein et al., 2008; Leupp et al., 2009; Hoffman and Baker, 2011).

4.4 Feeding dried distillers' grain with solubles to dairy cattle

Like beef cattle, dairy cows also can effectively utilize DDGS. However, the faster passage rate in dairy cows reduces fiber digestibility compared to beef cattle. Benchaar et al. (2013) reported a linear increase in DMI together with a linear decrease in NDF digestibility in lactating dairy cows fed incremental amounts of DDGS (from 0% to 30% of the diet DM). They further observed a linear reduction in enteric methane emissions as DDGS level increased from 0% to 30% of the diet DM, which can be associated with decreased NDF digestibility (Benchaar et al., 2013).

Dried distillers' grain with solubles is a good source of rumen undegradable protein, and about 55% of DDGS CP is considered undegradable in the rumen (Schingoethe et al., 2009). However, increased rumen undegradable protein compared to corn and other protein sources may be a consequence of the heat treatment generating Maillard reactions, which can ultimately decrease amino acid digestibility (Schingoethe et al., 2009). Usually, DDGS color is a good indicator of protein quality, and light golden-yellow color is preferable over the dark-brown counterpart. Lysine is the main limiting amino acid in diets containing DDGS. Paz and Kononoff (2014) observed a lower lysine intake and lower blood concentration of lysine in the subcutaneous abdominal vein in dairy cows fed 30% DDGS compared to 15% DDGS. They also observed a reduced milk protein concentration in cows fed 30% DDGS but milk protein yield was similar between 30% and 15% of DDGS in the diet DM (Paz and Kononoff, 2014). Interestingly, feeding rumen-protected lysine did not interact with dietary DDGS level (Paz and Kononoff, 2014).

A maximum inclusion rate between 20% and 30% of DDGS in lactating dairy cows diets has been recommended due to some adverse effects such as decreased milk fat concentration when replacing corn with DDGS (Schingoethe et al., 2009; Hoffman and Baker, 2011). In addition, studies showed that DDGS inclusion greater than 30% of the diet DM can increase P and N excretion, and reduced DMI and milk production (Kalscheur, 2005; Schingoethe et al., 2009; Benchaar et al., 2013).

4.5 Life-cycle analysis

It is not uncommon to see in the media negative press about the environmental impact of animal agriculture, especially ruminant production systems. However, the public is often misled and not presented with scientific-based information about the full production cycle. Therefore, LCA studies are important to provide a more realistic view of ruminant production systems placing animal agriculture as part of the solution instead of the 'major' contributor of environmental degradation and pollution (Liu et al., 2021). Specifically, LCA has become a powerful tool to access the environmental impact of agricultural products (Caffrey and Veal, 2013) despite being very complex as it must consider multiple steps of the production system and the lack of accurate information make LCA studies a challenging task.

Werth et al. (2021), using an LCA approach, showed that the transportation of DDGS, corn, and other ingredients from the feed mill can significantly contribute to the carbon footprint of cattle's diet in California. However, as the authors did not find information about the emissions of feedstuffs transportation, they created three simulations as follows: (A) all the trucks and trains returned with an empty load (100% empty return load); (B) half trucks and trains returned with an empty load (50% empty return load); and (C) none of the trucks and trains returned with an empty load (0% empty return load). The CO_2 equivalent ($CO2_{eq}$) for the three scenarios averaged 0.63, 0.58, and 0.52 kg of CO_2eq/kg of as-fed total mixed ration for A, B, and C, respectively (Werth et al., 2021). This study exemplifies that the LCA studies often depend on assumptions and created simulations, and data should be interpreted cautiously. Furthermore, the assessment of the environmental impact of feeding by-products such as DDGS depends on how much fossil fuel can be spared during processing and transportation.

In an LCA study, evaluating the emissions saving when feeding by-products of the corn-based ethanol production industry, Bremer et al. (2010) showed that a more pronounced greenhouse gas emission reduction was obtained when WDGS was fed to livestock compared to DDGS because of the energy required to dry DDGS. However, feeding WDGS has some limitations including product stability and transportation costs because more water is transported when WDGS is used on farms. Thus, feeding WDGS is limited to feedlots adjacent to ethanol plants. In Bremer et al. (2010) study, they also observed that the greenhouse gas emission saving varied according to the livestock class that the by-product was fed to, and a more substantial decrease was obtained in the net life cycle greenhouse gas emissions with feeding WDGS to beef cattle compared to dairy cattle or swine (Bremer et al., 2010).

Feeding ethanol by-products to livestock is a way to decrease the environmental impact and improve the economic viability of farms because traditional feedstuffs such as corn and soybean meal can be replaced. In the case of corn, it can be processed for human consumption or to produce biofuel which generates by-products including DDGS that is not suitable to be used as a food source, but it can be fed to livestock and ultimately be converted into high quality edible products (e.g. milk, meat, and eggs), a process known as 'upcycling'.

5 Conclusion

Animal agriculture is under increasing scrutiny due to environmental concerns, particularly greenhouse gas emissions and N and P pollution. However, the public is often misled with information that is not backed up by the scientific community and scientific data. For instance, the public is not generally aware of the importance of livestock systems to recycle nutrients from by-product feeds into meat, eggs, and milk that would otherwise be wasted in the environment. Therefore, animal agriculture has been effectively contributing to provide nutritious foods for humans and it is expected to contribute more during the upcoming decades due to the growing human population and demand for animal protein. The use of by-product feeds in livestock diets would likely increase as they can replace traditional feeds such as corn that is also used for humans, thus alleviating the feed/food competition. For optimizing the use of current and new by-product feeds, further research is needed not only focusing on production outcomes but also on animal health and environmental impact. In this scenario, modeling and LCA should be part of the evaluation processes of by-product feeds to provide producers and policy makers with information to make decisions to improve the economic, social, and environmental sustainability of livestock systems globally.

6 Where to look for further information

- NASEM. 2021. *Nutrient Requirements of Dairy Cattle*. 8th Edition. The National Academics Press.
- https://www.un.org/en/desa/world-population-projected-reach-98-billion-2050-and-112-billion-2100.
- https://www.ers.usda.gov/data-products/u-s-bioenergy-statistics/.
- https://ethanolrfa.org/markets-and-statistics/annual-ethanol-production.
- https://www.ers.usda.gov/data-products/u-s-bioenergy-statistics.
- https://www.feedipedia.org/.

7 References

Alexandratos, N. and Bruinsma, J. 2012. World agriculture towards 2030/2050: the 2012 revision. *ESA Working paper, Rome, FAO.* n. 12-03. p. 1–147. Available at: https://www.fao.org/3/ap106e/ap106e.pdf.

Association of American Feed Control Officials 2016. Oxford, UK: Association of American Feed Control.

Bampidis, V. A. and Robinson, P. H. 2006. Citrus by-products as ruminant feeds: a review. *Anim. Feed Sci. Technol.* 128(3–4):175–217. doi:10.1016/j.anifeedsci.2005.12.002.

Benchaar, C., Hassanat, F., Gervais, R., Chouinard, P. Y., Julien, C., Petit, H. V. and Massé, D. I. 2013. Effects of increasing amounts of corn drieddistillers grains with solubles in dairy cow diets on methane production,ruminal fermentation, digestion, N balance, and milk production. *J. Dairy Sci.* 96(4):2413–2427. doi:10.3168/jds.2012-6037.

Bernard, J. K. 2017. Considerations for using by-product feeds. *University of Georgia Cooperative Extension, Athens, GA.* v. 862. Available at: https://secure.caes.uga.edu/extension/publications/files/pdf/B%20862_3.PDF.

Boler, D. D., Overholt, M. F., Lowell, J. E., Dilger, A. C. and Stein, H. H. 2015. Effects of pelleting growing-finishing swine diets on growth, carcass, and bacon characteristics. In: 15th Annual Midwest Swine Nutrition Conference Proceedings, 23–30.

Bremer, V. R., Liska, A. J., Klopfenstein, T. J., Erickson, G. E., Yang, H. S., Walters, D. T. and Cassman, K. G. 2010. Emissions savings in the corn-ethanol life cyclefrom feeding coproducts to livestock. *J. Environ. Qual.* 39(2):472–482. doi:10.2134/jeq2009.0283.

Caffrey, K. R. and Veal, M. W. 2013. Conducting an agricultural life cycle assessment: challenges and perspectives. *Sci. World J.* 2013:472431. doi:10.1155/2013/472431.

Cromwell, G. L., Herkleman, K. L. and Stahly, T. S. 1993. Physical, chemical, and nutritional characteristics of distillers dried grains with solubles for chicks and pigs. *J. Anim. Sci.* 71(3):679–686.

de Vries, M. and de Boer, I. J. M. 2010. Comparing environmental impacts for livestock products: a review of life cycle assessments. *Livest. Sci.* 128(1–3):1–11.

Ergul, T., Martinez Amezcus, C., Parsons, C. M., Walters, B., Brannon, J. and Noll, S. L. 2003. Amino acid digestibility in corn distillers dried grains with solubles. *Poult. Sci.* 82(Suppl. 1):70.

Erickson, A. 2006. Corn wet milled feed products. *Corn Refiners Assoc.* Washington, DC. 4th Edition. p. 1-33. Available at: https://corn.org/wp-content/uploads/2018/10/Feed2006.pdf.

Faccenda, A., Zambom, M. A., Avila, A. S., Castagnara, D. D., Dri, R., Fischer, M. L., Tinini, R. C. R., Dessbesell, J. G., Almeida, A. E. and Almeida, K. V. 2021. Influence of the storage period on the nutritional and microbiological value of sun-dried brewer's grains. *Rev. Colomb. Cienc. Pecu* 34(4):254-266. doi:10.17533/UDEA.RCCP.V34N4A02.

Faccenda, A., Zambom, M. A., de Avila, A. S., Garcias, J., Eckstein, E. I., Fornari, J. L., de Almeida, K. V. and Santos, G. T. 2019. Nutrient digestibility and ruminal parameters of cattle fed dried brewers grains and Saccharomyces cerevisiae. *Livest. Sci.* 225:109–115. doi:10.1016/j.livsci.2019.05.003.

Gerrior, D., Delsoz Bahri, K., Kermanshahi-Pour, A., Eckelman, M. J. and Brar, S. K. 2022. Life cycle assessment and techno-economic analysis of a novel closed loop corn ethanol biorefinery. *Sustain. Prod. Consum.* 30:359–376.

Grasser, L. A., Fadel, J. G., Garnett, I. and Depeters, E. J. 1995. Quantity and economic importance of nine selected by-products used in California dairy rations. *J. Dairy Sci.* 78(4):962–971. doi:10.3168/jds.S0022-0302(95)76711-X.

FAOSTAT 2016. *Food and Agriculture Organization of the United Nations (FAO).* FAOSTAT database. Available at: http://faostat.fao.org/site/291/default.aspx.

Hoffman, L. A. and Baker, A. J. 2011. *Estimating the Substitution of Distillers' Grains for Corn and Soybean Meal in the US Feed Complex.* US Department of Agriculture. USDA Econ. Res. Serv. Rep. FDS-11-I-01.

Hünerberg, M., Little, S. M., Beauchemin, K. A., McGinn, S. M., O'Connor, D., Okine, E. K., Harstad, O. M., Kröbel, R. and McAllister, T. A. 2014. Feeding high concentrations of corn dried distillers' grains decreases methane, but increases nitrous oxide emissions from beef cattle production. *Agric. Syst.* 127:19–27. doi:10.1016/j.agsy.2014.01.005.

Hünerberg, M., McGinn, S. M., Beauchemin, K. A., Okine, E. K., Harstad, O. M. and McAllister, T. A. 2013. Effect of dried distillers' grains with solubles on enteric methane emissions and nitrogen excretion from finishing beef cattle. *Can. J. Anim. Sci.* 93(3):373–385. doi:10.4141/cjas2012-151.

Irwin, S. 2022. Trends in the operational efficiency of the U. S. ethanol industry : 2021 update. *Farmdoc Dly*. 12:1–5.

Jackowski, M., Niedźwiecki, Ł., Jagiełło, K., Uchańska, O. and Trusek, A. 2020. Brewer's spent grains–valuable beer industry by-product. *Biomolecules* 10(12):1–18.

Jeswani, H. K., Chilvers, A. and Azapagic, A. 2020. Environmental sustainability of biofuels: a review. *Proc. Math. Phys. Eng. Sci.* 476(2243):20200351. doi:10.1098/rspa.2020.0351.

Kalscheur, K. F. 2005. Impact of feeding distillers grains on milk fat, protein, and yield. In: *Proceedings of the Distillers Grains Technology Council, 10th Annual Symposium*, Louisville, KY.

Klopfenstein, T. J., Erickson, G. E. and Bremer, V. R. 2008. Board-invited review: use of distillers by-products in the beef cattle feeding industry. *J. Anim. Sci.* 86(5):1223–1231. doi:10.2527/jas.2007-0550.

Lardy, G., Anderson, V., Dahlen, C. and Carlson, Z. 2009. *Alternative Feeds for Ruminants.* North Dakota State University Extension, Fargo, ND.v. 1182. p. 1–27

Leupp, J. L., Lardy, G. P., Karges, K. K., Gibson, M. L. and Caton, J. S. 2009. Effects of increasing level of corn distillers dried grains with solubles on intake, digestion, and ruminal fermentation in steers fed seventy percent concentrate diets. *J. Anim. Sci.* 87(9):2906–2912. doi:10.2527/jas.2008-1712.

Liu, S., Proudman, J. and Mitloehner, F. M. 2021. Rethinking methane from animal agriculture. *CABI Agric. Biosci.* 2(1):1–13. doi:10.1186/s43170-021-00041-y.

Lumpkins, B. and Batal, A. 2005. The bio- availability of lysine and phosphorus in distillers dried grains with solubles. *Poult. Sci.* 84, 581–586.

Mahbub, N., Gemechu, E., Zhang, H. and Kumar, A. 2019. The life cycle greenhouse gas emission benefits from alternative uses of biofuel coproducts. *Sustain.Energy Technol. Assess.* 34:173–186.

McGinn, S. M., Chung, Y.-H., Beauchemin, K. A., Iwaasa, A. D. and Grainger, C. 2009. Use of corn distillers' dried grains to reduce enteric methane loss from beef cattle. *Can. J. Anim. Sci.* 89(3):409–413. doi:10.4141/CJAS08133.

Mirzaei-Aghsaghali, A. and Maheri-Sis, N. 2008. Nutritive value of some agro-industrial by-products for ruminants: a review. *World J. Zool.* 3:40–46.

Mottet, A., de Haan, C., Falcucci, A., Tempio, G., Opio, C. and Gerber, P. 2017. Livestock: on our plates or eating at our table? A new analysis of the feed/food debate. *Glob. Food Sec.* 14:1–8. doi:10.1016/j.gfs.2017.01.001.

NASEM 2021. *Nutrient Requirements of Dairy Cattle* (8th edn.). Washington, DC: The National Academics Press.

Paz, H. A. and Kononoff, P. J. 2014. Lactation responses and amino acid utilization of dairy cows fed low-fat distillers dried grains with solubles with or withoutrumen-protected lysine supplementation. *J. Dairy Sci.* 97(10):6519–6530.

Pedersen, C., Boersma, M. G. and Stein, H. H. 2007. Digestibility of energy and phosphorus in ten samples of distillers dried grains with solubles fed to growing pigs. *J. Anim. Sci.* 85(5):1168–1176. doi:10.2527/jas.2006-252.

Salim, H. M., Kruk, Z. A. and Lee, B. D. 2010. Nutritive value of corn distillers dried grains with solubles as an ingredient of poultry diets: a review. *Worlds Poult. Sci. J.* 66(3):411–432. doi:10.1017/S0043933910000504.

Schingoethe, D. J., Kalscheur, K. F., Hippen, A. R. and Garcia, A. D. 2009. The use of distillers products in dairy cattlediets. *J. Dairy Sci.* 92(12):5802–5813. doi:10.3168/jds.2009-2549.

Stein, H. 2007. Distillers dried grains with solubles (DDGS) in diets fed to swine. *Swine Focus*:1-8.

Stewart, L. 2022. Using cotton byproducts in beef cattle diets. University of Georgia Cooperative Extension, Athens, GA. v. 1311. Available at: https://secure.caes.uga.edu/extension/publications/files/pdf/b%201311_3.pdf.

Stein, H. H. and Shurson, G. C. 2009. BOARD-INVITED REVIEW: the use and application of distillers dried grains with solubles in swine diets. *J. Anim. Sci.* 87(4):1292–1303. doi:10.2527/jas.2008-1290.

Świątkiewicz, S. and Koreleski, J. 2008. The use of distillers dried grains with solubles (DDGS) in poultry nutrition. *Worlds Poult. Sci. J.* 64(2):257–266. doi:10.1017/S0043933908000044.

Swiatkiewicz, S., Swiatkiewicz, M., Arczewska-Wlosek, A. and Jozefiak, D. 2016. Efficacy of feed enzymes in pig and poultry diets containing distillers dried grains with solubles: a review. *J. Anim. Physiol. Anim. Nutr.* 100(1):15–26. doi:10.1111/jpn.12351.

Taheripour, F., Hertel, T. W., Tyner, W. E., Beckman, J. F. and Birur, D. K. 2010. Biofuels and their by-products: global economic and environmental implications. *Biomass Bioenergy* 34(3):278–289. doi:10.1016/j.biombioe.2009.10.017.

Thacker, P. A. 2006. Nutrient digestibility, performance and carcass traits of growing-finishing pigs fed diets containing dried wheat distiller's grains with solubles. *Can. J. Anim. Sci.* 86(4):527–529. doi:10.4141/A06-025.

Werth, S. J., Rocha, A. S., Oltjen, J. W., Kebreab, E. and Mitloehner, F. M. 2021. A life cycle assessment of the environmental impacts of cattle feedlot finishing rations. *Int. J. Life Cycle Assess.* 26(9):1779–1793. doi:10.1007/s11367-021-01957-3.

Westendorf, M. L. and Wohlt, J. E. 2002. Brewing by-products: their use as animal feeds. *Vet. Clin. North Am. Food Anim. Pract.* 18(2):233–252. doi:10.1016/S0749-0720(02)00016-6.

Zang, Y., Silva, L. H. P., Geng, Y. C., Ghelichkhan, M., Whitehouse, N. L., Miura, M. and Brito, A. F. 2021. Dietary starch level and rumen-protected methionine, lysine, and histidine: effects on milk yield, nitrogen, and energy utilization in dairy cows fed diets low in metabolizable protein. *J. Dairy Sci.* 104(9):9784–9800. doi:10.3168/jds.2020-20094.

Chapter 6

Optimizing slurry management

David Fangueiro, LEAF-Instituto Superior de Agronomia-ULisboa, Portugal; Jihane Elmahdi, Wageningen University and Research, The Netherlands; Jared Nyang'au, Aarhus University, Denmark; Stamatis Chrysanthopoulos, LEAF-Instituto Superior de Agronomia-ULisboa, Portugal; Jerke De Vries, Wageningen University and Research, The Netherlands; and Peter Sørensen, Aarhus University, Denmark*

1 Introduction

The rising demand for food has prompted an intensification of agricultural practices (Abdalla, 2002), leading to a revolution in food production with a focus on improving productivity (Wright et al., 2012; Eurostat, 2016). This has encouraged a growing reliance on mineral fertilizers (especially nitrogen (N), phosphorus (P) and potassium (K)). However, the extensive use of mineral fertilizers poses long-term environmental problems (Welch, 2002), including the greenhouse gases (GHG) involved in manufacturing synthetic fertilizers and the non-renewable nature of raw materials such as phosphate used to produce phosphorus (P) fertlizer (Eurostat, 2016). The European energy crisis has also had a significant impact on the availability and cost of synthetic fertilizers. In addition, these fertilizers do not contribute to levels of soil organic matter (SOM) which are increasingly seen as critical to long-term soil health and sustainable crop yields.

* Agricultural Biosystems Engineering Department, Wageningen University and Research, The Netherlands.

http://dx.doi.org/10.19103/AS.2023.0120.12

Animal manure is seen a potential solution to address the dual challenges of replacing synthetic fertilizers in optimizing crop yields whilst also building up soil organic matter at nominal expense. Animal manure includes not only nutrients but also organic matter (OM) and beneficial microorganisms that can improve soil health and thus crop yields (Köninger et al., 2021). This approach also aligns with recent EU strategies promoting recycling and a circular economy.

However, animal manures are also responsible for close to 10% of global GHG emissions and are the main source of ammonia emissions. If not properly managed, they can also negatively impact water quality due to N and P leaching and run-off from fields into watercourses. Global annual production of livestock manure is estimated to be about 80–140 Tg, with a N content exceeding synthetic N production (Oenema and Tamminga, 2005; Zhang et al., 2020). Cattle manure has the largest share at about 60% of total manure (Thangarajan et al., 2013). It is estimated that 50% of manure is deposited in pastures while 50% is collected: only half of the amount collected is recycled to agricultural land. Manure is one of the main sources of nitrous oxide (N_2O) and methane (CH_4) emissions in agriculture (Burney et al., 2010; Zhang et al., 2020). It has been estimated that, of the total N excreted by animals, only 15% is taken up by any subsequent crop (with a range of 0-60%). The other 85% of excreted N is lost via NH_3 volatilization, nitrification-denitrification, leaching and run-off (Oenema and Tamminga, 2005; Zhang et al., 2020). Manure also contains phosphorus that can be lost after soil application, resulting in aquatic eutrophication (Kleinman et al., 2020). These losses of P and N not only waste agronomically valuable nutrients but also contribute to serious environmental impacts such as climate change, soil acidification, eutrophication and harmful particulate matter formation in the atmosphere (Yuan et al., 2018).

Despite many advances in research, there is still a need for new, improved strategies for manure management to optimize its use and minimize emissions. One issue is a lack of a clear definition of manure in EU regulations: from 'waste products excreted by livestock' in the Nitrate directive (Directive 91/676/EEC) to 'any excrement and/or urine of farmed animals (...), with or without litter' in the Regulation on animal by-products (EC/1069/2009). Another factor to consider is that up to 90% of manure applied to soils has no or minimal processing (Foged et al., 2011).

Part of the reason for the lack of manure processing is the strong imbalance in production of manure in regions such as Europe (Eurostat, 2021). Particular areas in countries like Germany, Spain, France or Netherlands are hotspots of manure production. There has been a growing specialization of crop and livestock production with livestock production generally concentrated in specific areas and on specific livestock farms (sometimes operating on a large scale). In 2018, e.g. 80% of manure was produced by only 4% of European farms (Amann et al., 2018). This means both a problem for those livestock farms having to manage very large volumes of manure and for the wider agricultural sector in managing how this resource can be efficiently distributed to benefit crop production. This means that dealing with manure is as much about management as it is about technical solutions for treating manure (Niles et al., 2019; Varma et al., 2021). This is particularly the case given the risk of pollution swapping where, in solving one pollution problem, a particular treatment unintentionaly creates a new environmental problem. Indeed, techniques such as anaerobic digestion (AD), the use of covers for storage, slurry injection and acidification can all potentially result in pollution swapping (Emmerling et al., 2020). This highlights the importance of an integrated management approach.

These challenges necessitate the adoption of appropriate techniques and strategies for manure management to optimize the recovery of nutrients, mitigate the environmental consequences associated with its storage, transportation and application on soil and ensure its safe use (Amon et al., 2021; Fangueiro et al., 2023). This chapter examines how to achieve these objectives by looking at options at the various stages involved in slurry management, including collecting excreta within the animal house, storage, processing and field application. Each of these steps involves a multitude of potential treatments and technologies that can be employed to enhance use in terms of the agronomic/nutrient value of manure products reducing GHG emissions, odours and cost.

2 Current decision tools for optimizing manure management

A large number of tools are available for manure management. A list of the most relevant decision tools is shown in Table 1. These tools have been divided in four categories:

- Manure production and characteristics;
- Manure storage;
- Soil application of manure; and
- Managing the whole manure management chain.

In the first category, there are three examples of decision tools to estimate the amount of manure produced and/or the amount of nutrients contained in manure:

- Nutrient loading calculator calculates the amount of N, P, K and S that will be deposited in the form of manure, urine and waste feed depending on

Table 1 Most relevant tools available for manure management at several steps of the manure management continuum

Category	Name of tool	Link	Authors
Manure production and characteristics	Nutrient Loading Calculator	https://www.alberta.ca/nutrient-loading-calculator.aspx	Alberta government and Agri-Food Canada in collaboration with prairie provincial agricultural departments
	Manure Properties Calculation Tool	https://projects.luke.fi/manurestandards/en/results/	Project Manure Standards: Allan Kaasik, Christian Friis Børsting, Friederike Lehn
	COMPOSIM	http://www.rmt-fertilisationetenvironnement.org/moodle/mod/resource/view.php?id=802	Institut de l'élevage
Manure storage	Le Pré-Dexel et le Dexel	https://idele.fr/gestion-des-effluents-et-des-dejections-ged?tx_ideleregeff_feregeff%5Baction%5D=displayRessources&tx_ideleregeff_feregeff%5BcomeFrom%5D=themes&tx_ideleregeff_feregeff%5Bcontroller%5D=Screen&tx_ideleregeff_feregeff%5BwannaGoTo%5D=REGLEMENTATION_ZONE_VULNERABLE&cHash=4cb71ef3249f10af4d75226e3a5b9465	Institut de l'élevage
Soil application of manure	Manure Transportation Calculator	https://www.alberta.ca/manure-transportation-calculator.aspx	Alberta government
	Phosphorus Management Tool	https://www.alberta.ca/phosphorus-management-tool.aspx	developed by the Government of Alberta as part of the Alberta Phosphorus Watershed Project
	Manure Management Planner	https://www.alberta.ca/manure-management-planner.aspx	Purdue University for use in the United States. The programme was adapted by the Alberta government for use in Alberta

	Sorensen model	Sørensen, C.G., Jacobsen, B.H., Sommer, S.G., 2003. An Assessment Tool applied to Manure Management Systems using Innovative Technologies, Biosystems Engineering, 86, 315-325 https://doi.org/10.1016/S1537-5110(03)00137-5.	Claus G. Sørensen, Brian H. Jacobsen, Sven G. Sommer
	BeMestWijs management tool: Smart manure application and utilization:	https://www.wur.nl/en/article/Smart-manure-application-utilisation-a-practical-tool-for-issuing-advice-on-manure-application.htm	ir. G (Gerard) Migchels (wageningen) LTO Bedrijven LTO Noord DairyCampus
	Deere calculator coupled with HarvestLab 3000	https://www.deere.co.uk/en/tools/ag-turf/manure-sensing-calculator/	John Deer Company
	Ermes	https://www.ermes.pro/fonctionnalitesgenerales.html	IG tools
	EpandApp	https://www.mon-cultivar-elevage.com/content/regler-son-epandeur-sur-smartphone-avec-epandapp	Chambre Régionale d'Agriculture de Bretagne
	Manure Tracker	https://play.google.com/store/apps/details?id=ipcm.counter.prelim2&hl=pt_PT&gl=US	University of Wisconsin Integrated Pest Management
	Manure Valuator app	https://www.manuremanager.com/manure-app-calculates-value-14819/	University of Arkansas Division of Agriculture
	The Farm Crap App Pro	https://www.agricology.co.uk/resources/farm-crap-app-pro	SWARM Hub
Whole manure management chain	Manure Management N-flow tool	https://www.eea.europa.eu/publications/emep-eea-guidebook-2019/part-b-sectoral-guidance-chapters/4-agriculture/manure-management-n-flow-tool/view	Aether Ltd 2019 under contract to the EEA
	Manure Monitor	https://play.google.com/store/apps/details?id=com.movecreative.feedlotandroid&hl=pt_PT&gl=US	Move Creative

the feeding strategy considered. It makes it possible to adapt the herd to the available feed or the amount of feed needed for the existing herd.
- Manure Properties Calculation Tool allows the estimation of manure composition.
- COMPOSIM simulates the amount and characteristics of manure produced.

In the area of manure storage, there are two French decision tools:

- Pré-Dexel; and
- Dexel.

These make it possible to estimate the volume of storage needed in a specific farm and specifically the storage capacity needed to conform to the rules in nitrate vulnerable zones.

There are a large number of tools dedicated to the application of manure to soil. These include

- manure management planner; and
- BeMestWijs management tool.

These define where, when and how much manure can be applied. Both make it possible to simulate the effects of improvements/changes in management. Other tools include:

- Manure management planner: This analyses if it is possible to apply all the manure produced on a farm and enables long-term manure planning for up to 10 years.
- BeMestWijs management tool: This uses data from a number of management tools and databases to allow precision manure application.
- Manure transportation calculator: This analyses the costs associated with transport and application of manure.
- Deere calculator coupled with HarvestLab 3000: This allows variable-rate application of slurry based on continuous analysis of the composition of slurry using NIRS.
- Phosphorus management tool: This assesses the impact of manure application on P losses; it also gives information on best management practices to decrease P losses.
- ERMES: This uses geolocalization to track the location and amount of manure applied; an excellent tool to prepare manure application maps and assess the impacts of cumulative applications.

More recently, tools have been created as Apps to allow farmers and others easy access via smartphones. Among several Apps, we have selected four different tools with distinct objectives:

- Farm crap App determines the amount of crop-available nutrients (nitrogen (N), phosphorous (P) and potassium (K)) for different spreading rates of manure. This App helps to establish how much to spread in order to meet crop requirements with minimum waste, environmental impact and cost.
- Manure Valuator App calculates the value of manure based on the amount of fertilizer it replaces.
- Manure Tracker App automatically records manure application on fields using GPS (complementing the ERMES tool).
- Epand calculates the amount of manure to apply to reach an N target and can calibrate machinery used for manure application.

As noted earlier, avoiding problems such as pollution swapping requires a holistic approach. A final set of tools considers the whole manure management chain. These include:

- Manure management N-flow tool looks at N flows with special emphasis on GHG emissions (Sorensen et al., 2003). It compares different manure management strategies to identify the best solutions for specific farms. It also enables a system-oriented evaluation of labour requirements, machinery capacity and costs related to manure handling for any one approach.
- Manure Monitor is an App created to facilitate the recording of data related to manure management; it also includes an emergency response plan in case of manure-related pollution incidents.

3 Modifying animal slurry pH to enhance its value as a biobased fertilizer: (bio)-acidification and alkalinization

With over 140 million animals, the European Union is the second largest global producer of pigs, generating large quantities of slurry (Eurostat, 2022). Limited storage capacity in pig-breeding facilities and lack of a uniform manure regulation have resulted in untreated pig slurry being spread on fields. This practise has been associated with serious potential safety and environmental impacts, particularly in ammonia (NH_3) emissions (Hjorth et al., 2010), nutrient leaching to water courses (Fangueiro et al., 2014) and potential contamination of crops with pathogens (Rodrigues et al., 2021).

A promising and cost-effective approach is pH modification of swine slurry using additives. Due to its low dry matter content, slurry can be seen as an aqueous solution where an equilibrium exists between ammonium (NH_4^+) and unionized NH_3. This equilibrium is pH dependent and favours NH_4^+

concentration once animal slurry is acidified (Regueiro et al., 2016a). Modifying the pH of animal slurry can therefore cause significant changes to its physico-chemical and biological properties (Fangueiro et al., 2015). A typical range of initial animal manure pH lies between 7 and 8.4 (Regueiro et al., 2016). At this range, buffer components (weak acids or bases) are responsible for pH regulation. Slurry can exhibit high buffer capacity which can affect the amount of additive required to modify the pH (Sommer and Husted, 1995) .

A key benefit in altering slurry pH is mitigation of nitrogen (N) losses to the atmosphere via gaseous NH_3 (Regueiro et al., 2016; Overmeyer et al., 2021). Since NH_3 is a nitrogenous compound, the N fertilizer value of the slurry is also increased. Slurry sanitization, a crucial aspect related to the safety of biobased fertilizers, can also be attained through pH modification (Rodrigues et al., 2021). Alkalinization followed by stripping technologies also allows recovery of ammonia and the production of ammonium-based fertilizers (Overmeyer et al., 2020). There are, however, problems to overcome, including replacing inorganic with organic acid/alkali compounds to allow treated slurry to be used in organic agriculture (see European Regulation 2019/1009) and effective integration of this technique with others in an overall manure management strategy (Hjorth et al., 2010; Varma et al., 2021).

There are three strategies to modify the pH of animal slurry:

- Chemical acidification (Fangueiro et al., 2015);
- Biological acidification (Regueiro et al., 2022); and
- Alkalinization (Rodrigues et al., 2021).

Acidification is the most common strategy in reducing slurry pH. Acidification of slurry with inorganic acids, such as sulphuric acid (H_2SO_4), has been practised at a farm scale in Northern European countries (i.e. Denmark) for almost 20 years (Fangueiro et al. 2015; Regueiro et al., 2022). As alternatives to inorganic acids, organic acids (like acetic acid and citric acid) have also been assessed for their efficacy to reduce pH (Regueiro et al., 2016). A typical target pH is 5.5 since, at this range, the NH_4^+/NH_3 equilibrium favours NH_4^+ (Overmeyer et al., 2021).

Fangueiro et al. (2015) reported up to 70% and 88% NH_3 reduction with H_2SO_4 during in-house and storage acidification, respectively. Methane (CH_4) emissions from stored slurry can also be reduced by up to 99% by reducing its pH value to 5.5 (Fuchs et al., 2021). The fertilizer value of acidified slurry exhibits higher concentration of dissolved nutrients (Fangueiro et al., 2015). For instance, the solubility of phosphorus (P), e.g. is increased by acidification of cattle slurry (Pedersen et al., 2017). Swine slurry acidification to pH 5.5, e.g. increased the dissolution of total P (~70% as Ortho-P) (Christensen et al., 2009), facilitating P uptake by crops such as maize (Regueiro et al., 2020). Acidification is also important in slurry sanitization (Köninger et al., 2021). The EU regulation for fertilizers ((EU) 2019/1009) mandates a limit of 1000 CFU/g in five out of five tested samples for *Escherichia coli* or total coliforms, and the absence of *Salmonella* spp. in 25 g of fertilizer. Rodrigues et al. (2021) found effective swine slurry sanitization (<1000 CFU *E. coli*) was possible with H_2SO_4 at a target pH of 5.

There are, however, disadvantages with chemical acidification. The higher nutrient solubility that acidified slurry exhibits can increase nutrient leaching (Fangueiro et al., 2014). While acidification abates NH_3 losses, there are trade-offs in terms of N_2O emissions (Schreiber et al., 2023). In a study where untreated and acidified slurry were applied at a rate of 90 kg N ha^{-1}, 35% higher cumulative soil N_2O emissions were observed with acidified slurry compared to untreated slurry (Malique et al., 2021). A meta-analysis by Emmerling et al. (2020), however, concluded a decrease in CH_4, CO_2 and N_2O emissions from acidified slurry. Concentrated H_2SO_4 also poses significant safety risks for workers and is challenging to handle (Bastami et al., 2016; Prado et al., 2020). Cost is also an issue. de Vrieze et al. (2019) concluded that acidification with H_2SO_4 increased total net process costs by approximately 24%.

Biological-acidification (bio-acidification) of animal slurry is a microbially-driven process where the microorganisms present in slurry convert easily fermentable carbon sources into organic acids such as lactic acid (Regueiro et al., 2022). Bio-acidification with glucose and sucrose as substrates has exhibited promising results in pH modification and reduction in GHGs from animal slurry (Bastami et al., 2016; Prado et al., 2020; Regueiro et al., 2022). Bastami et al. (2016), e.g. observed CH_4 reduction of up to 99% using brewing sugar as labile C source to bio-acidify dairy slurry. Bio-acidification is able to achieve a target pH of 5.5 or lower (Prado et al., 2020). However, high doses of sucrose-rich additives are required to initiate and maintain the fermentation process, and this may not be economically viable since sucrose is often more valuable for other applications. Combining slurry acidification using H_2SO_4 followed by the addition of glucose and/or sucrose-rich additives has been explored as an alternative (Overmeyer et al., 2021; Regueiro et al., 2022). Pre-acidification of swine slurry with H_2SO_4 to pH 5.5 followed by glucose addition (2% and 4%) appeared to maintain pH below 5 more consistently than glucose (2%) treatment alone (Regueiro et al., 2022).

Slurry alkalinization, a strategy much less studied, is the opposite of acidification since it increases slurry pH. Additives used to increase animal slurry pH include lime products such as calcium hydroxide ([$Ca(OH)_2$]) as well as other hydroxide salts like potassium hydroxide (KOH) and sodium hydroxide (NaOH) (Rodrigues et al., 2021). In principle, alkalinization shifts the NH_4^+/NH_3 equilibrium towards the latter, causing NH_3 losses to the atmosphere and consequently reducing the fertilizer value of the slurry. However, alkalinization at pH 12 can inactivate pathogenic microorganisms (Anderson et al., 2015).

Table 2 Agro-industrial by-products as alternative additives to modify the pH of animal slurry

Additive	**Slurry type used**	**Tested amount**	**Additional benefit**	**References**
Apple pulp Brewers grain Dairy washings Dairy waste Grass silage Maize silage Sugar beet molasses	Dairy cattle slurry	70;150 70;150 150 100 70;150 70;150 30;50;70 (g kg^{-1})	Reduce NH_3 (up to 67%) and CH_4 (15-70%) emissions during storage	(Kavanagh et al., 2021)
Brewing sugar Brewing sugar & Actiferm EM	Dairy cattle slurry	10% (w.w.) 10% & 5% (v.w.)	Reduce CH_4 (85-99%) emissions during storage	(Bastami et al., 2016)
Brown juice (coagulated & fermented; different combinations with H_2SO_4 and glucose)	Pig slurry	20-50 % (w.w.)	pH reduction up to 3.9 during storage Increased lactic acid production with fermented brown juice Negligible CH_4 and CO_2 emissions	(Regueiro et al., 2022)
Cheese whey Rice bran	Dairy slurry	100-250 (g 0.5 kg^{-1} slurry)	Reduce NH_3 emissions (58%) with cheese whey P and Mg enrichment in slurry treated with rice bran	(Prado et al., 2020)
Acidic whey Flushing milk	Cattle manure slurry	20-50% (v.w.)	pH reduction during storage (1 pH unit for flushing milk and up to pH 4.5 with acidic whey)	(Sepperer et al., 2021)
Banana peel	Pig slurry	20-120 (g L^{-1} slurry)	pH reduction up to 5.4 Dissolution of particulate P	(Moyo et al., 2022)
Cheese whey	Dairy cattle slurry	10-80% (w.w.)	pH reduction below 5 during storage Reduce total GHG's (up to 54%) and NH_3 (up to 90%) during storage Increase CH_4 (up to 53%) production during AD	(Gioelli et al., 2022)

Rodrigues et al. (2021) succeeded in sanitizing pig slurry with KOH addition by reaching pH 9.5, though they observed higher NH_3 emissions compared to acidified slurry. Alkalinization has also been tested in dairy sludge and anaerobically digested biosolids (Case and Jensen, 2019).

An alternative approach is to use agro-industrial by-products as additives to reach a target pH and enhance the value of slurry as biobased fertilizer (Chojnacka et al., 2020; Prado et al., 2020). Table 2 summarizes the range of alternative additives that have been used to modify slurry pH, primarily via bio-acidification with a particular focus on reducing GHGs and NH_3 emissions.

Acidified manure has also been considered as a potential feedstock in AD to produce energy (biogas) from organic residues such as animal manure (Chojnacka et al., 2020). Two main approaches have been considered:

- Acidification of raw slurry prior to AD; and
- Co-digestion of raw and acidified slurry.

Using the first approach, Sutaryo et al. (2013) acidified raw sow slurry with H_2SO_4 to pH 5.5. They observed a significant reduction in CH_4 yields compared to untreated slurry (256.6 L kg VS^{-1} and 372.7 L kg VS^{-1}, respectively). This was attributed to the inhibition of sulphur reduction in the acidified slurry. However, utilization in AD of cattle slurry bio-acidified with cheese whey resulted in a significant increase in CH_4 production (+53%) compared to non-acidified slurry (Gioelli et al., 2022). The feedstock ratio between acidified and non-acidified slurry fraction is crucial in the second approach. Moset et al. (2016) found that the 10% inclusion of acidified slurry in a thermophilic reactor increased CH_4 yields by 10%; however, CH_4 yields were reduced with over 30% acidified cattle slurry. Co-digestion of acidified pig slurry (30% inclusion) reduced CH_4 yields (Moset et al., 2012, 2016).

Solid-liquid separation of manure is a common treatment on livestock farms (Varma et al., 2021). However, solid and liquid fractions can release much higher NH_3 and GHG emissions compared to non-separated raw slurry during storage (Prado et al., 2022). This requires acidification of slurry prior to solid-liquid separation, which can also increase P concentration in the acidified liquid fraction (Fangueiro et al., 2009). Acidification of a separated liquid fraction to pH 5.5 required less H_2SO_4 compared to raw slurry but resulted in significant pH increase during storage (Overmeyer et al., 2021). de Vrieze et al. (2019) evaluated 15 different methods to enhance P recovery, of which nine involved acidification of swine manure or digestate with H_2SO_4. Most of these scenarios involved manure acidification prior to centrifugation to assist P transfer from the solid to the liquid fraction.

In conclusion, pH modification of pig and other types of slurry has both strengths and weaknesses. Slurry acidification can reduce NH_3 and CH_4

emissions, though pollution swapping is possible, particularly with N_2O emissions. Little attention has been paid on increasing slurry's initial pH, probably due to an expected increase in NH_3 emissions. The fate of pathogens and nutrient enrichment of slurry (when using KOH as additive) are strong arguments to explore alkalinization further. A recent development is combining additives to mitigate the use of H_2SO_4 (Fuchs et al., 2021). Alternative additives from agro-industrial activities could replace chemical acids/bases but more research is needed to evaluate their impact on the environment, crop fertilization and economic feasibility. Integration of slurry acidification in a manure management system requires careful design, including economic assessment for large-scale application.

4 Improving manure management systems to minimize trade-offs

A main reason behind failure of current technologies to effectively limit all environmental impact is the trade-offs taking place between different emissions. This requires a holistic manure management strategy that considers key aspects like emissions, nutrient availability and practicability for farmers.

Measures to reduce environmental impacts can be implemented at various steps of the manure management chain. Reducing crude protein in diets, e.g. can reduce NH_3 emissions in animal housing, during storage and following soil application (Hou et al., 2015). Solid-liquid separation has been found to effectively reduce CH_4 and N_2O emissions (Fangueiro et al., 2008; Aguirre-Villegas et al., 2014; Aguirre-Villegas et al., 2019; Emmerling et al., 2020; Kupper et al., 2020). However, separation and emission reduction technologies can require costly and complex technologies and may not deal with emissions before separation (Vu et al., 2016).

Source separation is gaining interest based on simple principles like gravity to separate urine from faeces in animal houses (Aarnink et al., 2007; De Vries et al., 2013; Koger et al., 2014; Loussouarn et al., 2014; Vu et al., 2016). In principle, separating faeces and urine at source should ensure rapid removal of faeces from the animal house, separate urea (in urine) from urease (faeces), and separate ammonium (in urine) from a carbon-rich source (faeces). This could potentially reduce emissions of NH_3, N_2O and CH_4 (Fig. 1) simultaneously. Compared to traditional slurry systems, source separation in pig housing systems was found to decrease NH_3 emissions by 46 to 70% and CH_4 emissions by up to 80% (Aarnink et al., 2007; Aarnink and Ogink, 2007; De Vries et al., 2015b). A source separation system can also achieve high separation efficiencies in terms of nutrients compared to traditional liquid/solid separation from slurry (mainly N from P) (Vu et al., 2016). Since most research is on pigs, more research is needed on source separation systems for dairy manure.

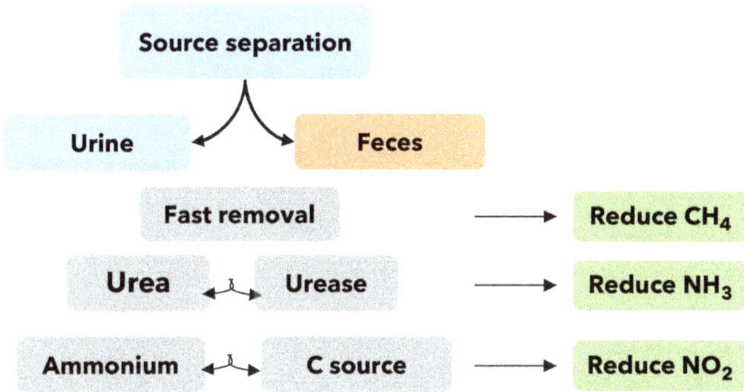

Figure 1 Potential effect of source separation of manure on emissions and involved mechanisms. C refers to carbon.

Emissions during storage are a particular concern. Strategies to manage emissions (and, ideally, valorize slurry) include AD, acidification, composting, drying, use of different natural additives (zeolite, biochar and straw) and storage covers. AD has been found to reduce CH_4 emissions during storage and alter nutrient availability (increase mineral N and reduce soluble P) (Clemens et al., 2006; Chadwick et al., 2011 ; De Vries et al., 2012; Hou et al., 2015; Dennehy et al., 2017 ; Hou et al., 2018 ; Maldaner et al., 2018 ; Aguirre-Villegas et al., 2019; Cao et al., 2019 ; Ershadi et al., 2020 ; Li et al., 2020). Acidification and use of covers are effective tools to reduce NH_3 emissions during storage and following field application (Ten Hoeve et al., 2016; Hou et al., 2017; Yuan et al., 2018; Fangueiro et al., 2015; Emmerling et al., 2020; Wagner et al., 2021; Keskinen et al., 2022). Better application methods and timing, and avoiding disturbance after application, can mitigate gaseous emissions and nutrient leaching (Webb et al., 2010; Chadwick et al., 2011; Emmerling et al., 2020; Erwiha et al., 2020; Dittmer et al., 2020; De Vries et al., 2015b). However, as noted earlier, there are usually trade-offs in using any one measure which need to be addressed by a more integrated approach.

Most studies of integrated systems using a range of techniques focused on limiting trade-offs between NH_3 and GHG emissions, or between environmental impacts like global warming, acidification and eutrophication (Aguirre-Villegas et al., 2014; Holly et al., 2017; Yuan et al., 2018; Aguirre-Villegas et al., 2019). These studies quantify NH_3 and GHG emissions at different stages of the chain, and sometimes energy used in manure processing and nutrient availability for crops.

Aguirre-Villegas et al. (2014), e.g. modelled the effect of solid-liquid separation of slurry, AD and both techniques combined, on NH_3, global warming potential, fossil fuel depletion and nitrogen availability from a whole manure management chain (barn, storage and field application). Slurry was stored in open basin and applied to field using broadcasting. AD combined

Table 3 Function of the single strategies, their pollution swapping and related problems, and proposed integrated solutions

Established single strategies	Function	Problem (pollution swapping)	Proposed integrated solution	References
Liquid/solid separation	-Manage nutrients -Easier handling -Reduce NH_3 and GHG emissions during storage	-Emissions in the animal house not tackled -Poor separation of nutrients (P and N stay in both fractions) -Soluble organic matter is available in liquid fraction causing CH_4 emission -Increase emissions of NH_3 and N_2O after field application especially from the liquid fraction	-Source separation (faeces/urine) separates nutrients efficiently and at the same time reduces NH_3, N_2O and CH_4 in the animal house	De Vries et al., 2013; Kupper et al., 2020; Emmerling et al., 2020
Anaerobic digestion (AD)	-Harvest CH_4/produce renewable energy reduces uncontrolled CH_4 emissions -Increase nutrient availability for plants (N and P)	-Increases emissions of NH_3 during storage and N_2O, NH_3, and NO_3^- after field application -Mono-digesting slurry is not cost effective	-Combining AD with drying of digestate or zeolite addition or acidification with sealed storage and adapted field application will control the emissions in all the stages after AD -Faster AD process of source separated faeces & possible higher yield per ton faeces -AD of source segregated faeces instead of slurry will need less digester size leading to reduced costs	De Vries et al., 2015a; Vu et al., 2016; Holly et al., 2017; Emmerling et al., 2020

Drying of slurry, manure digestates	-Reduce microbial activity leading to reduced emissions in storage -Easier handling/ transport	-Massive NH_3 emission during drying -High energy required for slurry drying	-Combining with zeolite addition before drying and sealed storage will control the NH_3 emissions -Combining with source separation will control ammoniacal nitrogen loss because most ammoniacal nitrogen stays in the urine. -Combining with AD will provide renewable energy for drying -Drying faeces only will require less energy than slurry (lower volume and lower moisture content)	Liu et al., 2019; Aguirre-Villegas et al., 2019; De Vries et al., 2015a
Covered storage	-Reduce emissions during storage	-High emissions prior to storage (animal house) -N_2O and CH_4 emissions upon exposure in addition to NH_3 and after field application	-Combining with source separation will control the emissions in the animal house -Combining with AD will reduce CH_4 production during storage and volatilization after field application -Acidification of separated faeces and urine to pH 5.5 will control NH_3 emissions after field application -Combining with adapted application methods will control N_2O after field application	De Vries et al., 2015a; Yuan et al., 2018; Keskinen et al., 2022; Wagner et al., 2021
Injection application field	-Reduce NH_3 emissions	-Increase N_2O emissions -Excess N or P applied leading to leaching and run-off (unbalanced N/P ratio on animal manures)	-Combining with source separation, AD, additives or drying, and sealed outside storage will lead to separated treated urine and faeces with more balanced N/P ratios and less prone to emissions -Combining with adapted application rate, time and reduced intervention will result in the control of all emissions and leaching	Nyord et al., 2013; De Vries et al., 2015a; Wagner et al., 2021; Keskinen et al., 2022

with solid-liquid separation had a lower impact in studied categories except NH_3 emissions. NH_3 emissions increased by 44% compared to untreated slurry, whilst a 40% and 2% increase occurred with single use of AD and solid-liquid separation, respectively. The combination of AD with solid-liquid separation also reduced nitrogen availability more than AD and solid-liquid separation used separately. These technologies are known to reduce GHG emissions but increase NH_3 emissions (Holly et al., 2017; Emmerling et al., 2020). An optimized management system thus needs to incorporate a NH_3 emissions mitigation strategy. Aguirre-Villegas et al. (2019) found AD combined with liquid-solid separation achieved a higher reduction of GHGs during storage than each of the technologies used separately. However, N_2O emissions during field application increased net GHG emissions.

Yuan et al. (2018) compared four pig slurry management systems integrating AD, separation and wastewater treatment and field application using life cycle assessment (LCA). Outdoor storage was uncovered and the slurry was surface applied. Results showed that even the best-performing system, which consisted of in-house separation and field application, still had high environmental impacts, in eutrophication and acidification. This study also highlights the importance of designing management systems considering all objectives and including interventions from the animal house to the field. Including techniques for optimized storage and field application could have reduced eutrophication and acidification potential through reduced NH_3 emissions. This study also suggested that the best management system could be further optimized by improving P removal, reducing P losses after soil application and eutrophication potential.

Table 3 summarizes potential ways to integrate treatment technologies to reduce all emissions simultaneously with minimum energy required. For each technology, the pollution-swapping problem is highlighted, and an integrated approach is proposed.

As noted, manure management involves potentially challenging trade-offs between environment, economic, community and farmer priorities (Niles et al., 2022). Integrated manure management systems would be more effective in balancing these trade-offs if they included an initial system design step. De Vries et al. (2015a) used engineering design methods to develop an integrated system covering the whole chain, including setting clear objectives and system boundaries to prevent pollution swapping and reduce the environmental impact of the whole chain. They highlighted the importance of starting with source segregation of urine and faeces since this was found to have no pollution swapping effect, could significantly reduce GHG emissions (up to 82%), acidification and particulate matter formation (up to 49%) with no significant effect on eutrophication (De Vries et al., 2013). When they modelled the effect of such an integrated system compared to conventional management, they

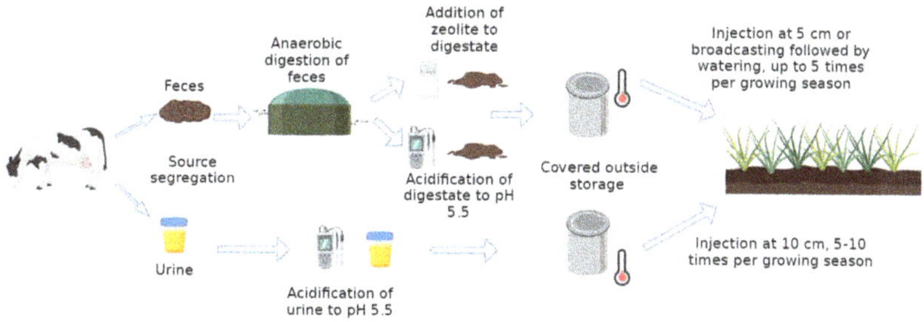

Figure 2 Graphical illustration of an integrated system to manage dairy manure from animal house to field.

found that an increase of up to 70% in N use efficiency, a decrease of excess P application to soil, and a reduction of GHG and NH_3 emissions with minimal energy use (De Vries et al., 2015b).

Steps to limit CH_4 emissions include the following:

1 Define factors regulating methane emissions in manure management.
2 Define functions that can be designed into the system to control these factors (C availability, temperature, oxygen and pH).
3 Define how these factors can be managed (e.g. remove C, reduce temperature, increase oxygen and lower pH).
4 Define techniques to achieve this (e.g. grooved floor, mono-digestion and adding acid).
5 Define possible interactions of this technique with other process, e.g. dealing with NH_3 or N_2O emissions.
6 Iterate between these steps.

The final result should be an integrated manure management system covering the whole chain (from the animal house up to field application). Figure 2 illustrates an example of an integrated system. A good integrated system is also one that can be verified by appropriate measurement and testing to follow manure fractions through all the steps in the chain. Measurement should be able to track changes in composition after each step/treatment and measure all emissions in the animal house, during storage and following field application.

5 Combining manure management with anaerobic digestion

One strategy to optimize manure management is to combine it with AD. AD is a bioenergy technique in which organic materials are broken down in an oxygen-free environment into biogas and digestate as a by-product. Digestate from the AD process is characterized by a high NH_4^+/total N ratio due to the mineralization of organically bound N and is stabilized due to the conversion of less decomposable organic matter to biogas. This reduces N immobilization after manure application and increases available N for plant utilization (Sørensen et al., 2012).

Substrate physical and chemical properties change during AD, affecting physico-chemical properties such as the dry matter content, NH_4^+/total N ratio, viscosity and pH (Barampouti et al., 2020). Digestate dry matter and viscosity decreases, enhancing soil infiltration. Low dry matter reduces contact time with the soil surface, reducing ammonia volatilization and improving nutrient value of the digestate (Hafner et al., 2018). Decomposition and recovery of labile carbon from substrates as biogas (CH_4 and CO_2) reduces N immobilization and increases plant-available nitrogen (Webb et al., 2013).

Organic matter content in manure is degraded during complex biochemical reactions in AD, resulting in thinner manure with better soil infiltration and less microbial immobilization of nutrients in the soil. Volatile solids in manure are recovered as CH_4 and CO_2 during AD, avoiding GHG emissions. Stabilized digestates decrease the microbial demand for oxygen after application, reducing heterotrophic denitrification and N_2O emissions (Chadwick et al., 2011). However, the AD increases manure pH, increasing the risk of ammonia volatilization (Sommer and Husted, 1995). This requires measures during storage and digestate application to minimize this risk.

Input substrate characteristics influence biogas yield and digestate quality (Hagos et al., 2017). The AD process may involve mono-digestion with only one substrate or co-digestion with multiple substrates digested together. Due to prior carbon degradation in the animal's digestive system, mono-digestion of manure for biogas production yields relatively low amounts of biogas. Mono-digestion is also characterized by nutritional imbalance and a lack of diverse microorganisms required to optimize the AD process (Hagos et al., 2017).

Co-digestion of manure and other substrates, such as food waste, industrial food wastes and agricultural crop residues, maximizes biogas yield (Moset et al., 2017; Sarker et al., 2019). It allows system optimization and improved economic viability of biogas plants, as well as the ability to recycle and reuse nutrients from other waste streams (Sayara and Sánchez, 2019; Sørensen et al., 2019). As a result, there has been an increasing shift towards co-digestion of the slurry with solid organic biomass (such as straw-rich litter, grass and straw) to produce biogas, for instance in Denmark (Møller et al., 2022).

The co-digestion of manure with solid biomasses has a high biogas potential. However, most substrates are lignocellulosic with high fibre content, limiting their degradation during AD. Digestion of recalcitrant biomass poses problems with pumping and agitation, which require investments in adapting biogas plants to handle the high dry matter content in solid biomass (Møller and Nielsen, 2016). These limitations result in low biogas yields and affect digestate characteristics such as dry matter content which influence nitrogen use efficiency during field application. These challenges require adapting AD systems.

Co-digestion of lignocellulosic biomasses with livestock slurry/manure results in a digestate with high dry matter content and viscosity due to the low digestibility of the lignin (Barampouti et al., 2020). This poses agronomic and environmental challenges during digestate storage and application (Olesen et al., 2021; Webb et al., 2013). High dry matter content increases the ammonia volatilization per unit area during surface application given the reduced infiltration rate of digestate (Søgaard et al., 2002). Losses are further exacerbated by high digestate pH which reduces fertilizer value. Møller et al. (2022) estimated an increase in ammonia emissions of 0.14 to 0.3 kg NH_3-N ton^{-1} biomass digested due to higher pH. Nyang'au et al. (2023) also reported higher ammonia volatilization from digestates with higher dry matter content following application to winter wheat compared to digestates with low dry matter content, which they attributed to reduced infiltration rates and higher pH.

Olesen et al. (2021) and Pedersen et al. (2021) found that adhesive characteristics and increased surface area of digestate might limit soil infiltration compared to untreated manure, resulting in higher NH_3 volatilization, outweighing the benefits of reduced dry matter content. NH_3 losses from digestates could be minimized by utilizing low-emission application techniques such as digestate injection into soil (Møller et al., 2022; Sørensen et al., 2019). However, this is difficult in some crops, such as in winter cereals, where injection can cause significant damage to crops. Alternative technologies, such as solid-liquid separation, can be used to increase digestate performance, especially those with relatively higher dry matter content (Sørensen and Thomsen, 2005).

The application of digestates with a high fraction of undecomposed organic material, characterized by a high C/N ratio, promotes soil N immobilization. Microbial assimilation of N linked to respiration of manure C typically starts immediately after manure application and reaches a peak after 1-4 weeks, after which mineralization starts (Sørensen, 2004). Immobilization reduces readily available nitrogen to crops, reducing fertilizer value.

With increasing utilization of recalcitrant substrates as inputs, only 40-70% of the biomass is converted into biogas, resulting in a digestate with a significantly high degradable organic matter (Romio et al., 2021). Incorporating such digestates increases oxygen demand in the soil, inducing denitrification. Due to denitrification, loss of N from manure application in the form of N_2 and N_2O is usually high compared to synthetic fertilizers (Velthof et al., 2003). Denitrification losses are significantly influenced by manure distribution, soil water content and soil type (Webb et al., 2013). Assessing the benefits of incorporating manure into AD should include a mass flow approach at the whole farm level, including factors such as NH_3 volatilization, reduced carbon sequestration after digestate application and methane emissions (Chadwick et al., 2011; Møller et al., 2022).

6 Pre- and post-treatment of biomass for anaerobic digestion

During the AD process, hydrolysis and methanogenesis are limiting stages due to the recalcitrant nature of biomass and the formation and accumulation of undesirable intermediary products such as volatile fatty acids (Rajagopal et al., 2013; Siegert and Banks, 2005). These limitations result in less stable biogas yields with digestate characterized by partially-decomposed organic matter with reduced nutrient availability. Ways of improving the AD process include incorporating a pre-treatment step for lignocellulosic biomass before AD or incorporating a post-treatment stage as illustrated in Fig. 3.

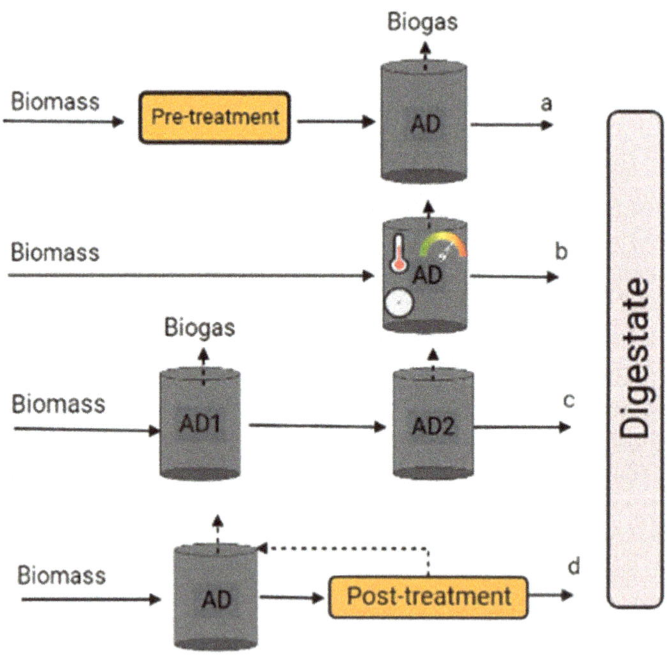

Figure 3 Schematic diagram of how anaerobic digestion process can be optimized for higher biogas yield and enhanced nutrient availability, (a) illustrates the incorporation of a pre-treatment step before AD, (b) optimization of operation conditions, i.e. temperature and HRT, (c) extending the HRT by adjusting reactors configuration to have primary and secondary AD steps and (d) a post-treatment step, i.e. solid-liquid separation.

6.1 Pre-treatment of biomass

Pre-treatment of biowastes facilitates degradation by enzymes and microbes during AD by increasing reaction kinetics and nutrient solubilization (Sarker et al., 2019; Zheng et al., 2014). Pre-treatment also helps to overcome lignin recalcitrance and crystalline cellulose barriers during digestion, optimizing process efficiency and increasing biogas production and digestate quality (Meegoda et al., 2018; Sarker et al., 2019). Several studies have reported positive effects of pre-treatment techniques on biogas yield (Harris and McCabe, 2015; Hernández-Beltrán et al., 2019; Sayara and Sánchez, 2019).

Pre-treatment techniques include physical, chemical and biological methods. Physical pre-treatment techniques e.g. include grinding, milling, extrusion, sonication, high pressure, microwave digestion, ozonolysis and electrokinetic treatments. These increase the accessible surface area of the substrate and decrease the crystallinity of cellulose during AD (Kumari and Singh, 2018; Sayara and Sánchez, 2019). The increase in surface area and release of intracellular components increases the degradation rate and accelerates the AD process, increasing methane yield.

Nyang'au et al. (2023), e.g. found that an electrokinetic pre-treatment step ($P < 0.05$) increased NH_4^+-N/total N in digestates before the second AD step, but the effect levelled off after the secondary digestion step (Fig. 4). Electrokinetic and ultrasonication pre-treatments tended to increase the mineral fertilizer equivalent of ammonium-N in digestates (Table 4). They concluded that these pre-treatment steps could improve digestate fertilizing properties such as lower dry matter content and viscosity alongside increasing biogas yields.

In selecting an appropriate pre-treatment technique to incorporate into the biogas digester, cost and energy consumption must be adequately evaluated (Meegoda et al., 2018). Pre-treatment also needs to significantly improve cellulose and hemicellulose solubilization at low cost (Kim et al., 2014). Other factors include increasing biogas yield, improving digestate fertilizer properties and reducing biowaste treatment time (Nyang'au et al., 2023). High biogas recovery after pre-treatment can reduce labile organic matter in digestates, increasing stability and the ratio of available N.

6.2 Post-treatment of biomass

Due to regulations limiting nitrogen application up to 170 kg N ha^{-1} in many EU countries (EU nitrates directive), digestates often have to be transported to other areas for use. The high water content of digestates increases costs of transportation and storage, potentially exceeding the fertilizer value of the digestate (Plana and Noche, 2016). To deal with these challenges and improve

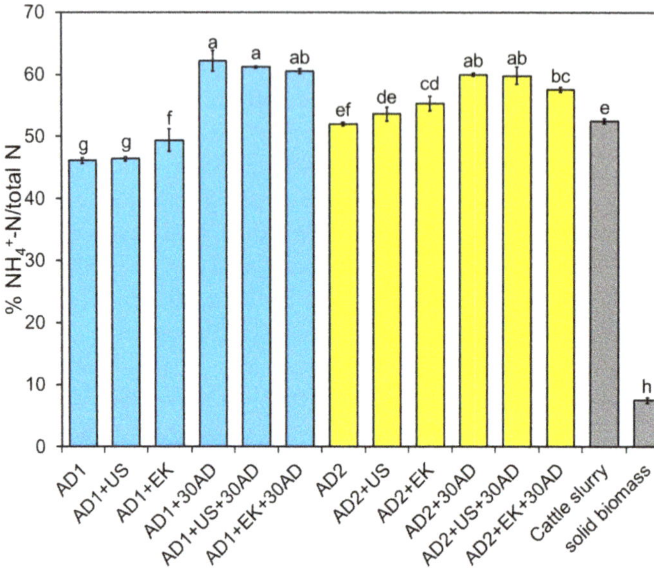

Figure 4 Effect of pre-treatment techniques, electrokinetic (EK) and ultrasonication (US), and two-step-AD (+30AD) on percentage NH_4^+-N/total N in two digestates (AD1 and AD2) sourced from two biogas plants with contrasting HRT in a primary AD step. Letters indicate significant differences among the treatments estimated with Tukey's test ($P < 0.05$). Error bars indicate standard deviation (n=3). Reprinted from Nyang'au et al. (2023) with permission.

digestate NUE, post-treatment technologies such as solid-liquid separation can be adopted for more tailored field applications.

Separation of digestate results in a liquid fraction with a high proportion of ammonium nitrogen and potassium, and a solid fraction rich in phosphorus and organic matter which is lighter and cheaper to transport. The liquid fraction could replace nitrogen-based mineral fertilizer, while the solid fraction of the digestate can be utilized in regions with P-deficient soils (Fangueiro et al., 2011; Romero-Güiza et al., 2016). The liquid fraction can also be further treated to meet legislation requirements such as RENURE criteria relating to N recovery from manure (Reuland et al., 2021) or by recovering nutrients such as N through ammonia stripping. The low dry matter content of the liquid fraction makes it less viscous and able to infiltrate soil faster, reducing the risk of NH_3 volatilization and improving NUE after surface application (Sørensen and Thomsen, 2005). The liquid fraction also has a low C/N ratio and a high potential N fertilizer value (Webb et al., 2013). Sørensen and Thomsen (2005) observed improved NUE after surface application of separated liquid fraction of slurry in winter wheat, though only techniques including rapid incorporation of the solid fraction and surface banding or injection of the liquid fraction improved overall NUE.

Table 4 Recovery of labelled ammonium-N, mineral fertilizer equivalence (MFE) of ammonium-N in digestates calculated based on ^{15}N uptake in winter wheat and spring barley in above-ground biomass relative to ^{15}N uptake in mineral fertilizer and estimated ammonia losses in winter wheat after surface banding application of digestates pre-treated by electrokinetic (EK) and ultrasonication (US) in a two-step AD

	Secondary AD step	Winter wheat		Spring barley		
		% Recovery	% MFE	% Recovery	% MFE	Estimated % NH_4^+-N losses
AD1	-	31.1	51.7 (1.4)^{b}	36.0	82.4 (1.0)^{a}	30.7^{ab}
AD1+EK	-	32.2	53.6 (0.7)^{b}	36.4	83.2 (5.9)^{a}	29.7^{abc}
AD1+US	-	31.6	52.5 (2.4)^{b}	37.2	85.1 (4.7)^{a}	32.6^{a}
AD1^{+AD30}	+	37.7	62.6 (1.8)^{a}	36.4	88.3 (1.3)^{a}	20.7^{bcd}
AD1+EK	+	37.6	62.6 (1.3)^{a}	35.7	81.5 (3.1)^{a}	19.0^{cd}
AD1+US	+	38.5	63.9 (0.1)^{a}	35.1	80.2 (3.1)^{a}	16.3^{d}

Means followed by different letters within each column are significantly different ($P < 0.05$). For % MFE, values correspond to averages and standard errors in brackets ($n = 4$). The ammonia loss is estimated from the difference in MFE between spring barley and winter wheat, assuming no ammonia loss by injection to spring barley. Data from Nyang'au et al. (2023).

7 Optimization of anaerobic digestion operations to optimize digestate quality

There are a huge range of variables which affect AD performance in terms of biogas or digestate yield and quality (Sarker et al., 2019; Veluchamy et al., 2019). However, optimization of AD is often based on biogas and methane yield with limited attention given to the digestate quality (Logan and Visvanathan, 2019). This results in a digestate with undesirable properties such as high dry matter content. Extending biomass hydraulic retention time (HRT) using staged reactors e.g. can result in both higher biogas yield and high nutrient mineralization in digestate (Feng et al., 2017).

In Denmark, e.g. there has been an increasing trend of incorporating secondary reactors into biogas plants to both optimize biogas yields and utilization of biomass such as straw (Møller and Nielsen, 2016). Increasing the residence time of substrates in digesters enhances degradation, resulting in more biogas recovery, nutrient solubilization and stabilized digestate. Applying stabilized digestate from longer HRT to soil results in less C mineralization without affecting soil carbon retention (Nyang'au et al., 2022). Nyang'au et al. (2023) found that incorporating a secondary digestion step for biowastes to increase HRT could, on average, increase winter wheat N uptake from 39% to 42% following digestate surface application and from 75% to 86% following digestate injection in soil. Through ^{15}N-labelling of ammonium-N in digestates, they also found that the extra digestion step significantly reduced ammonia losses through volatilization (Table 4).

There are typically two temperature regimes used in AD:

- Mesophilic; and
- Thermophilic.

Mesophilic digestion operates at around 35-45°C and and, whilst requiring less energy, is associated with a slower degradation rate and less biogas production. Thermophilic digestion occurs at higher temperatures (around 50-55°C) and is characterized by faster degradation, potentially increasing biogas yield and nutrient solubilization (Hartmann and Ahring, 2006). Hossain et al. (2021) e.g. found that thermophilic temperatures marginally increased inorganic N release of digestate to soil compared to digestion at mesophilic temperature. Thermophilic conditions could provide additional benefits by destroying pathogens such *Escherichia coli* and *Salmonella* spp., weed seeds and other foreign material before field application (Smith et al., 2005; Sørensen et al., 2019).

8 References

Aarnink, A. and Ogink, N. (2007). Environmental impact of daily removal of pig manure with a conveyer belt system. In: *International Symposium on Air Quality and Waste Management for Agriculture*. Broomfield, CO.

Aarnink, A. J. A., Veld, J. H. I. T. and Hol, A. I. V. (2007). *Kempfarm vleesvarkensstal: milieu emissies en Investeringskosten*. Wageningen University and Research.

Abdalla, C. W. (2002). The industrialization of agriculture; implications for public concern and environmental consequences of intensive livestock operations. *Penn State Environmental Law* 10: 175–191.

Aguirre-Villegas, H. A., Larson, R. and Reinemann, D. J. (2014). From waste-to-worth: Energy, emissions, and nutrient implications of manure processing pathways. *Biofuels, Bioproducts and Biorefining* 8(6): 770–793.

Aguirre-Villegas, H. A., Larson, R. A. and Sharara, M. A. (2019). Anaerobic digestion, solid-liquid separation, and drying of dairy manure: measuring constituents and modeling emission. *Science of the Total Environment* 696: 134059.

Amann, M., Anderl, M., Borken-Kleefeld, J., Cofala, J., Heyes, C., Höglund Isaksson, L., Kiesewetter, G., Klimont, Z., Moosmann, L. and Rafaj, P. (2018). *Progress Towards the Achievement of the EU's Air Quality and Emissions Objectives*. IIASA, Laxenburg, Austria.

Amon, B., Stouman Jensen, L. S., Groenestein, K. and Sutton, M. (2021). Sustainable nitrogen management for housed livestock, manure storage and manure processing. In: Baines, R. (Ed.). *Reducing greenhouse gas emissions from livestock production*. Burleigh Dodds Science Publishing, Cambridge, 46 pages.

Anderson, C., Malambo, D. H., Gonzalez Perez, M. E., Nobela, H. N., de Pooter, L., Spit, J., Hooijmans, C. M., de Vossenberg, Jv., Greya, W., Thole, B., van Lier, J. B. and Brdjanovic, D. (2015). Lactic acid fermentation, urea and lime addition: Promising faecal sludge sanitizing methods for emergency sanitation. *International Journal of Environmental Research and Public Health* 12(11): 13871–13885.

Barampouti, E. M., Mai, S., Malamis, D., Moustakas, K. and Loizidou, M. (2020). Exploring technological alternatives of nutrient recovery from digestate as a secondary resource. *Renewable and Sustainable Energy Reviews* 134: 110379.

Bastami, M. S. B., Jones, D. L. and Chadwick, D. R. (2016). Reduction of methane emission during slurry storage by the addition of effective microorganisms and excessive carbon source from brewing sugar. *Journal of Environmental Quality* 45(6): 2016–2022.

Burney, J. A., Davis, S. J. and Lobell, D. B. (2010). Greenhouse gas mitigation by agricultural intensification. *Proceedings of the National Academy of Sciences of the United States of America* 107(26): 12052–12057.

Cao, L., Keener, H., Huang, Z., Liu, Y., Ruan, R. and Xu, F. (2020). Effects of temperature and inoculation ratio on methane production and nutrient solubility of swine manure anaerobic digestion. *Bioresource Technology* 299: 122552.

Case, S. D. C. and Jensen, L. S. (2019). Nitrogen and phosphorus release from organic wastes and suitability as bio-based fertilizers in a circular economy. *Environmental Technology* 40(6): 701–715.

Chadwick, D., Sommer, S., Thorman, R., Fangueiro, D., Cardenas, L., Amon, B. and Misselbrook, T. (2011). Manure management: implications for greenhouse gas emissions. *Animal Feed Science and Technology* 166–167: 514–531.

Clemens, J., Trimborn, M., Weiland, P. and Amon, B. (2006). Mitigation of greenhouse gas emissions by anaerobic digestion of cattle slurry. *Agriculture, Ecosystems and Environment* 112(2-3): 171-177.

Chojnacka, K., Moustakas, K. and Witek-Krowiak, A. (2020). Bio-based fertilizers: A practical approach towards circular economy. *Bioresource Technology* 295: 122223.

Christensen, M. L., Hjorth, M. and Keiding, K. (2009). Characterization of pig slurry with reference to flocculation and separation. *Water Research* 43(3): 773-783.

De Vries, J. W., Aarnink, A. J. A., Groot Koerkamp, P. W. G. and De Boer, I. J. M. (2013). Life cycle assessment of segregating fattening pig urine and feces compared to conventional liquid manure management. *Environmental Science and Technology* 47(3): 1589-1597.

De Vries, J. W., Groenestein, C. M., Schröder, J. J., Hoogmoed, W. B., Sukkel, W., Groot Koerkamp, P. W. G. and De Boer, I. J. M. (2015). Integrated manure management to reduce environmental impact: II. Environmental impact assessment of strategies. *Agricultural Systems* 138: 88-99.

De Vries, J. W., Vinken, T. M. W. J., Hamelin, L. and De Boer, I. J. M. (2012). Comparing environmental consequences of anaerobic mono- and co-digestion of pig manure to produce bio-energy - A life cycle perspective. *Bioresource Technology* 125: 239-248.

De Vrieze, J., Colica, G., Pintucci, C., Sarli, J., Pedizzi, C., Willeghems, G., Bral, A., Varga, S., Prat, D., Peng, L., Spiller, M., Buysse, J., Colsen, J., Benito, O., Carballa, M. and Vlaeminck, S. E. (2019). Resource recovery from pig manure via an integrated approach: A technical and economic assessment for full-scale applications. *Bioresource Technology* 272: 582-593.

Dennehy, C., Lawlor, P. G., Jiang, Y., Gardiner, G. E., Xie, S., Nghiem, L. D. and Zhan, X. (2017). Greenhouse gas emissions from different pig manure management techniques: A critical analysis. *Frontiers of Environmental Science and Engineering* 11(3): 1-16.

Dittmer, K. M., Darby, H. M., Goeschel, T. R. and Adair, E. C. (2020). Benefits and tradeoffs of reduced tillage and manure application methods in a Zea mays silage system. *Journal of Environmental Quality* 49(5): 1236-1250.

Directive 91/676/EEC of the Council of the European Communities of 12 December 1991 concerning the protection of waters against pollution caused by nitrates from agricultural sources.

Emmerling, C., Krein, A. and Junk, J. (2020). Meta-analysis of strategies to reduce NH3 emissions from slurries in European agriculture and consequences for greenhouse gas emissions. *Agronomy* 10(11): 1633.

Ershadi, S. Z., Dias, G., Heidari, M. D. and Pelletier, N. (2020). Improving nitrogen use efficiency in crop-livestock systems: a review of mitigation technologies and management strategies, and their potential applicability for egg supply chains. *Journal of Cleaner Production* 265: 121671.

Erwiha, G. M., Ham, J., Sukor, A., Wickham, A. and Davis, J. G. (2020). Organic fertilizer source and application method impact ammonia volatilization. *Communications in Soil Science and Plant Analysis* 51(11): 1469-1482.

Eurostat (2022). *Key Figures on the European Food Chain* (2022 edn.). 7 December.

Eurostat (2016). Agri-environmental indicator - mineral fertiliser consumption [WWW Document]. Available at: https://ec.europa.eu/eurostat/statistics-explained/index

.php/Agri-environmental_indicator_-_mineral_fertiliser_consumption (accessed 4 January 2023).

Eurostat (2021). Glossary: Fertiliser. Available at: https://ec.europa.eu/eurostat/statistics-explained/index.php/Glossary:Fertiliser.

Fangueiro, D., Coutinho, J., Chadwick, D., Moreira, N. and Trindade, H. (2008). Effect of cattle slurry separation on greenhouse gas and ammonia emissions during storage. *Journal of Environmental Quality* 37(6): 2322–2331.

Fangueiro, D., Ribeirom,H., Vasconcelos, E., Coutinho, J. and Cabral, F. (2009). Treatment by acidification followed by solid–liquid separation affects slurry and slurry fractions composition and their potential of N mineralization. *Bioresource Technology* 100(20): 4914–4917.

Fangueiro, D., Surgy, S., Napier, V., Menaia, J., Vasconcelos, E. and Coutinho, J. (2014). Impact of slurry management strategies on potential leaching of nutrients and pathogens in a sandy soil amended with cattle slurry. *Journal of Environmental Management* 146: 198–205.

Fangueiro, D., Hjorth, M. and Gioelli, F. (2015). Acidification of animal slurry–a review. *Journal of Environmental Management* 149: 46–56.

Fangueiro, D., Merino, P., Pantelopoulos, A., Pereira, J. L. S., Amon, B. and Chadwick, D. R. (2023). The implications of animal manure management on ammonia and greenhouse gas emissions. In: Bartzanas, T. (Ed.). *Technology for Environmentally Friendly Livestock Production Smart Animal Production*. Springer, Cham. https://doi.org/10.1007/978-3-031-19730-7_5.

Feng, L., Wahid, R., Ward, A. J. and Møller, H. B. (2017). Anaerobic co-digestion of cattle manure and meadow grass: effect of serial configurations of continuous stirred tank reactors (CSTRs). *Biosystems Engineering* 160: 1–11.

Foged, H., Flotats, X., Blasi, A. B., Palatsi, J., Magri, A. and Schelde, K. M. (2011). Inventory of manure processing activities in Europe. Technical Report No. I concerning 'Manure Processing Activities in Europe' to the European Commission, Directorate-General Environment.

Fuchs, A., Dalby, F. R., Liu, D., Kai, P. and Feilberg, A. (2021). Improved effect of manure acidification technology for gas emission mitigation by substituting sulfuric acid with acetic acid. *Cleaner Engineering and Technology* 4: 100263.

Gioelli, F., Grella, M., Scarpeci, T. E., Rollè, L., Pierre, F. D. and Dinuccio, E. (2022). Bio-acidification of cattle slurry with whey reduces gaseous emission during storage with positive effects on biogas production. *Sustainability* 14(19): 12331.

Hafner, S. D., Pacholski, A., Bittman, S., Burchill, W., Bussink, W., Chantigny, M., Carozzi, M., Génermont, S., Häni, C., Hansen, M. N., Huijsmans, J., Hunt, D., Kupper, T., Lanigan, G., Loubet, B., Misselbrook, T., Meisinger, J. J., Neftel, A., Nyord, T., Pedersen, S. V., Sintermann, J., Thompson, R. B., Vermeulen, B., Vestergaard, A. V., Voylokov, P., Williams, J. R. and Sommer, S. G. (2018). The ALFAM2 database on ammonia emission from field-applied manure: description and illustrative analysis. *Agricultural and Forest Meteorology* 258: 66–79.

Hagos, K., Zong, J., Li, D., Liu, C. and Lu, X. (2017). Anaerobic co-digestion process for biogas production: progress, challenges and perspectives. *Renewable and Sustainable Energy Reviews* 76: 1485–1496.

Harris, P. W. and McCabe, B. K. (2015). Review of pre-treatments used in anaerobic digestion and their potential application in high-fat cattle slaughterhouse wastewater. *Applied Energy* 155: 560–575.

Hartmann, H. and Ahring, B. K. (2006). Strategies for the anaerobic digestion of the organic fraction of municipal solid waste: An overview. *Water Science and Technology* 53(8): 7-22.

Hernández-Beltrán, J. U., Hernández-De Lira, I. O., Cruz-Santos, M. M., Saucedo-Luevanos, A., Hernández-Terán, F. and Balagurusamy, N. (2019). Insight into pretreatment methods of lignocellulosic biomass to increase biogas yield: Current state, challenges, and opportunities. *Applied Sciences* 9(18): 3721.

Hjorth, M., Christensen, K. V., Christensen, M. L. and Sommer, S. G. (2010). Solid-liquid separation of animal slurry in theory and practice. A review. *Agronomy for Sustainable Development* 30(1): 153–180.

Holly, M. A., Larson, R. A., Powell, J. M., Ruark, M. D. and Aguirre-Villegas, H. (2017). Greenhouse gas and ammonia emissions from digested and separated dairy manure during storage and after land application. *Agriculture, Ecosystems and Environment* 239: 410–419.

Hossain, M. K., Islam, M. R., Jahiruddin, M., Sorensen, P., Møller, H. B. and Islam, M. S. (2021). Effect of anaerobic digestion temperature and manure type on N and S mineralization. *Communications in Soil Science and Plant Analysis* 52(20): 2431–2444.

Hou, Y., Velthof, G. L., Case, S. D. C., Oelofse, M., Grignani, C., Balsari, P., Zavattaro, L., Gioelli, F., Bernal, M. P., Fangueiro, D., Trindade, H., Jensen, L. S. and Oenema, O. (2018). Stakeholder perceptions of manure treatment technologies in Denmark, Italy, the Netherlands and Spain. *Journal of Cleaner Production* 172: 1620-1630.

Hou, Y., Velthof, G. L., Lesschen, J. P., Staritsky, I. G. and Oenema, O. (2017). Nutrient recovery and emissions of ammonia, nitrous oxide, and methane from animal manure in Europe: effects of manure treatment technologies. *Environmental Science and Technology* 51(1): 375–383.

Hou, Y., Velthof, G. L. and Oenema, O. (2015). Mitigation of ammonia, nitrous oxide and methane emissions from manure management chains: A meta-analysis and integrated assessment. *Global Change Biology* 21(3): 1293–1312.

Kavanagh, I., Fenton, O., Healy, M. G., Burchill, W., Lanigan, G. J. and Krol, D. J. (2021). Mitigating ammonia and greenhouse gas emissions from stored cattle slurry using agricultural waste, commercially available products and a chemical acidifier. *Journal of Cleaner Production* 294: 126251.

Keskinen, R., Termonen, M., Salo, T., Luostarinen, S. and Räty, M. (2022). Slurry acidification outperformed injection as an ammonia emission-reducing technique in boreal grass cultivation. *Nutrient Cycling in Agroecosystems* 122(2): 139–156.

Kim, T. H., Oh, K. K., Ryu, H. J., Lee, K. and Kim, T. H. (2014). Hydrolysis of hemicellulose from barley straw and enhanced enzymatic saccharification of cellulose using acidified zinc chloride. *Renewable Energy* 65: 56–63.

Kleinman, P. J. A., Spiegal, S., Liu, J., Holly, M., Church, C. and Ramirez-Avila, J. (2020). Managing animal manure to minimize phosphorus losses from land to water. In: Waldrip, H. M., Pagliari, P. H. and He, Z. *Animal Manure*, pp. 201–228.

Koger, J. B., O'Brien, B. K., Burnette, R. P., Kai, P., van Kempen, M. H. J. G., van Heugten, E. and van Kempen, T. A. T. G. (2014). Manure belts for harvesting urine and feces separately and improving air quality in swine facilities. *Livestock Science* 162: 214–222.

Köninger, J., Lugato, E., Panagos, P., Kochupillai, M., Orgiazzi, A. and Briones, M. J. I. (2021). Manure management and soil biodiversity: towards more sustainable food systems in the EU. *Agricultural Systems* 194: 103251.

Kumari, D. and Singh, R. (2018). Pretreatment of lignocellulosic wastes for biofuel production: A critical review. *Renewable and Sustainable Energy Reviews* 90: 877–891.

Kupper, T., Häni, C., Neftel, A., Kincaid, C., Bühler, M., Amon, B. and VanderZaag, A. (2020). Ammonia and greenhouse gas emissions from slurry storage - a review. *Agriculture, Ecosystems and Environment* 300: 106963.

Li, B., Dinkler, K., Zhao, N., Sobhi, M., Merkle, W., Liu, S., Dong, R., Oechsner, H. and Guo, J. (2020). Influence of anaerobic digestion on the labile phosphorus in pig, chicken, and dairy manure. *Science of the Total Environment* 737: 140234.

Logan, M. and Visvanathan, C. (2019). Management strategies for anaerobic digestate of organic fraction of municipal solid waste: current status and future prospects. *Waste Management and Research* 37(1_suppl): 27–39.

Loussouarn, A., Lagadec, S., Robin, P. and Hassouna, M. (2014). Raclage en «V»: Bilan environnemental et zootechnique lors de sept années de fonctionnement à Guernévez. In: *46èmes Journées de la Recherche Porcine*. Paris, France, pp. 199–204.

Maldaner, L., Wagner-Riddle, C., VanderZaag, A. C., Gordon, R. and Duke, C. (2018). Methane emissions from storage of digestate at a dairy manure biogas facility. *Agricultural and Forest Meteorology* 258: 96–107.

Malique, F., Wangari, E., Andrade-Linares, D. R., Schloter, M., Wolf, B., Dannenmann, M., Schulz, S. and Butterbach-Bahl, K. (2021). Effects of slurry acidification on soil N2O fluxes and denitrification. *Journal of Plant Nutrition and Soil Science* 184(6): 696–708.

Meegoda, J. N., Li, B., Patel, K. and Wang, L. B. (2018). A review of the processes, parameters, and optimization of anaerobic digestion. *International Journal of Environmental Research and Public Health* 15(10): 2224.

Møller, H. B. and Nielsen, K. J. (2016). Biogas taskforce: udvikling og effektivisering af biogasproduktionen i Danmark. DCA-Nationalt Center for Fødevarer og Jordbrug.

Møller, H. B., Sørensen, P., Olesen, J. E., Petersen, S. O., Nyord, T. and Sommer, S. G. (2022). Agricultural biogas production–climate and environmental impacts. *Sustainability* 14(3): 1849.

Moset, V., Fontaine, D. and Møller, H. B. (2017). Co-digestion of cattle manure and grass harvested with different technologies. Effect on methane yield, digestate composition and energy balance. *Energy* 141: 451–460.

Moset, V., Cerisuelo, A., Sutaryo, S. and Møller, H. B. (2012). Process performance of anaerobic co-digestion of raw and acidified pig slurry. *Water Research* 46(16): 5019–5027.

Moset, V., Ottosen, L. D. M., Xavier, C. and Møller, H. B. (2016). Anaerobic digestion of sulfate-acidified cattle slurry: one-stage vs. two-stage. *Journal of Environmental Management* 173: 127–133.

Moyo, L. B., Simate, G. S. and Mutsatsa, T. (2022). Biological acidification of pig manure using banana peel waste to improve the dissolution of particulate phosphorus: A critical step for maximum phosphorus recovery as struvite. *Heliyon* 8(8): e10091.

Niles, M. T., Wiltshire, S., Lombard, J., Branan, M., Vuolo, M., Chintala, R. and Tricarico, J. (2022). Manure management strategies are interconnected with complexity across US dairy farms. *PLoS ONE* 17(6): e0267731.

Niles, M. T., Horner, C., Chintala, R. and Tricarico, J. (2019). A review of determinants for dairy farmer decision making on manure management strategies in high-income countries. *Environmental Research Letters* 14(5): 053004.

Nyang'au, J. O., Moller, H. B. and Sorensen, P. (2022). Nitrogen dynamics and carbon sequestration in soil following application of digestates from one- and two-step anaerobic digestion. *Science of the Total Environment* 851(1): 158177.

Nyang'au, J. O., Møller, H. B. and Sørensen, P. (2023). Effects of electrokinetic and ultrasonication pre-treatment and two-step anaerobic digestion of biowastes on the nitrogen fertiliser value by injection or surface banding to cereal crops. *Journal of Environmental Management* 326(A): 116699.

Nyord, T., Liu, D., Eriksen, J. and Adamsen, A. P. S. (2013). Effect of acidification and soil injection of animal slurry on ammonia and odour emission. *15th RAMIRAN Conference*, Versailles, France.

Oenema, O. and Tamminga, S. (2005). Nitrogen in global animal production and management options for improving nitrogen use efficiency. *Science in China Series C Life Sciences* 48(S2): 871–887.

Olesen, J. E., Møller, H. B., Petersen, S. O., Sørensen, P., Nyord, T. and Sommer, S. G. (2021). Sustainable biogas-climate and environmental effects of biogas production. DCA Report. Aarhus Universitet, DCA – Danish Centre for Food and Agriculture, p. 85.

Overmeyer, V., Holtkamp, F., Clemens, J., Büscher, W. and Trimborn, M. (2020). Dynamics of different buffer systems in slurries based on time and temperature of storage and their visualization by a new mathematical tool. *Animals: An Open Access Journal from MDPI* 10(4): 724.

Overmeyer, V., Kube, A., Clemens, J., Büscher, W. and Trimborn, M. (2021). One-time acidification of slurry: what is the most effective acid and treatment strategy? *Agronomy* 11(7): 1319.

Pedersen, I. F., Rubæk, G. H. and Sørensen, P. (2017). Cattle slurry acidification and application method can improve initial phosphorus availability for maize. *Plant and Soil* 414(1-2): 143–158.

Pedersen, J., Andersson, K., Feilberg, A., Delin, S., Hafner, S. and Nyord, T. (2021). Effect of exposed surface area on ammonia emissions from untreated, separated, and digested cattle manure. *Biosystems Engineering* 202: 66–78.

Plana, P. V. and Noche, B. (2016). A review of the current digestate distribution models: Storage and transport. *WIT Transactions on Ecology and the Environment* 202: 345–357.

Prado, J., Chieppe, J., Raymundo, A. and Fangueiro, D. (2020). Bio-acidification and enhanced crusting as an alternative to sulphuric acid addition to slurry to mitigate ammonia and greenhouse gases emissions during short term storage. *Journal of Cleaner Production* 263: 121443.

Prado, J., Ribeiro, H., Alvarenga, P. and Fangueiro, D. (2022). A step towards the production of manure-based fertilizers: disclosing the effects of animal species and slurry treatment on their nutrients content and availability. *Journal of Cleaner Production* 337: 130369.

Rajagopal, R., Massé, D. I. and Singh, G. (2013). A critical review on inhibition of anaerobic digestion process by excess ammonia. *Bioresource Technology* 143: 632–641.

Regueiro, I., Coutinho, J., Balsari, P., Popovic, O. and Fangueiro, D. (2016). Acidification of pig slurry before separation to improve slurry management on farms. *Environmental Technology* 37(15): 1906–1913.

Regueiro, I., Coutinho, J. and Fangueiro, D. (2016). Alternatives to sulfuric acid for slurry acidification: impact on slurry composition and ammonia emissions during storage. *Journal of Cleaner Production* 131: 296–307.

Regueiro, I., Siebert, P., Liu, J., Müller-Stöver, D. and Jensen, L. S. (2020). Acidified animal manure products combined with a nitrification inhibitor can serve as a starter fertilizer for maize. *Agronomy* 10(12): 1941.

Regueiro, I., Gómez-Muñoz, B., Lübeck, M., Hjorth, M. and Jensen, L. S. (2022). Bio-acidification of animal slurry: efficiency, stability and the mechanisms involved. *Bioresource Technology Reports* 19: 101135.

Regulation 1069/2009 of the European Parliament and of the Council of 21 October 2009 laying down health rules as regards animal by-products and derived products not intended for human consumption and repealing Regulation (EC) No 1774/2002 (Animal by-products Regulation).

Reuland, G., Sigurnjak, I., Dekker, H., Michels, E. and Meers, E. (2021). The potential of digestate and the liquid fraction of digestate as chemical fertiliser substitutes under the RENURE criteria. *Agronomy* 11(7): 1374.

Rodrigues, J., Alvarenga, P., Silva, A. C., Brito, L., Tavares, J. and Fangueiro, D. (2021). Animal slurry sanitization through PH adjustment: process optimization and impact on slurry characteristics. *Agronomy* 11(3): 517.

Romero-Güiza, M. S., Mata-Alvarez, J., Chimenos Rivera, J. M. and Astals Garcia, S. (2016). Nutrient recovery technologies for anaerobic digestion systems: an overview. *Revista ION* 29(1): 7–26.

Romio, C., Kofoed, M. V. W.and Moller, H. B. (2021). Digestate post-treatment strategies for additional biogas recovery: a review. *Sustainability* 13(16): 9295.

Sarker, S., Lamb, J. J., Hjelme, D. R. and Lien, K. M. (2019). A review of the role of critical parameters in the design and operation of biogas production plants. *Applied Sciences* 9(9): 1915.

Sayara, T. and Sánchez, A. (2019). A review on anaerobic digestion of lignocellulosic wastes: pretreatments and operational conditions. *Applied Sciences* 9(21): 4655.

Schreiber, M., Bazaios, E., Ströbel, B., Wolf, B., Ostler, U., Gasche, R., Schlingmann, M., Kiese, R. and Dannenmann, M. (2023). Impacts of slurry acidification and injection on fertilizer nitrogen fates in grassland. *Nutrient Cycling in Agroecosystems* 125(2): 171–186.

Sepperer, T., Petutschnigg, A. and Steiner, K. (2021). Effect of flushing milk and acidic whey on pH and nitrogen loss of cattle manure slurry. *Atmosphere* 12(9): 1222.

Siegert, I. and Banks, C. (2005). The effect of volatile fatty acid additions on the anaerobic digestion of cellulose and glucose in batch reactors. *Process Biochemistry* 40(11): 3412–3418.

Smith, S. R., Lang, N. L., Cheung, K. H. M. and Spanoudaki, K. (2005). Factors controlling pathogen destruction during anaerobic digestion of biowastes. *Waste Management* 25(4): 417–425.

Søgaard, H. T., Sommer, S. G., Hutchings, N. J., Huijsmans, J. F. M., Bussink, D. W. and Nicholson, F. (2002). Ammonia volatilization from field-applied animal slurry–the ALFAM model. *Atmospheric Environment* 36(20): 3309–3319.

Sommer, S. G. and Husted, S. (1995). A simple model of pH in slurry. *The Journal of Agricultural Science* 124(3): 447–453.

Sørensen, C. G., Jacobsen, B. H. and Sommer, S. G. (2003). An assessment tool applied to manure management systems using innovative technologies. *Biosystems Engineering* 86(3): 315–325. https://doi.org/10.1016/S1537-5110(03)00137-5.

Sørensen, P. (2004). Immobilisation, remineralisation and residual effects in subsequent crops of dairy cattle slurry nitrogen compared to mineral fertiliser nitrogen. *Plant and Soil* 267(1-2): 285-296.

Sørensen, P., Bechini, L. and Jensen, L. S. (2019). *Manure management in organic farming*. In: Köpke, U. (ed.) *Improving organic crop cultivation*, Burleigh Dodds Science Publishing: Cambridge, UK.

Sørensen, P., Khan, A. R., Møller, H. B. and Thomsen, I. K. (2012). Effects of anaerobic digestion of organic manures on N turnover and N utilization. *Proceedings of the 17th Nitrogen Workshop-Innovations for Sustainable Use of Nitrogen Resources*, pp. 80-81.

Sørensen, P. and Thomsen, I. K. (2005). Separation of pig slurry and plant utilization and loss of nitrogen-15-labeled slurry nitrogen. *Soil Science Society of America Journal* 69(5): 1644-1651.

Sommer, S. G. and Husted, S. (1995). The chemical buffer system in raw and digested animal slurry. *The Journal of Agricultural Science* 124(1): 45-53.

Sutaryo, S., Ward, A. J. and Møller, H. B. (2013). Anaerobic digestion of acidified slurry fractions derived from different solid-liquid separation methods. *Bioresource Technology* 130: 495-501.

Ten Hoeve, M., Gómez-Muñoz, B., Jensen, L. S. and Bruun, S. (2016). Environmental impacts of combining pig slurry acidification and separation under different regulatory regimes - A life cycle assessment. *Journal of Environmental Management* 181: 710-720.

Thangarajan, R., Bolan, N. S., Tian, G., Naidu, R. and Kunhikrishnan, A. (2013). Role of organic amendment application on greenhouse gas emission from soil. *Science of the Total Environment* 465: 72-96.

Varma, V. S., Parajuli, R., Scott, E., Canter, T., Lim, T. T., Popp, J. and Thoma, G. (2021). Dairy and swine manure management-Challenges and perspectives for sustainable treatment technology. *Science of the Total Environment* 778: 146319.

Velthof, G. L., Kuikman, P. J. and Oenema, O. (2003). Nitrous oxide emission from animal manures applied to soil under controlled conditions. *Biology and Fertility of Soils* 37(4): 221-230.

Veluchamy, C., Gilroyed, B. H. and Kalamdhad, A. S. (2019). Process performance and biogas production optimizing of mesophilic plug flow anaerobic digestion of corn silage. *Fuel* 253: 1097-1103.

Vu, P. T., Melse, R. W., Zeeman, G. and Groot Koerkamp, P. W. G. (2016). Composition and biogas yield of a novel source segregation system for pig excreta. *Biosystems Engineering* 145: 29-38.

Wagner, C., Nyord, T., Vestergaard, A. V., Hafner, S. D. and Pacholski, A. S. (2021). Acidification effects on in situ ammonia emissions and cereal yields depending on slurry type and application method. *Agriculture* 11(11): 1053.

Webb, J., Pain, B., Bittman, S. and Morgan, J. (2010). The impacts of manure application methods on emissions of ammonia, nitrous oxide and on crop response–a review. *Agriculture, Ecosystems and Environment* 137(1-2): 39-46.

Webb, J., Sørensen, P., Velthof, G., Amon, B., Pinto, M., Rodhe, L., Salomon, E., Hutchings, N., Burczyk, P. and Reid, J. (2013). An assessment of the variation of manure nitrogen efficiency throughout Europe and an appraisal of means to increase manure-N efficiency. *Advances in Agronomy*: 371-442.

Welch, R. M. (2002). The impact of mineral nutrients in food crops on global human health. *Plant and Soil* 247(1): 83-90.

Wright, I. A., Tarawali, S., Blummel, M., Gerard, B., Teufel, N. and Herrero, M. (2012). Integrating crops and livestock in subtropical agricultural systems. *Journal of the Science of Food and Agriculture* 92(5): 1010-1015.

Yuan, Z., Pan, X., Chen, T., Liu, X., Zhang, Y., Jiang, S., Sheng, H. and Zhang, L. (2018). Evaluating environmental impacts of pig slurry treatment technologies with a life-cycle perspective. *Journal of Cleaner Production* 188: 840-850.

Zhang, X., Fang, Q., Zhang, T., Ma, W., Velthof, G. L., Hou, Y., Oenema, O. and Zhang, F. (2020). Benefits and trade-offs of replacing synthetic fertilizers by animal manures in crop production in China: A meta-analysis. *Global Change Biology* 26(2): 888-900.

Zheng, Y., Zhao, J., Xu, F. Q. and Li, Y. (2014). Pretreatment of lignocellulosic biomass for enhanced biogas production. *Progress in Energy and Combustion Science* 42: 35-53.

Chapter 7

Optimizing livestock manure as a biofertilizer and bioenergy source

V. Riau, L. Morey, R. Cáceres, M. Cerrillo and A. Bonmatí, Institute of Agrifood Research and Technology (IRTA), Spain; and A. Robles, BETA Tech Center (UVIC-UCC), Spain

1 Introduction

In 2020, 11.2 million tonnes of mineral fertilizer was used in the European agricultural sector. Ten million tonnes was mineral nitrogen (N) and 1.2 million tonnes was mineral phosphorus (P) (Eurostat, 2022). More than 45% of the total N applied on agricultural fields comes from mineral fertilizers (Hendriks et al., 2022).

Energy from natural gas is typically used to manufacture nitrogenous fertilizers which means its cost is highly dependent on oil prices. The N-based fertilizer industry in Europe depends to a large extent on gas from Russia, which also plays a key role in the global production of P rock-based fertilizers. Moreover, phosphorus extraction is mainly performed outside the EU, leading to high production and transportation costs, which are also related to oil prices. The current geopolitical situation is thus leading to an increase in the price of N and P mineral fertilizer that will have a direct impact on the use of fertilizers in agriculture in the EU (Eurostat, 2022). The current worldwide energy crisis

http://dx.doi.org/10.19103/AS.2023.0120.11

and dependence on non-renewable energy resources has therefore driven interest in the production of biobased fertilizers to secure a more sustainable and resilient agricultural sector (Hendriks et al., 2022).

In this context, livestock manure should be considered as a resource rather than waste as it has an abundance of available nutrients for crops, carbon for bioenergy production and can promote soil carbon stocks. Global livestock production in 2018 was around 1450 million heads, comprising cattle (965 million), pigs (242 million) and chickens (237 million), generating approximately 55 billion tonnes of manure. In terms of nutrient content, this represents around 125 million tonnes of N, only 27% of which is processed in some way. An estimated 12–14 million tonnes per year of P are estimated to be excreted by livestock worldwide (Nagarajan et al., 2023).

Circular agriculture focuses on using minimal amounts of external inputs, closing nutrient loops, regenerating soils and minimizing impacts on the environment, thus aiding the transition towards sustainable and resilient energy and farming systems (Kristinn et al., 2021). Using manure as a source of macro- and micronutrients in agriculture potentially enables crop and livestock production that does not deplete non-renewable sources and does not harm the environment, since it reduces dependence on mineral fertilizers (Prado et al., 2022). In Europe, it is estimated that a circular approach to food systems could reduce the use of chemical fertilizers by 80% (Kristinn et al., 2021).

However, if not managed properly, manure application to soil can lead to detrimental environmental effects from ammonia emissions, pollution of surface water by run-off and groundwater via NO_3^- (nitrate) leaching. Consequently, manure and by-product application to soil in the EU is strictly regulated (maximum 170 kg N ha^{-1} $year^{-1}$ in Nitrate Vulnerable Zones), due to the environmental risks associated with nitrate pollution (European Commission, 1991). To overcome all these limitations, the European Parliament and the EU Council approved the new Fertilizing Product Regulation 1009/2019 (effective as of 16 July 2022), which facilitated access to the EU Single Market for biobased fertilizers. These biobased fertilizers can be awarded the 'CE mark', making it easier to commercialize and, consequently, promote production (Rizzioli et al., 2023). In addition, the possible application of the RENURE criteria (REcovered Nitrogen from manURE) aims to diminish the legal limitations on the application of livestock manure in agriculture. Under these criteria, the application of manure-based fertilizers with a ratio $N_{mineral}/N_{total} > 90\%$, or a ratio of total organic carbon/$N_{total} \leq 3$ would be less restrictive, and the production and use of biobased fertilizers would be promoted (Prado et al., 2022; Hendriks et al., 2022; Brienza et al., 2021).

However, many farmers still prefer the use of mineral fertilizers. This is not only due to logistic/regulative barriers and costs in using alternatives, but also because they are sometimes reluctant to use manure-derived materials that have a wide range of physical and chemical characteristics, and with unbalanced and usually unknown nutrient content. In addition, the investment cost and lack of incentives often makes it difficult to install valorization technologies to process manure at farm level (Khoshnevisan et al., 2021; Tur-Cardona et al., 2018). Therefore, further research is required to establish a systematic framework based on regional requirements to develop an integrated and circular manure nutrient management plan with minimum environmental risks, maximum profit and broad acceptance. A promising solution for the future sustainable development of the agri-food sector is livestock manure management using a biorefinery approach to produce manure-derived fertilizers and/or bioenergy (Awasthi et al., 2022). Technology selection would depend on environmental policies, the investment costs and incentives provided, and the requirement/market of an end-product in a particular country/city (Khoshnevisan et al., 2021).

This chapter provides an overview of the main technologies for manure valorization into bioenergy and biofertilizer products. Some technologies are already well developed and widely implemented, such as solid-liquid (S/L) separation, anaerobic digestion and composting. Others are more innovative and less developed, including struvite precipitation, ammonia stripping/scrubbing as well as bioelectrochemical and membrane systems. These technologies can be classified as

- biological (composting, anaerobic digestion and bioelectrochemical systems);
- mechanical (S/L separation); and
- physico-chemical (struvite precipitation, membrane filtration and stripping/scrubbing).

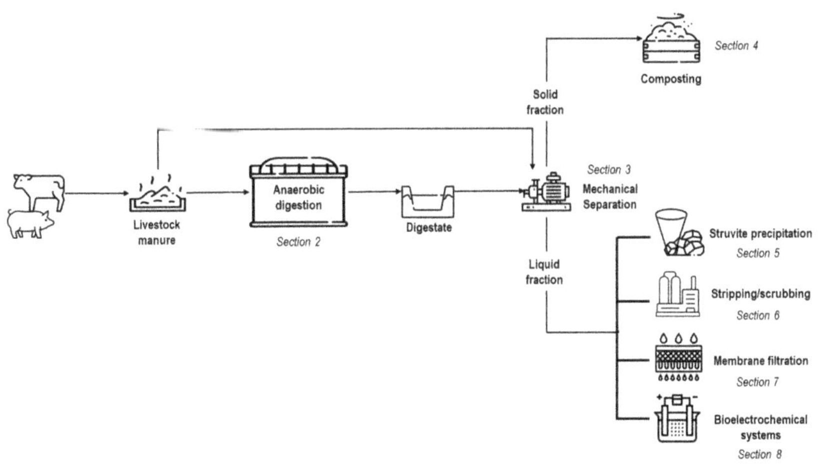

Figure 1 Overview of the different technologies for the production of biobased fertilizers and bioenergy included in this chapter.

However, this chapter has been structured following the scheme shown in Fig. 1, describing the different technologies in the order they are usually implemented along the treatment chain.

2 Anaerobic digestion

Anaerobic digestion (AD) is a well-established and widely implemented technology in the European agricultural sector for treating livestock manure since it generates a renewable fuel in the form of biogas, improves manure fertilizer quality, and reduces odours, pathogens as well as greenhouse gas emissions (Awasthi et al., 2022; Torrellas et al., 2018). The biochemical steps and the factors affecting the AD process have been extensively studied over the last few decades and are well-documented for innumerable feedstocks. This section is therefore mainly focused on the optimization of manure-based bioenergy and biofertilizer production in full-scale AD biogas plants. Figure 2 shows an overview of the main inputs and outputs of the biogas and biomethane production process in the livestock sector.

According to the European Biogas Association, there were almost 15,000 agro-industrial biogas plants in the European Union by the end of 2019 (https://www.europeanbiogas.eu/). However, the AD of manure is only 13–65% efficient when converting organic matter into biogas (Nagarajan et al., 2023; Khoshnevisan et al., 2021), mainly due to the high lignocellulosic content of manure and its low C/N ratio, which results in low methane yields. It has been reported that a suitable C/N ratio from AD systems ranges from 20 to 30, with an optimal of 25 (Zhu et al., 2021). Moreover, biogas yield is strongly dependent on the type of livestock manure and is highly variable among the different studies carried out, ranging from 107 to 438 $LCH_4/kg\ VS_{removed}$ (Bumharter et al., 2023; Sawatdeenarunat et al., 2016).

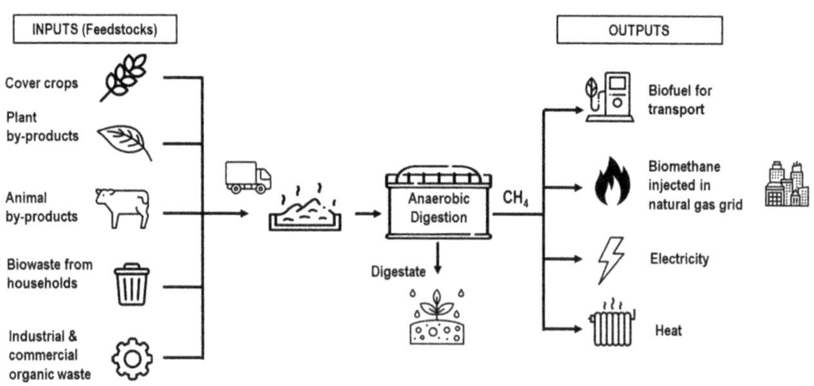

Figure 2 Schematic overview of inputs and outputs of the biogas and biomethane production from livestock manure (modified from EBA's statistical report 2021).

In order to increase bioenergy production from manure, pre-treatment methods such as thermochemical, microwave, ultrasonication and mechanical techniques can be applied to promote the breakdown of the recalcitrant structure of lignocellulosic material and improve the hydrolysis step (Khoshnevisan et al., 2021). However, anaerobic co-digestion of manure, together with other organic substrates containing a high labile carbon content, is the main approach used by agro-industrial biogas plants to improve process efficiency and biogas production (Zhu et al., 2021). It offers multiple advantages compared with single digestion, i.e. better nutrient balance, improved C:N ratio and a robust microbial consortia. Many agricultural feedstocks have been investigated as potential co-substrates over the last few years, including catch crops (Riau et al., 2021), sugar beets (Aboudi et al., 2016), maize silage (Varol and Ugurlu, 2017) or rice straw (Tian et al., 2023). Khoshnevisan et al. (2021) recently published a complete list of synergistic substrates for the anaerobic co-digestion of different livestock manures.

The biogas produced via the AD process has been used in a wide range of applications. To date, the most common industrial-scale application has been heat and electricity generation using a combined heat and power engine, with electrical and thermal conversion efficiencies around 40% and 50%, respectively (Horváth et al., 2016; Sawatdeenarunat et al., 2016). A recent trend is biogas upgrading to remove undesirable compounds, such as H_2S, CO_2, siloxanes, ammonia, moisture and particles from biogas. This can produce biomethane with a CH_4 content greater than 95% (Słupek et al., 2020). The European Biogas Association reported the number of biomethane plants increased from 483 plants in 2018 to 729 units in 2020 (https://www.europeanbiogas.eu/). Biomethane can be used as a fuel for transport, in the form of compressed natural gas, liquid natural gas or injected into a natural gas grid (Fig. 2). Several technologies for biogas purification have been developed in recent years, including adsorption, refrigeration with condensation, membrane technologies and biological methods (Budzianowski, 2016). Apart from its use as a renewable energy resource for heat and electricity generation, biogas can also be subjected to biological and thermochemical processes to produce, for instance, biohydrogen or biomethanol (Sawatdeenarunat et al., 2016).

Apart from biogas production, thermochemical manure treatments such as gasification to produce syngas (H_2 + CO), combustion to generate heat and electricity, or hydrothermal liquefaction/carbonization that can directly convert wet manure into bio-oil or hydrochar, have been developed to obtain manure-based bioenergy. However, the implementation of these technologies at farm level is still very limited. Drawbacks that are yet to be overcome include feedstock composition, undesirable compounds in syngas requiring expensive post-treatment or the configuration of the conversion system, among others (Khoshnevisan et al., 2021). Therefore, further studies and thorough knowledge of these processes in terms of operating conditions are needed to optimize the thermal conversion of livestock manure.

Moreover, in addition to biogas production, the AD of manure also generates a digestate product with good fertilizing value and similar behaviour to mineral fertilizers, since most of the nitrogen, phosphorus, potassium and calcium contained in the input substrates remain in the digestate. The remaining organic carbon also helps to maintain or even increase soil organic matter, which is particularly valuable in marginal or degraded soils (Ehmann et al., 2018). Hence, using anaerobic digestates as organic fertilizer is a common practice that partially reduces the costs derived from both mineral fertilizer and digestate disposal (Kovačić et al., 2022).

Digestate is usually separated into a solid and a liquid fraction in order to reduce the water content and improve transportability, thus reducing management costs (see Section 3). The solid fraction is usually subjected to a composting process to obtain a more stable product and transform (at least partially) N-ammonium into N-nitrate (Cáceres et al., 2015) (see Section 4). The SF can also be valorized into syngas, bio-oil, bio-hydrogen, methanol or hydrochar (Khoshnevisan et al., 2021). Treatment of the liquid fraction has generated considerable interest over the last few years as a feedstock for biobased inorganic fertilizer production such as struvite (Cerrillo et al., 2015) and ammonium sulphate (Sigurnjak et al., 2019) (see Sections 5 and 6, respectively). However, the existing digestate treatments still show considerable optimization potential to increase the effectiveness of nutrient recovery. In this sense, novel technologies to recover nutrients and increase biogas production from digestate, such as membrane systems and bioelectrochemical , have also aroused interest over the last few years (Cerrillo et al., 2023, 2021) (see Sections 7 and 8).

By combining AD and nutrient recovery technologies, energy, nutrients and carbon recovery can be achieved in a more sustainable way. Hence, the integration of these technologies within a single biorefinery is a promising approach that leads to more environmental and economic benefits: reducing nutrient spreading, connecting livestock farmers with industrial sectors, and considering livestock manure as a valuable feedstock for the biobased economy (Khoshnevisan et al., 2021; Sawatdeenarunat et al., 2016).

Livestock manure can be a key feedstock for a lignocellulosic biorefinery, in which materials and chemicals with lignin as a by-product are produced from the cellulose/hemicellulose portion of lignocellulosic biomass (Takkellapati et al., 2018). Hence, Fig. 2 can be transformed into Fig. 3 in order to consider manure AD as a key process within a biorefinery.

The feasibility of AD as a key technology to move to more resilient and sustainable livestock systems has been demonstrated, not only due to the production of biogas as a precursor of environmentally friendly and renewable

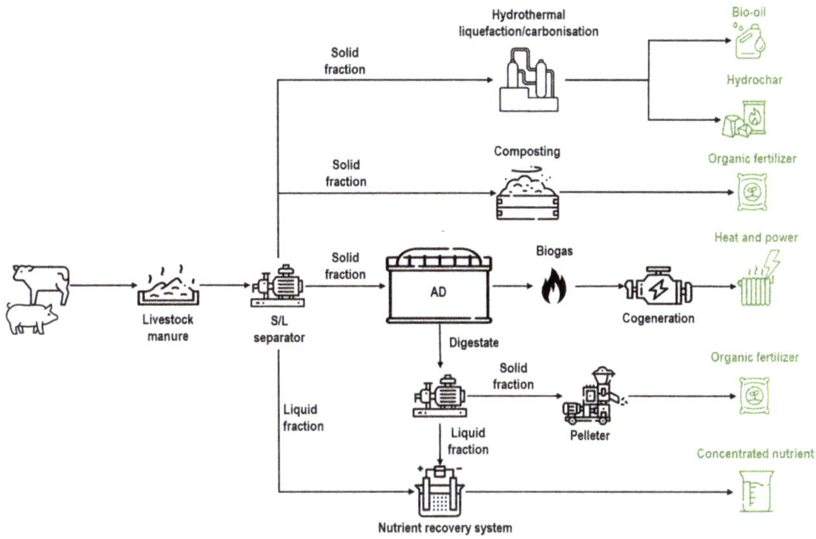

Figure 3 Example of a livestock manure-based biorefinery. Source: modified from Khoshnevisan et al. (2021).

bioenergy and bio-products, but also by generating a valuable fertilizer product that can be further treated to increase nutrient recovery and carbon sequestration in soils.

3 Mechanical separation

Solid/liquid separation is a technology commonly used as a first step in manure and anaerobic digestate processing. After the S/L separation, a liquid fraction (LF) with soluble components, such as mineral N, potassium and other ions, is obtained. A solid fraction (SF) containing most of the organic matter and a significant proportion of P is also generated. Therefore, this technology allows nutrient and carbon distribution between the LF and SF, thus reducing management costs and enabling nutrient transport over longer distances (Kovačić et al., 2022; Jorgensen and Jensen, 2009). Separation mainly occurs via natural settling or mechanical methods (Bonmatí-Blasi et al., 2020; Prenafeta-Boldú and Parera, 2020). During natural settling, particles precipitate to the bottom of the manure pit, while the LF is extracted from the top (Lyons et al., 2021). This is a common process in the field of urban wastewater treatment, but it does not have many applications in manure management as it requires long retention times and large ponds, free of turbulence. A more effective separation can be obtained using mechanical methods which are able to achieve high removal efficiencies and a solid stream with high dry matter content and most of the nutrients (Prenafeta-Boldú and Parera, 2020).

The most common mechanical separation technologies can be divided into three main groups (Bonmatí-Blasi et al., 2020; Prenafeta-Boldú and Parera, 2020; Møller et al., 2000):

- Gravity (pore diameter 250–750 μm), which includes statics, vibratory and rotatory sieves. In this case, solids cross the sieve via gravity, so no additional energy is required to achieve separation, which makes it the most energy-saving separation technology, and is typically used together with other separation systems.
- Pressure (pore diameter 300–650 μm), by screw pressing and filter belts; additional pressure is applied in order to increase the separation efficiency through the sieve.
- Centrifugation, using decanting centrifuges.

Over the last few decades, many studies have focused on comparing the efficiency and performance of different separation technologies (Prenafeta-Boldú and Parera, 2020; Møller et al., 2000; Flotats Ripoll et al., 2011). The percentage of dry matter and the nutrient content recovered in the SF are typically used as indicators of separation efficiency (Lyons et al., 2021; Møller et al., 2000). The most common methods for measuring efficiency are the simple separation efficiency index (SSEI) and the reduced separation efficiency index (RSEI). The SSEI is calculated by considering the nutrient and solids recovery, in terms of mass, as a proportion of the input material, so it indicates the proportion of a certain compound in the SF. An SSEI index of 0.5 would mean 50% of the component has been moved to the SF. On the contrary, the RSEI considers the difference in the masses of both fractions, showing the increase or decrease in concentration in the SF relative to the starting product (Lyons et al., 2021; Cocolo et al., 2012).

In order to determine the nutrient distribution between the two fractions, it is necessary to know the flow rate of the input manure and the output flow rate of both the SF and LF, as well as quantifying their nutrient content. Therefore, the factors affecting the efficiency of nutrient recovery in the SF must be considered in the decision-making process, and include the characteristics of the separation equipment, the size of the farm, the type of livestock, the investment and maintenance costs, ease of handling, adaptability to different types of manure, processing capacity, level of odour reduction (Møller et al., 2002) and the use of additives such as coagulant-flocculants or polyelectrolytes, which improve the separation efficiency, especially during centrifugation (Flotats Ripoll et al., 2011) (Table 1). These additives should not affect the quality and subsequent use of the separated SF as a fertilizer or organic amendment (Beggio et al., 2022; Prenafeta-Boldú and Parera, 2020).

Table 1 Reference values of separation efficiencies in terms of mass and nitrogen recovery in the solid fraction (%), depending on the type of livestock and separation technology

Separation technology		Fattening (>7% DM)		Sows (>3% DM)		Cattle (>5% DM)	
		Mass	Nitrogen	Mass	Nitrogen	Mass	Nitrogen
Gravity	Without additive	15	20	10	15	15	20
	With additive	20	25	15	20	20	25
Pressure	Without additive	15	20	10	15	20	25
	With additive	25	30	15	20	25	30
Centrifugation	Without additive	14	28	10	30	-	-
	With additive	20	50	15	45	20	50

Translated from Prenafeta-Boldú and Parera (2020).

Tables 1 and 2 show data from a review of manure treatment technologies carried out by the Government of Catalonia and the Institute of Agrifood Research and Technology (IRTA) in Spain (Prenafeta-Boldú and Parera, 2020). Table 1 compares the efficiency of the three main groups of separation techniques as a function of the type of livestock and the use of coagulant or flocculant; the highest efficiency is achieved using a centrifuge together with an additive. However, it should be noted that the associated costs in comparison to pressure technologies are significantly higher. Investment costs can range from €10 000 for sieves to €150 000 for a decanter centrifuge, while operational costs are mostly influenced by the use of additives (Flotats Ripoll et al., 2011). In addition, Table 2 shows a comparison of two of the most common separation technologies regarding nutrient (N, P) and metal content in the different fractions.

Fertilizer characteristics of the SF and LF, and hence the availability of nutrients for plants, are affected by the separation technique employed (Fangueiro et al., 2015) (Table 1). Furthermore, the inclusion of a previous process (e.g. anaerobic digestion or acidification) will also greatly influence nutrient speciation, removal efficiencies and the characteristics of the two streams obtained (Regueiro et al., 2016; Møller et al., 2000).

Prior to the solid-liquid separation, acidification minimizes NH_3 emissions during the solid-liquid separation and increases the P content of the liquid fraction (Cocolo et al., 2012; Regueiro et al., 2016). Both fractions can be used directly as a fertilizer, but it should be noted that the composition varies after

Table 2 Comparison of the nutrient content (nitrogen, phosphorus and potassium) and heavy metals (copper and zinc) in the different fractions, after a solid-liquid separation using a centrifuge with coagulants and polymers and a sieve and/or screw press separation system

	[a]Centrifuge + additive			[b]Sieve + screw press		
Parameter	Manure (kg/m^3)	Liquid fraction (kg/m^3)	Solid fraction (kg/t)	Manure (kg/m^3)	Liquid fraction (kg/m^3)	Solid fraction (kg/t)
Total nitrogen	3.37	1.93	9.24	4.30	4.10	6.07
Organic nitrogen	1.12	0.27	5.86	1.29	1.14	3.30
P	0.56	0.07	3.16	1.46	1.29	4.10
K	0.24	0.20	0.20	0.24	0.23	0.23
Cu (mg/kg)	4.88	0.66	30.43	15.16	14.70	23.27
Zn (mg/kg)	26.00	3.25	153.26	124.54	119.43	200.30

Translated from Prenafeta-Boldú and Parera (2020).

[a] Samples collected from one livestock farm with a centrifuge treating a mixture of pig and dairy manure (n = 6).

[b] Samples collected from seven livestock farms with a sieve and/or screw press separation system treating pig manure (n = 17).

separation and thus agronomic management should be modified accordingly (Table 2) (Prenafeta-Boldú and Parera, 2020).

Regarding the SF, the organic matter content (between 20% and 30%) provides a readily available carbon source (Camilleri-Rumbau et al., 2021) and, due to its high C/N ratio, it is more appropriate to be used as an organic amendment or in basal dressing; whereas the LF, with a lower C/N ratio (dry matter content of around 2–6%), is more suitable to be applied as a top dressing fertilizer, in fertigation or in soils with high phosphorus content, since the N/P ratio is higher than in the raw manure (Kovačić et al., 2022; Prenafeta-Boldú and Parera, 2020).

In parallel to management as fertilizer products, both fractions should be further processed in order to maximize nutrient recovery and/or to produce bioenergy. The solid fraction can be treated with consolidated technologies such as anaerobic digestion (see Section 2) and composting (see Section 4), or with other biomass valorization processes such as combustion, pelletization or pyrolysis. In addition, since solid-liquid separation does not guarantee high recovery of the nutrients still available in the obtained LF, a further post-treatment of the diluted LF is usually required (Camilleri-Rumbau et al., 2021). In this case, apart from the use of consolidated technologies such as anaerobic digestion or nitrification/denitrification (that cannot be considered a technology to recover but to remove nitrogen), other technologies such as stripping/absorption, struvite precipitation, membrane filtration or bioelectrochemical systems, in many cases combined with each other, can be implemented to maximize nutrient recovery from the LF, mainly in the form of struvite and ammonium sulphate (see Sections 5–8). These treatments result in different products that should be adequately managed taking into account their characteristics and chemical composition.

4 Composting

Composting can be defined as a nature-based solution for transforming solid organic wastes into compost, which is an organic amendment for agricultural soils, gardening and soil remediation (Cáceres et al., 2022); other specific compost products can be used as a growing medium for containerized plant production in certain proportions. The transformation of organic matter and other nutrients during composting is achieved by aerobic microorganisms under controlled conditions. Composting is usually carried out by piling up the solid organic matter (or a mixture with a bulking agent). The pile of organic material (composting pile) can be placed outside, with a cover structure, or composting can be achieved indoors within special tunnels (closed systems).

The primary field/lab parameters that should be measured to monitor the process are: temperature, oxygen concentration within the composting pile and moisture of the material. The temperature of the pile rises, as composting releases heat as a result of aerobic microbial activity (Haug, 1993; Cáceres et al., 2015a). Although oxygen availability is key in composting, water availability is necessary to maintain microbial activity, and optimize organic matter breakdown and compost maturity. Oxygen within the composting matrix decreases due to microbial consumption and the release of CO_2 as a product of respiration; this exhausted air should be renewed by fresh air. Therefore, the actions to manage the process are mainly aeration and irrigation of the material.

There are basically two types of composting systems that can be used in manure composting (i.e. static and dynamic composting) and they are distinguished by the way in which air is provided to the solid material. Air renewal (and also homogenization of the composting matrix) can be achieved by turning the material, and this is the key concept of dynamic composting. A static composting system consists of blowing air from the bottom of the composting piles through pipes connected to a blower (Cáceres et al., 2006). Static systems can be used outdoors, employing a simple configuration; but they are typically used in tunnel systems, which are closed structures where emitted gas can also be collected and treated to avoid the release of undesirable gases into the atmosphere.

Raw materials in livestock composting are mainly: manure from cattle or dairy cows, the solid fraction of pig or cattle slurry, droppings from pig or cattle manure (from organic livestock production), poultry manure and droppings from other animals under livestock production. These products are characterized by a high moisture content and high N and P content. Solid excrement is usually mixed with urine, which is why the manure has a high water content. A high moisture content (50–70%, fresh weight basis) is necessary to trigger the composting process, which is facilitated by a quick temperature rise. Other products, like the solid fraction of cattle or pig slurry, could have a very high moisture content (depending on the physical structure of the material) and should be mixed with bulking agents to diminish the water content of the mixture. When the solid fraction is obtained from slurry, the structure of the material and the moisture content sometimes hinders piling up the material making it necessary to separate the slurry combining two systems (such as gravity and a screw press).

Solid manure is characterized by the presence of bedding materials used for animal welfare (straw or wood-based materials such as sawdust). Bulking agents are normally added to other raw materials during composting; however, for livestock manure, the addition of such materials is not usually necessary, although in some cases it would be convenient in order to increase the C/N ratio and avoid certain ammonia and greenhouse gas production. The emission of such gases are issues of concern, but mitigating this release can be achieved by raising the C/N ratio of the manure or the mixtures by adding bulking agents. At lab level, it has been demonstrated that the addition of biochar can diminish the gases released into the atmosphere (Czekala et al., 2016; Janczak et al., 2017).

The solid fraction of cattle or pig/cattle slurry normally has a lower N content than the raw slurry, which is due to part of the N in the slurry remaining in the liquid portion, after separation. In general, the solid fraction of cattle slurry is less concentrated than any other kind of manure (Table 3).

The chemical and physical characteristics of the solid fraction of cattle or pig slurry can be significantly improved by adding a bulking agent: the C/N increases, the moisture content diminishes, the organic matter increases and the N decreases. On the other hand, the physico-chemical parameters (pH and EC) diminish. Table 3 shows the effect of the bulking agent (pine debris) addition, improving the air capacity of the mixture. The study by Cáceres et al., tested the composting of just separated cattle slurry using dynamic composting, separated cattle slurry using a static method for composting and separated cattle slurry with pine debris also composted dynamically. Results show that the evolution of several chemical parameters and physico-chemical parameters differed, in particular, regarding electrical conductivity, ammonia-N and nitrate-N (the different evolution of this N species had a clear consequence on pH as nitrification naturally drops the pH) (Fig. 4, Cáceres et al., 2006).

Over the last few years, composting has been studied as a process of processes (Cáceres et al., 2018). In particular, the fate of N during composting is of interest due to the enrichment of N in the compost, the loss of this element in the form of NH_3 at the beginning of the process and the loss of N_2O due to other processes during composting (Cáceres et al., 2018). Specifically, it has been demonstrated that nitrification, which is a process that can occur in soil and other liquid treatment processes, is also possible during composting under certain conditions (Cáceres et al., 2006, 2016).

Table 3 Characterization of the solid fraction of cattle slurry (SF-CS) without and with pile debris (PD) at 2/1, v/v rate (Cáceres et al., 2006)

Parameter	Units	SF-CS	SF-CS/PD (2/1, v/v)
Gravimetric moisture	%, fb	82.0	72.1
Organic matter	%, db	92.4	95.5
Norg	%, db	1.4	1.1
C/Norg	-	39	50
pH	-	7.7	7.2
Electrical conductivity	dS/m	0.80	0.7
Total porosity	% vol.	95.9	94.8
Volumetric moisture	% vol.	31.9	21.9
Air capacity	% vol.	64.0	72.9

Adapted from: Cáceres et al. (2006).
db, dry basis; Fb, fresh basis; Norg, organic N.

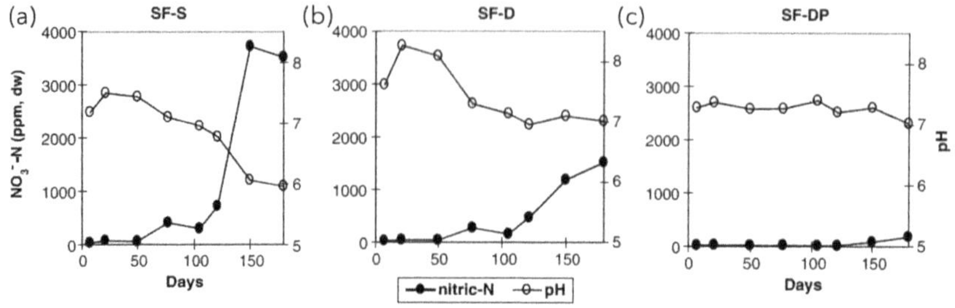

Figure 4 Time course of nitrate-N and pH during the composting of the solid fraction of cattle slurry: (a) solid fraction using a static method (SF-S), (b) solid fraction using a dynamic method (SF-D) and (c) solid fraction with pine debris (2/1, v/v) using a dynamic method. Source: Figures extracted from Cáceres et al. (2006).

Composts based on manure are, in general, rich in plant nutrients but also in other undesirable elements for plant growth, such as sodium which increases the electrical conductivity of the product (Cáceres et al., 2015b). Composts of the solid fraction of pig or cattle slurry are suitable for many applications and the fate of N (i.e. organic N, ammonium N and nitrate N) should be studied for each climate condition (Cáceres et al., 2022), and in vulnerable areas, it is a simple and advisable way to manage slurries. Separators (see Section 3) can be purchased or rented, and the liquid can be applied in cultures near the farms, while the solid fraction can be transformed into compost on the farm (*on farm composting*) and in this way the compost (reduced by 50% in volume or weight with respect to the initial mixture) can be used in other locations. Composting can also be used in combination with anaerobic digestion. The solid part of the digestate can be composted in order, for example, to transform N-ammonia, which is the prominent mineral form in such a by-product, into N-nitrate. Furthermore, organic matter can be transformed even further under aerobic conditions before applying the product (digestate) to the soil.

Varying actions can be applied to promote the composting of different animal manures by using, for instance, alternative bulking agents. The Catalan Government recently published a manual for on-farm composting of manure, which includes the characterization of different types of manures and bulking agents, including aspects such as using the generated leachates or the commercialization of compost (Cáceres and Parera, 2022). Indeed, important amounts of leachates can be generated depending on the management of the composting piles and the climatic conditions. Hence, new approaches to the transformation (via nitrification) of these effluents and their use in horticultural production have been proposed (Cáceres et al., 2015c).

5 Struvite precipitation

Struvite precipitation has been described as the most promising and universal solution for phosphorus recovery from livestock manure and anaerobic digestates. Struvite, or magnesium ammonium phosphate (MAP, $MgNH_4PO_4{\cdot}6H_2O$), is a white crystalline element (Fig. 5) composed of ammonia nitrogen (NH_4^+), phosphorus (phosphate, PO_4^{3}) and magnesium (Mg^{2+}) in equimolar relation. Struvite precipitation is achieved under alkaline conditions when the concentration of Mg^{2+}, NH_4^+ and PO_4^{3-} exceeds the solubility product, according to the following reaction (Cerrillo et al., 2015):

$$Mg^{2+} + NH_4^+ + H_2PO_4^- + 6H_2O \rightarrow MgNH_4PO_4 \cdot 6H_2O + 2H^+$$

Struvite precipitation is dependent on many factors, a principal parameter being pH, as it has a direct effect on the concentration of free ions available for reaction and on struvite solubility. A wide range of optimal pH values have been reported for struvite precipitation in the literature; ranging from 7 to 11 with an optimal at 8.5–9.3 depending, for instance, on the feedstock characteristics and operational conditions (Nagarajan et al., 2023). Moreover, since struvite production also generates H^+, the continuous addition of alkali (generally using NaOH or $Mg(OH)_2$) is needed to keep the pH in the optimal range and to favour struvite precipitation. Besides pH, the N/P and Mg/P molar ratios, the physical characteristics of the reactor (stirring and settling properties), the temperature and the ionic strength of competitive cations such as Ca^{2+} and Na^+, are other factors affecting struvite precipitation (Tao et al., 2016; Cerrillo et al., 2015).

On the other hand, high NH_4^+ concentrations are usually found in anaerobic digestates in comparison with PO_4^{3-}, but as Mg^{2+} is insufficient, supplements such as $Mg(OH)_2$, $MgCl_2$ or $MgSO_4$ have to be added to the reactor in order

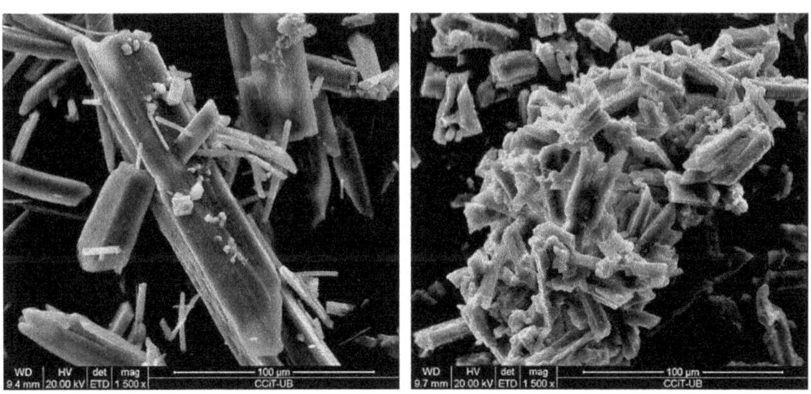

Figure 5 Struvite crystals obtained from pig slurry (left) and synthetic wastewater (right). (Bonmatí ©).

to induce struvite formation and achieve optimal molar ratios (Nagarajan et al., 2023; Lorick et al., 2020; Cerrillo et al., 2015). It should be noted that the co-precipitation caused by Ca^{2+} (as calcium phosphate), which is usually highly concentrated in raw dairy manure, would negatively affect the recovery yield of struvite. In this sense, the highest recovery rates have been reported from studies on the treatment of the liquid fraction of manure-derived digestates after a mechanical solid-liquid separation (Brown et al., 2018). However, in many cases, the major obstacle for P recovery as struvite is due to 55–65% of the total amount of P from manure digestate being concentrated in the solid fraction (see Section 3), so P should be previously recovered in the liquid fraction as soluble inorganic phosphorus. Therefore, it may be necessary to carry out pre-treatment to maximize phosphorus recovery. According to the literature, acid treatment would be the best pre-treatment for the release of phosphorus (Shi et al., 2018).

In a meta-analyses of different livestock treatment technologies, struvite was shown to be the most promising for P recovery, achieving the highest recovery among all the evaluated technologies with a rate of 83.0% (Shi et al., 2022). In this sense, struvite precipitation has been extensively studied and applied to anaerobically digested dairy and pig manure (Wagner and Karthikeyan, 2022; Cerrillo et al., 2021; Tao et al., 2016; Cerrillo et al., 2015) poultry manure (Yetilmezsoy and Sapci-Zengin, 2009), or the liquid fraction of pig slurry (Huang et al., 2011), reporting PO_4^{3-} recovery efficiencies of up to 95%. The typical struvite precipitation reactor consists of a stirred crystallization reactor composed of a reaction zone and a sedimentation zone (Fig. 6). The reactor could also include an aeration system to favour CO_2 stripping and pH increase, and prevent mechanical mixing (Cerrillo et al., 2015). However, other innovative approaches for P recovery as struvite, such as bioelectrochemical and membrane systems, with different configurations, have been gaining attention over the last few years (Cerrillo et al., 2023; Cerrillo et al., 2021).

In relation to its fertilizer quality, struvite is considered to be an effective and clean slow-release fertilizer, as it is sparingly soluble under neutral and alkaline conditions but readily soluble in acid (Carreras-Sempere et al., 2021; Cerrillo et al., 2021). Nagarajan et al. (2023) reported that 1 kg of MAP per day is enough to fertilize 2.6 hectares of arable land. The value of struvite as fertilizer has only recently been understood and it is now the focus of increasing research attention (Robles-Aguilar et al., 2019, 2020). The advantages of struvite as a solid fertilizer include ease of application, low transportation costs, reduced bulk storage, and the avoidance of odour and pathogen contamination issues (Tao et al., 2016). In addition, P recovery in the form of struvite reduces the use of P rock-based mineral fertilizers, allowing the release of nutrients at an appropriate rate for crop uptake due to its low solubility, while reducing nutrient run-off and leaching (Nkoa, 2014). Struvite has also been applied to nutrient

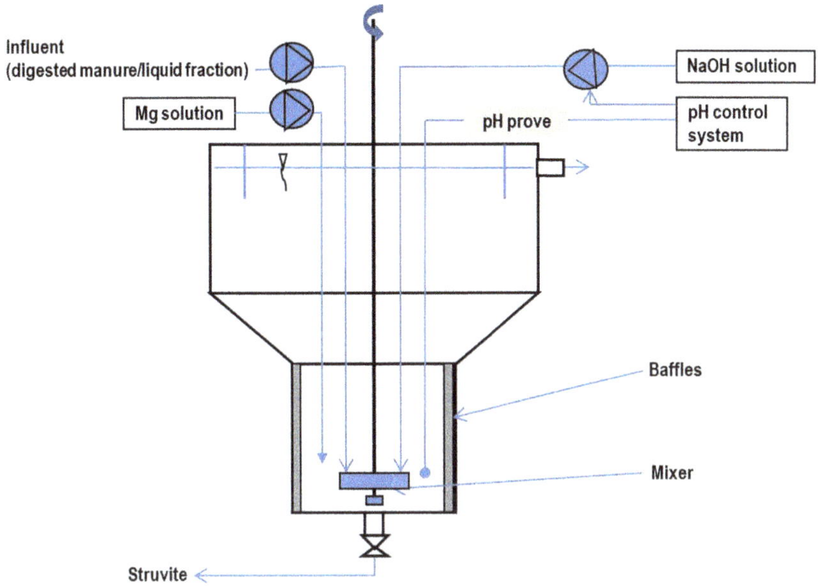

Figure 6 Schematic diagram of a crystallization lab-scale reactor. Source: Cerrillo et al. (2015).

solutions in horticulture (without an additional P source), attaining similar plant performance (Carreras-Sempere et al., 2021, 2022).

Struvite precipitation has been implemented in more than 40 industrial plants around the world (Rizzioli et al., 2023). However, this process is typically applied for wastewater and municipal sludge digestate (Pikaar et al., 2022; Sena and Hicks, 2018), with some commercial full-scale systems already on the market (Ostara Pearl®, NuReSys®, PHOSPAQ™, Airprex®) (Rizzioli et al., 2023; Bonmatí-Blasi et al., 2020). However, there is limited information for agricultural digestate, due to the low P content, which has a direct bearing on process efficiency. In addition, P recovery from agricultural residues is not often applied as there is not a specific regulation focused on this nutrient, as is the case for N by means of the Nitrate Directive. However, since agricultural residues are considered to be a critical raw material, some countries such as The Netherlands, Denmark, Belgium and Germany have already established special authorization for the use of struvite recovered from manure as a fertilizer (Corona et al., 2021). An example of a prototype demonstration in an operational environment (TRL7) is the REVAWASTE® system (CARTIF company, Spain), at demonstrative scale, that treats agricultural residues using the AD process, including a crystallization unit that allows nutrient recovery of up to 95.4% (Rizzioli et al., 2023).

6 Stripping/scrubbing

Sigurnjak et al. (2019), described the possible pathways to recover ammonia via stripping-scrubbing at full scale in Europe. Among the different possibilities, we will discuss the so-called end-of-pipe pathway, where livestock manure is firstly separated into a liquid and a solid fraction and, since ammonia represents the major nitrogen compound of the liquid fraction, this fraction is further treated in a stripping-scrubbing unit to produce ammonium sulphate or ammonium nitrate (Rizzioli et al., 2023).

Ammonia stripping can be achieved using air, steam or biogas to separate the gaseous NH_3 from the liquid phase (Adghim et al., 2023). The NH_3-saturated stripping gas subsequently comes into contact with an acid solution, usually against the flow, to capture ammonia (Fig. 7). When sulphuric acid is used, a $(NH_4)_2SO_4$ solution is formed. Nitric acid can also be used to capture ammonia that will produce NH_4NO_3, which is more interesting for fertilization, but is more expensive (Bonmatí et al., 2020).

The amount of ammonia that can be stripped from liquid waste or absorbed in the acidic solution is based on two thermodynamic equilibria, which are pH and temperature dependent (Bonmatí-Blasi et al., 2020):

- Ammonia dissociation equilibrium in the liquid: $NH_{3\,(l)} \leftrightarrow NH_3\,(g)$; and
- Ammonia gas/liquid equilibrium: $NH_4^+ \leftrightarrow NH_3 + H^+$.

For instance, at room temperature (about 20°C), ammonia stripping requires a high pH to move the ammonia equilibrium towards free ammonia. However, if the temperature increases, a lower pH is needed. As previously mentioned, pH

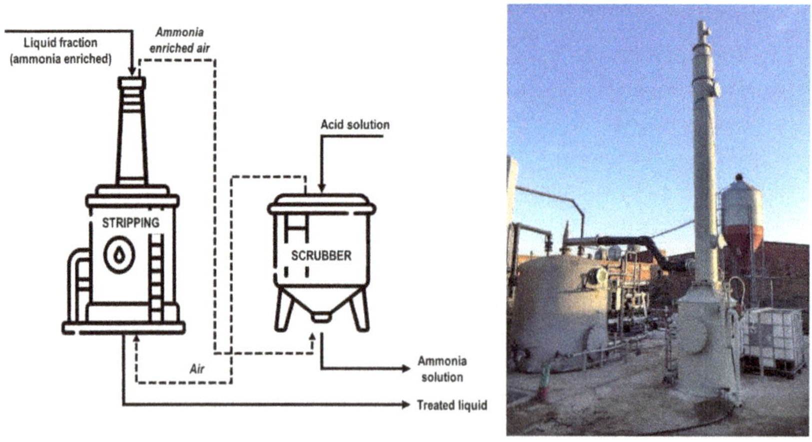

Figure 7 (left) Schematic diagram of the stripping/scrubbing process and (right) image of equipment installed at an existing biogas plant in Catalonia, Spain (own picture).

is usually set between 9 and 10 by means of base addition (lime is frequently used due to its low cost) or previous CO_2 stripping. Nevertheless, if air stripping is performed at high temperature, the high manure buffering capacity could enable maintaining the pH at the required value, and the amount of alkali could be reduced. On the other hand, pH values below 5 highly favour ammonia scrubbing, with almost no temperature effect (Laureni et al., 2013; Bonmatí and Flotats, 2003). Thus, one of the main limiting factors for ammonia air stripping, which is usually performed at high temperature, is the availability of an economical thermal energy source. In this regard, by combining anaerobic digestion and a stripping/absorption process, the biogas generated could provide the input thermal energy needed for stripping at high temperature (Shi et al., 2018; Bonmatí and Flotats, 2003).

In terms of efficiency, up to 95% of ammonia reduction can be achieved by setting optimal conditions and almost complete ammonia recovery is possible via absorption in acid solutions with only a little acid stoichiometric excess (Bonmatí-Blasi et al., 2020). Shi et al. (2018) compared several studies on NH_3 stripping from pig manure, dairy manure, chicken manure and food waste. Among the different conditions and feedstocks, the higher efficiencies (greater than 90%) were achieved when treating pig manure and keeping the pH above 9.5. Moreover, when combined with prior anaerobic digestion, the performance of the stripping process can be altered, as the biological process increases the ammonia content, part of the organic nitrogen is hydrolyzed to ammonia, which slightly increases the pH and reduces alkali requirement (Bonmatí-Blasi et al., 2020). Bonmatí and Flotats (2003) studied the differences between the stripping performance when using digested or fresh pig manure. For fresh slurry, despite working at 80°C, a high initial pH (11.5) was required to achieve ammonia removal at approximately 63%. On the other hand, for digested pig manure, 90% of ammonia was removed without pH modification and there was significantly less organic matter contaminating the recovered ammonia salt. It was also shown that substrates characterized by low organic matter content resulted in ammonia stripping efficiencies greater than 80% (Laureni et al., 2013).

The ammonium salt produced after the scrubber can be used as a biobased fertilizer, thus representing a valid alternative to fossil-based fertilizers, as it contains total N entirely in mineral form, similar to synthetic mineral N fertilizers, produced via the Haber–Bosch process. However, the high variability of N concentrations in these products is still the biggest challenge for their recognition as N fertilizers (Sigurnjak et al., 2019). As mentioned previously, the current European legislative framework prohibits application of these end-products above 170 kg N ha^{-1} y^{-1}, which reduces the market opportunities for the installation owners and hinders the development of a circular economy. Fortunately, the new Fertilizing Product Regulation 1009/2019 and the technical proposal of manure-derived RENURE criteria could open new alternatives for the use of NPK biobased fertilizers (Brienza et al., 2021).

Brienza et al. (2021), analysed the performance of vacuum NH_3 stripping and a scrubbing system that allowed the recovery of 57% of ammonia present in the digestate, in the form of an ammonium sulphate solution. The characteristics of the recovered product (22% solution) would fit the proposed RENURE criteria thus enabling the reuse of N derived from manure as inorganic biobased N fertilizer in the European market to replace mineral N fertilizer (Sigurnjak et al., 2019). However, future studies should further investigate the agronomical and environmental performance of ammonium nitrate and ammonium sulphate under varying conditions and over multi-year crop rotations.

Some examples of commercially available stripping technologies are the AMFER® stripping system (Colsen industry, The Netherlands), the Biogas Bree system (Bree, Belgium), the 'Detricon' (Gistel, Belgium) and the 'NITROStripp' system (Chiari, Italy). The operational costs associated with this process range between €2 and 7/kg N removed, primarily depending on the reagent used for pH regulation, the temperature and energy consumption (Rizzioli et al., 2023).

7 Membrane filtration

Membrane filtration is a physical process in which a membrane acts as a filter thus separating a liquid stream into a solid fraction, also known as concentrate or retentate, and into a liquid fraction, called a permeate, that are not thermally, chemically or biologically modified. Membrane filtration is usually applied to the manure or digestate liquid fraction after mechanical separation (Rizzioli et al., 2023). However, depending on the separation system, a second separation process is sometimes needed in order to avoid damaging the membrane modules due to the high solid content of the liquid fraction (see Section 3). Membrane filtration has gained interest in recent years in the agri-food sector, especially in surplus nutrient areas with high livestock density, as it concentrates manure nutrients in small volumes that could be exported as fertilizers to other agricultural regions.

Membranes are typically classified depending on their pore size, applied pressure, molecular weight cut-off and permeability. These mainly include hydraulic pressure membrane processes such as microfiltration (MF; pore size: 0.1–5 μm), ultrafiltration (UF; pore size: 0.005–0.1 μm), nanofiltration (NF; pore size: 0.001–0.01 μm), reverse osmosis (RO; pore size: 0.0001–0.001 μm), and other emerging membrane technologies such as forward osmosis (FO), membrane distillation (MD) and electrodialysis (ED) (Pinnekamp and Friedrich, 2006).

Masse et al. (2007) performed a critical review of the current state of research on the use of membrane filtration for manure concentration and treatment. MF and UF membranes basically act as very efficient solid-liquid separators that can isolate nutrients associated with particles such as phosphorus. Furthermore, MF or UF membranes can successfully eliminate particles from digestate to achieve over 80% removal of organic matter, when combined with paper filtration, and UF up to 95.7% of suspended solids can be removed from digested swine manure (Zhang et al., 2020).

Ammonia and potassium retention requires NF or RO and represents one of the major challenges of using RO filtration for animal manure treatment. Konieczny et al. (2011) recovered water from pig slurry by implementing an integrated system consisting of centrifugation followed by UF and NF. Guo and Jin (2014) investigated anaerobically digested cattle manure treatments with RO pre-treated with UF, the concentration of NH_4^+ in the final permeate was around 63 mg/L from the initial 836 mg/L. Rejection of other species such as K^+, Na^+ and Cl^- was observed to be as high as 90%.

Table 4 shows a summary of these membrane processes and operational characteristics.

However, depending on the type of membrane used, there are several technical challenges for the application of membrane processes at an industrial level, such as membrane fouling, contaminant enrichment, ammonia volatilization and high maintenance costs, (Zhang et al., 2020). Fouling seems to be one of the main problems affecting NF and RO since it impairs membrane performance and shortens membrane lifetime, thereby restraining the productivity of nutrient recovery, due to the high concentrations of various foulants, such as humic acids, suspended solids, colloids and inorganic salts, which are nutrients in digested effluent (Zhang et al., 2020; Velthof, 2011). Thus, the maximum amount of solids and organic matter should be removed from the liquid fraction prior the application of RO by applying pre-treatments that include techniques such as UF, dissolved air flotation, low pressure membrane filtration and manure/digestate coagulation-flocculation, among others (Zhang et al., 2020; Masse et al., 2007).

Figure 8 shows the flowchart of a pig manure processing plant located in The Netherlands, which includes anaerobic co-digestion, solid-liquid separation and further treatment of the liquid fraction by ultrafiltration, followed by reverse osmosis trying to overcome RO membrane fouling.

However, these pre-treatment techniques can reduce the nutrient content in the digestate, and subsequent nutrient concentration, and require additional investment for residual treatment. In this sense, forward osmosis, membrane distillation and electrodialysis are three emerging membrane-based processes considered suitable to overcome the challenges of nutrient recovery, which could potentially represent a paradigm shift in nutrient management (Zhang et al., 2020). In FO, the driving force of the process is the chemical potential difference across the membrane, instead of hydraulic pressure, which forces

Table 4 Membrane process classification and operational conditions

Membrane process	Pore size (μm)	Size of the particles, colloids or molecules to be separated	Driving force (pressure difference, bar)	Application
Microfiltration	0.1–5	Solids > 0.1 μm	0.1–3	Separation of solid matter from suspensions
Ultrafiltration	0.005–0.1	200 000–20 000 D[a]	0.5–10	Separation of macromolecules or colloids, disinfection
Nanofiltration	0.001–0.01	20 000–200 D[a]	2–40	Separation of dissolved organic molecules and polyvalent inorganic ions
Reverse osmosis	0.0001–0.001	<200 D*[a]	5–70 Up to 120 in special cases	Separation of organic molecules and all ions

[a]Dalton, numerically equivalent to the molecular weight in g/mol (modified from Pinnekamp and Friedrich, 2006)

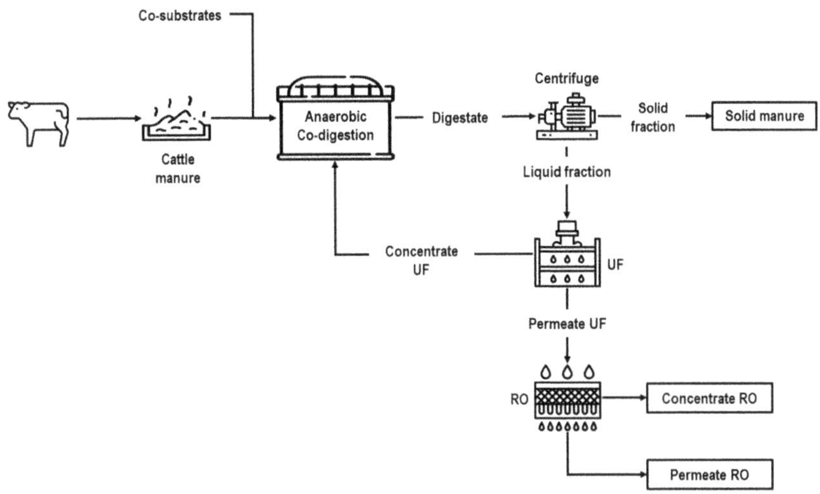

Figure 8 Treatment scheme of a manure treatment plant located in The Netherlands. Source: Modified from Velthof, 2011. https://edepot.wur.nl/192069.

the liquid to flow from the low to high concentrated compartment. FO has been demonstrated to be less sensible and more reversible to fouling than RO (Blandin et al., 2016).

Using hydrophobic membranes for MD, the driving force is based on the volatility and vapour pressure of the different compounds that are to be separated. Since the process is driven by temperature gradients, residual heat can be used, and the use of expensive pressure pumps is unnecessary. Recently, hydrophobic membranes permeable to gases have been applied to ammonia recovery from manure with recovery efficiencies of more than 97%. Furthermore, hydrophobic membranes can be combined with other treatment technologies such as anaerobic digestion (see Section 2) to increase biogas production, or with bioelectrochemical systems to recover ammonia (see Section 8) (Cerrillo et al., 2023; Cerrillo et al., 2021).

Finally, ED is an electro-membrane process in which ions are transported through an ion exchange membrane under the influence of a potential gradient. In ED, the direct current field forces the migration of cations and anions toward cathodes and anodes, respectively (Zhang et al., 2020). The ion separation is achieved via ion-exchange membranes that include cation-selective, anion-selective and bipolar membranes that serve as a barrier to fraction these ions (Zhang et al., 2020). Therefore, the excess nitrogen and phosphorus, which are found as ions in nutrient-rich streams, such as swine liquid fraction, can be removed using this technique (Fukumoto and Haga, 2004).

8 Bioelectrochemical systems

Bioelectrochemical systems (BESs) are an innovative and highly versatile technology for the treatment of livestock manure to obtain biofertilizers and recover bioenergy. These devices are bioreactors that include an anode and a cathode, and that use microorganisms to catalyse oxidation and/or reduction reactions. Several organic substrates have been used in BESs, e.g. livestock manure or digested livestock manure (Cerrillo et al., 2016). The organic matter of the substrate is oxidized by exoelectrogenic bacteria that grow on a biofilm on the bioanode, and the produced electrons circulate through an external electrical circuit to the cathode.

BESs have different applications in the field of livestock manure treatment. First, electron movement through the external circuit produces electricity, which can be used directly to power small devices via the chemical energy contained in the organic substrate as an energy source, and has even been successfully boosted up to 99 V (Koffi and Okabe, 2020). In this case, the BES is called a microbial fuel cell (MFC).

Second, the movement of electrons can also be used to force the migration of ions through a membrane in a dual-chambered MFC. In this case, an anode compartment and a cathode compartment are divided by a cation or an anion exchange membrane. This way, the electron circulation from the anode to the cathode generates a charge imbalance in the cell, which is compensated by ion migration through the membrane, from one compartment to the other. Using a cationic exchange membrane in a dual-chambered BES, ammonium can be recovered from livestock manure, which is introduced into the anode compartment, in a clean solution placed in the cathode compartment (Sotres et al., 2015). The livestock manure treated in the anode compartment will have a lower organic and nitrogen content, while the cathodic solution, rich in ammonium, can be used as a fertilizer. Since the migration of ammonia is related to the amount of charge that is circulating between the anode and the cathode, ammonium recovery can be boosted with the aid of a small energy input, achieving a flux through the membrane of up to 119 gN m^{-2} day^{-1} (Rodríguez Arredondo et al., 2019). In a recent review, it has been reported that the energy consumption for ammonia recovery in BESs is in the range of 1.17–2.7 kWh kg_N^{-1}, generally only considering the external energy supplied to enhance ammonium migration (Cerrillo et al., 2023).

BESs have evolved for the simultaneous recovery of ammonium and phosphate, in the form of a concentrated solution or via salt precipitation as struvite, using a combination of different types of membranes or number of chambers (Fig. 9) (Haddadi et al., 2014; Cerrillo et al., 2021b, 2023; Cusick et al., 2014). For the recovery of phosphorus from livestock manure, a first step may involve solubilization from ferrous phosphate mineral, in order to allow

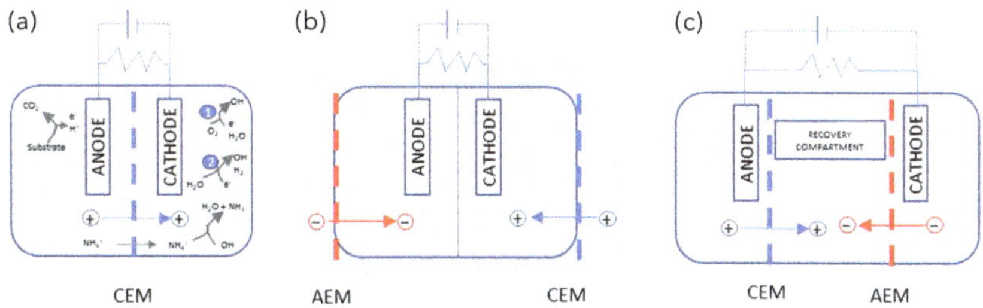

Figure 9 Three examples of typical configurations of BESs for nutrient recovery. + and – stand for ions charged positively (i.e. NH_4^+) and negatively (i.e. PO_4^{3-}), respectively. (a) Dual-chambered BESs, (b) submersed BESs and (c) three-chambered cell with recovery compartment. The main reactions taking place in BESs are shown in (a) by way of example, where ❶ and ❷ stand for MFC and MEC cathodic reactions, respectively (Cerrillo et al., 2023).

migration in the form of phosphate through the anionic membranes (Cerrillo et al., 2021b).

Third, by applying a small external voltage to BESs, certain non-spontaneous reactions can be favoured, which is the case for the production of biohydrogen or biomethane in microbial electrolysis cells (MECs). Hydrogen is produced via the combination of electrons that travel to the cathode with protons (eq. 1) (Cheng and Logan, 2011), while for the production of methane, carbon dioxide has to be introduced into the biocathode compartment, where bioelectromethanogenesis reactions occur (Zeppilli et al., 2020). Methane obtained from CO_2 in BESs can be achieved via two different mechanisms (Villano et al., 2011): (i) indirectly, through the microbially catalysed and/or intermediate abiotic electrochemical production of hydrogen in the cathodic compartment (Equations 1and 2) or (ii) directly, via reducing CO_2 to methane by taking the necessary electrons from the cathode (eq. 3).

$$2H^+ + 2e^- \rightarrow H_2 \,(1)$$
$$4H_2 + CO_2 \rightarrow CH_4 + 2H_2O \,(2)$$
$$CO_2 + 8H^+ + 8e^- \rightarrow CH_4 + 2H_2O \,(3)$$

Bioelectromethanogenesis application in BESs is of great interest for bioenergy production, as biomethane can be produced from carbon dioxide captured from different sources while treating livestock manure. One possible source of carbon dioxide is biogas obtained from anaerobic digestion, so MECs can be applied to biogas upgrading, increasing biomethane production thanks to the conversion of carbon dioxide, which differs from conventional biogas-upgrading technologies. A tubular MEC of 12 L, continuously fed by a gas mixture composed of CO_2 at 30% and N_2 at 70% to simulate the CO_2 content of a biogas, has been reported to achieve a methane production of 1.25 L day^{-1}. This study reported a low energy consumption of 0.33 kWh m^{-3} CO_2, resulting in a lower value than commercially available technologies on the market, such as 0.8 kWh m^3 CO_2 for water scrubbing for biogas upgrading (Zeppilli et al., 2020).

The most important advantage of BESs is that they can be used for a combined purpose, such as biomethane production while simultaneously recovering ammonia (Zeppilli et al., 2017, 2019; Cerrillo et al., 2021a), or hydrogen production while recovering struvite (Song et al., 2021). Furthermore, it is a technology with great possibilities when combined with AD, either to recover nutrients or to increase biogas production. BESs can be submersed in anaerobic digesters (Zhang and Angelidaki, 2015), become a post-treatment for digestate (Cerrillo et al., 2016a), or be placed in a recirculation loop of the anaerobic digester to increase the stability of the system and biogas production (Cerrillo et al., 2016b). A recovery of 40% of the nitrogen in pig slurry in a combined AD-MEC system and an increase of 55% in methane production when the MEC was integrated in a recirculation loop in the AD has been reported (Cerrillo et al., 2016b).

Despite the great versatility and potential of BESs for biofertilizers and bioenergy recovery, most of the experience to date has been at lab or small pilot scale. More research is needed to improve nutrient and energy recovery efficiencies of BESs, with a low construction cost, in order to boost scaling up and commercialization of this technology.

9 Case study: farm for the future

Figure 10 shows an example of a circular approach based on a mixed farming system (ruminant production + fodder crop production) to be implemented in cattle farms, which includes not only some of the technologies described in this chapter, such as anaerobic digestion, solid/liquid separation or composting, but also strategies to increase biogas production and agronomical practices that allow reducing the environmental impact derived from the use of manure-based fertilizers in croplands.

The case study was built up in the framework of a LIFE project called '*Farms for the future: innovation for sustainable manure management from farm to soil* (LIFE12 ENV/ES/000647)'. The main objective of this study was to investigate the nutrient extraction efficiency from soil of three different catch crop (CC) species, as well as the viability of using them as co-substrates for anaerobic co-digestion with dairy manure, with emphasis on the effect of ensiling on the anaerobic biodegradability and biogas potential. Digestate from the anaerobic co-digestion was used to fertilize the main crop, thus closing nutrient cycles. The study was published in the *Waste Management Journal*, and is available in open access (Riau et al., 2021).

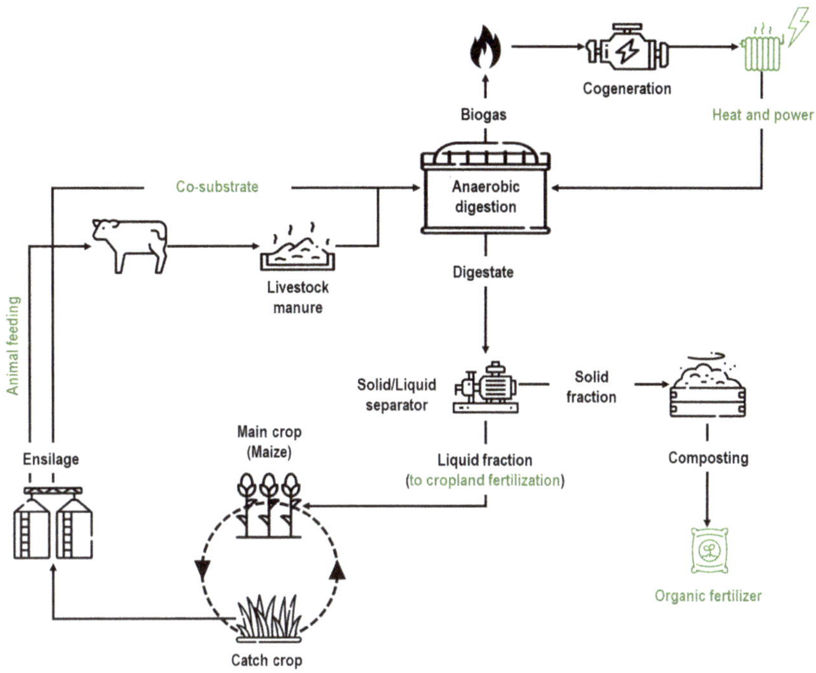

Figure 10 Circular approach in a cattle farm. Source: Riau et al. (2021).

CCs can be grown in the period between two main crops that have been fertilized with manure, anaerobic digestates and/or its liquid fraction, as a strategy to retain soil mineral N as well as P and heavy metals, thus reducing leaching and run-off losses and improving soil quality, by adding organic matter and reducing fertilizer requirements of the following growing season.

In this study, three CCs species were tested: *Lolium multiflorum* (ryegrass), *Brassica napus* (forage rape) and *Avena strigosa* (black oat). The CCs were sown once the main crop was harvested, by the end of September or the beginning of October, and were grown with minimum cultivation and harvesting effort as well as without using fertilizers and irrigation water. Finally, CCs were harvested in March. This rotation was repeated during three consecutive years. As an example, the experimental design of one rotation is summarized in Table 5. As a control of the process, a plot without CCs was also tested.

Despite spontaneous herbage growth, N uptake from soil via catch crops was always significantly higher (42–264%) than those cases in which CC was not sown, although highly variable among years. Thus, CC may reduce the risk of nitrate leaching from soil. Over different years, the highest N and P extractions were achieved by forage rape. On the other hand, it is known that anaerobic digestion does not affect the removal of dissolved Cu and Zn in dairy manure. However, Cu and Zn bioavailability is increased during the process, thus

Table 5 Experimental design of field trials with maize and catch crop rotations (plots: 131 m^2; three repetitions each; three consecutive seasons)

Maize fertilization (N dosage = 170 kg N ha^{-1})	Year 1			Year 2
	Ja F M	A M Jn J A	S O N D	Ja F
	Catch crop	Main crop	Catch crop	
Liquid fraction of digested manure	Black oat	Maize	Black oat	
Liquid fraction of digested manure	Ryegrass	Maize	Ryegrass	
Liquid fraction of digested manure	Forage rape	Maize	Forage rape	
Liquid fraction of digested manure	-	Maize	-	

making anaerobic digestate less favourable for its application in agriculture. Nevertheless, this could be solved by introducing CCs into crop rotation. Zn and Cu uptake from all catch crops studied were always significantly higher than those obtained by spontaneous herbages (27–197%).

As mentioned in Section 1, once harvested, CC biomass can be processed as a co-substrate in the anaerobic digestion process, together with livestock manure, in order to increase biogas production since, on one side, manure provides buffering capacity and a wide range of nutrients, while on the other, CCs, with high carbon content, balance the carbon to nitrogen (C/N) ratio of the feedstock, thereby decreasing the risk of ammonia inhibition. However, since CCs are usually seasonally sown (autumn) and harvested (spring), it is necessary to find a storage method that guarantees the availability of this feedstock for biogas production during the whole year. This is why CCs were ensiled after harvesting, as it has been demonstrated that ensiling can maintain the energy content of crops (over 90%), ensuring good nutritional value when used as animal feed. Moreover, during ensiling, the different biochemical processes, such as hydrolysis and acidification, directly or indirectly affect biogas production by altering the feedstock properties. A faster pH decrease produces more water-soluble carbohydrates in the silage mass resulting in more biogas. Therefore, ensiling not only allows maintaining crop characteristics but, as a feedstock for biogas production, increases the anaerobic biodegradability of lignocellulosic materials thus improving process efficiency. In fact, the anaerobic biodegradability of the ensiled ryegrass, forage rape and black oat was increased by 10%, 14% and 10%, respectively, which resulted in an increase of the methane yield by 40%, 46% and 50%, in terms of LCH_4 per tonne of waste and 19%, 25% and 27%, in terms of methane production per volatile solid added.

Table 6 Organic loading rate, operational conditions and methane yields obtained at steady-state conditions for each semi-continuous experiment

		Ryegrass		Forage rape		Black oat	
	Units	Control	Co-dig	Control	Co-dig	Control	Co-dig
OLR	kgCOD $m^{-3} \cdot day^{-1}$	1.8 ± 0.1	2.4 ± 0.4	1.2 ± 0.2	1.4 ± 0.4	1.3 ± 0.2	1.9 ± 0.4
VS removal	%	40 ± 19	56 ± 7	42 ± 15	57 ± 10	52 ± 10	57 ± 6
Methane content	%	61 ± 15	63 ± 0	59 ± 5	60 ± 2	65 ± 1	63 ± 1
Methane yield	$m^3\ CH_4\ t^{-1}_{waste}$	8.8 ± 2.6	12.6 ± 1.0	8.1 ± 2.0	10.9 ± 3.0	8.7 ± 1.2	12.9 ± 2.8
Average productivity	$m^3\ CH_4\ m^{-3} \cdot day^{-1}$	0.25 ± 0.04	0.33 ± 0.07	0.22 ± 0.04	0.26 ± 0.07	0.25 ± 0.04	0.38 ± 0.07

Mean value ± standard deviation.

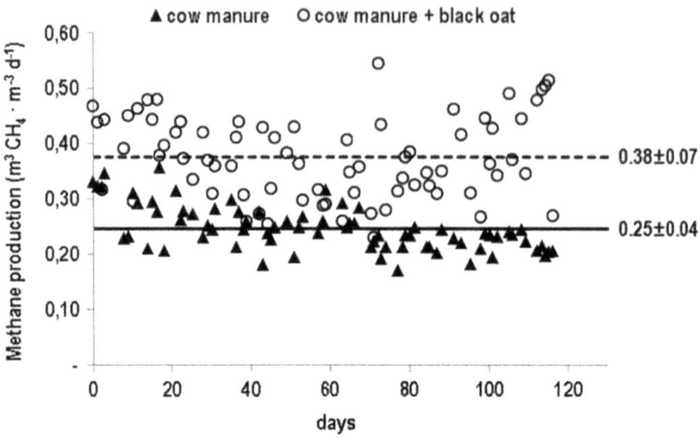

Figure 11 Evolution of volumetric methane production, expressed as m^3 CH_4 m^{-3} day^{-1} at standard temperature and pressure in both the control and the co-digestion reactors (black oat as a co-substrate). Numbers and lines represent mean values ± standard deviations. Source: Riau et al. (2021).

Hence, anaerobic co-digestion with livestock manure was carried out using the ensiled CCs as co-substrates and was compared with a single anaerobic digestion of manure as a control process. Table 6 summarizes the main control parameters of the anaerobic co-digestion process for the three CCs tested. As it can be seen, using CCs as co-substrates increased the organic content in the influent and therefore, the organic loading rate (ORL) rose to between 17% and 46%, while the hydraulic retention time remained constant at 40 days. Generally, the overall process performance was enhanced; for ryegrass, methane yield increased by 43% when increasing the organic load rate by 33%, in comparison with the control digester, whereas organic matter degradation was improved by 16% in terms of volatile solid (VS) removal. Similarly, co-digestion using forage rape as a co-substrate showed a 35% increase of methane yield and the organic matter degradation increased by 15% in terms of VS added. However, it must be taken into account that OLR only increased by 17% in this case. With regards to co-digestion when using black oat, methane yield was increased by 48% while VS removal was only 5%. In this case, the OLR increased significantly (46%) in comparison with the other experiments.

As shown in Fig. 11, using black oat, for instance, the volumetric methane production over time was always higher than in the control digester. Calculations in terms of electric self-sufficiency made on the biogas plant, which provided the liquid fraction for fertilization and the inoculum for the anaerobic digestion tests, foresee an increase in energy production by 42% (from 3.3 to 4.7 mill kWh/a) when CCs are introduced as co-substrates, in comparison to the current plant operation.

10 Conclusion and future trends

The recovery of biobased fertilizers and bioenergy from animal manure to partially replace synthetic mineral fertilizers and fossil fuels should be considered as a key strategy to move towards more sustainable and resilient agriculture and livestock production. This has driven the development of numerous technologies over the last few decades. Some technologies have already been consolidated and widely implemented, such as anaerobic digestion or composting, for bioenergy and biofertilizer production, respectively, but they can still be optimized in a particular context and included as relevant technologies in a biorefinery approach. Other well-known technologies, like solid/liquid separation, maximize nutrient recovery and improve further treatment of the derived fractions, but its efficiency is highly variable and dependent on many factors.

Treatment of the liquid fraction to produce biobased inorganic fertilizers, such as struvite or ammonium sulphate, have gained interest in recent years and innovative technologies have been developed that are able to generate products with similar characteristics to those of mineral fertilizers. As a source of bioenergy, apart from biogas production, thermochemical manure treatments such as gasification to produce syngas, combustion to generate heat and electricity, or hydrothermal liquefaction/carbonization that can directly convert wet manure into bio-oil or hydrochar, have been developed to obtain manure-based bioenergy. However, the operation on an industrial scale is still quite limited and further optimization studies should be performed to overcome the different limiting factors hampering the application of these innovative technologies.

The technologies for nutrient and bioenergy recovery from manure should not be considered as single treatments but as a combination of processes, ideally in a biorefinery framework and with a circular approach, to close nutrient and energy cycling as much as possible along the manure treatment chain.

11 Where to look for further information

11.1 Further reading

- Nutrient Recovery and Reuse (NRR) in European agriculture. A review of the issues, opportunities, and actions (2016). RISE Foundation. https://risefoundation.eu/wp-content/uploads/2020/07/2016_RISE_NRR_Exec_Summary_EN.pdf.
- Schoumans, O. F., Bouraoui, F., Kabbe, C., Oenema, O. and van Dijk, K. C. Phosphorus management in Europe in a changing world. Ambio 2015. https://doi.org/10.1007/s13280-014-0613-9.

11.2 Websites

- European Biogas Association. https://www.europeanbiogas.eu/.
- Eurostat. https://ec.europa.eu/eurostat.
- EIP-Agri. https://ec.europa.eu/eip/agriculture/en/node.
- EIP-AGRI. Focus group on Circular Horticulture https://eu-cap-network.ec.europa.eu/sites/default/files/publication/2023-05/eip-agri_fg_circular_horticulture_final_report_2019_en.pdf.

11.3 Major international research projects

- Circular Agronomics. https://www.circularagronomics.eu/.
- Nutri2cycle. https://www.nutri2cycle.eu/.
- Fertimanure. https://www.fertimanure.eu/.
- Lex4bio. https://www.lex4bio.eu/.
- LIFE Agriclose: https://agriclose.eu/en/.
- Organic Plus: https://organic-plus.net/.

12 References

Aboudi, K., Álvarez-Gallego, C. J. and Romero-García, L. I. (2016). Biomethanization of sugar beet byproduct by semi-continuous single digestion and co-digestion with cow manure. *Bioresour. Technol.* 200, 311–319. https://doi.org/10.1016/j.biortech.2015.10.051.

Awasthi, S. K., Kumar, M., Sarsaiya, S., Ahluwalia, V., Chen, H., Kaur, G., Sirohi, R., Sindhu, R., Binod, P., Pandey, A., Rathour, R., Kumar, S., Singh, L., Zhang, Z., Taherzadeh, M. J. and Awasthi, M. K. (2022). Multi-criteria research lines on livestock manure biorefinery development towards a circular economy: from the perspective of a life cycle assessment and business models strategies. *J. Cleaner Prod.* 341. https://doi.org/10.1016/j.jclepro.2022.130862.

Blandin, G., Vervoort, H., Le-Clech, P. and Verliefde, A. R. D. (2016). Fouling and cleaning of high permeability forward osmosis membranes. *J. Water Process Eng.* 9, 161–169. https://doi.org/10.1016/J.JWPE.2015.12.007.

Beggio, G., Peng, W., Lü, F., Cerasaro, A., Bonato, T. and Pivato, A. (2022). Chemically enhanced solid–liquid separation of digestate: suspended solids removal and effects on environmental quality of separated fractions. *Waste Biomass Valor.* 13(2), 1029–1041. https://doi.org/10.1007/s12649-021-01591-y.

Bonmatí-Blasi, A., Cerrillo-Moreno, M. and Riau-Arenas, V. (2020). Systems based on physical-chemical processes: nutrient recovery for cycle closure. In: I. Management Association (Ed.). *Waste Management: Concepts, Methodologies, Tools, and Applications* (pp. 526–558). IGI Global. https://doi.org/10.4018/978-1-7998-1210-4.ch026.

Brienza, C., Sigurnjak, I., Meier, T., Michels, E., Adani, F., Schoumans, O., Vaneeckhaute, C. and Meers, E. (2021). Techno-economic assessment at full scale of a biogas refinery plant receiving nitrogen rich feedstock and producing renewable energy and biobased fertilizers. *J. Cleaner Prod.* 308, 127408. https://doi.org/10.1016/j.jclepro.2021.127408.

Brown, K., Harrison, J. and Bowers, K. (2018). Struvite precipitation from anaerobically digested dairy manure. *Water Air Soil Pollut.* 229(7). https://doi.org/10.1007/s11270-018-3855-5.

Budzianowski, W. M. (2016). A review of potential innovations for production, conditioning and utilization of biogas with multiple-criteria assessment. *Renew. Sustain. Energy Rev.* 54, 1148–1171. https://doi.org/10.1016/j.rser.2015.10.054.

Bumharter, C., Bolonio, D., Amez, I., García Martínez, M. J. and Ortega, M. F. (2023). New opportunities for the European Biogas industry: a review on current installation development, production potentials and yield improvements for manure and agricultural waste mixtures. *J. Cleaner Prod.* 388, 135867. https://doi.org/10.1016/j.jclepro.2023.135867.

Cáceres, R., Flotats, X. and Marfà, O. (2006). Changes in the chemical and physicochemical properties of the solid fraction of cattle slurry during composting using different aeration strategies. *Waste Manage.* 26(10), 1081–1091. https://doi.org/10.1016/j.wasman.2005.06.013.

Cáceres, R., Coromina, N., Malińska, K. and Marfà, O. (2015a). Evolution of process control parameters during extended co-composting of green waste and solid fraction of cattle slurry to obtain growing media. *Bioresour. Technol.* 179, 398–406. https://doi.org/10.1016/j.biortech.2014.12.051.

Cáceres, R., Cunill, C. and Marfà, O. (2015b). Composts que es produeixen a Catalunya: caracterització I viabilitat del seu ús com a substrat. *Quad. AGRARIS ICEA* 39, 7–34.

Cáceres, R., Magri, A. and Marfà, O. (2015c). Nitrification of leachates from manure composting under field conditions and their use in horticulture. *Waste Manage.* 44, 72–81. http://doi.org/10.1016/j.wasman.2015.07.039.

Cáceres, R., Malińska, K. and Marfà, O. (2018). Nitrification within composting: a review. *Waste Manage.* 72, 119–137. https://doi.org/10.1016/j.wasman.2017.10.049.

Cáceres, R., Biel, C. and Ortiz, C. (2022). Composting as a nature-based solution for coping with circularity challenges in horticulture: the use of nitrified compost in sustainable horticulture. *Acta Hortic.* (1355), 149–156. https://doi.org/10.17660/ActaHortic.2022.1355.19.

Cáceres, R. and Parera, J. (2022). Guia per al compostatge en granja de dejeccions ramaderes. Manual for the on-farm composting of livestock droppings. *Departament d'Acció Climàtica*, Alimentació i Agenda Rural (DACC). Barcelona. , 108 pages. Available (free (catalán language)) in the Website at: https://agriclose.eu/wp-content/uploads/2023/04/Guia_compostatge_AGRICLOSE-1.pdf

Camilleri-Rumbau, M. S., Briceño, K., Fjerbæk Søtoft, L., Christensen, K. V., Roda-Serrat, M. C., Errico, M. and Norddahl, B. (2021). Treatment of manure and digestate liquid fractions using membranes: opportunities and challenges. *Int. J. Environ. Res. Public Health* 18(6), 3107. https://doi.org/10.3390/ijerph18063107.

Carreras-Sempere, M., Caceres, R., Viñas, M. and Biel, C. (2021). Use of recovered struvite and ammonium nitrate in fertigation in tomato (Lycopersicum esculentum) production for boosting circular and sustainable horticulture. *Agriculture* 11(11). https://doi.org/10.3390/agriculture11111063.

Carreras-Sempere, M., Biel, C., Viñas, M., Guivernau, M. and Caceres, R. (2022). The use of recovered struvite and ammonium nitrate in fertigation in a Horticultural rotation: agrononomic and microbiological assessment. *Environ.Technol. (UK).* https://doi.org/10.1080/09593330.2022.2154172.

Cerrillo, M., Palatsi, J., Comas, J., Vicens, J. and Bonmatí, A. (2015). Struvite precipitation as a technology to be integrated in a manure anaerobic digestion treatment plant - removal efficiency, crystal characterization and agricultural assessment. *J. Chem. Technol. Biotechnol.* 90(6), 1135–1143. https://doi.org/10.1002/jctb.4459.

Cerrillo, M., Viñas, M. and Bonmatí, A. (2016a). Removal of volatile fatty acids and ammonia recovery from unstable anaerobic digesters with a microbial electrolysis cell. *Bioresour. Technol.* 219, 348–356. https://doi.org/10.1016/j.biortech.2016.07.103.

Cerrillo, M., Viñas, M. and Bonmatí, A. (2016b). Overcoming organic and nitrogen overload in thermophilic anaerobic digestion of pig slurry by coupling a microbial electrolysis cell. *Bioresour. Technol.* 216, 362–372. https://doi.org/10.1016/j.biortech.2016.05.085.

Cerrillo, M., Burgos, L. and Bonmatí, A. (2021a). Biogas upgrading and ammonia recovery from livestock manure digestates in a combined electromethanogenic biocathode–hydrophobic membrane system. *Energies* 14(2). https://doi.org/10.3390/en14020503.

Cerrillo, M., Burgos, L., Noguerol, J., Riau, V. and Bonmatí, A. (2021b). Ammonium and phosphate recovery in a three chambered microbial electrolysis cell: towards obtaining struvite from livestock manure. *Processes* 9(11), 1916. https://doi.org/10.3390/pr9111916.

Cerrillo, M., Riau, V. and Bonmatí, A. (2023). Recent advances in bioelectrochemical systems for nitrogen and phosphorus recovery using membranes. *Membranes* 13(2), 186. https://doi.org/10.3390/membranes13020186.

Cheng, S. and Logan, B. E. (2011). High hydrogen production rate of microbial electrolysis cell (MEC) with reduced electrode spacing. *Bioresour. Technol.* 102(3), 3571–3574. https://doi.org/10.1016/j.biortech.2010.10.025.

Cocolo, G., Curnis, S., Hjorth, M. and Provolo, G. (2012). Effect of different technologies and animal manures on solid-liquid separation efficiencies. *J. Agric. Eng.* 43(2). https://doi.org/10.4081/jae.2012.e9.

Corona, F., Hidalgo, D., Martín-Marroquín, J. M. and Meers, E. (2021). Study of pig manure digestate pre-treatment for subsequent valorisation by struvite. *Environ. Sci. Pollut. Res. Int.* 28(19), 24731–24743. https://doi.org/10.1007/s11356-020-10918-6.

Cusick, R. D., Ullery, M. L., Dempsey, B. A. and Logan, B. E. (2014). Electrochemical struvite precipitation from digestate with a fluidized bed cathode microbial electrolysis cell. *Water Res.* 54, 297–306. Available at: https://doi.org/10.1016/j.watres.2014.01.051.

Czekala, W., Malinska, K., Cáceres, R., Janczak, D., Dach, J. and Lewicki, A. (2016). Co-composting of poultry manure mixtures amended with biochar - the effect of biochar on temperature and C-CO2 emission. *Bioresour. Technol.* 200, 921–927. https://doi.org/10.1016/j.biortech.2015.11.019.

Directive of the Council of 12 December 1991 Concerning the Protection of Waters against Pollution Caused by Nitrates from Agricultural Sources (91/676/EC).

Ehmann, A., Thumm, U. and Lewandowski, I. (2018). Fertilizing potential of separated biogas digestates in annual and perennial biomass production systems. *Front. Sustain. Food Syst.* 2. https://doi.org/10.3389/fsufs.2018.00012.

Eurostat (2022). Agri-Environmental Indicator - Mineral Fertilizer Consumption - Statistics Explained. Available at: europa.eu.

Fangueiro, D., Coutinhob, J., Borges, L. and Vasconcelos, E. (2015). Recovery efficiency of nitrogen from liquid and solid fractions of pig slurry obtained using different separation technologies. *J. Plant Nutr. Soil Sci.* 178(2), 229–236. https://doi.org/10.1002/jpln.201400261.

Flotats Ripoll, X., Foged, H., Bonmatí Blasi, A., Palatsi Civit, J., Magrí Aloy, A. and Schelde, K. M. (2011). Manure processing technologies. Technical Report No. II concerning. Manure Processing Activities in Europe to the European Commission, Directorate-General Environment. Project reference. EnvironmentalistB.1/ETU/2010/0007, 184 pp. Available at: http://hdl.handle.net/2117/18944.

Fukumoto, Y. and Haga, K. (2004). Advanced treatment of swine wastewater by electrodialysis with a tubular ion exchange membrane. *Anim. Sci. J.* 75(5), 479–485. https://doi.org/10.1111/j.1740-0929.2004.00216.x.

Guo, X. and Jin, X. (2014). Purification of UF-treated anaerobically digested manure wastewater by two-pass reverse osmosis. *Desalin. Water Treat.* 52(16–18), 3027-3034. https://doi.org/10.1080/19443994.2013.797671.

Haddadi, S., Nabi-Bidhendi, G. and Mehrdadi, N. (2014). Nitrogen removal from wastewater through microbial electrolysis cells and cation exchange membrane. *J. Environ. Health Sci. Eng.* 12(1), 48. https://doi.org/10.1186/2052-336X-12-48.

Haug, R. T. (1993). *The Practical Handbook of Compost Engineering*. ISBN: 0-87371-373-7. 715 pp. https://doi.org/10.1201/9780203736234.

Hendriks, C. M. J., Shrivastava, V., Sigurnjak, I., Lesschen, J. P., Meers, E., Noort, Rv., Yang, Z. and Rietra, R. P. J. J. (2022). Replacing mineral fertilizers for bio-based fertilizers in potato growing on sandy soil: A case study. *Appl. Sci.* 12(1), 341. https://doi.org/10.3390/app12010341.

Horváth, I. S., Tabatabaei, M., Karimi, K. and Kumar, R. (2016). Recent updates on biogas production - a review. *Biofuel Res. J.* 3(2), 394–402. https://doi.org/10.18331/BRJ2016.3.2.4.

Huang, H., Xu, C. and Zhang, W. (2011). Removal of nutrients from piggery wastewater using struvite precipitation and pyrogenation technology. *Bioresour. Technol.* 102(3), 2523–2528. https://doi.org/10.1016/j.biortech.2010.11.054.

Janczak, D., Malińska, K., Czekała, W., Cáceres, R., Lewicki, A. and Dach, J. (2017). Biochar to reduce ammonia emissions in gaseous and liquid phase during composting of poultry manure with wheat straw. *Waste Manage.* 66, 36–45. https://doi.org/10.1016/j.wasman.2017.04.033.

Jorgensen, K. and Jensen, L. S. (2009). Chemical and biochemical variation in animal manure solids separated using different commercial separation technologies. *Bioresour. Technol.* 100(12), 3088–3096. https://doi.org/10.1016/j.biortech.2009.01.065.

Khoshnevisan, B., Duan, N., Tsapekos, P., Awasthi, M. K., Liu, Z., Mohammadi, A., Angelidaki, I., Tsang, D. C., Zhang, Z., Pan, J., Ma, L., Aghbashlo, M., Tabatabaei, M. and Liu, H. (2021). A critical review on livestock manure biorefinery technologies: sustainability, challenges, and future perspectives. *Renew. Sustain. Energy Rev.* 135, 110033. https://doi.org/10.1016/j.rser.2020.110033.

Koffi, N. J. and Okabe, S. (2020). High voltage generation from wastewater by microbial fuel cells equipped with a newly designed low voltage booster multiplier (LVBM). *Sci. Rep.* 10(1), 18985. https://doi.org/10.1038/s41598-020-75916-7.

Konieczny, K., Kwiecińska, A. and Gworek, B. (2011). The recovery of water from slurry produced in high density livestock farming with the use of membrane processes. *Sep. Purif. Technol.* 80(3), 490–498. https://doi.org/10.1016/j.seppur.2011.06.002.

Kovačić, Đ., Lončarić, Z., Jović, J., Samac, D., Popović, B. and Tišma, M. (2022). Digestate management and processing practices: a review. *Appl. Sci.* 12(18), 9216. https://doi.org/10.3390/app12189216.

Lorick, D., Macura, B., Ahlström, M., Grimvall, A. and Harder, R. (2020). Effectiveness of struvite precipitation and ammonia stripping for recovery of phosphorus and nitrogen from anaerobic digestate: a systematic review. *Environ. Evid.* 9(1), 27. https://doi.org/10.1186/s13750-020-00211-x.

Lyons, G. A., Cathcart, A., Frost, J. P., Wills, M., Johnston, C., Ramsey, R. and Smyth, B. (2021). Review of two mechanical separation technologies for the sustainable management of agricultural phosphorus in nutrient-vulnerable zones. *Agronomy* 11(5), 836. https://doi.org/10.3390/agronomy11050836.

Masse, L., Massé, D. I. and Pellerinb, Y. (2007). The use of membranes for the treatment of manure: a critical literature review. *Biosyst. Eng.* 98(4), 371–380. https://doi.org/10.1016/J.BIOSYSTEMSENG.2007.09.003.

Møller, H. B., Lund, I. and Sommer, S. G. (2000). Solid-liquid separation of livestock slurry: efficiency and cost. *Bioresour. Technol.* 74, 223–229. https://doi.org/10.1016/S0960-8524%2800%2900016-X.

Møller, H. B., Sommer, S. G. and Ahring, B. K. (2002). Separation efficiency and particle size distribution in relation to manure type and storage conditions. *Bioresour. Technol.* 85(2), 189–196. https://doi.org/10.1016/s0960-8524(02)00047-0.

Nagarajan, A., Goyette, B., Raghavan, V., Bhaskar, A. and Rajagopal, R. (2023). Nutrient recovery via struvite production from livestock manure-digestate streams: towards closed loop bioeconomy. *Process Saf. Environ. Prot.* 171, 273–288. https://doi.org/10.1016/j.psep.2023.01.006.

Nkoa, R. (2014). Agricultural benefits and environmental risks of soil fertilization with anaerobic digestates: a review. *Agron. Sustain. Dev.* 34(2), 473–492. https://doi.org/10.1007/s13593-013-0196-z.

Pikaar, I., Guest, J., Ganigué, R., Jensen, P., Rabaey, K., Seviour, T., Trimmer, J., van der Kolk, O., Vaneeckhaute, C. and Verstraete, W. (2022). Resource recovery from water: principles and application. In: IWA Publishing (Ed.). ISBN electronic: 9781780409566. https://doi.org/10.2166/9781780409566.

Pinnekamp, J. and Friedrich, H. (2006). Membrane technology for waste water treatment. *Municipal Water Waste Management*. FiW-Verlag. ISBN 978-3-939377-01-6 2.

Prado, J., Ribeiro, H., Alvarenga, P. and Fangueiro, D. (2022). A step towards the production of manure-based fertilizers: disclosing the effects of animal species and slurry treatment on their nutrients content and availability. *J. Cleaner Prod.* 337. https://doi.org/10.1016/j.jclepro.2022.130369.

Prenafeta-Boldú, F. X. i Parera, J. (2020). Guia de les tecnologies de tractament de les dejeccions ramaderes a Catalunya, Pesca i Alimentació (DARP), Barcelona, 72 pàgines. *Departament d'Agricultura, Ramaderia*. Available at: https://www.agrodigital.com/wp-content/uploads/2020/12/Guiadeyeccionesc.pdf.

Regueiro, I., Coutinho, J., Balsari, P., Popovic, O. and Fangueiro, D. (2016). Acidification of pig slurry before separation to improve slurry management on farms. *Environ. Technol.* 37(15), 1906–1913. https://doi.org/10.1080/09593330.2015.1135992.

Regulation (EU) 2019/1009 of the European Parliament and of the Council of 5 June 2019 laying down rules on the making available on the market of EU fertilising products and amending Regulations (EC) 1069/2009) and (EC) No 1107/2009 and repealing

Regulation (EC)(2003/2003). Available at: https://eur-lex.europa.eu/eli/reg/2019/1009/oj

Riau, V., Burgos, L., Camps, F., Domingo, F., Torrellas, M., Antón, A. and Bonmatí, A. (2021). Closing nutrient loops in a maize rotation. Catch crops to reduce nutrient leaching and increase biogas production by anaerobic co-digestion with dairy manure. *Waste Manage.* 126, 719–727. https://doi.org/10.1016/j.wasman.2021.04.006.

Rizzioli, F., Bertasini, D., Bolzonella, D., Frison, N. and Battista, F. (2023). A critical review on the techno-economic feasibility of nutrients recovery from anaerobic digestate in the agricultural sector. *Sep. Purif. Technol.* 306, 122690. https://doi.org/10.1016/j.seppur.2022.122690.

Robles-Aguilar, A. A., Pang, J., Postma, J. A., Schrey, S. D., Lambers, H. and Jablonowski, N. D. (2019). The effect of pH on morphological and physiological root traits of *Lupinus angustifolius* treated with struvite as a recycled phosphorus source. *Plant Soil* 434(1–2), 65–78. https://doi.org/10.1007/s11104-018-3787-2.

Robles-Aguilar, A. A., Schrey, S. D., Postma, J. A., Temperton, V. M. and Jablonowski, N. D. (2020). Phosphorus uptake from struvite is modulated by the nitrogen form applied. *J. Plant Nutr. Soil Sci.* 183(1), 80–90. https://doi.org/10.1002/jpln.201900109.

Rodríguez Arredondo, M., Kuntke, P., ter Heijne, A. and Buisman, C. J. (2019). The concept of load ratio applied to bioelectrochemical systems for ammonia recovery. *J. Chem. Technol. Biotechnol.* 94(6), 2055-2061. Available at: https://doi.org/10.1002/jctb.5992.

Sawatdeenarunat, C., Nguyen, D., Surendra, K. C., Shrestha, S., Rajendran, K., Oechsner, H., Xie, L. and Khanal, S. K. (2016). Anaerobic biorefinery: current status, challenges, and opportunities. *Bioresour. Technol.* 215, 304–313. https://doi.org/10.1016/j.biortech.2016.03.074.

Sena, M. and Hicks, A. L. (2018). Life cycle assessment review of struvite precipitation in wastewater treatment. *Resour. Conserv. Recycl.* 139, 194–204. https://doi.org/10.1016/J.RESCONREC.2018.08.009.

Shi, L., Simplicio, W. S., Wu, G., Hu, Z., Hu, H. and Zhan, X. (2018). Nutrient recovery from digestate of anaerobic digestion of livestock manure: a review. *Curr. Pollut. Rep.* 4(2), 74–83. https://doi.org/10.1007/s40726-018-0082-z.

Shi, S., Tong, B., Wang, X., Luo, W., Tan, M., Wang, H. and Hou, Y. (2022). Recovery of nitrogen and phosphorus from livestock slurry with treatment technologies: A meta-analysis. *Waste Manage.* 144, 313–323. https://doi.org/10.1016/j.wasman.2022.03.027.

Słupek, E., Makoś, P. and Gębicki, J. (2020). Theoretical and economic evaluation of low-cost deep eutectic solvents for effective biogas upgrading to bio-methane. *Energies* 13(13), 3379. https://doi.org/10.3390/en13133379.

Song, Y.-H., Hidayat, S., Effendi, A. J. and Park, J. (2021). Simultaneous hydrogen production and struvite recovery within a microbial reverse-electrodialysis electrolysis cell. *J. Ind. Eng. Chem.* 94, 302–308. https://doi.org/10.1016/j.jiec.2020.10.043.

Sotres, A., Cerrillo, M., Viñas, M. and Bonmatí, A. (2015). Nitrogen recovery from pig slurry in a two-chambered bioelectrochemical system. *Bioresour. Technol.* 194, 373–382. https://doi.org/10.1016/j.biortech.2015.07.036.

Takkellapati, S., Li, T. and Gonzalez, M. A. (2018). An overview of biorefinery-derived platform chemicals from a cellulose and hemicellulose biorefinery. *Clean Technol. Environ. Policy* 20(7), 1615–1630. https://doi.org/10.1007/s10098-018-1568-5.

Tao, W., Fattah, K. P. and Huchzermeier, M. P. (2016). Struvite recovery from anaerobically digested dairy manure: a review of application potential and hindrances. *J. Environ. Manage.* 169, 46-57. https://doi.org/10.1016/j.jenvman.2015.12.006.

Tian, P., Gong, B., Bi, K., Liu, Y., Ma, J., Wang, X., Ouyang, Z. and Cui, X. (2023). Anaerobic co-digestion of pig manure and rice straw: optimization of process parameters for enhancing biogas production and system stability. *Int. J. Environ. Res. Public Health* 20(1), 804. https://doi.org/10.3390/ijerph20010804.

Torrellas, M., Burgos, L., Tey, L., Noguerol, J., Riau, V., Palatsi, J., Antón, A., Flotats, X. and Bonmatí, A. (2018). Different approaches to assess the environmental performance of a cow manure biogas plant. *Atmos. Environ.* 177, 203–213. https://doi.org/10.1016/j.atmosenv.2018.01.023.

Tur-Cardona, J., Bonnichsen, O., Speelman, S., Verspecht, A., Carpentier, L., Debruyne, L., Marchand, F., Jacobsen, B. H. and Buysse, J. (2018). Farmers' reasons to accept bio-based fertilizers: A choice experiment in seven different European countries. *J. Cleaner Prod.* 197, 406–416. https://doi.org/10.1016/j.jclepro.2018.06.172.

Varol, A. and Ugurlu, A. (2017). Comparative evaluation of biogas production from dairy manure and co-digestion with maize silage by CSTR and new anaerobic hybrid reactor. *Eng. Life Sci.* 17(4), 402–412. https://doi.org/10.1002/elsc.201500187.

Velthof, G. L. (2011). *Synthesis of the Research within the Framework of the Mineral Concentrates Pilot*. Available at: https://edepot.wur.nl/192069. 72 pp. Wageningen, Alterra. Alterra report 2224.

Villano, M., Monaco, G., Aulenta, F. and Majone, M. (2011). Electrochemically assisted methane production in a biofilm reactor. *J. Power Sources* 196(22), 9467–9472. https://doi.org/10.1016/j.jpowsour.2011.07.016.

Wagner, E. and Karthikeyan, K. G. (2022). Precipitating phosphorus as struvite from anaerobically-digested dairy manure. *J. Cleaner Prod.* 339, 130675. https://doi.org/10.1016/j.jclepro.2022.130675.

Yetilmezsoy, K. and Sapci-Zengin, Z. (2009). Recovery of ammonium nitrogen from the effluent of UASB treating poultry manure wastewater by MAP precipitation as a slow release fertilizer. *J. Hazard. Mater.* 166(1), 260–269. https://doi.org/10.1016/j.jhazmat.2008.11.025.

Zeppilli, M., Mattia, A., Villano, M. and Majone, M. (2017). Three-chamber bioelectrochemical system for biogas upgrading and nutrient recovery. *Fuel Cells* 17(5), 593–600. https://doi.org/10.1002/fuce.201700048.

Zeppilli, M., Simoni, M., Paiano, P. and Majone, M. (2019). Two-side cathode microbial electrolysis cell for nutrients recovery and biogas upgrading. *Chem. Eng. J.* 370, 466–476. https://doi.org/10.1016/j.cej.2019.03.119.

Zeppilli, M., Cristiani, L., Dell'Armi, E. and Majone, M. (2020). Bioelectromethanogenesis reaction in a tubular microbial electrolysis cell (MEC) for biogas upgrading. *Renew. Energy* 158, 23–31. Available at: https://doi.org/10.1016/j.renene.2020.05.122.

Zhang, Y. and Angelidaki, I. (2015). Counteracting ammonia inhibition during anaerobic digestion by recovery using submersible microbial desalination cell. *Biotechnol. Bioeng.* 112(7), 1478–1482. https://doi.org/10.1002/bit.25549.

Zhang, Z., Xu, Z., Song, X., Zhang, B., Li, G., Huda, N. and Luo, W. (2020). Membrane processes for resource recovery from anaerobically digested livestock manure effluent: opportunities and challenges. *Curr. Pollut. Rep.* 6(2), 123–136. https://doi.org/10.1007/s40726-020-00143-7.

Zhu, Q. L., Wu, B., Pisutpaisal, N., Wang, Y. W., Ma, K. D., Dai, L. C., Qin, H., Tan, F. R., Maeda, T., Xu, Y. S., Hu, G. Q. and He, M. X. (2021). Bioenergy from dairy manure: technologies, challenges and opportunities. *Sci. Total Environ.* 790, 148199. https://doi.org/10.1016/j.scitotenv.2021.148199.

Chapter 8

Safe and sustainable use of bio-based fertilizers in agricultural production systems

April Leytem, Robert Dungan, Mindy Spiehs and Dan Miller, United States Department of Agriculture, USA*

1 Introduction

Development of circular agricultural production systems necessitates the reuse/recycling/upcycling of agricultural byproducts (e.g. animal manure, crop residues, food waste and food processing waste) back into the production cycle thereby lessening the overall footprint of agriculture. Historically, many agricultural byproducts have been classified as 'wastes', as they were not the primary product of production. However, the goal of circularity is to utilize these 'wastes' as a resource instead of simply discarding them. Many agricultural byproducts contain significant amounts of nitrogen (N), phosphorus (P) and potassium (K) as well as a wide variety of micronutrients which makes them an excellent nutrient source for crop growth. Because of this, the development of bio-based fertilizers from a variety of agricultural byproducts is of increasing interest. In addition to supplying nutrients, bio-based fertilizers can improve soil health via the addition of organic carbon (C) which may improve soil structure, water holding capacity and water infiltration and support a healthy microbiome. Although these products contain valuable N and P for agricultural production, the conversion efficiency into edible food or other end products

* Corresponding author: april.leytem@usda.gov, 01-208-423-6530, USDA-ARS NWISRL, 3793N 3600E, Kimberly, ID, 83341, USA.

http://dx.doi.org/10.19103/AS.2023.0120.16
Published by Burleigh Dodds Science Publishing Limited, 2024.

tends to be low with typical use efficiencies of 30% or less (Bordirsky et al., 2012; UNEP and WHRC, 2007; FAO, 2006) leading to potential environmental losses that impact air and water quality as well contribute greenhouse gasses (GHG) to the atmosphere which are associated with climate change. Efficiently integrating these bio-based fertilizers back into agricultural production is dependent on the characteristics of the product, availability, quality control, transportation costs/logistics, environmental regulation and public acceptance (Westerman and Bicudo, 2005). Additionally, valorization of these products is challenging as the nutrient contents tend to be diluted compared to synthetic fertilizer, less consistent and in some cases harder to handle and apply than traditional fertilizers. However, the additional benefits of providing a wider variety of micronutrients, compared to synthetic fertilizer, along with additions of organic C that may improve overall soil health and productivity, need to be taken into consideration as well.

In addition to concerns related to nutrient pollution, other environmental contaminants that may be associated with bio-based fertilizers include heavy metals, pathogens, antibiotics and chemicals of emerging concern (CEC; e.g. pesticides, hormones, polychlorinated biphenyls [PCBs], polycyclic aromatic hydrocarbons [PAHs] and microplastics; Tran et al., 2018; O'Connor et al., 2022; Tian et al., 2022). Bio-based fertilizers originating from biosolids, manures and other organic wastes can contain heavy metals (O'Connor et al., 2021). Heavy metals are toxic in high concentrations and are persistent pollutants in the environment where they can bioaccumulate and biomagnify through the food chain (Sardar et al., 2013). Pathogens may be present in many organic byproducts (livestock manure, manure compost, biosolids) used as bio-based fertilizers. The application of these fertilizers to agricultural crops, in particular those grown for fresh produce, has resulted in foodborne illness outbreaks (Alegbeleye et al., 2018; Callejón et al., 2015). The use of antibiotics in animal production and the presence of antibiotics in biosolids have raised concerns about the release of these drugs into the environment, particularly with land application of these byproducts, and the increasing prevalence of antibiotic resistant bacteria (ARB) and antibiotic resistance genes (ARGs) in the environment. Pesticide residues are commonly found in food wastes which may pose a threat to the quality of composts and digestates generated from these byproducts (Nguyen et al., 2020). The persistence and phytotoxic effects of these residues are unknown but may cause damage to soil quality and plant growth when applied to cropland as bio-based fertilizers at high rates (Boudh and Singh, 2019). There are a host of other CEC that are introduced into agricultural production and they remain persistent within the system. These chemicals have the potential to damage soil health and can pose health risks to both animals and humans. Given the large variety of agricultural byproducts that can serve as bio-based

Published by Burleigh Dodds Science Publishing Limited, 2024.

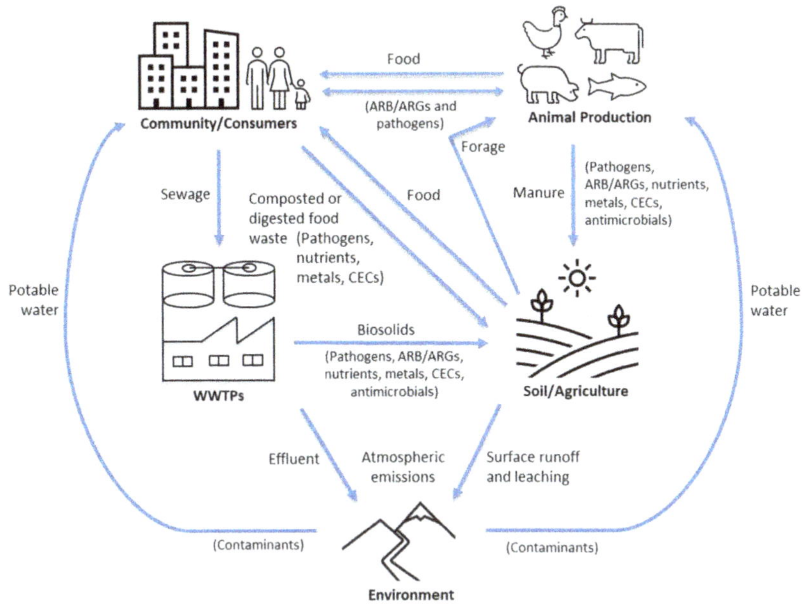

Figure 1 The routes by which biological and chemical contaminants in bio-based fertilizers are spread in environmental ecosystems. The contaminants make their way into the environment when manure, food waste and biosolids are applied to soils, either through direct application, animal excretion or disposal. The contaminants can then be transported in surface runoff or by leaching to surface and ground water, respectively, or emitted to the atmosphere. Humans and animals can be continuously exposed to contaminants via ingestion of food and water, while humans can also be exposed via contact in recreational waters.

fertilizers available for recycling within the agricultural system, it is important not only to consider the benefits in terms of C and nutrients, but also to assess any potential tradeoffs with respect to contaminants. Figure 1 illustrates some of the potential pathways of contaminant flow within circular systems.

2 Risk factors associated with utilizing bio-based fertilizers in agricultural production

2.1 Nutrient losses to the environment

Estimates of the conversion of N used in food production to that consumed by humans range from 10% to 20% (Bordirsky et al., 2012; UNEP and WHRC, 2007), with greater than 50% of N applied to cropland lost to downstream and downwind environments (Davidson et al., 2012). It is estimated that only 15% to 30% of applied phosphorus (P) is taken up by harvested crops (FAO, 2006). With these low nutrient use efficiency rates, the integration of agricultural byproducts as bio-based fertilizers into agricultural production needs to be carefully managed to reduce nutrient losses to the environment.

In terms of nutrients generated, livestock manure is one of the largest sources of bio-based fertilizers available. Globally, it is estimated that livestock production generates 131 and 23 Tg of manure N and P, respectively, per year (Liu et al., 2017; Zhang, 2017). Manure nutrient production is expected to increase as human and livestock populations increase and the dietary intake of meat increases in developing countries (Herrero and Thornton, 2013). While the quantity of manure nutrients generated over time are increasing, the growing disconnect between livestock and crop production on the agricultural landscape has left most modern livestock producers with a surplus of on-farm nutrients which contribute to environmental degradation in many regions. Overall, the greatest concerns have been related to air quality degradation due to emissions of NH_3, water quality impacts from nitrate (NO_3) leaching and P losses via runoff and climate impacts and ozone depletion from nitrous oxide (N_2O) emissions (Holly et al., 2018).

2.1.1 Reactive nitrogen losses to the environment

Reactive N (N_r) is essential to the growth of plants and animals and is typically the most limiting nutrient in agricultural production. The current rate of N_r loss to the environment, due to agricultural production, is more than 10 times the rate that occurred at the end of the 1800s (UNEP and WHRC, 2007) largely due to industrial-scale plant fertilizer production using the Haber-Bosch process in the twentieth century. Nitrogen losses are of increasing concern due to their negative consequences on human, animal and environmental health. The primary forms of N_r loss are ammonia (NH_3), nitrate (NO_3), nitrous oxide (N_2O) and N oxides (NO and NO_2 or NO_x).

Food production, including cropland and livestock production, is estimated to generate up to 90% of global NH_3 emissions (Ma et al., 2021; Zhan et al., 2021). Liu et al. (2022) estimated that, in 2010, approximately 58 Tg of N was lost as NH_3, with approximately 67% of that being generated from livestock production and land application of manures. Ammonia in the atmosphere is an important driver for the formation of fine particulate matter ($PM_{2.5}$) which is both an air quality and human health concern. In addition, the deposition of this N into terrestrial ecosystems can cause soil acidification, lead to eutrophication of sensitive waterbodies, affect biodiversity, lead to secondary N_2O emissions and cause changes in terrestrial carbon sinks (Liu et al., 2019; Elser et al., 2009; Zhan et al., 2017; Wang et al., 2017; Sutton et al., 2014). Liu et al. (2022) reported that global agricultural NH_3 emissions and N deposition increased by 78% and 72% from 1980 to 2018, respectively. Understanding the sources of NH_3 emissions and temporal and spatial patterns of N deposition will be essential for developing abatement strategies.

Reductions in NH_3 emissions from livestock via improved feed efficiency and enhancing NH_3-N capture from livestock manure and recycling that N back into cropping systems can help improve the overall N use efficiency and circularity of agricultural systems. In addition, development of enhanced efficiency bio-based fertilizers that are stable and release N slowly, as well as improved N management (better matching N to crop needs over time, immediate incorporation of fertilizers into soils, etc.), can also help minimize NH_3 volatilization losses in cropping systems.

Nitrate is one of the most widespread groundwater contaminants in the world, and its consumption has been linked with a variety of human health ailments. The main sources of NO_3 pollution of groundwater are from intensive use of N fertilizers (both synthetic and livestock manures) in agricultural production, disposal of sanitary and industrial wastes, leakage from septic systems and landfills and NO_x air stripping waste from air pollution control devices (Bhatnagar and Sillanpää, 2011; Abascal et al., 2022). Transport of NO_3 from ground waters to surface waters as well as runoff of NO_3 to surface waters has led to eutrophication in N limited systems, resulting in hypoxia and anoxia, loss of biodiversity and harmful algal blooms damaging fisheries and pristine marine environments (BijaySingh and Craswell, 2021). Optimizing N application rates and timing can be effective approaches to reducing NO_3 losses to ground and surface waters. In addition, development of more stable bio-based fertilizers with slow N release characteristics can reduce losses by better matching N release with plant uptake.

Nitrous oxide is a potent GHG as well as a stratospheric ozone depleting substance with a lifetime of 116 years (Prather et al., 2015). The largest global source of anthropogenic N_2O emissions is agriculture, representing 52% of total emissions (Tian et al., 2020). Tian et al. (2020) reported that the largest increases in anthropogenic N_2O emissions in the last four decades were due to direct emissions from agricultural soils receiving N additions (71%). There is also considerable positive feedback between climate change and increasing N_2O emissions. As the recent growth in N_2O emissions have exceeded some of the highest projected emission scenarios, mitigating these emissions is urgent (Gidden et al., 2019). Reducing excess N application to croplands and adopting precision fertilizer application methods provide the greatest immediate opportunities for the abatement of N_2O emissions.

Improving the connectivity between agricultural byproducts and agricultural production can improve N use efficiency and reduce N_r losses through more efficient recycling of nutrients. Application of bio-based fertilizers following the 4R strategy of applying the right fertilizer at the right rate and time with the right application method can enhance overall system efficiencies and thereby reduce their environmental impact.

2.1.2 Phosphorus losses

Globally, P deficiency is considered a major limitation for crop production, particularly in low-input agricultural systems (Raghothama, 2005). However, in many developed countries, over application of P as fertilizer (synthetic and bio-based fertilizers) has become a significant source for water pollution. In areas with concentrated livestock production, managing manure P to reduce negative environmental impacts is a challenge. Although manure P availability is greater or similar to synthetic fertilizer P use, manure P is not efficiently used in crop production. The inefficient use of manure P is due to several factors: uneven distribution of manure by grazing animals, incomplete collection and inadequate storage of manure from housed animals, poor timing of manure application, high cost of transporting manure, manure N:P ratios that do not necessarily match N:P needs of crops resulting in overapplication of P in some instances and the relatively low price of synthetic P fertilizer. Due to the high moisture content and bulky nature of manures, they are generally land applied within close proximity of where they are produced which had resulted in the buildup of P in soils surrounding livestock farms. This excessive P application has led to P surpluses in croplands, decreased P use efficiency and increasing P losses to surface waters.

Elevated P concentrations in receiving waters can lead to eutrophication, which can be costly to remediate. For example, in England and Wales, it has been estimated that damages due to agricultural losses of P are near £19 million (Bateman et al., 2011). In some countries, direct discharge of P in wastewater to surface waters is still common. For example, in Thailand, P containing wastewaters discharged directly to surface waters from dairy and swine farms were estimated to add 554 and 261 T P y^{-1} (Prathumchai et al., 2018). These discharges have a direct negative impact on surface water quality in these regions. Phosphorus surplus has been shown to increase with increasing livestock density in studies at continental, national and regional scale, especially at livestock densities above 2 LSU ha^{-1} (Liu et al., 2017; Nesme et al., 2015). One of the main causes of these surpluses is the large amount of P imported in feed coupled with low P use efficiency of most livestock. Therefore, there is often a clear relationship between livestock density and P balances at the farm level. In several countries (e.g. the United States, Netherlands, Norway, Denmark, Finland), manure P generated can meet or even exceed that needed for sustainable crop production in some regions (Smit et al., 2015; Hanserud et al., 2016; Yang et al., 2016; Parchomenko and Borsky, 2018; Svanbäck et al., 2019). Despite this large potential for within country P recycling, areas with the largest amounts of manure P are not necessarily co-located with the highest P needs. Therefore, hot spots occur where manure P exceeds that regionally needed for crop production. In addition, fertilizer P is still often used in these production systems leading to a further buildup of soil P.

Development of technologies to capture and concentrate P from agricultural byproduct streams will be essential to enable more sustainable recycling of P back through the production system. Additionally, cost-effective methods for redistribution of byproduct P from areas of surplus to areas of deficit will need to be developed. In some instances, particularly in livestock dense regions, regulations, economic incentives and technical solutions for enhanced relocation of byproduct P from areas with surplus to areas with deficit will be crucial. Support for processing and trading of byproduct-based P fertilizers can help reduce nutrient imbalances between areas with excess nutrient and those needing this valuable nutrient.

2.2 Heavy metals

Heavy metal contamination of byproducts used as bio-based fertilizers is a concern as these metals can bioaccumulate in agricultural crops and subsequently endanger the health of the consumer (Polechónska et al., 2018). Two of the most common byproducts used as bio-based fertilizers are manures and sewage sludges which may be applied raw or composted. On average, sludges have the highest values of heavy metal concentrations, compared to composts and manure, particularly when wastewater treatment plants also collect industrial effluents (Lopes et al., 2011). Cajamarca et al. (2019) characterized several agro-industrial, livestock and food wastes used to create bio-based fertilizers and found that some material contained high levels of trace metals, in particular cadmium (Cd). Livestock manures can contain elevated levels of arsenic (As), copper (Cu) and zinc (Zn) as these are found in feed additives to improve overall health and growth of livestock (Bloem et al., 2017). Repeated application of these bio-based fertilizers to cropland can lead to buildup of metals in soils which can potentially enter the food chain. Monte Carlo simulations indicated that repeated application of bio-based fertilizer likely increased the concentrations of Zn, Cd and As in soil compared with soil background levels (Gong et al., 2019). Concern over heavy metal contamination is ubiquitous where sludges, manures, composts and other bio-based fertilizers are applied to agricultural lands resulting in legislation in many countries regulating application (Lopes et al., 2011; Shi et al., 2018; Gong et al., 2019; Nunes et al., 2021).

In China, it was reported that livestock manure was one of the predominant sources of trace metals entering agricultural soils, accounting for approximately 55%, 69% and 51% of the total Cd, Cu and Zn inputs, respectively (Luo et al.,

2009; Shi et al., 2018). Tan and Tran (2021) reported that Cu and Zn accumulated in soil, water and rice with application of sewage sludge at varying levels with concentrations exceeding permissible Vietnamese standards (QCVN 03: 2008) and US EPA 503 criteria. A risk assessment focused on reuse of organic waste as bio-based fertilizers, indicated that Zn was the main contributor to total risk due to its high concentration in bio-based fertilizers and high bio-transfer potential (Lopes et al., 2011). While others reported that health risk index values for Cd, cobalt (Co) and lead (Pb) suggested that these metals had the probability to cause health problems in people who consume vegetables grown with bio-based fertilizers (Ugulu et al., 2021). Others have reported heavy metal contents of bio-based fertilizers being below regulatory thresholds (Nekvapil et al., 2021; Golovko et al., 2022). In addition to health concerns, elevated levels of heavy metals in agricultural soils can result in decreased germination, altered metabolism, growth reduction, reduced biomass production and reduced yield for sensitive plants (Sethy and Ghosh, 2013; Goyal et al., 2020). While phytoremediation of metals from soils is feasible, there are few remediation strategies available, therefore prevention is key.

2.3 Pathogens

Pathogens can be present in food wastes, manures and other byproducts that are used as bio-based fertilizers. This contamination can pose potentially serious risks to human health as the application of these bio-based fertilizers to croplands can result in foodborne and waterborne outbreaks. The risks will vary from low to high, depending upon a number of factors, such as animal health, microbial concentration, land application method and environmental conditions (Venglovsky et al., 2009; Adegoke et al., 2016). Some pathogens commonly found in manures and food wastes are *Campylobacter*, *Escherichia coli* O157:H7, *Clostridium*, *Listeria*, *Salmonella*, hepatitis E virus, *Cryptosporidium parvum*, *Giardia lamblia* and *Shigella* (Kraus et al., 2003; Pepper et al., 2006; Manyi-Loh et al., 2016; O'Connor et al., 2022). The level of these zoonotic pathogens can exceed thousands per gram of material, with infection causing temporary illness or mortality, especially in high-risk individuals (Hutchison et al., 2005; Klein et al., 2010; Létourneau et al., 2010). Exposure of humans to pathogens can occur through occupational and recreational exposures, ingestion of contaminated food and water or aerogenic routes (Matthews, 2006; Dungan, 2010). According to the Centers for Disease Control and Prevention (CDC), *Escherichia coli* O157:H7, *Listeria monocytogenes* and *Salmonella* spp. are three major foodborne pathogens (CDC, 2020). These bacteria have been associated with numerous outbreaks in the United States (Ratnam et al., 1988; Nightingale et al., 2004; CDC, 2020).

Application of contaminated bio-based fertilizers to soils, particularly surface application, can result in the transport of pathogens to surface or ground waters (Abu-Ashour et al., 1994; Jamieson et al., 2002; Tyrrel and Quinton, 2003; Bloem et al., 2017). The overland transport of microorganisms is called horizontal movement, while the leaching of microorganisms through soil and other porous subsurface strata is referred to as vertical movement. Unless a soil is saturated or contains an impermeable barrier, vertical movement of microorganisms will occur (Mawdsley et al., 1995). Some physical and chemical properties that influence the vertical movement of microorganisms are soil type, water content and water flow, microbe and soil particle surface properties, cell motility, pH, plant roots, temperature and presence of micro- and meso-faunal organisms (Mawdsley et al., 1995; Unc and Goss, 2004).

Rapid horizontal transport of microorganisms to surface waters can occur when either the rainfall intensity exceeds the soil's infiltration rate or when the soil becomes so saturated that no rainfall can percolate (Tyrrel and Quinton, 2003). Factors that influence the level of microbiological contamination in runoff from agricultural lands are organism die-off rates, quantity and type of amendment applied, sloping terrain, rainfall intensity and water infiltration rate (Evans and Owens, 1972; Doran and Linn, 1979; Baxter-Potter and Gilliland, 1988; Abu-Ashour and Lee, 2000; Jenkins et al., 2006; Ramos et al., 2006). Methods to mitigate the offsite transport of microorganisms in runoff from amended soils include use of vegetative filter strips (Coyne et al., 1995; Fajardo et al., 2001) or vegetative treatment systems with a settling basin for solids collection and a vegetated area (Koelsch et al., 2006; Berry et al., 2007).

Alternatively, bio-based fertilizers can be treated prior to land application, thus reducing subsequent risks associated with pathogens (Lund et al., 1996; Tiquia et al., 1998). Various physical, chemical and biological treatment technologies could be used to reduce or eliminate the presence of pathogens (Heinonen-Tanski et al., 2006). While there are advantages and disadvantages with these methods, some can provide additional benefits, such as the production of compost that can be used to enhance the properties of agricultural soils (Tester, 1990) or biogas for energy generation (Holm-Nielsen et al., 2009). There are a wide variety of technologies available to treat byproducts; however, the only processes with a documented record of cost-effective pathogen reduction are composting and anaerobic digestion (Sobsey et al., 2006; Martens and Böhm, 2009; Gurtler et al., 2018; Jiang et al., 2020; Thakalie and MacRae, 2021).

2.4 Antibiotics, antibiotic resistant bacteria and antibiotic resistance genes

In many countries, antibiotics are used therapeutically (high doses) in livestock production to treat specific diseases or subtherapeutically (low doses) by incorporating into feed to improve growth efficiency (Sarmah et al., 2006). It is also common practice to simultaneously administer multiple classes of antibiotics to livestock at the production facility (Song et al., 2007). Because not all antibiotics are absorbed in the gut of animals, they are excreted via urine and feces in unaltered form and as metabolites (Halling-Sørensen et al., 1998; Boxall et al., 2004). It has been estimated that as much as 80% of orally ingested antibiotics can be excreted in urine and feces (Elmund et al., 1971; Levy, 1992; Halling-Sørensen et al., 2002). Several classes of veterinary pharmaceuticals and antibiotics, including coccidiostats, ionophores, lincosamides, macrolides, sulfonamides and tetracyclines, have been detected in surface waters adjacent to livestock operations (Campagnolo et al., 2002; Hao et al., 2006; Song et al., 2007). In addition, the practice of land applying livestock manure as a bio-based fertilizer provides for the introduction of antibiotics over large areas in the environment, resulting in frequent detection of these compounds in soils and waters worldwide (Hamscher et al., 2002; Christian et al., 2003; Kemper, 2008).

Antibiotics can select for ARB in the animal gut, which are then released into the environment via excreted feces. ARGs are the genetic code ARB use to produce proteins responsible for antibiotic resistance. ARGs can be distributed to similar, distantly related and pathogenic bacteria through horizontal gene transfer mechanisms (Alekshun and Levy, 2007) and are considered an emerging contaminant (Pruden et al., 2006). Prophylactic and therapeutic uses of antibiotics in food-animal production has the potential to contribute to drug resistant bacteria in vivo, during manure storage, and in soils receiving manure solids or wastewater as a fertilizer (Binh et al., 2008; Negreanu et al., 2012; Chantziaras et al., 2014). Use of antimicrobials in livestock production may also intensify the resistance of pathogens to antibiotics, reducing the ability to treat infected individuals (Boxall et al., 2003; Bahe et al., 2006).

The continued use of large quantities of antibiotics in animal production and human health raises concerns about the release of these drugs and the increasing prevalence of ARB and ARGs in the environment when livestock manure and municipal biosolids are used as bio-based fertilizers. While use of antibiotic drugs can enrich ARB, the interplay between antibiotic drug use, ARB/ARGs and land use practices in agroecosystems is poorly understood (Williams-Nguyen et al., 2016). Manure application can transfer ARB and ARGs to soils, as well as antibiotic residues and other xenobiotic compounds, resulting in the expansion of antibiotic resistance reservoirs when compared to that of native soils (Heuer and Smalla, 2007; Cytryn, 2013; Amarakoon et al., 2016; Dungan et al., 2019). The detection of ARGs in soils, manures and agriculturally impacted environments is well documented in the scientific literature, but the risk of elevated ARG levels on public health is not well understood. Low concentrations of antibiotics and their metabolites can enter the food chain when plants are grown on fields which have received bio-based fertilizers contaminated by antibiotics (Bloem et al., 2017). This could potentially contribute to antibiotic resistance in humans and animals when edible crops are contaminated by traces of antibiotics, as it has been shown that even low-level antibiotic concentrations can select for antibiotic resistance in bacteria (Sandegren, 2014).

To minimize the effects of antibiotics, ARB and ARG in bio-based fertilizers, removal of these contaminants prior to land application may be necessary. Removal of antibiotics during anaerobic digestion has had varying success, even within the same class of antibiotics, while composting has been shown to significantly reduce antibiotics in nearly all cases (Youngquist et al., 2016). The removal of ARB and ARGs with digestion and composting varies with treatment effectiveness dependent on temperature (Beneragama, et al., 2013; Diehl and La Para, 2010; Guan et al., 2007; Resende et al., 2014). Alternatively, bioremediation using algae (Guo and Chen, 2015; Yu et al., 2017; Waseem et al., 2017) and fungal (Naghdi et al., 2018) species have successfully removed antimicrobials in water treatment facilities.

2.5 Chemicals of emerging concern

The trace levels of emerging or recently detected pollutants in soil and water receiving bio-based fertilizers is a growing concern for human health and the environment because at high concentrations, these chemicals can affect animal, human and environmental health. This category of emerging contaminants consists of pharmaceuticals, pesticides, hormones, disinfectants and their metabolites (Chaturvedi et al., 2021; Rathi et al., 2021). Hormones may be naturally occurring endogenous hormones, or natural or synthetic exogenous hormones used as growth promoters. Hormones found in animal waste are a concern as they may act as endocrine disruptors and influence the production of natural hormones, providing the blockage, minimization, stimulation or their inhibition (Matthiessen et al., 2002). Hormones have been found in high concentrations in pig farming effluent (Honorio et al., 2019; Fine et al., 2003). Hutchins et al. (2007) reported estrogens and estrogen metabolites in swine sow, finisher and nursery lagoons, as well as dairy, beef and poultry lagoons. Some hormones have even been detected in groundwater near livestock wastewater lagoons (Fine et al., 2003; Bartelt-Hunt et al., 2011).

Pesticides and microplastics are commonly found in food wastes due to their presence in soils and on foods (Nguyen et al., 2020; Tian et al., 2022). Some pesticides adsorb strongly to soil particles and can accumulate in soils and potentially be taken up by plants (Li et al., 2018). Pesticide residues that persist in bio-based fertilizers can damage soil quality as well as plant health

(Boudh and Singh, 2019). Composts containing residual herbicides were found to inhibit seedling vigor, emergence rate and plant growth in sensitive plants when mixed with soil at rates of 10–20% (Chang et al., 2017). Some plant growth regulator herbicides do not break down easily and can pass through animal digestive systems and composting facilities and have the potential to cause significant damage to plants and crops.

A wide range of microplastics composed of polyethylene, polypropylene, polyamide, polyester, polystyrene, polyethylene terephthalate, polyvinyl chloride, polyurethane, polyvinylidene chloride and polyacrylonitrile have been identified in the environment. Polymer-coated, slow-release fertilizers, plastic mulching for weed and pathogen control and other cultivation practices are important sources contributing to microplastics in soil. Composting, which usually reduces CEC in manures and agricultural wastes, accelerates microplastic formation when larger plastic contaminants become more fragmented. Although ingestion by earthworms and other soil fauna would be the most direct biological effect, microplastic's capacity to serve as particles capable of concentrating other contaminants, such as antibiotics and pharmaceutically active compounds, may be an unrecognized contributor to CEC risk.

PCBs are a widely recognized environmental and food contaminant, with human exposure primarily coming from consumption of animal-derived food such as meat, dairy and eggs, and fishery products (Weber et al., 2018). Animal feeds and feed additives are major sources of dioxins and PCB contamination for food of animal origin. Animals that are exposed to contaminated soil, such as beef cattle and veal from suckler cow herds, sheep and free-range chickens, have the highest risk of exposure through uptake of contaminated soil, grass, silage and hay. Application of sewage sludge and contaminated sediment deposits on agricultural land caused elevated PCB levels in meat from sheep and beef cattle grazing the contaminated areas (Weber et al., 2018). Management recommendations include not grazing contaminated areas, feeding non-contaminated feed sources during the finishing phase prior to slaughter, increasing the cutting height of plants grown on contaminated soil to reduce exposure to contaminated soil and mixing non-contaminated feed with contaminated feed sources to dilute the PCB in the whole feed. PAHs are classified as carcinogenic, mutagenic, teratogenic and immunotoxic to micro- and macro-organisms (Patel et al., 2020). PAHs are commonly found in aquatic and terrestrial environments and are strongly adsorbed to soil particles where they can enter the food chain due to plant uptake of contaminated soils (O'Connor et al., 2022).

Assuming increasing use of bio-based fertilizers for food production in the future, it would be beneficial to have concentration limits for CEC. Risk estimation based on risk quotients (RQs) indicated generally low environmental risks associated with application of bio-based fertilizer to soils for food crop production. However, the toxicity and potential concentration and build-up in soils of CEC mixtures needs to be considered when estimating the risks from application of bio-based fertilizers on agricultural land or in other production systems.

3 Case study: intensive dairy production in the northwest United States

3.1 Background

Intensive livestock production regions generate large volumes of manure that can be recycled back through crops, creating more sustainable circular systems. However, the large volume of manure and the potential contaminants found within can also create challenges. As an example, we will discuss several aspects of an intensive dairy production region and issues regarding sustainability/circularity of using manure as a bio-based fertilizer within the region. Idaho is the third largest dairy producing state in the United States with 70% of the lactating herd or ~894 000 animals (including mature and young stock) located in South Central Idaho, also known as the 'Magic Valley' region (Fig. 2). This region has approximately 429 000 ha of cropland resulting in an animal density of ~2 AU ha^{-1} which has been shown to be the stocking rate at which regional nutrient surpluses can occur (Liu et al., 2017; Nesme et al., 2015). All the manure generated within the region is land applied, with a majority applied either as a solid (some of which is composted) or a liquid (following solid separation) which is blended with irrigation water and applied through the irrigation system.

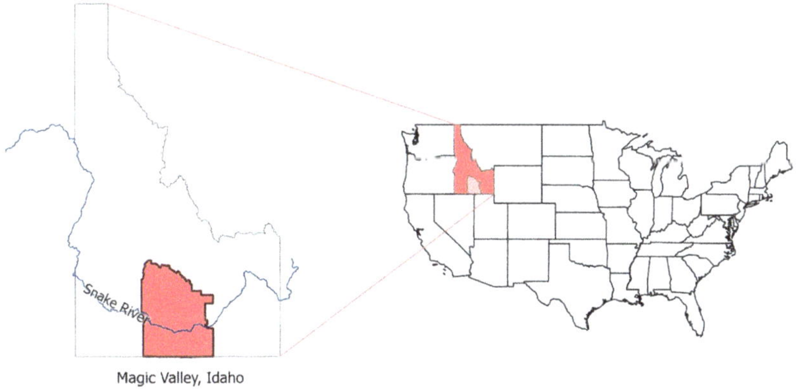

Figure 2 The Magic Valley region of South Central Idaho. This region is the third largest dairy producing region in the United States with ~894 000 animals (including mature and young stock) and approximately 429 000 ha of cropland resulting in an animal density of ~2 AU ha^{-1}.

This region has a semi-arid climate with an annual average precipitation of only 270 mm, which mainly occurs during the non-growing season. Therefore, all cropland in the region is irrigated with water diverted from the Snake River. This water is managed by the Twin Falls Canal Company (TFCC) south of the Snake River and the Northside Canal Company to the north. The TFCC routes irrigation water through 180 km of main canals and over 1600 km of smaller channels and laterals. Irrigation water flows by gravity from the Snake River throughout the 82 000 ha watershed (Fig. 3). Natural channels or coulees often convey water to laterals and collect runoff and return flow from fields. These return flows ultimately convey water back to the Snake River and can carry nutrients and other contaminants that may impair water quality. This irrigation tract has served as a benchmark watershed for the USDA Agricultural Research Service since 2004 and provides a unique system in which to evaluate the impact of land management practices on water quality in the region. In this section, we discuss the impacts of the recycling of manure products on nutrient balances and losses as well as the impact on antibiotic resistance in the regional environment.

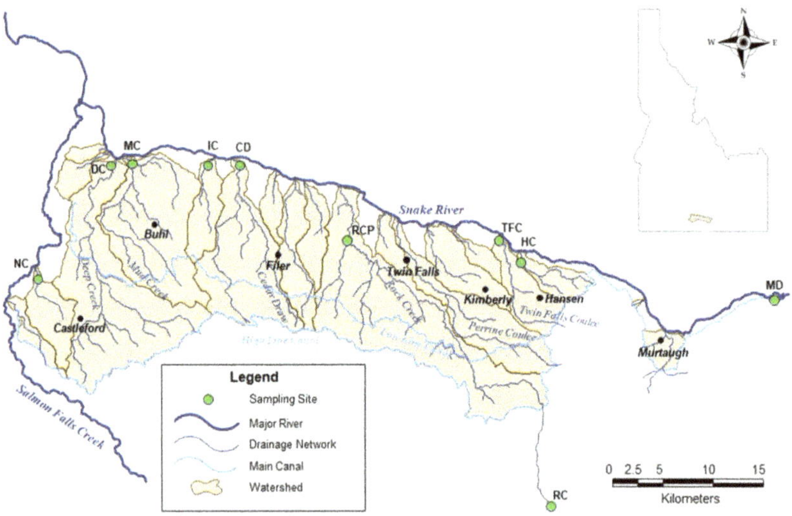

Figure 3 The Twin Falls Canal Company irrigation system including sampling sites for studies located within the watershed. The sampling sites within the Upper Snake Rock Watershed in South Central Idaho: N coulee (NC), Deep Creek (DC), Mud Creek (MC), I Coulee (IC), Cedar Draw (CD), Rock Creek Poleline (RCP), Twin Falls Coulee (TFC), Hansen Coulee (HC), Milner Dam (MD) and Rock Creek (RC). Both MD and RC were upstream sites and used to determine background concentrations. The NC and other return streams flow in a northerly direction into Salmon Falls Creek and Snake River, respectively. The Snake River flows to the west.

3.2 Impact of intensive dairy production on regional nutrient balances and losses of reactive nitrogen

Dairy production in the Magic Valley region of Idaho is characterized by high animal density at the regional (~2 AU ha^{-1}) as well as the farm level, where typical densities are closer to 4–20 AU ha^{-1}. Due to the import of nutrients via feed, on-farm surpluses of approximately 11 kg P cow^{-1} $year^{-1}$ and 174 kg N cow^{-1} $year^{-1}$ (including unaccounted for N) have been reported (Hristov et al., 2006; Spears et al., 2003). Unaccounted for N in these budgets are associated with losses of volatile N in housing and manure storage, which are ~50% of N excreted in some cases, resulting in N deficits for on-farm forage production (Leytem et al., 2018). Modeling of representative farms in the region has indicated that 70% of farms cannot meet crop N needs with manure only and need to import fertilizer N, while 80% of dairies produced more P on-farm than can be used by growing crops (Dell et al., 2022). These farm gate P surpluses have led to a buildup of soil P on many producer fields as transportation of manure longer distances is cost prohibitive. The Snake River, which receives runoff from agricultural fields in the region, has many segments of the river which have been identified as impaired with respect to P, resulting in total maximum daily loads (TMDLs) being set at 0.075 mg L^{-1} for total P (TP; IDEQ, 2010). In response, the Idaho State Department of Agriculture (ISDA) has required dairies in the state to develop a nutrient management plan to regulate, in particular, the amount of P being land-applied and evaluate the risk of potential P losses to both surface and groundwater via use of a P Site Index (Leytem et al., 2017).

Leytem et al. (2021) investigated the potential to recycle nutrients generated by dairy cattle in the region within the current agricultural system to better balance nutrients produced with regional crop demands, potentially ameliorating negative environmental impacts. They reported that manure N and P accounted for 45% and 55% of total regional inputs, respectively, with the balance consisting of fertilizer and biological N fixation (15%). There was a regional N surplus of 24 172 MT N $year^{-1}$, when accounting for all N inputs, compared to crop removal, and accounting for losses of both manure and fertilizer N as NH_3 (Fig. 4a). Overall, N losses as NH_3 were estimated to be approximately 44 000 MT N $year^{-1}$, representing 47% of N removed by crops each year. There was a regional P surplus of 8913 MT P $year^{-1}$ (Fig. 4b). However, when assessing P application per ha of cropland reported as receiving manure, the average surplus was 146 kg P ha^{-1} $year^{-1}$. For both N and P, there was a positive relationship between nutrient surpluses and dairy cattle populations ($r^2 > 0.9$). If manure N could be distributed across the regional cropland, then manure N could supply 44% of crop demand. However, it is important to keep in mind that 100% of the manure N is not available the year it is applied

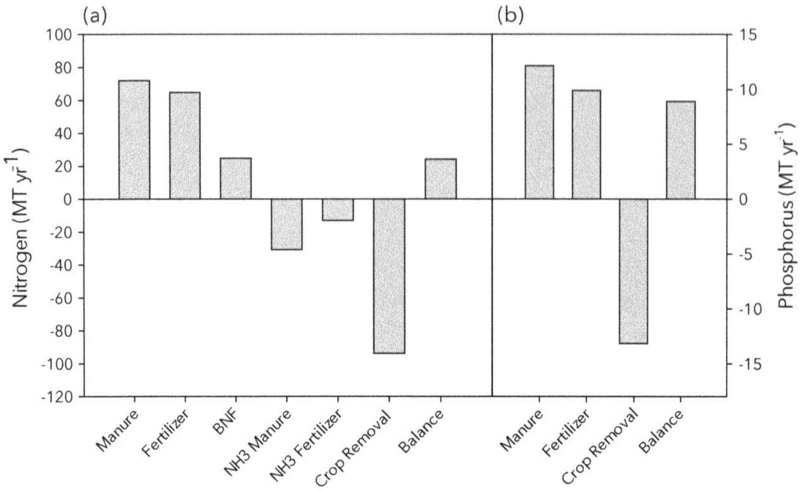

Figure 4 Regional nitrogen (a) and phosphorus (b) balances within the Magic Valley of South Central Idaho, an intensive dairy production region. Balances include N and P applied with manure and fertilizer and removed through crop uptake. Biological nitrogen fixation (BNF) was included as a source of N and losses of N as, ammonia (NH_3), from manure and fertilizer were also included in the balance.

and can vary substantially due to manure type, soils and climate. Average N mineralization rates for this region for solid manure (which represents roughly 67% of total manure) range between 15% and 30% (Leytem, unpublished data) each year. However, over time this N may mineralize and become available for plant uptake or losses to the environment. If N lost as NH_3 could be captured and re-used, then manure N could supply 77% of crop uptake. If manure P were distributed across the regional cropland, it would account for 92% of crop uptake and could replace synthetic fertilizer on most of the acreage. Utilizing the available road system, manure would need to be transported approximately 12.9 km to efficiently cycle it through the regional cropping system. However, due to fields that already have excess P and are not suitable for additional P applications, this transportation distance is likely greater. At present, the average maximum manure transport distance is 6–7 km from the dairy, therefore cost-effective methods for redistribution of manure P from areas of surplus to areas of deficit would need to be developed.

Losses of reactive N in this region have the potential to negatively impact the environment. Regionally, 44 000 MT N year^{-1} is lost as NH_3-N from agricultural production and there is potential for redeposition of N in areas where it could have negative impacts on N sensitive ecosystems. The Magic Valley is surrounded by native rangeland and is upwind of Yellowstone and Grand Teton National Parks, both of which are sensitive to N inputs. Estimating transport and deposition of this NH_3-N has been a new area of investigation in the region.

Utilizing a regional NH_3 monitoring network, estimates of N fluxes indicated a deposition rate of 10 to 40 kg N ha^{-1} $year^{-1}$ in areas impacted by agriculture and dairy production (Leytem et al., unpublished) while the surrounding natural environment may be sensitive to inputs as low as 2 kg ha^{-1} $year^{-1}$. These additions of N can lead to water quality issues and ecosystem impacts in downwind regions. With a regional N surplus of 24 172 MT N $year^{-1}$, water quality is also a concern. The USGS evaluated nitrate trends in groundwater over the periods 2000 to 2020 and found increasing trends in regions dominated by intensive agriculture/dairy production, although a majority of wells tested have NO_3 + NO_2 concentrations less than 10 mg L^{-1}, which is below the EPA threshold for drinking water (Skinner, 2022). Lentz et al. (2018) reported that NO_3-N in shallow groundwater in the region increased 1.4-fold from the late 1960s to early 2000s. Even though most well NO_3 concentrations are below the current U.S. EPA threshold, there is concern about increasing trends over time and what that might mean for groundwater quality in the future.

In the Magic Valley region, given the large annual inputs of N, there is potential of significant N_2O losses which can have negative impacts on climate change. To evaluate these potential losses, plot scale research has been conducted within the region to determine the effects of manure/fertilizer application, crop rotation and climate on N_2O losses. In general, fluxes of N_2O-N were greatest when both soil moisture and temperature were high (0.35 m^3 m^{-3} and 25°C), with a few large emission pulses accounting for the majority of growing season N_2O losses (Dungan et al., 2017, 2021; Leytem et al., 2019). In general, the addition of stacked solid manure had greater N_2O emissions than fertilizers or composted manures (Dungan et al., 2017). This enhanced loss of N_2O carried over beyond the year of application with higher losses from treatments receiving manure application 2 years previously. Growing season N_2O emissions also increased significantly with increasing manure application rate (Leytem et al., 2019). The amount of N_2O-N lost as a percentage of the total N applied across multiple studies ranged from <0.01% to 0.41% (Table 1). Regionally, losses of N_2O from application of manure and fertilizer would contribute approximately 1179 MT N_2O-N $year^{-1}$ (assuming 1% loss) to the atmosphere contributing to climate change.

Despite the higher N_2O emissions from manure application, soil organic carbon (SOC) was found to have increased in the soil profile (0-91 cm), while fertilizer treatments had a slight net loss of SOC (Bierer et al., 2021). This increase in SOC can help offset GHG emissions; however, SOC levels begin to decrease once manure applications are terminated, thus the offset will certainly decrease with time since manure was last applied. In addition to providing a potential GHG offset, studies have shown that manure can improve soil health by slowing down or reversing declining organic matter levels. The effect of manure application on soil health was evaluated in the top 30 cm of soil using 12 different biological

Table 1 Emission factors for nitrous oxide losses (N_2O) within the magic valley of southern Idaho

Reference	Treatment	%N_2O-N of TN applied
Dungan et al., 2017	Urea[a]	0.21
	SuperU[b]	0.09
	Compost[c] (34 Mg ha^{-1} annually)	0.09
	Fall manure[d] (52 Mg ha^{-1} annually)	0.09
	Spring manure (52 Mg ha^{-1} annually)	0.12
Leytem et al., 2019	Low manure application rate (18 Mg ha^{-1} annually)	0.13
	Low manure application rate (36 Mg ha^{-1} biennially)	0.15
	High manure application rate (52 Mg ha^{-1} annually)	0.18
	Urea	0.24
Dungan et al., 2021	Urea	<0.01
	Super U	<0.01
	Compost (34 Mg ha^{-1} annually)	0.06
	Fall manure (52 Mg ha^{-1} annually)	0.13
	Spring manure (52 Mg ha^{-1} annually)	0.41

Emission factors are presented as the loss of N_2O-N as a percentage of the total amount of N applied and represent the average (or cumulative) losses over multiple years.
[a] Urea fertilizer only.
[b] SuperU [(stabilized granular urea with urease {*N*-[*n*-butyl]-thiophosphoric triamide) and nitrification (dicyandiamide) inhibitors}].
[c] Composted dairy manure scraped from open lots and composted in windrows.
[d] Solid manure scraped from open lot dairies applied in the fall (after harvest) or in the spring (prior to planting) with target application rates of 18, 36 and 52 Mg ha^{-1} annually (dry weight basis).

and chemical indicators (Dungan et al., 2022). These metrics are commonly used to quantify organic matter pools and biological nutrient cycling via enzyme activities and N transformation rates. Compared to synthetic fertilizer, application of manure had a significant positive effect on most indicators, which increased with increasing manure application rate. While these results suggest that manured soil is healthier than soil receiving synthetic fertilizer, the caveat is that applying high rates of manure for many years has resulted in the gradual increase of N and P in the top 30 cm compared to synthetic fertilizer. As previously discussed, this buildup of N and P can pose a threat to water quality.

3.3 Impact of intensive agricultural production on determinants of antibiotic resistance in the environment

While land application of manure can improve soil fertility and health, there are growing concerns over the impact of this practice on the development and dissemination of antibiotic resistance in the environment (Fig. 5). An

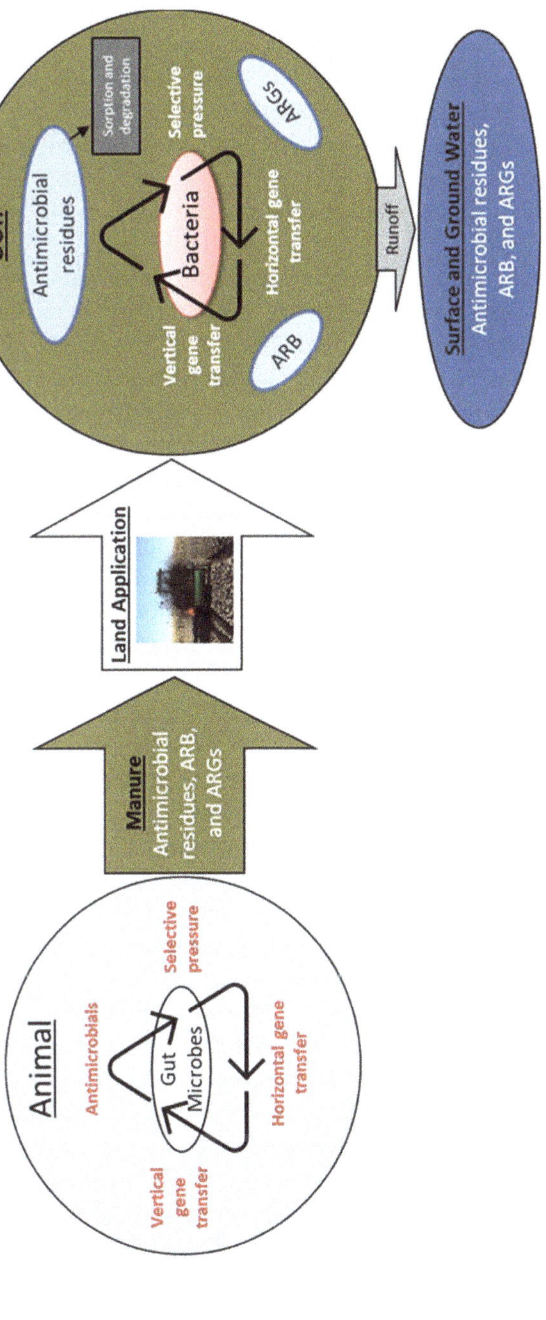

Figure 5 A conceptual diagram showing the potential fate of antimicrobial residues and mechanisms by which antibiotic resistance can be acquired by bacteria within the animal gut and in soil following the land application animal manure. Animal manure application is a major pathway by which antimicrobials, antibiotic resistance genes and bacteria are introduced into the soil environment. Figure adapted from Chee-Sanford et al. (2009).

investigation into the effect of solid dairy manure application on the abundance of ARGs during crop production was performed at plot scale (McKinney et al., 2018). Some ARGs detected at higher relative abundances (normalized to the 16S rRNA gene) in soil with manure application were those encoding resistance to sulfonamide and tetracycline antibiotics. In general, it was found that (1) manure application significantly increased ARG abundance in the top 30 cm of soil compared to plots receiving synthetic fertilizer only, (2) gene abundance increased with increasing manure application rate and (3) subsequent annual applications of manure did not increase the gene abundance above that of the first application.

Dairy wastewater is also commonly applied to agricultural soils; thus, it represents another mechanism by which antibiotic resistance determinants can be disseminated in the environment. To improve knowledge on this topic, a small-scale plot study was established to determine the effect of straight or diluted dairy wastewater on the abundance of ARGs (Dungan et al., 2018). The relative abundance of most ARGs increased dramatically after wastewater irrigation, compared to irrigation of control plots with canal water, and high gene levels were maintained throughout the 6-month study period. The results from this study suggest that increased ARG abundance was by addition of intracellular and extracellular genes present in the wastewater, not by enrichment of ARB. However, it was not known if the presence of antibiotic residues and other xenobiotic compounds may have affected gene selection and persistence in this study. What was clear from these results was that wastewater irrigation dramatically increased ARG abundance in soils receiving straight or diluted wastewater. The increase of the ARG reservoir is a potential cause for concern, as it could facilitate acquired resistance in bacteria that are pathogenic to humans and food-producing animals.

Moving beyond plot-scale studies, research was conducted to determine the abundance of several ARGs in agricultural and non-agricultural soils under various land use practices in the region, including cropland, forestland, inactive cropland, pastureland, rangeland, recreational and residential lands (Fig. 6, Dungan et al., 2019). Soil samples were obtained from 96 sites within seven counties in South Central Idaho. ARGs were detected in many of the soils (15 to 58 out of 96 samples), with a sulfonamide resistance gene being detected the most frequently (60% of samples). All the genes were detected more frequently in the cropland soils (46 sites) and at statistically greater relative abundances than in soils from the other land use categories. When the cropland gene data were separated by sites that had received dairy manure, dairy wastewater and/or biosolids (27 sites), it was revealed that the genes were found at statistically greater abundances (7- to 22-fold higher on average) than in soils that were not treated. This study provided evidence that agricultural production, especially use of manure and biosolids, is contributing to the

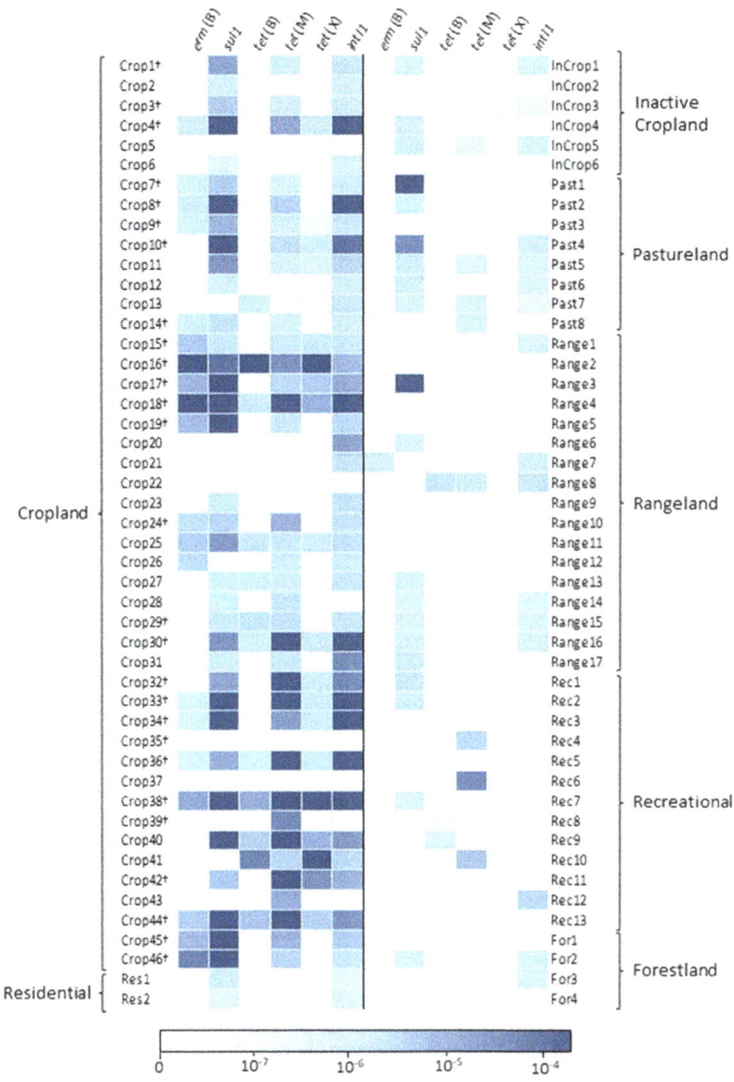

Figure 6 A heat map illustrating the relative abundance of selected antibiotic resistance genes in soils by land use category. The tetracycline resistance genes cover three known mechanisms of resistance including efflux [*tet*(B)], ribosomal protection [*tet*(M)] and enzymatic [*tet*(X)]; *sul1* represents a main determinant of sulfonamide resistance in Gram-negative bacteria; *erm*(B) confers resistance to macrolide-lincosamide-streptogramin B (MLSB) antibiotics and *intI1* is a class 1 integron-integrase gene and is linked to antibiotic resistance genes. The color scale at the bottom indicates the gene abundance from no detection (white) to $\geq 10^{-4}$ gene copies/16S rRNA gene copies (dark blue). †Indicates cropland soils that have a history of receiving dairy manure, dairy wastewater and/or biosolids.

expansion of antibiotic resistance-related determinants in regional cropland soils.

Land application of manures and biosolids containing antibiotic residues can pose a risk for transport of these residues to surface water via runoff (Joy et al., 2013; Soni et al., 2015; Bartelt-Hunt et al., 2009; Dolliver and Gupta, 2008; Kay et al., 2004). Antibiotics associated with clinical uses can also reach the environment from municipal wastewater treatment plant effluent discharges (Batt et al., 2006; Michael et al., 2013). Regardless of source, antibiotic contamination is recognized as being widespread in water resources. The spatial and temporal occurrence of 21 antibiotics in eight irrigation return flows (IRFs) within the TFCC irrigation tract was investigated (Dungan et al., 2017). Seven antibiotics of veterinary and human origins were detected at frequencies ranging from 3.1% to 62.5%. Monensin, which is exclusively used in cattle and poultry production, was the most frequently detected antibiotic and it was present in all IRFs (Fig. 7). This result suggests that monensin, and other veterinary antibiotics, are entering return flows in runoff from fields that had received livestock manure or wastewater. Some antibiotics were also detected at the site of irrigation water diversion from the Snake River. Therefore, even cropped soils that are not treated with manure, wastewater or biosolids are still receiving low-level antibiotics during irrigation events. The occurrence of antibiotics at low concentrations may pose a health risk to humans and animals due to the proliferation of antibiotic resistance in the environment (Kemper, 2008).

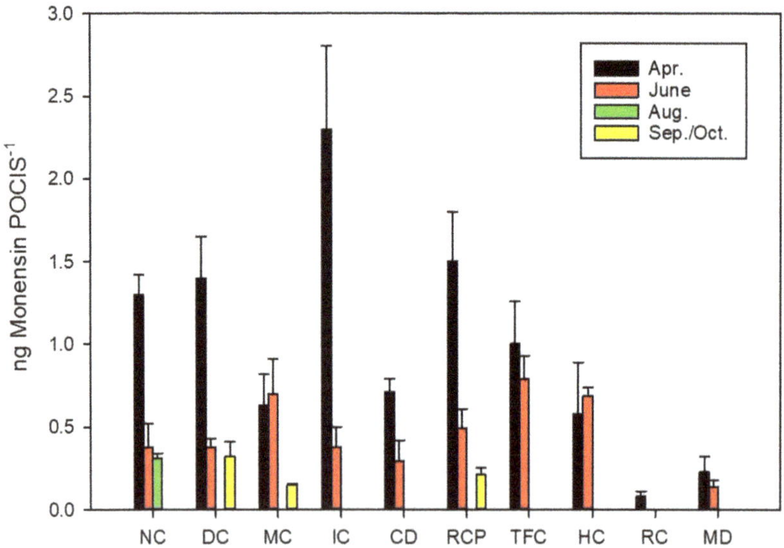

Figure 7 Mass of monensin extracted from polar organic compound integrative samplers (POCIS) that were deployed at the sampling sites within the Upper Snake Rock watershed in South Central Idaho. Columns represent means ± SD (n = 3).

Surface waters can be a dominant route by which ARGs and ARB are disseminated in the regional watershed. A study was conducted to determine the abundance of selected ARGs in the IRFs that transport irrigation water from agricultural fields back to the Snake River (Dungan and Bjorneberg, 2020). The ARGs were recovered at all IRF sampling sites with detections ranging from 55 to 81 out of 81 water sampling events. When compared to the average annual relative gene abundances in the canal water samples, they were found to be at statistically greater levels. The fact that most IRFs contained higher levels than found in the canal water indicates that IRFs can be a source of ARGs that ultimately discharge into the Snake River.

The antimicrobial susceptibility among *E. coli* and enterococcal isolates that were obtained from the regional watershed were determined (Dungan and Bjorneberg, 2021). *E. coli* and enterococci are important indicators for understanding the impact of fecal pollution on water resources but reports on antimicrobial resistance among these organisms in IRFs are lacking. Environmental isolates can be a potentially important source of AMR and return flows may be one way resistance genes are transported out of agroecosystems. For *E. coli*, 75% of the isolates were pan-susceptible, while 13% of isolates were resistant to tetracycline and fewer numbers being resistant to 13 other antimicrobials. For the enterococcal species, only 9% of isolates were pan-susceptible and the single highest resistance was to lincomycin (75%). Furthermore, 17 *E. coli* and 13 enterococcal isolates were found to be multidrug resistant to up to seven different drug classes. These results indicate that the IRFs are polluted with material of fecal origin, which would not be surprising given that this is a mixed-use watershed and livestock manures are commonly applied to cropland soils. However, some of the *E. coli* and enterococcal isolates could be from naturalized extraintestinal populations that were released from sediments and soils. Regardless of fecal bacteria source, a wide variety of resistance patterns were found among many of the isolates, suggesting the potential for horizontal transfer of ARGs in the aquatic environment.

4 Conclusion

The valuable nutrients and C found in many agricultural byproducts have the potential to be recycled as bio-based fertilizers within the greater agricultural system, thereby enhancing the circularity and sustainability of this sector. While the potential for beneficial use of these products is great, there are potential risks associated with their use that must be carefully considered. Balancing nutrients in areas with intensive livestock production, avoiding health risks associated with bioaccumulation and biomagnification of heavy metals as well as reducing exposure to pathogens, antibiotics and other CEC are imperative for these products to be used safely and sustainably. Careful risk assessment and management/treatment may overcome most barriers for reuse of these valuable products enabling greater use throughout the agricultural sector.

5 Future trends in research

To further improve the recycling and reuse of agricultural byproducts as bio-based fertilizer in food production systems, further research is needed to address several areas of concern. Improved nutrient use efficiency and reduction of losses, in particular N and P, to the environment are greatly needed. Technologies that allow capture and stabilization of N and P from livestock manure over a range of farm sizes are needed to more effectively recycle these nutrients back to cropland that is deficient in these nutrients. This also includes methods to condense nutrients to reduce the cost of transporting them from regions of surplus to those of deficit. Improved models for estimating crop N uptake to better match N additions are necessary to enhance overall N use efficiencies. Improved models are also needed to estimate N and P losses to the environment to better predict the effects of management strategies on air and water quality as well as losses of N_2O that may impact climate change.

In livestock systems, alternatives to the use of metals (Cu and Zn) and antibiotics in diets as growth promoters are necessary to reduce their potential negative impacts when manure is used as a fertilizer in the landscape. Effective treatment technologies are needed to remove metals, kill pathogens, remove or inactivate antibiotics, and other CECs from agricultural byproducts that may be used as bio-based fertilizers are needed to ensure that these products can be safely and sustainably used within the food production system. Improved risk assessments to establish critical levels of these potential contaminants are also needed to provide guidance for reuse of byproducts as bio-based fertilizers.

6 References

Abascal, E., Gómez-Coma, L., Ortiz, I. and Ortiz, A. (2022). Global diagnosis of nitrate pollution in groundwater and review of removal technologies, *Sci. Total Environ.* 810, 152233. doi: 10.1016/j.scitotenv.2021.152233.

Abu-Ashour, J. and Lee, H. (2000). Transport of bacteria on sloping soil surfaces by runoff, *Environ. Toxicol.* 15(2), 149–153.

Abu-Ashour, J., Joy, D. M., Lee, H., Whiteley, H. R. and Zelin, S. (1994). Transport of microorganisms through soil, *Water Air Soil Pollut.* 75(1–2), 141–158.

Adegoke, A. A., Awolusi, O. O. and Stenström, T. A. (2016). Organic fertilizers: public health intricacies. In: Larramendy, M. L. and Soloneski, S. (Eds), *Organic Fertilizers-From Basic Concepts to Applied Outcomes*. IntechOpen, Croatia, pp. 343–374.

Published by Burleigh Dodds Science Publishing Limited, 2024.

Alegbeleye, O. O., Singleton, I. and Sant'Ana, A. S. (2018). Sources and contamination routes of microbial pathogens to fresh produce during field cultivation: a review, *Food Microbiol.* 73, 177-208.

Alekshun, M. N. and Levy, S. B. (2007). Molecular mechanisms of antibacterial multidrug resistance, *Cell* 128(6), 1037-1050.

Amarakoon, I. D., Surac, S., Zvomuya, F., Cessna, A. J., Larney, F. J. and McAllister, T. A. (2016). Dissipation of antimicrobials in a seasonally frozen soil after beef cattle manure application, *J. Environ. Qual.* 45(5), 1644-1651.

Bahe, A., Classen, J. and Williams, B. (2006). Antibiotic resistance and the environment: Current science and policy considerations. In Rice, J. M. Caldwell, D. F. and Humenik, F. J. (Eds). *Animal Agriculture and the Environment, National Center for Manure & Animal Waste Management White Papers*. American Society of Agricultural and Biological Engineers, St Joseph, pp. 89-108.

Bartelt-Hunt, S. L., Snow, D. D., Damon, T., Shockley, J. and Hoagland, K. (2009). The occurrence of illicit and therapeutic pharmaceuticals in wastewater effluent and surface waters in Nebraska, *Environ. Pollut.* 157(3), 786-791.

Bartelt-Hunt, S. L., Snow, D. D., Damon-Powell, T. and Misebach, D. (2011). Occurrence of steroid hormones and antibiotics in shallow groundwater impacted by livestock waste control facilities, *J. Contam. Hydrol.* 123(3-4), 94-103.

Bateman, A., der Horst, D., Boardman, D., Kansal, A., and Carliell-Marquet, C.(2011). Closing the phosphorus loop in England: the spatio-temporal balance of phosphorus capture from manure versus crop demand for fertilizer, *Res. Conserv. Recyl* 55, 1146-1153.

Batt, A. L., Snow, D. D. and Aga, D. S. (2006). Occurrence of sulfonamide antimicrobials in private water wells in Washington County, Idaho, USA, *Chemosphere* 64(11), 1963-1971.

Baxter-Potter, W. R. and Gilliland, M. W. (1988). Bacterial pollution in runoff from agricultural lands, *J. Environ. Qual.* 17(1), 27-34.

Beneragama, N., Iwasaki, M., Lateef, S. A., Yamashiro, T., Ihara, I. and Umetsu, K. (2013). The survival of multidrug-resistant bacteria in thermophilic and mesophilic anaerobic co-digestion of dairy manure and waste milk, *Anim. Sci. J.* 84(5), 426-433.

Berry, E. D., Woodbury, B. L., Nienaber, J. A., Eigenberg, R. A., Thurston, J. A. and Wells, J. E. (2007). Incidence and persistence of zoonotic bacterial and protozoan pathogens in a beef cattle feedlot runoff control vegetative treatment system, *J. Environ. Qual.* 36(6), 1873-1882.

Bhatnagar, A. and Sillanpää, M. (2011). A review of emerging adsorbents for nitrate removal from water, *Chem. Eng. J.* 168(2), 493-504.

Bierer, A. M., Leytem, A. B., Dungan, R. S., Moore, A. D. and Bjorneberg, D. L. (2021). Soil organic carbon dynamics in semi-arid irrigated cropping systems, *Agronomy* 11(3), 484. doi: 10.3390/agronomy11030484.

BijaySingh, and Craswell, E. (2021). Fertilizers and nitrate pollution of surface and ground water: an increasingly pervasi`ve global problem, *SN Appl. Sci.* 3(4), 518. doi: 10.1007/s42452-021-04521-8.

Binh, C. T. T., Heuer, H., Kaupenjohann, M. and Smalla, K. (2008). Piggery manure used for soil fertilization is a reservoir for transferable antibiotic resistance plasmids, *FEMS Microbiol. Ecol.* 66(1), 25-37.

Bloem, E., Albihn, A., Elving, J., Hermann, L., Lehmann, L., Sarvi, M., Schaaf, T., Schick, J., Turtola, E. and Ylivainio, K. (2017). Contamination of organic nutrient sources with potentially toxic elements, antibiotics and pathogen microorganisms in relation to P fertilizer potential and treatment options for the production of sustainable fertilizers: a review, *Sci. Total Environ.* 607-608, 225–242.

Bodirsky, B. L., Popp, A., Weindl, I., Dietrich, J. P., Rolinski, S., Scheiffele, L., Schmitz, C. and Lotze-Campen, H. (2012). N_2O emission from the global agricultural nitrogen cycle: current state and future scenarios, *Biogeosciences* 9(10), 4169–4197.

Boudh, S. and Singh, J. S. (2019). Pesticide contamination: environmental problems and remediation strategies. In: Bharagave, R. N. and Chowdhary, P. (Eds). *Emerging and Eco-friendly Approaches for Waste Management*. Springer, Singapore, pp. 245–269.

Boxall, A. B. A., Kolpin, D. W., Sorensen, B. H. and Tolls, J. (2003). Are veterinary medicines causing environmental risks?, *Environ. Sci. Technol.* 37(15), 286A–294A.

Boxall, A. B. A., Fogg, L. A., Blackwell, P. A., Kay, P., Pemberton, E. J. and Croxford, A. (2004). Veterinary medicines in the environment, *Rev. Environ. Contam. Toxicol.* 180, 1–91.

Cajamarca, S. M. N., Martins, D., da Silva, J., Fontenelle, M., Guedes, Í., de Figueiredo, C. and Pacheco Lima, C. (2019). Heterogeneity in the chemical composition of biofertilizers, potential agronomic use, and heavy metal contents of different agro-industrial wastes, *Sustainability* 11(7), 1995. doi: 10.3390/su11071995. Available at: www.mdpi.

Callejón, R. M., Rodríguez-Naranjo, M. I., Ubeda, C., Hornedo-Ortega, R., Garcia-Parrilla, M. C. and Troncoso, A. M. (2015). Reported foodborne outbreaks due to fresh produce in the United States and European Union: trends and causes, *Foodb. Pathog. Dis.* 12(1), 32–38.

Campagnolo, E. R., Hohnson, K. R., Karpati, A., Rubin, C. S., Kolpin, D. W., Meyer, M. T., Esteban, J. E., Currier, R. W., Smith, K., Thu, K. M. and McGeehin, M. (2002). Antimicrobial residues in animal water and water resources proximal to large-scale swine and poultry feeding operations, *Sci. Total Environ.* 299(1–3), 89–95.

Center for Disease Control (2020). Foodborne germs and illnesses. Available at: https://www.cdc.gov/foodsafety/foodborne-germs.html (accessed October 17, 2022).

Chang, R. X., Michel Jr., F. C., Gan, J. J., Wang, Q., Wang, Z. Z. and Li, Y. M. (2017). Effect of single and combined herbicides in compost on growth of sensitive crops: green bean, cucumber, and tomato, *Compost Sci. Util.* 25 (sup1), S23–S30.

Chantziaras, I., Boyen, F., Callens, B. and Dewulf, J. (2014). Correlation between veterinary antimicrobial use and antimicrobial resistance in food-producing animals: a report on seven countries, *J. Antimicrob. Chemother.* 69(3), 827–834.

Chaturvedi, P., Shukla, P., Giri, B. S., Chowdhary, P., Chandra, R., Gupta, P. and Pandey, A. (2021). Prevalence and hazardous impact of pharmaceutical and personal care products and antibiotics in environment: a review on emerging contaminants, *Environ. Res.* 194, 110664.

Chee-Sanford, J. C., Mackie, R. I., Koike, S., Krapac, I. G., Lin, Y.-F., Yannarell, A. C., Maxwell, S. and Aminov, R. I. (2009). Fate and transport of antibiotic residues and antibiotic resistance genes following land application of manure waste, *J. Environ. Qual.* 38, 1086–1108. doi: 10.2134/jeq2008.0128.

Christian, T., Schneider, R. J., Färber, H. A., Skutlarek, D., Meyer, M. T. and Goldbach, H. E. (2003). Determination of antibiotic residues in manure, soil, and surface waters, *Acta Hydrochim. Hydrobiol.* 31(1), 36–44.

Coyne, M. S., Gilfillen, R. A., Rhodes, R. W. and Belvins, R. L. (1995). Soil and fecal coliform trapping by grass filter strips during simulated rains, *J. Soil Water Conserv.* 50, 405–408.

Cytryn, E. (2013). The soil resistome: the anthropogenic, the native, and the unknown, *Soil Biol. Biochem.* 63, 18-23.

Davidson, E. A., David, M. B., Galloway, J. N., Goodale, C. L., Haeuber, R., Harrison, J. A., Howarth, R. W., Jaynes, D. B., Lowrance, R. R., Nolan, B. T., Peel, J. L., Pinder, R. W., Porter, E., Snyder, C. S., Townsend, A. R. and Ward, M. H. (2012). Excess nitrogen in the U.S. environment: trends, risks, and solutions. Issues in Ecology, report #15. Ecological Society of America, Washington, DC.

Dell, C. J., Baker, J. M., Spiegal, S., Porter, S. A., Leytem, A. B., Flynn, K. C., Rotz, C. A., Bjorneberg, D. L., Bryant, R. B., Hagevoort, G. R., Williamson, J. C., Slaughter, A. and Kleinman, P. J. A. (2022). Challenges and opportunities for manureshed management across U.S. dairy systems: case studies from four regions, *J. Environ. Qual.* 51(4), 521-539.

Diehl, D. L. and La Para, T. M. (2010). Effect of temperature on the fate of genes encoding tetracycline resistance and the integrase of class 1 integrons within anaerobic and aerobic digesters treating municipal wastewater solids, *Environ. Sci. Technol.* 44(23), 9128-9133.

Dolliver, H. and Gupta, S. (2008). Antibiotic losses in leaching and surface run-off from manure-amended agricultural land, *J. Environ. Qual.* 37(3), 1227-1237.

Doran, J. W. and Linn, D. M. (1979). Bacteriological quality of runoff water from pastureland, *Appl. Environ. Microbiol.* 37(5), 985-991.

Dungan, R. S. (2010). Board-Invited Review: fate and transport of bioaerosols associated with livestock operations and manures, *J. Anim. Sci.* 88(11), 3693-3706.

Dungan, R. S. and Bjorneberg, D. L. (2020). Antibiotic resistance genes, class 1 integrons, and IncP-1/IncQ-1 plasmids in irrigation return flows, *Environ. Pol.L.* 257, 113568, 1-8.

Dungan, R. S. and Bjorneberg, D. L. (2021). Antimicrobial resistance in Escherichia coli and enterococcal isolates from irrigation return flows in a high-desert watershed, *Front. Microbiol.* 12, 660697.

Dungan, R. S., Leytem, A. B., Tarkalson, D. D., Ippolito, J. A. and Bjorneberg, D. L. (2017a). Greenhouse gas emissions from an irrigated cropping system as influenced by nitrogen source and timing, *Soil Sci. Soc. Am. J.* 81(3), 537-545.

Dungan, R. S., Snow, D. D. and Bjorneberg, D. L. (2017b). Occurrence of antibiotics in an agricultural watershed in south-central Idaho, *J. Environ. Qual.* 46(6), 1455-1461.

Dungan, R. S., McKinney, C. W. and Leytem, A. B. (2018). Tracking antibiotic resistance genes in soil irrigated with dairy wastewater, *Sci. Total Environ.* 635, 1477-1483.

Dungan, R. S., Strausbaugh, C. A. and Leytem, A. B. (2019). Survey of selected antibiotic resistance genes in agricultural and non-agricultural soils in south-central Idaho, *FEMS Microbiol. Ecol.* 95(6). doi: 10.1093/femsec/fiz071.

Dungan, R. S., Leytem, A. B. and Tarkalson, D. D. (2021). Greenhouse gas emissions from an irrigated cropping rotation with dairy manure utilization in a semiarid climate, *Agron. J.* 113(2), 1222-1237.

Dungan, R. S., McKinney, C. W., Acosta-Martinez, V. and Leytem, A. B. (2022). Response of soil health indicators to long-term dairy manure in a semiarid irrigated cropping system, *Soil Sci. Soc. Am. J.* 86(6), 1597-1610. doi: 10.1002/saj2.20462.

Elmund, G. K., Morrison, S. M., Grant, D. W. and Nevins, S. M. Sr. (1971). Role of excreted chlorotetracycline in modifying the decomposition process in feedlot waste, *Bull. Environ. Contam. Toxicol.* 6(2), 129-132.

Elser, J. J., Andersen, T., Baron, J. S., Bergström, A. K., Jansson, M., Kyle, M., Nydick, K. R., Steger, L. and Hessen, D. O. (2009). Shifts in lake N:P stoichiometry and nutrient limitation driven by atmospheric nitrogen deposition, *Science* 326(5954), 835–837.

Evans, M. R. and Owens, J. D. (1972). Factors affecting the concentration of faecal bacteria in land-drainage water, *J. Gen. Microbiol.* 71(3), 477–485.

Fajardo, J. J., Bauder, J. W. and Cash, S. D. (2001). Managing nitrate and bacteria in runoff from livestock confinement areas with vegetative filter strips, *J. Soil Water Conserv.* 56, 185–191.

FAO (2006). *Plant Nutrition for Food Security: A Guide for Integrated Nutrient Management*. FAO Fertilizer and Plant Nutrition Bulletin 16. Food and Agriculture Organization of the United Nations, Rome.

Fine, D. D., Breidenbach, G. P., Price, T. L. and Hutchins, S. R. (2003). Quantitation of estrogens in ground water and swine lagoon samples using solid-phase extraction, pentafluorobenzyl/trimethylsilyl derivatizations and gas chromatography-negative ion chemical ionization tandem mass spectrometry, *J. Chromatogr. A* 1017(1–2), 167–185.

Gidden, M. J., Riahi, K., Smith, S. J., Fujimori, S., Luderer, G., Kriegler, E., van Vuuren, D. P., van den Berg, M., Feng, L., Klein, D., Calvin, K., Doelman, J. C., Frank, S., Fricko, O., Harmsen, M., Hasegawa, T., Havlik, P., Hilaire, J., Hoesly, R., Horing, J., Popp, A., Stehfest, E. and Takahashi, K. (2019). Global emissions pathways under different socioeconomic scenarios for use in CMIP6: a dataset of harmonized emissions trajectories through the end of the century, *Geosci. Model Dev.* 12(4), 1443–1475.

Golovko, O., Ahrens, L., Schelin, J., Sörengård, M., Bergstrand, K. J., Asp, H., Hultberg, M. and Wiberg, K. (2022). Organic micropollutants, heavy metals and pathogens in anaerobic digestate based on food waste, *J. Environ. Manage.* 313, 114997.

Gong, Q., Chen, P., Shi, R., Gao, Y., Zheng, S. A., Xu, Y., Shao, C. and Zheng, X. (2019). Health assessment of trace metal concentrations in organic fertilizer in northern China, *Int. J. Environ. Res. Public Health* 16(6), 1031.

Goyal, D., Yadav, A., Prasad, M., Singh, T.B., Shrivastav, P., Ali, A., Dantu, P. K. and Mishra, S. (2020). Effect of heavy metals on plant growth: an overview. In: Naeem, M., Ansari, A. and Gill, S. (Eds). *Contaminants in Agriculture*. Springer Nature, Switzerland.

Guan, J., Wasty, A., Grenier, C. and Chan, M. (2007). Influence of temperature on survival and conjugative transfer or multiple antibiotic-resistant plasmids in chicken manure and compost microcosms, *Poult. Sci.* 86(4), 610–613.

Guo, R. X. and Chen, J. Q. (2015). Application of alga-activated sludge combined systems (AASCS) as a novel treatment to remove cephalosporins, *Chem. Eng. J.* 260, 550–556.

Gurtler, J. B., Doyle, M. P., Erickson, M. C., Jiang, X., Millner, P. and Sharma, M. (2018). Composting to inactivate foodborne pathogens for crop soil application: a review, *J. Food Prot.* 81(11), 1821–1837.

Halling-Sørensen, B., Nielsen, S. N., Lanzky, P. F., Ingerslev, F., Holten Lützhøft, H. C. and Jørgensen, S. E. (1998). Occurrence, fate and effects of pharmaceutical substances in the environment – a review, *Chemosphere* 36(2), 357–393.

Halling-Sørensen, B., Sengeløv, G. and Tjørnelund, J. (2002). Toxicity of tetracyclines and tetracycline degradation products to environmentally relevant bacteria, including selected tetracycline-resistant bacteria, *Arch. Environ. Contam. Toxicol.* 42(3), 263–271.

Published by Burleigh Dodds Science Publishing Limited, 2024.

Hamscher, G., Sczesny, S., Höper, H. and Nau, H. (2002). Determination of persistent tetracycline residues in soil fertilized with liquid manure by high-performance liquid chromatography with electrospray ionization tandem mass spectrometry, *Anal. Chem.* 74(7), 1509–1518.

Hanserud, O. S., Brod, E., Øgaard, A. F., Müller, D. B. and Brattebø, H. (2016). A multi-regional soil phosphorus balance for exploring secondary fertilizer potential: the case of Norway, *Nutr. Cycl. Agroecosyst.* 104(3), 307–320.

Hao, C., Lissemore, L., Nguyen, B., Kleywegt, S., Yang, P. and Solomon, K. (2006). Determination of pharmaceuticals in environmental waters by liquid chromatography/electrospray ionization/tandem mass spectrometry, *Anal. Bioanal. Chem.* 384(2), 505–513.

Heinonen-Tanski, H., Mohaibes, M., Karinen, P. and Koivunen, J. (2006). Methods to reduce pathogen microorganisms in manure, *Livest. Sci.* 102(3), 248–255.

Herrero, M. and Thornton, P. K. (2013). Livestock and global change: emerging issues for sustainable food systems, *Proc. Natl Acad. Sci. U. S. A.* 110(52), 20878–20881.

Heuer, H. and Smalla, K. (2007). Manure and sulfadiazine synergistically increased bacterial antibiotic resistance in soil over at least two months, *Environ. Microbiol.* 9(3), 657–666.

Holly, M. A., Kleinman, P. J., Bryant, R. B., Bjorneberg, D. L., Rotz, C. A., Baker, J. M., Boggess, M. V., Brauer, D. K., Chintala, R., Feyereisen, G. W., Gamble, J. D., Leytem, A. B., Reed, K. F., Vadas, P. A. and Waldrip, H. M. (2018). Short communication: identifying challenges and opportunities for improved nutrient management through the USDA's Dairy Agroecosystem Working Group, *J. Dairy Sci.* 101(7), 6632–6641.

Holm-Nielsen, J. B., Al Seadi, T. and Oleskowicz-Popiel, P. (2009). The future of anaerobic digestion and biogas utilization, *Bioresour. Technol.* 100(22), 5478–5484.

Honorio, J. F., Veit, M. T. and Tavares, C. R. G. (2019). Alternative adsorbents applied to the removal of natural hormones from pig farming effluents and characterization of the biofertilizer, *Environ. Sci. Poll. Res. Int.* 26(28), 28429–28435.

Hristov, A. N., Hazen, W. and Ellsworth, J. W. (2006). Efficiency of use of imported nitrogen, phosphorus, and potassium and potential for reducing phosphorus imports on Idaho dairy farms, *J. Dairy Sci.* 89(9), 3702–3712.

Hutchins, S. R., White, M. V., Hudson, F. M. and Fine, D. D. (2007). Analysis of lagoon samples from different concentrated animal feeding operations for estrogens and estrogen conjugates, *Environ. Sci. Technol.* 41(3), 738–744.

Hutchison, M. L., Walters, L. D., Avery, S. M., Munro, F. and Moore, A. (2005). Analyses of livestock production, waste storage, and pathogen levels and prevalences in farm manures, *Appl. Environ. Microbiol.* 71(3), 1231–1236.

Idaho Department of Environmental Quality. Issued permits and water quality certification. Available at: https://www.deq.idaho.gov/permits/issued-permits-and-water-quality-certifications/ (accessed 17 October 2022).

Jamieson, R. C., Gordon, R. J., Sharples, K. E., Stratton, G. W. and Madani, A. (2002). Movement and persistence of fecal bacteria in agricultural soils and subsurface drainage waters: a review, *Can. Biosyst. Eng.* 44, 1.1–1.9.

Jenkins, M. B., Endale, D. M., Schomberg, H. H. and Sharpe, R. R. (2006). Fecal bacteria and sex hormones in soil and runoff from cropped watersheds amended with poultry litter, *Sci. Total Environ.* 358(1–3), 164–177.

Jiang, Y., Xie, S. H., Dennehy, C., Lawlor, P. G., Hu, Z. H., Wu, G. X., Zhan, X. M. and Gardiner, G. E. (2020). Inactivation of pathogens in anaerobic digestion systems for converting biowastes to bioenergy: a review, *Renew. Sustain. Energy Rev.* 120, 109654.

Joy, S. R., Bartelt-Hunt, S. L., Snow, D. D., Gilley, J. E., Woodbury, B. L., Parker, D. B., Marx, D. B. and Li, X. (2013). Fate and transport of antimicrobials and antimicrobial resistance genes in soil and runoff following land application of swine manure slurry, *Environ. Sci. Technol.* 47(21), 12081–12088.

Kay, P., Blackwell, P. A. and Boxall, A. B. (2004). Fate of veterinary antibiotics in a macroporous tile drained clay soil, *Environ. Toxicol. Chem.* 23(5), 1136–1144.

Kemper, N. (2008). Veterinary antibiotics in the aquatic and terrestrial environment, *Ecol. Indic.* 8(1), 1–13.

Klein, M., Brown, L., Tucker, R. W., Ashbolt, N. J., Stuetz, R. M. and Roser, D. J. (2010). Diversity and abundance of zoonotic pathogens and indicators in manures of feedlot cattle in Australia, *Appl. Environ. Microbiol.* 76(20), 6947–6950.

Koelsch, R., Lorimor, J. and Mankin, K. (2006). 'Vegetative treatment systems for open lot runoff: Review of literature. In Rice, J. M., Caldwell, D. F. and Humenik, F. J. (Eds). *Animal Agriculture and the Environment, National Center for Manure & Animal Waste Management White Papers*. American Society of Agricultural and Biological Engineers, St Joseph, MI, pp. 576–608.

Kraus, H., Weber, A., Appel, M., Enders, B., Isenberg, H. D., Schiefer, H. G., Slenczka, W., von Graevenitz, A. and Zahner, H. (2003). *Zoonoses: Infectious Diseases Transmissible from Animals to Humans*. ASM Press, Washington, DC.

Lentz, R. D., Carter, D. L. and Haye, S. V. (2018). Changes in groundwater quality and agriculture in forty years on the Twin Falls irrigation tract in southern Idaho, *J. Soil Water Conserv.* 73(2), 107–119.

Létourneau, V., Duchaine, C., Côté, C., Letellier, A., Topp, E. and Massé, D. (2010). Presence of zoonotic pathogens in physico-chemically characterized manure from hog finishing houses using different production systems, *Bioresour. Technol.* 101(11), 4048–4055.

Levy, S. B. (1992). *The Antibiotic Paradox. How Miracle Drugs Are Destroying the Miracle*. Plenum Press, New York.

Leytem, A., Bjorneberg, D. and Tarkalson, D. (2017). The phosphorus site index: A systematic approach to assess the risk of nonpoint source pollution of Idaho waters by agricultural phosphorus. Available at: https://agri.idaho.gov/main/wp-content/uploads/2018/12/Phosphorus-Site-Index-reference-2017-revised.pdf (accessed 14 October 2022).

Leytem, A. B., Bjorneberg, D. L., Rotz, C. A., Moraes, L. E., Kebreab, E. and Dungan, R. S. (2018). Ammonia emissions from dairy lagoons in the western U.S., *Trans. ASABE* 61(3), 1001–1015.

Leytem, A. B., Moore, A. D. and Dungan, R. S. (2019). Greenhouse gas emissions from an irrigated crop rotation utilizing dairy manure, *Soil Sci. Soc. Am. J.* 83(1), 137–152.

Leytem, A. B., Williams, P., Zuidema, S., Martinez, A., Chong, Y. L., Vincent, A., Vincent, A., Cronan, D., Kliskey, A., Wulfhorst, J. D., Alessa, L. and Bjorneberg, D. (2021). Cycling phosphorus and nitrogen through cropping systems in an intensive dairy production region, *Agronomy* 11(5), 1005.

Li, H., Sun, Z., Qiu, Y., Yu, X., Han, X. and Ma, Y. (2018). Integrating bioavailability and soil aging in the derivation of DDT criteria for agricultural soils using crop species sensitivity distributions, *Ecotoxicol. Environ. Saf.* 165, 527–532.

Liu, M., Huang, X., Song, Y., Tang, J., Cao, J., Zhang, X., Zhang, Q., Wang, S., Xu, T., Kang, L., Cai, X., Zhang, H., Yang, F., Wang, H., Yu, J. Z., Lau, A. K. H., He, L., Huang, X., Duan, L., Ding, A., Xue, L., Gao, J., Liu, B. and Zhu, T. (2019). Ammonia emission control in China would mitigate haze pollution and nitrogen deposition, but worsen acid rain, *Proc. Natl Acad. Sci. U. S. A.* 116(16), 7760–7765.

Liu, Q., Wang, J., Bai, Z., Ma, L. and Oenema, O. (2017). Global animal production and nitrogen and phosphorus flows, *Soil Res.* 55(6), 451–462.

Liu, L., Xu, W., Zhong, B., Lu, X., Zhong, B., Guo, Y., Lu, X., Zhao, Y., He, W., Wang, S., Zhang, X., Liu, X. and Vitousek, P. (2022). Exploring global changes in agricultural ammonia emissions and their contribution to nitrogen deposition since 1980, *PNAS* 119(14), e2121998119. doi: 10.1073/pnas.2121998119.

Lopes, C., Herva, M., Franco-Uría, A. and Roca, E. (2011). Inventory of heavy metal content in organic waste applied as fertilizer in agriculture: evaluating the risk of transfer into the food chain, *Environ. Sci. Pollut. Res. Int.* 18(6), 918–939.

Lund, B., Jensen, V. F., Have, P. and Ahring, B. (1996). Inactivation of virus during anaerobic digestion of manure in laboratory scale biogas reactors, *Anton. Leeuw.* 69(1), 25–31.

Luo, L., Ma, Y., Zhang, S., Wei, D. and Zhu, Y. G. (2009). An inventory of trace element inputs to agricultural soils in China, *J. Environ. Manage.* 90(8), 2524–2530.

Ma, R., Li, K., Guo, Y., Zhang, B., Zhao, X., Linder, S., Guan, C., Chen, G., Gan, Y. and Meng, J. (2021). Mitigation potential of global ammonia emissions and related health impacts in the trade network, *Nat. Commun.* 12(1), 6308.

Manyi-Loh, C. E., Mamphweli, S. N., Meyer, E. L., Makaka, G., Simon, M. and Okoh, A. I. (2016). An overview of the control of bacterial pathogens in cattle manure, *Int. J. Environ. Res. Public Health* 13(9), International, J. Environ. Res, *Pub Health* 13, 843.

Martens, W. and Böhm, R. (2009). Overview of the ability of different treatment methods for liquid and solid manure to inactivate pathogens, *Bioresour. Technol.* 100(22), 5374–5378.

Matthews, K. R. (2006). *Microbiology of Fresh Produce*. ASM Press, Washington, DC.

Matthiessen, P., Allen, Y., Bamber, S., Craft, J., Hurst, M., Hutchinson, T., Feist, S., Katsiadaki, I., Kirby, M., Robinson, C., Scott, S., Thain, J. and Thomas, K. (2002). The impact of oestrogenic and androgenic contamination on marine organisms in the United Kingdom-summary of the EDMAR programme. Endocrine disruption in the marine environment, *Mar. Environ. Res.* 54(3–5), 645–649.

Mawdsley, J. L., Bardgett, R. D., Merry, R. J., Pain, B. F. and Theodorou, M. K. (1995). Pathogens in livestock waste, their potential movement through soil and environmental pollution, *Appl. Soil Ecol.* 2(1), 1–15.

McKinney, C. W., Dungan, R. S., Moore, A. and Leytem, A. B. (2018). Occurrence and abundance of antibiotic resistance genes in agricultural soil receiving dairy manure, *FEMS Microbiol. Ecol.* 94(3), 1–10.

Michael, I., Rizzo, L., McArdell, C. S., Manaia, C. M., Merlin, C., Schwartz, T., Dagot, C. and Fatta-Kassinos, D. (2013). Urban wastewater treatment plants as hotspots for the release of antibiotics in the environment: a review, *Water Res.* 47(3), 957–995.

Naghdi, M., Taheran, M., Brar, S. K., Kermanshahi-Pour, A., Verma, M. and Surampalli, R. Y. (2018). Removal of pharmaceutical compounds in water and wastewater using fungal oxidoreductase enzymes, *Environ. Pollut.* 234, 190–213.

Negreanu, Y., Pasternak, Z., Jurkevitch, E. and Cytryn, E. (2012). Impact of treated wastewater irrigation on antibiotic resistance in agricultural soils, *Environ. Sci. Technol.* 46(9), 4800–4808.

Nekvapil, F., Ganea, I. V., Ciorîţă, A., Hirian, R., Ogresta, L., Glamuzina, B., Roba, C. and Cintă Pinzaru, S. (2021). Wasted biomaterials from crustaceans as a compliant natural product regarding microbiological, antibacterial properties and heavy metal content for reuse in blue bioeconomy: A preliminary study, *Materials (Basel)* 14(16), 4558.

Nesme, T., Senthilkumar, K., Mollier, A. and Pellerin, S. (2015). Effects of crop and livestock segregation on phosphorus resource use: A systematic, regional analysis, *Eur. J. Agron.* 71, 88–95.

Nguyen, T. T., Rosello, C., B´elanger, R.and Ratti, C. (2020). Fate of residual pesticides in fruit and vegetable waste (FVW) processing, *Foods* 9(10), 1468.

Nightingale, K. K., Schukken, Y. H., Nightingale, C. R., Fortes, E. D., Ho, A. J., Her, Z., Grohn, Y. T., McDonough, P. L. and Wiedmann, M. (2004). Ecology and transmission of Listeria monocytogenes infecting ruminants and in the farm environment, *Appl. Environ. Microbiol.* 70(8), 4458–4467.

Nunes, N., Ragonezi, C., Gouveia, C. S. S. and Pinheiro de Carvalho, M. Â. A. (2021). Review of sewage sludge as a soil amendment in relation to current international guidelines: A heavy metal perspective, *Sustainability* 13(4), 2317.

O'Connor, J., Hoang, S. A., Bradney, L., Dutta, S., Xiong, X., Tsang, D. C. W., Ramadass, K., Vinu, A., Kirkham, M. B. and Bolan, N. S. (2021). A review on the valorisation of food waste as a nutrient source and soil amendment, *Environ. Pollut.* 272, 115985.

O'Connor, J., Mickan, B. S., Siddique, K. H. M., Rinklebe, J., Kirkham, M. B. and Bolan, N. S. (2022). Physical, chemical, and microbial contaminants in food waste management for soil application: a review, *Environ. Pollut.* 300, 118860.

Parchomenko, A. and Borsky, S. (2018). Identifying phosphorus hot spots: A spatial analysis of the phosphorus balance as a result of manure application, *J. Environ. Manage.* 214, 137–148.

Patel, A. B., Shaikh, S., Jain, K. R., Desai, C. and Madamwar, D. (2020). Polycyclic aromatic hydrocarbons: sources, toxicity, and remediation approaches, *Front. Microbiol.* 11, 562813.

Pepper, I. L., Brooks, J. P. and Gerba, C. P. (2006). Pathogens in biosolids, *Adv. Agron.* 90, 1–41.

Polechónska, L., Klink, A., Dambiec, M. and Rudecki, A. (2018). Evaluation of *Ceratophyllum demersum* as the accumulative bioindicator for trace metals, *Ecol. Indic.* 93, 274-281.

Prather, M. J., Hsu, J., DeLuca, N. M., Jackman, C. H., Oman, L. D., Douglass, A. R., Fleming, E. L., Strahan, S. E., Steenrod, S. D., Søvde, O. A., Isaksen, I. S., Froidevaux, L. and Funke, B. (2015). Measuring and modeling the lifetime of nitrous oxide including its variability, *J. Geophys. Res. Atmos.* 120(11), 5693–5705.

Prathumchai, N., Polprasert, C. and Englande Jr., A. J. (2018). Phosphorus distribution and loss in the livestock sector – the case of Thailand, *Resour. Conserv. Recycl.* 136, 257-266.

Pruden, A., Pei, R. T., Storteboom, H. and Carlson, K. H. (2006). Antibiotic resistance genes as emerging contaminants: studies in northern Colorado, *Environ. Sci. Technol.* 40, 7445–7450.

Published by Burleigh Dodds Science Publishing Limited, 2024.

Raghothama, K. G. (2005). Phosphorus and plant nutrition: an overview. In: Sims, J. T. and Sharpley, A. N. (Eds). *Phosphorus: Agriculture and the Environment, Agronomy Monograph NO. 46 ASA, CSSA, SSSA*. Madison Book Company, Madison, pp. 355-378.

Ramos, M. C., Quinton, J. N. and Tyrrel, S. F. (2006). Effects of cattle manure on erosion rates and runoff water pollution by faecal coliforms, *J. Environ. Manage.* 78(1), 97-101.

Rathi, B. S., Kumar, P. S. and Show, P. L. (2021). A review on effective removal of emerging contaminants from aquatic systems: current trends and scope for further research, *J. Hazard. Mater.* 409, 124413.

Ratnam, S., March, S. B., Ahmed, R., Bezanson, G. S. and Kasatiya, S. (1988). Characterization of Escherichia coli serotype O157:H7, *J. Clin. Microbiol.* 26(10), 2006-2012.

Resende, J. A., Silva, V. L., Rocha de Oliveira, T. L., de Oliveira Fortunato, S., da Costa Carneiro, J., Otenio, M. H. and Diniz, C. G. (2014). Prevalence and persistence of potentially pathogenic and antibiotic resistant bacteria during anaerobic digestion treatment of cattle manure, *Bioresour. Technol.* 153, 284-291.

Sandegren, L. (2014). Selection of antibiotic resistance at very low antibiotic concentrations, *Ups. J. Med. Sci.* 119(2), 103-107.

Sardar, K., Ali, S., Hameed, S., Afzal, S., Fatima, S., Shakoor, M. B., Bharwana, S. A. and Tauqeer, H. M. (2013). Heavy metals contamination and what are the impacts on living organisms, *Greener J Environ. Manag. Public Saf* 2(4), 172-179.

Sarmah, A. K., Meyer, M. T. and Boxall, A. B. A. (2006). A global perspective on the use, sales, exposure pathways, occurrence, fate and effects of veterinary antibiotics (VAs) in the environment, *Chemosphere* 65(5), 725-759.

Sethy, S. K. and Ghosh, S. (2013). Effect of heavy metals on germination of seeds, *J. Nat. Sci. Biol. Med.* 4(2), 272-275.

Shi, T., Ma, J., Wu, X., Ju, T., Lin, X., Zhang, Y., Li, X., Gong, Y., Hou, H., Zhao, L. and Wu, F. (2018). Inventories of heavy metal inputs and outputs to and from agricultural soils: a review, *Ecotoxicol. Environ. Saf.* 164, 118-124.

Skinner, K. D. (2022)-5060. *Trends in Groundwater Levels, and Orthophosphate and Nitrate Concentrations in the Middle Snake River Region, South-Central Idaho*. Scientific Investigations Report. United State Geologic Survey, Reston, VA.

Smit, A. L., van Middelkoop, J. C., van Dijk, W. and van Reuler, H. (2015). A substance flow analysis of phosphorus in the food production, processing and consumption system of the Netherlands, *Nutr. Cycl. Agroecosyst.* 103(1), 1-13.

Sobsey, M. D., Khatib, L. A., Hill, V. R., Alocilja, E., Pillai, S. (2006). 'Pathogen in animal wastes and the impacts of waste management practices on their survival, transport and fate'. In Rice, J. M., Caldwell, D. F. and Humenik, F. J. (Eds). Animal Agriculture and the Environment, National Center for Manure & Animal Waste Management White Papers. American Society of Agricultural and Biological Engineers, St. Joseph, MI, pp. 609-666.

Song, W., Huang, M., Rumbeiha, W. and Li, H. (2007). Determination of amprolium, carbadox, monensin, and Tylosin in surface water by liquid chromatography/tandem mass spectrometry, *Rapid Commun. Mass SPectrom.* 21(12), 1944-1950.

Soni, B., Bartelt-Hunt, S. L., Snow, D. D., Gilley, J. E., Woodbury, B. L., Marx, D. B. and Li, X. (2015). Narrow grass hedges reduce Tylosin and associated antimicro-bial resistance genes in agricultural runoff, *J. Environ. Qual.* 44(3), 895-902.

Spears, R. A., Kohn, R. A. and Young, A. J. (2003). Whole-farm nitrogen balance on western dairy farms, *J. Dairy Sci.* 86(12), 4178–4186.

Sutton, M. A., Mason, K. E., Sheppard, L. J., Sverdrup, H., Haeuber, R. and Hicks, W. K. (2014). Nitrogen deposition, critical loads and biodiversity. Springer, Dordretch, Netherlands.

Svanbäck, A., McCrackin, M. L., Swaney, D. P., Linefur, H., Gustafsson, B. G., Howarth, R. W. and Humborg, C. (2019). Reducing agricultural nutrient surpluses in a large catchment-Links to livestock density, *Sci. Total Environ.* 648, 1549–1559.

Tan, L. V. and Tran, T. (2021). Heavy metal accumulation in soil and water in pilot scale rice field treated with sewage sludge, *ChemEngineering* 5(4), 77.

Tester, C. F. (1990). Organic amendment effects on physical and chemical properties of a sandy soil, *Soil Sci. Soc. Am. J.* 54(3), 827–831.

Thakali, A. and MacRae, J. D. (2021). A review of chemical and microbial contamination in food: what are the threats to a circular food system?, *Environ. Res.* 194, 110635.

Tian, H., Xu, R., Canadell, J. G., Thompson, R. L., Winiwarter, W., Suntharalingam, P., Davidson, E. A., Ciais, P., Jackson, R. B., Janssens-Maenhout, G., Prather, M. J., Regnier, P., Pan, N., Pan, S., Peters, G. P., Shi, H., Tubiello, F. N., Zaehle, S., Zhou, F., Arneth, A., Battaglia, G., Berthet, S., Bopp, L., Bouwman, A. F., Buitenhuis, E. T., Chang, J., Chipperfield, M. P., Dangal, S. R. S., Dlugokencky, E., Elkins, J. W., Eyre, B. D., Fu, B., Hall, B., Ito, A., Joos, F., Krummel, P. B., Landolfi, A., Laruelle, G. G., Lauerwald, R., Li, W., Lienert, S., Maavara, T., MacLeod, M., Millet, D. B., Olin, S., Patra, P. K., Prinn, R. G., Raymond, P. A., Ruiz, D. J., van der Werf, G. R., Vuichard, N., Wang, J., Weiss, R. F., Wells, K. C., Wilson, C., Yang, J. and Yao, Y. (2020). A comprehensive quantification of global nitrous oxide sources and sinks, *Nature* 586(7828), 248–256.

Tian, L., Jinjin, C., Ji, R., Ma,Y. and Yu, X. (2022). 'Microplastics in agricultural soils: sources, effects, and their fate', *Curr. Opin, Environ. Sci. Health* 25, 100311.

Tiquia, S. M., Tam, N. F. Y. and Hodgkiss, I. J. (1998). Salmonella elimination during composting of spent pig litter, *Bioresour. Technol.* 63(2), 193–196.

Tran, N. H., Reinhard, M. and Gin, K. Y. H. (2018). Occurrence and fate of emerging contaminants in municipal wastewater treatment plants from different geographical regions-a review. *Water Res.* 133, 182–207.

Tyrrel, S. F. and Quinton, J. N. (2003). Overland flow transport of pathogens from agricultural land receiving faecal wastes, *J. Appl. Microbiol.* 94 (Suppl.), 87S–93S.

Ugulu, I., Akhter, P., Khan, Z. I., Akhtar, M. and Ahmad, K. (2021). Trace metal accumulation in pepper (*Capsicum annuum* L.) grown using organic fertilizers and health risk assessment from consumption, *Food Res. Int.* 140, 109992.

Unc, A. and Goss, M. J. (2004). Transport of bacteria from manure and protection of water resources, *Appl. Soil Ecol.* 25(1), 1–18.

UNEP and WHRC (2007). *Reactive Nitrogen in the Environment: Too Much or Too Little of a Good Thing*. United Nations Environment Programme, Paris.

Venglovsky, J., Sasakova, N. and Placha, I. (2009). Pathogens and antibiotic residues in animal manures and hygienic and ecological risks related to subsequent land application, *Bioresour. Technol.* 100(22), 5386–5391.

Wang, R., Goll, D., Balkanski, Y., Hauglustaine, D., Boucher, O., Ciais, P., Janssens, I., Penuelas, J., Guenet, B., Sardans, J., Bopp, L., Vuichard, N., Zhou, F., Li, B., Piao, S., Peng, S., Huang, Y. and Tao, S. (2017). Global forest carbon uptake due to nitrogen and phosphorus deposition from 1850 to 2100, *Glob. Change Biol.* 23(11), 4854–4872.

Waseem, H., Williams, M. R., Stedtfeld, R. D. and Hashsham, S. A. (2017). Antimicrobial resistance in the environment, *Water Environ. Res.* 89(10), 921-941.

Weber, R., Herold, C., Hollert, H., Kamphues, J., Blepp, M. and Ballschmiter, K. (2018). Reviewing the relevance of dioxin and PCB sources for food from animal origin and the need for their inventory, control and management, *Environ. Sci. Eur.* 30(1), 42.

Westerman, P. W. and Bicudo, J. R. (2005). Management considerations for organic waste use in agriculture, *Bioresour. Technol.* 96(2), 215–221.

Williams-Nguyen, J., Sallach, J. B., Bartelt-Hunt, S., Boxall, A. B., Durso, L. M., McLain, J. E., Singer, R. S., Snow, D. D. and Zilles, J. L. (2016). Antibiotics and antibiotic resistance in agroecosystems: state of the science, *J. Environ. Qual.* 45(2), 394-406.

Yang, Q., Tian, H., Li, X., Ren, W., Zhang, B., Zhang, X. and Wolf, J. (2016). Spatiotemporal patterns of livestock manure nutrient production in the conterminous United States from 1930 to 2012, *Sci. Total Environ.* 541, 1592–1602.

Youngquist, C. P., Mitchell, S. M. and Cogger, C. G. (2016). Fate of antibiotics and antibiotic resistance during digestion and composting: a review, *J. Environ. Qual.* 45(2), 537-545.

Yu, Y., Zhou, Y., Wang, Z., Torres, O. L., Guo, R. and Chen, J. (2017). Investigation of the removed mechanisms of antibiotic ceftazidime by green algae and subsequent microbic impact assessment, *Sci. Rep.* 7, 1–11.

Zhan, X., Bo, Y., Zhou, F., Liu, X., Paerl, H. W., Shen, J., Wang, R., Li, F., Tao, S., Dong, Y. and Tang, X. (2017). Evidence for the importance of atmospheric nitrogen deposition to eutrophic lake Dianchi, China, *Environ. Sci. Technol.* 51(12), 6699-6708.

Zhan, X., Adalibieke, W., Cui, X., Winiwarter, W., Reis, S., Zhang, L., Bai, Z., Wang, Q., Huang, W. and Zhou, F. (2021). Improved estimates of ammonia emissions from global croplands, *Environ. Sci. Technol.* 55(2), 1329–1338.

Zhang, B., Tian, H., Lu, C., Dangal, S. R. S., Yang, J. and Pan, S. (2017). Global manure nitrogen production and application in cropland during 1860–2014: a 5 arcmin gridded global dataset for Earth system modeling, *Earth Syst. Sci. Data* 9(2), 667–678.

Part 3

Co-products

Chapter 9

Producing biogas from livestock manure and agricultural biomass

R. Fragoso, University of Lisbon, Portugal; D. Hidalgo, CARTIF Technology Center – Circular Economy Area, Spain; and M. Paterson, Kuratorium für Technik und Bauwesen in der Landwirtschaft e.V. (KTBL), Germany*

1 Introduction

The commitment to achieve climate neutrality by 2050, and a net-zero emissions economy in Europe, is a key starting point on the path to long-term change towards sustainability. It will require a substantial acceleration in current efforts across the EU to achieve these goals (EEA, 2021). This has driven the European Commission to establish the 'European Green Deal', a strategy to achieve a climate-neutral Europe by 2050 (EC, 2021a). There is a need to reduce greenhouse gas (GHG) emissions, including from agriculture. Methane (CH_4) emissions from enteric fermentation (from ruminants) and nitrous oxide (N_2O) emissions from soils account for more than 80% of total agricultural GHG emissions, whilst methane from manure management is the third most important source of emissions, representing 10% of the total (EC, 2021b).

* LEAF, TERRA, ISA, University of Lisbon, Portugal.

http://dx.doi.org/10.19103/AS.2023.0120.18

Methane is the second most important anthropogenic GHG gas after carbon dioxide, but over a 100-year timescale, its warming effect is 28–36 times greater (IEA, 2021).

Controlled methanation in anaerobic digesters in the agricultural sector is therefore key to reducing GHGs and thus justifies support for agricultural biogas production in the European Commission's methane strategy (EC, 2020a). The EU is seen as having the most advanced biogas market in the world, with more advanced legislation in this field than in any other region (Korbag et al., 2020). Biogas and biomethane[1] are expected to play an important role in achieving EU energy goals in three main ways (Meyer et al., 2018; Paterson et al., 2016):

- Storage capacity and demand-based (flexible) utilization of these gases as energy sources;
- An already established role in a wide range of applications (e.g. combined heat and power (CHP) production and transport); and
- The very diverse biomasses (livestock manure, energy crops, residues and waste) that can be used for their production.

In 2019, primary energy production from biogas [from anaerobic digestion (AD) and thermal processes] across the 28 members of the European Union reached 16.6 Mtoe, around the same level as in 2017. The implementation of regulations which were less supportive of using food-type energy crops to produce biogas contributed to this stagnation in growth in 2019 and was aggravated by limited capacity allocated to biogas tenders and less attractive biogas electricity payment terms. The member states posting the most positive output growth were those that encouraged biomethane injection and recovered energy from fermentable waste, e.g. France, Germany, Denmark and Greece (EurObserv'ER, 2020).

A recent analysis shows that the contribution of biogas at global level could increase to 95 Mtoe in 2030 and up to 150 Mtoe by 2040. Although energy crop residues are included in these predictions, energy crop feedstocks grown specifically to produce biogas and biomethane are not included because their sustainability has been questioned (IEA, 2020). This means growth potential is still high, even if the implementation of a more attractive regulatory framework is required (regarding substrates for AD), with greater political determination to replace reliance on fossil gas as an energy source and also stricter regulations to measure and control GHG emissions to support activities which help reduce these emissions. These expectations run counter to the recent lack of growth in the EU given European regulations against the use of food-type energy crops in biogas production and less attractive feed-in tariffs (EurObserv'ER, 2020).

Energy crops (maize, grass, etc.) have been widely used for AD as they can enhance the digestion process and are easily available in agricultural areas.

[1] Biogas (contains 50-75 % CH_4) purified and upgraded into biomethane, which meets the quality of natural gas (>97 % CH_4).

There has been recent controversy as to whether the cultivation of energy crops competes with the cultivation of crops for food and feed production ('food or tank' debate). Public and political debate about competition for arable land has increased interest in using residues and waste in biogas plants, with cultivated biomass used primarily for food and feed production. To adapt to new regulations, the biogas sector has set its sights on a new growth model more dependent on the recovery of fermentable waste (with emphasis on livestock manure, agricultural residues and agro-industrial waste) rather than relying on energy crops.

In Europe (EU 27 and UK), the total amount of animal manure produced is 1400 million tonnes per year (Köninger et al., 2021). This amount of manure leads to the emission of 73 million t CO_2 eq./year of GHG, which represents 50% of N_2O emissions and 15–25% of CH_4 emissions from agriculture in the EU (Fangueiro et al., 2021). In Germany, e.g. the storage and spreading of farm manure results in more than 250 000 t of methane or 6.3 million t of CO_2 eq. annually, without including direct and indirect nitrous oxide emissions. Based on the assumptions of Rösemann et al. (2021), methane emissions from conventional slurry storage and land spreading in livestock farming can be reduced by about 80–90% by using AD and storing digestate in a way that avoids emissions.

Since manure management usually includes application to fertilize crops, it can also lead to soil and water pollution. Manure can be a pathway for the introduction of heavy metals, pathogens and micropollutants (e.g. pharmaceutical compounds) into the soil and then to groundwater through leaching (Ghirardini et al., 2020). Manure leaching from field into watercourses also promotes eutrophication which damages water quality and biodiversity.

In dealing with waste such as manure, EU waste prevention planning aims to decouple economic production and social welfare from resource conservation, energy use and environmental conservation (EC, 2020b). Ambitious targets for the treatment of agricultural waste (a ready-to-use carbon source) using AD can contribute to EU waste reduction targets as well as reducing GHG emissions and reliance on fossil fuels (Achinas and Euverink, 2020). To address these issues, manure management has become more focused on technologies that allow the recovery of nutrients through the production of biobased fertilizers (e.g. nitrogen-rich liquid fertilizer). One such technology is AD which is outlined in Section 2.

2 Anaerobic digestion

Anaerobic digestion (AD) is a natural biological process that occurs in the absence of oxygen and converts organic matter into biogas, mainly methane (50–70%) and carbon dioxide (30–50%) with other trace gases (KTBL, 2013; Mata-Alvarez et al., 2014). Biogas from AD has a calorific value in the range

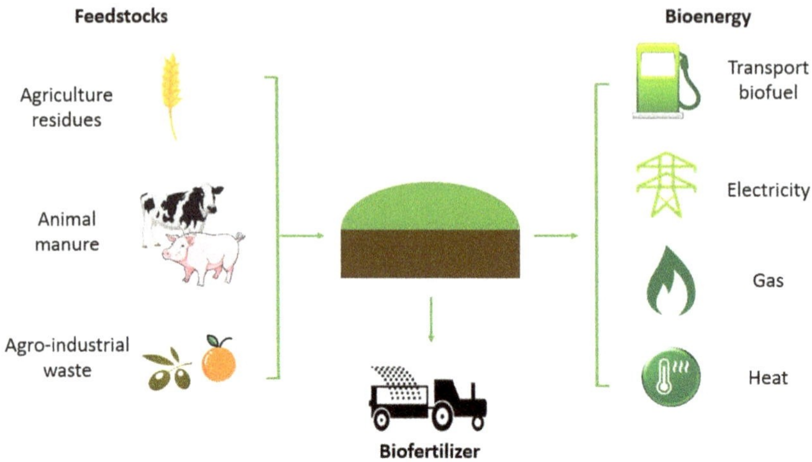

Figure 1 Schematic overview of the biogas production chain.

of 21–24.5 MJ/m^3 and can be converted to around 1.9–3.2 kWhe (KTBL, 2013; Caposciutti et al., 2020). Figure 1 provides a simple overview of the biogas production chain, from common types of feedstock used as inputs to the main outputs, different types of bioenergy sources and digestate that can be used as biofertilizer.

AD provides an opportunity to create a circular bioeconomy by creating high-value products: biogas and biomethane as energy sources and biofertilizer (Awasthi et al., 2019). An additional potential benefit is that, even in small amounts, contaminants such as antibiotics, disinfectants or solvents, herbicides or salts of heavy metals can inhibit the degradation process in the digester (Paterson et al., 2016). This requires careful selection and feedstock and process monitoring but means the process can be an effective way of identifying and removing contaminants that might otherwise pose safety and pollution risks.

AD technology offers several advantages in mitigating potential environmental impacts from manure management. One of the main advantages is avoiding GHG emissions from open manure storage by capturing the CH_4 into the biogas. Methane emissions from liquid manure storage and spreading can be reduced by around 80–90% through AD plants (with emission-avoiding storage of the digestate). No nitrous oxide emissions occur due to the absence of oxygen in the AD process. Ammonia emissions are also negligible if digestate is stored correctly.

Biogas produced by AD is renewable energy and displaces the use of fossil fuels and, consequently, emissions of associated CO_2-eq. AD also produces digestate as a by-product which has more stabilized and available organic matter than manure and can replace inorganic fertilizers (reducing the emissions associated with conventional fertilizer production and application). Biogas fits in with the goal of a circular economy as it recovers energy and nutrients from waste. Decentralized AD deployment can also contribute significantly to sustainable development in rural areas and provide additional revenue to farmers. Biogas contributes to improving the security of the energy supply, whilst local production improves flexibility in managing the energy supply.

Despite its benefits, several factors potentially limit the feasibility of AD. According to Einarsson and Persson (2017), the potential for using AD in different parts of the EU is affected by parameters such as maximum substrate removal rate and transportation distance, minimum viable plant size and minimum/maximum C/N ratio and dry matter content of substrate mix. According to these authors, the efficiency of biomethane production can vary by 50%, depending on these factors. The authors highlight the importance of overcoming these constraints by optimizing substrate C/N ratio and dry matter concentration, either by complementing them with suitable co-substrates or through further development of biogas technology (e.g. pre-treatment of substrates and recirculation of effluent). Other relevant factors affecting the profitability of biogas production include demand for biogas (power and heat or biofuel), energy prices, available subsidies (e.g. feed-in tariffs) and the cost of capital.

One issue is manure availability and transport logistics, as there needs to be enough substrate to allow profitable operation of biogas plant. Larger biogas plant sizes are normally more cost-effective but may involve higher transport costs to guarantee sufficient substrate supply from a wider network of farms or other sources.

Feedstock composition also has a major influence on the potential of AD to produce biogas efficiently. The presence, e.g. of inhibitory compounds and poor nutrient balance (low C/N), can disrupt the process (KTBL, 2013). Water content is also a key aspect, as it has an impact on transportation logistics, process yield and building costs of the AD plant.

To overcome these potential constraints, it is common to adopt a co-digestion regime (AcoD), integrating one or more co-substrates in the digestion process. The selection of co-substrate and blending ratio should aim at promoting synergies (e.g. balanced C/N ratio, macro- and micronutrient equilibrium, etc.) and diluting inhibitory compounds in feedstock. Introducing co-substrates in biogas plants using diluted substrates such as pig slurry increases carbon content, leading to a more balanced C/N ratio, enhancing production efficiency and improving economic viability. Meyer et al. (2018) carried out a study on the energy potential of biogas from animal manure for the EU-28 through 2030, suggesting that the key to success was the co-digestion of different substrates, such as straw from cereal production, residual grass from permanent and rotational grasslands. Using excess grass from permanent grassland and meadows such as recirculating nutrients from permanent grass areas to areas of crop cultivation.

3 Recent research on co-digestion

Zhou et al. (2021) have performed a meta-analysis reviewing 124 peer-reviewed articles focussed on the AcoD of livestock manure. They found that cattle manure was the most researched manure (ca 50%), followed by pig (ca 25%) and chicken manure (ca 17%). Regarding co-substrates contributing to achieving a balanced C/N ratio, crop straw was the most commonly tested substrate (51%), probably due to its availability. Commonly tested food-processing by-products included sugar beet pulp, vinasse (from sugar refining) and apple pomace, among others. Other frequently tested co-substrates were fruit and vegetable waste, horse manure, microalgae and sewage sludge. Mesophilic processing (at intermediate temperatures of 37–42°C) was adopted in 85% of the studies reviewed, probably due to a better net energy balance.

Duarte et al. (2021) compared the energy performance of livestock manure (pig and cattle) used in mono-digestion and in co-digestion with agro-industrial wastes. Agro-industrial waste included non-marketable pears; peels from oranges, carrots and pineapple; waste sardine oil from the fish canning industry; and coffee grounds (resulting from the water-pressured extraction to produce soluble coffee and coffee drinks). They used an energy performance indicator (EPI) to show the percentage increase in specific methane yield associated with the shift to co-digestion. They found that adopting co-digestion could lead to an increase in EPI value from 30% to 250%, clearly indicating better process stability, probably due to a better C/N ratio and buffer capacity compared to mono-digestion trials.

Aravani et al. (2022) have assessed the availability of agricultural and livestock wastes in China and Greece, including crop residues, animal manure and agro-industrial wastes (e.g. from olive oil extraction). They concluded that in Greece, the average quantity of agricultural residues, animal manure and agro-industrial wastes is 10, 26 and 13.2 Mt/year, respectively. According to this study, if all these wastes could be used, theoretically, up to 77 TWh could be recovered (11 TWh from agricultural wastes and 66 TWh from livestock manure), covering the energy needs of the country.

Reparaz et al. (2021) studied the economic feasibility of AD plants on pig farms, both to reduce GHG emissions from traditional manure management and to supply biogas to replace natural gas, using the region of Castilla y León (Spain) as case study. The economic model included internalization of CO_2 costs based on the Emission Trading System used in Europe. The authors concluded that this approach could meet 68% of the electricity demand in municipalities with more than 5000 inhabitants.

4 Biomass availability and biomethane potential: livestock manure, agricultural residues and agro-industrial waste

One of the key success factors in the development of AD plants is the availability of affordable biomass to be used as feedstock and its transport logistics and cost (Paterson et al., 2016). The availability of appropriate biomass near an AD plant significantly enhances operational efficiency and consequent viability. It is crucial to integrate statistical and geospatial data to map available biomass that can be used as feedstock for biogas production, e.g. livestock manure, agricultural residues and agro-industrial waste. This section presents data on the availability and biogas potential of these biomasses.

4.1 Livestock manure

In Europe (EU 27 and UK), the total amount of animal manure produced per year is 1400 million tonnes, 75% of which is from cattle, with pig and chicken accounting for around 12% each. As it can be seen from Fig. 2, manure production is concentrated in Germany, Spain, France, Italy, Poland and the United Kingdom, which together represent about 68% of the total manure (Köninger et al., 2021). The distribution of manure by animal type (cattle, pig and chicken) is very dependent on regional specialization in livestock production.

It has been estimated by Scarlat et al. (2018) that around 70% of the manure produced is available in practice for AD. Meyer et al. (2018) established three scenarios of availability: high, moderate and low and took into consideration the animal species (pig, poultry and cattle). For pig and poultry, availability was, respectively, 90%, 80% and 70%; whereas for cattle, due to differences in farming system and disposal practices, it was 70%, 60% and 50%. A German study shows that currently, about 30% of the farm manure produced in Germany is used to generate energy in biogas plants. Around two-thirds of the technical potential is therefore currently still unexploited. It is estimated that the current utilization rate of farm manure for AD can be increased by an additional 35% (Majer et al., 2019). Einarsson and Persson (2017) have estimated that around 200 million metric dry tonnes per year of crop residues and manure are available for biogas production, with cattle manure being the largest contributor at 84 million metric dry tonnes per year.

Farm manure, especially slurry, has a low energy density due to high water content and low specific gas yield. Relatively large fermentation volumes are therefore needed, which increases costs. The use of manure and other residues may also require sanitary treatment to deal with pathogens and contaminants, which also adds cost. The biomethane potential of livestock manure depends on factors such as animal species, farm practices regarding water and slurry/manure management (e.g. cleaning practices and solid/liquid separation

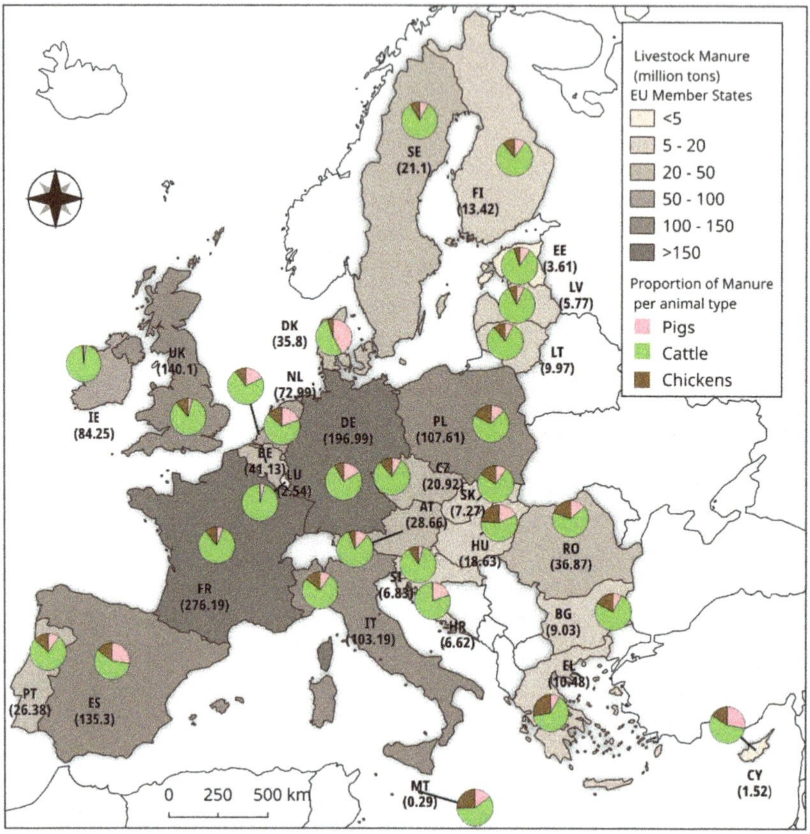

Figure 2 Annual manure production (million tonnes) in the European Union and UK and distribution according to main animal types (Period: 2016-2019) (Köninger et al., 2021; https://doi.org/10.1016/j.agsy.2021.103251 R), under the CC BY license (http://creativecommons.org/licenses/by/4.0/).

Table 1 Biomethane potential of livestock manure

Manure	Dry matter (% of fresh matter)	Organic dry matter (% of dry matter)	Biomethane potential (L_N CH_4/kg odm)	Reference
Cattle slurry	9	80	230	
Cattle dung	25	85	250	
Horse dung	31	82	220	KTBL 2021
Pig slurry	5	75	250	
Poultry dung	50	80	280	

N - standard conditions; odm, organic dry matter.

practices). Table 1 shows gas yield reference values for livestock manure and the resulting average biomethane yield calculated by the Kuratorium für Technik und Bauwesen in der Landwirtschaft e.V (KTBL) in Germany (KTBL, 2021).

4.2 Agricultural residues and agro-industrial waste

Agricultural residues include the organic materials generated as by-products from crop production (e.g. rice straw and corn stover). Agro-industrial waste results from crop processing; for example olive processing to produce olive oil generates large amounts of olive pomace. The type of residues/waste produced depends on the sector. For the fruit sector, agricultural residues include pruning and leaves whilst processing mainly leads to pomace (the pulpy residue left after juice extraction). Harvesting of cereals leads to straw production in fields, and processing mainly generates bran (the hard outer layers of cereal grains). In the livestock sector, farming mainly produces manure, whereas slaughtering wastes include fatty tissue, skin and bone. Agro-industrial waste type and availability can vary significantly according to geographical region, not only because of different crops grown but because processing practices are often diverse.

Monforti et al. (2015) have provided an estimate of potential agricultural residues from several crops (wheat, barley, oat, rye, maize, rapeseed, rice and sunflower) in the 27 EU member states. The analysis took into consideration collection rates (40% of the residues present on the ground for wheat, barley, oat and rye residues and 50% for maize, rapeseed, rice and sunflower residues) and subtracted the amount of residue diverted to other uses (e.g. straw for animal bedding or horticulture). The resulting spatial distribution of residues is shown in Fig. 3.

Bedoić et al. (2019) have estimated the production of waste along the food production chain (cultivation, processing and consumption) in the 28 EU member states for four sectors (fruit, vegetable, cereals and animal) from 2010 to 2016. Examples of commodities in the study included olives, tomatoes, onions and carrots (for the vegetable sector); barley, maize, oats, rice and wheat (for the cereal sector); cattle, dairy cows, pigs and chickens (for the animal sector). Key parameters for estimating the amount of waste of each commodity were the quantities produced (Eurostat data), exported and imported (FAOSTAT) and consumed commodity (per capita per year) for each EU 28 country. For the animal sector, production of waste depended on animal breeding, slaughtering and consumption of meat and meat products. The study showed that from 2010 to 2016 in the EU 28 the estimated quantity of food production waste was around 18.4 billion tonnes, with vegetable, animal, cereal and fruit sectors representing 44%, 31%, 22% and 2%, respectively. Figure 4 shows the geographical distribution (expressed in kg/km^2) of the agricultural residues and agro-industrial waste reviewed in Bedoić et al. (2019).

As the analysed commodities for the fruit sector were mainly citrus fruit, the highest quantities of waste were produced in Mediterranean countries such as Italy (around 40 t/km^2) and Greece (50 t/km^2). For the vegetable sector, the

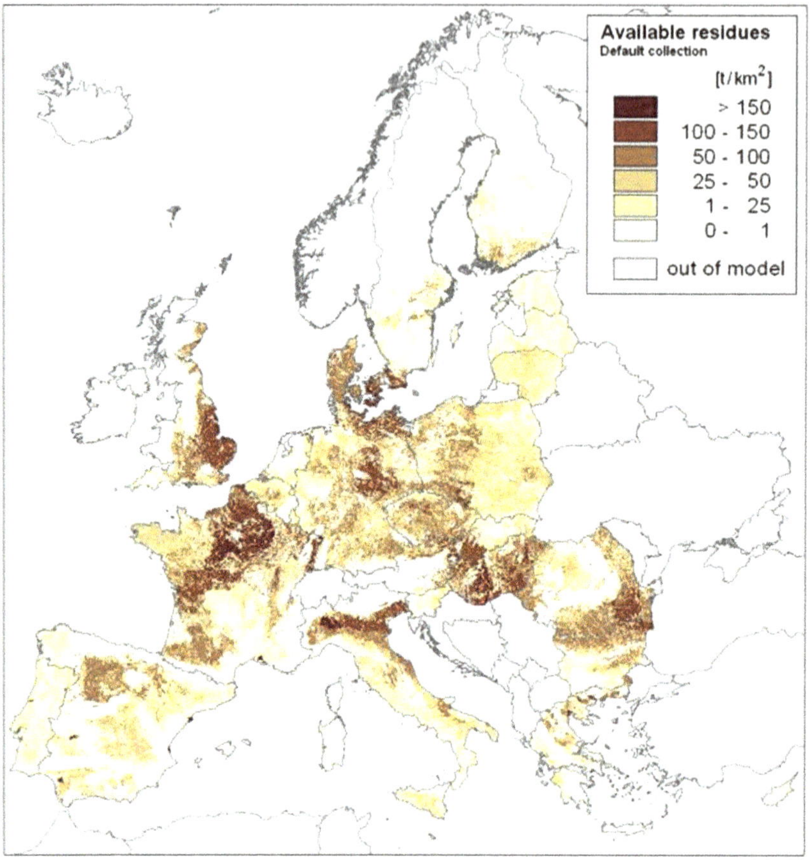

Figure 3 Agricultural residues available for energy use in EU-27 as calculated in tonnes of dry matter per square kilometre (Monforti et al., 2015; http://dx.doi.org/10.1016/j.rser.2014.12.033), under the CC BY-NC-ND license (http://creativecommons.org/licenses/by-nc-nd/4).

Netherlands and Belgium showed the highest values, respectively, with 2600 t/km^2 and 2525 t/km^2. Hungary (ca. 360 t/km^2), Denmark (ca. 330 t/km^2), Belgium (ca. 225 t/km^2) and Germany (ca. 220 t/km^2) present the highest quantities of waste resulting from the cereal sector. For the animal sector, the highest waste production has been estimated for the Netherlands (ca. 2200 t/km^2) and Belgium (ca. 1500 t/km^2). Denmark, Ireland and Luxembourg generate between 1000 t/km^2 and 1200 t/km^2 and Germany and France have shown values of 450 t/km^2 and 700 t/km^2, respectively.

Tamburini et al. (2020) estimated the potential of biogas production from agro-industrial waste, manure and agricultural residues from the Emilia Romagna region (Italy). Agro-industrial waste produced in the region includes tomato pomace, potato residues, vegetable and fruit waste, grape pomace,

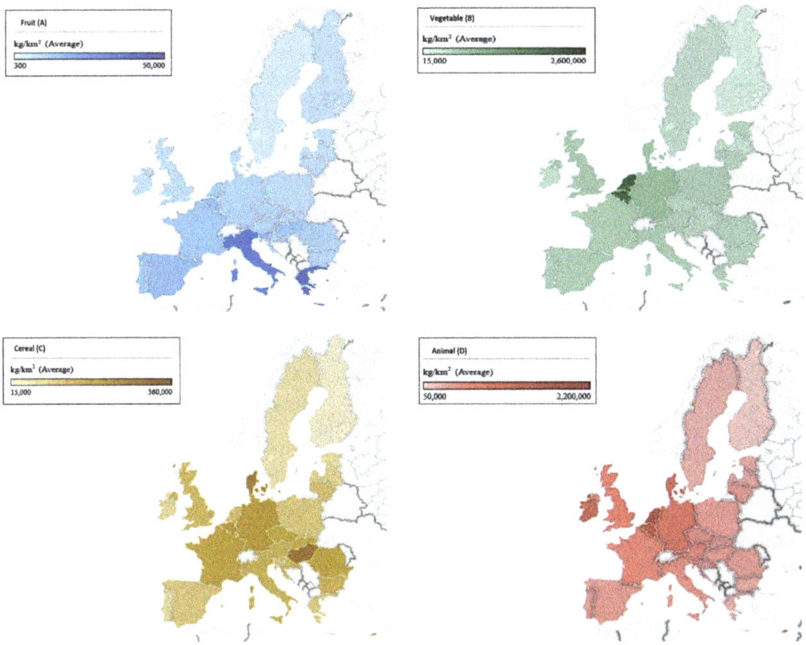

Figure 4 Average quantity of agriculture residues and agro-industrial waste from fruit (a), vegetable (b), cereal (c) and animal (d) sectors per area in the period 2010–2016. Adapted from Bedoić et al. (2019) (https://doi.org/10.1016/j.scitotenv.2019.05.219), CC BY license (http://creativecommons.org/licenses/by/4.0/).

olive oil pomace and milk whey. The total amount of manure produced in the region was 16 793 329 t/year, whereas total agro-industrial waste was 2 182 792 t/year and total agricultural residues were 1 335 420 t/year. Considering the area of the region (22 453 km^2), these values correspond to 747, 97 and 59 t/km^2.year, respectively.

These studies suggest that the volume of agricultural residues and agro-industrial waste produced could support the deployment of AD as a strategy for promoting renewable energy production and regional development. However, more studies are needed in order to assess the accessibility of these feedstocks and the amount diverted to other uses. The collection of reliable statistical data on these feedstocks is challenging. For this reason, it would be very useful to have an open-access database with statistics of production of manure, agriculture residues and agro-industrial wastes available at regional/national levels for European countries.

In order to estimate the potential biogas and biomethane production in a certain region, besides the type and amount of feedstock, it is essential to know the biomethane potential of the feedstock. The biomethane potential of agricultural residues and agro-industrial waste show a wide range of values

Table 2 Biomethane potential of several agricultural residues and agro-industrial wastes

Biomass	Biomethane potential (L_N CH_4/kg odm)	Reference
Alfafa silage	320	KTBL (2021)
Alfafa	300-500	Scarlat et al. (2018)
Apple peel	407	Suhartini et al. (2021)
Coffee spent grounds	271-325	Garcia et al. (2019)
Corn stover	427	Dar et al. (2021)
Food leftovers	450-600	KTBL (2021)
Grass silage	320	KTBL (2021)
Grass	300-450	Scarlat et al. (2018)
Maize silage	355	KTBL (2021)
	250-450	Scarlat et al. (2018)
Grain cleaning	360	KTBL (2021)
Onion peels	400	Garcia et al. (2019)
Orange peel	455-486	Suhartini et al. (2021)
Organic Biowaste	150-400	KTBL (2021)
Potato peels	267-320	Garcia et al. (2019)
Rice straw	263	Suhartini et al. (2021)
Slaughterhouse waste	550-650	Scarlat et al. (2018)
Straw	200-250	KTBL (2021); Scarlat et al. (2018)
Sugar beet	230-380	KTBL (2021); Scarlat et al. (2018)
Sunflower husk	200	Garcia et al. (2019)
Sunflower silage	300	KTBL (2021)

N - standard conditions; odm, organic dry matter.

depending on their composition (e.g. carbohydrate, protein and fat content). Table 2 shows the average biomethane yield to be expected from several types of agricultural residues and agro-industrial waste. The biomethane potential of energy crops was included because their residues can still be used as feedstock for AD (KTBL, 2021).

5 Types of anaerobic digestion plant and their economics

Development of agriculture-based AD plants is generally of two types:

- On-farm plants, either based on mono-digestion of manure or co-digestion of manure with other agricultural residues available at farm and sometimes agro-industrial waste from processing on the farm; and

- Centralized plants, which co-digest manures collected from nearby farms together with agro-industrial waste from local food-processing operations.

AD plants can involve different uses of produced biogas:

- Generating electricity and/or heat; and
- Upgrading to biomethane used as a natural gas substitute.

Electricity and/or heat production can be achieved by using the biogas in a combined heat and power (CHP) plant. The electricity produced can be fed into the grid or usedfor lighting and power on a farm, whilst the thermal energy can be used for heating on farm or fed into a nearby district-heating system. Surplus heat of a CHP can also be converted into cooling energy through an absorption refrigeration process. Biogas can be used in a boiler to generate heat that can be used for heating, producing steam or for drying processes (e.g. to dry crops).

Another option is to upgrade biogas into biomethane using one methods such as pressure washing, pressure swing adsorption, amine scrubbing and treatment by membrane technology (Hidalgo et al., 2020). This process requires a relatively high level of technical effort for processing and feeding-in the biogas into the gas grid. This type of plant is usually more expensive than an AD plant with a CHP utilization. The biomethane can be supplied to the natural gas grid for use at a local or regional scale (e.g. to feed a centralized CHP unit with a boiler). Biomethane can also be pressurized for use as a vehicle fuel for a local bus fleet (KTBL, 2013). However, this is only viable for biogas plants above a certain size (economy of scale) or the gas from several AD plants must be bundled via a pipeline network.

The investment in an agricultural biogas plant depends heavily on factors such as the size of the plant, available substrates and how biogas is used. AD plants typically need to conform to extensive legal requirements regarding construction, operation, environmental impact and safety. Table 3 shows indicative guideline values for assessing the economic viability of agricultural biogas plants in Germany.

6 Case studies: small-scale mono-digestion and co-digestion anaerobic digestion plants

This section and the following section present case studies using data collected from the EU project BioEnergyFarm2, the EU Biogas3 project and IEA Bioenergy Task 37 Project (https://task37.ieabioenergy.com).

Table 3 Economic guideline values for exemplary agricultural biogas plants (KTBL, 2022, supplemented)

Cost item	Guideline values		
Electrical installed power (kW_{el})	75	75	250
Rated power capacity (kW_{el})	69	69	125
Electricity fed into grid (kWh/year)	608 000	608 000	1 090 000
Substrate share	100% FM	70% FM30% EC	70% FM30% EC
Amount of farm manure	8580 t cattle slurry/year 600 t cattle manure/year	2670 t cattle slurry/year	5100 t cattle slurry/year
Specific investment (based on installed capacity)	€14 300/kW_{el}	€10 050/kW_{el}	€5160/kW_{el}
Operating costs/running costs (fixed/variable)	€150 200/year	€160 900/year	€260 600/year
Electricity production costs	24.8 ct/kWh	20.4 ct/kWh	23.8 ct/kWh
Revenues (electricity and heat)	€142 600/year[a](20% heat output)	€48 600/year[b](20% heat output)	€198 600/year[c](35% heat output)
Plant construction	Usually delivered as a turnkey installation by a plant builder (with/without in-house involvement)		

FM, farm manure; EC, energy crops.
[a]German feed-in rate for small slurry plants with a slurry content of at least 80% by mass (EEG 2021).
[b]German feed-in rate for biomass plants without participation in tender up to 100 kW_{el} (EEG 2021).
[c]German feed-in rate for biomass plants in the context of tenders (EEG 2021).

6.1 Small-scale mono-digestion anaerobic digestion plants

Microscale AD systems (5–50 kW) have been deployed at the farm level both to treat manure and to meet the farm's energy needs. The most common are mono-digestion systems with the biogas converted into electrical energy and heat. The digestate is used as organic fertilizer. These systems mitigate methane emissions, reduce on-farm GHG emissions, minimize odour issues associated with manure storage, reduce pathogen risks from manure and improve nitrogen bioavailability in digestate.

There are many examples of microscale AD plants around Europe, and Table 4 presents three case studies. One of these case studies is from Bioelectric: one of the main suppliers of this kind of solution. The digestor is a continuous stirring tank reactor (CSTR) made of stainless-steel insulated panels operated under mesophilic conditions (35°C–42°C). Electrical and automation equipment is housed in a 20-foot shipping container. The key benefits of this plug-and-play solution are easy installation and deployment and full automation.

Another example is an Austrian biogas plant designed to meet a farm's electricity demand with heat used for digester heating and to support local district heating. The CHP runs when electricity is needed on the farm; otherwise, a gas boiler consumes the gas for heat production.

Table 4 Example of microscale mono-digestion AD plants in Europe

Manufacturer	Franz Bernecker & Müller Abfallprojekte GmbH	Bioelectric	Gosmer Biogas
Location	Austria	Belgium	Denmark
Type of feedstock	Cattle slurry and cattle dung	Cattle slurry	Pig slurry
Amount of biomass treated (t/year)	1170 t (730 t cattle slurry, 440 t cattle dung)	2900	9100
Amount of biogas produced (m³/year)	50 000	69 600	120 000
Electric power of the gas engine (kW)	6	19.4	30 (2 × 15 kW)
Generated electric energy (kWh/year)	52 000	155 200	150 000
Generated thermal energy (kWh/year)	148 000	335 000	360 000 kWh from engine heat + 130 000 kWh from a gas boiler
Investment cost (k€)	100	150 (net yearly income 37.4 k€)	700

Source: Data were collected from available documents produced in the EU project BioEnergyFarm2 (https://www.bioenergyfarm.eu).

In the Danish biogas plant shown in Table 4, liquid pig manure is initially processed in a low-tech solid-liquid separator system. The solid fraction, rich in phosphorus, is sold as a fertilizer, while the liquid fraction (3–6% dry matter) is fed into the digester. The digestate is stored in a secondary digester with no mixing, and the sedimented fraction is pumped into the primary digester, enriching the volatile solid and consequently enhancing the biogas yield. The heat produced is used in sow stables and in the farmhouse.

Most common examples of manure-based AD systems have installed electric power above 50 kW. Table 5 provides examples of small mono-digestion AD plants for electricity and/or heat production.

In the example from Germany, the biogas plant consists of a digester with a concrete cover and a storage tank which is covered with a double-membrane gasholder roof. For better insulation and more convenient loading of the solid feeder, both tanks are partially built into the ground. The thermal energy is used for heating the digester, the farmhouse and the milking parlour.

In the Spanish biogas plant, the main objective is to use the pig slurry produced in the farm for the production of thermal energy, replacing 90 000 l of fuel previously needed to heat the farm.

The example from the Netherlands is of a Microferm farm-scale digestor, which is a tower-type digester. The benefit of this system is the plug-and-play principle that makes it easier to deploy; however, it has the disadvantage of a high investment cost.

None of the earlier presented case studies (Table 5) include a biogas upgrading system. In fact, that is more common in larger AD plants. However, there are some farms that adopted this conceptual approach, such as the

Table 5 Case studies of small mono-digestion AD plants for electricity and/or heat production

Manufacturer	NQ Anlagentechnik GmbH	Biovec	HoSt B.V.
Location	Germany	Spain	Netherlands
Type of feedstock	Cattle slurry and cattle dung	Pig slurry	Cow manure
Amount of biomass treated (t/year)	7580 (6800 t cattle slurry, 780 t cattle dung)	2000	15 000
Amount of biogas produced (m^3/year)	321 300	150 000	525 000
Electric power of CHP (kW)	75	170 kW (boiler, only thermal)	66
Generated electric energy (kWh/year)	630 000	-	500 000
Generated thermal energy (kWh/year)	749 400	900 000	1500
Investment cost (k€)	700	220	n.a.

n.a., no information available. *Source*: Data were collected from available documents produced in the aim of EU project BioEnergyFarm2 (https://www.bioenergyfarm.eu).

200 cows' dairy farm located in the Netherlands that digests 10000 m³ of manure/year, producing up to 480000 m³ of biogas per year which is upgraded using amine scrubbing technology. The 300000 m³ of biomethane per year can be sold to the natural gas grid or used as a transport fuel.

6.2 Small-scale co-digestion anaerobic digestion plants

There are many examples of farm-scale co-digestion AD plants. Table 6 shows some examples. Many current co-digestion biogas plants still use energy crops, though the trend is to use other feedstocks.

The example from France is of a co-digestion farm AD plant producing thermal energy. Co-digesting pig slurry (4000 t/year) with leftovers (50 t/year) and grass silage (100 t/year) produces 100000 m³ of biogas per year. The biogas is burned in a boiler (110 kW) for heating the livestock buildings, replacing the propane previously used. The farmer has also installed an ammonia striping unit to produce concentrated liquid fertilizer from the digestate.

Another example from France is a farm with 80 dairy cows and 170 ha of crop cultivation (colza, wheat, barley and corn silage) that implemented a Silogaz® type digestor. Feedstock includes cattle manure (4000 t/year), waste from a food-processing industry (1000 t/year) and green water (550 t/year). The 255000 m³ of biogas produced per year generates 660 MWh of thermal energy (used in the AD and for heating the milking parlour and farmhouse) and 500 MWh of electricity.

In the Italian case study, the thermal energy from the CHP heats the digester and the farmhouse and is used to dry hay.

In the Dutch AD plant, cattle slurry is first pumped into the hydrolysis unit and then co-digested with glycerin. However, the farmer has found that enhancement of process performance was not enough to compensate for the energy spent in manure hydrolysis.

In the Austrian case study, the main purpose of the plant is to treat pig slurry produced on the farm. Co-substrates are energy crops and residues from vegetable processing, which are delivered by neighbouring farmers and mixed in a 30:70% ratio. Co-substrates are stored in concrete silos. Leachates are also collected and discharged into the slurry storage tank. The farmers use the resulting digestate as fertilizer. This plant has a two-stage continuous stirred tank reactor system with a yearly input of raw materials of 11000 t of energy crops/agricultural residues and 7300 t of manure plus leachates. It has a total electrical capacity of 1 MW and a thermal capacity of 1034 kW. The 4.02 million m³ of biogas produced per year are burned in two CHP engines and converted into 8030 MWh/year of electrical energy and 8223 MWh/year of thermal energy. The sale of electrical energy amounted to 7468 MWh per year,

Table 6 Case studies of small co-digestion AD plants for electricity and/or heat production

Manufacturer	Rota[a]	Fermtech systems[a]	AAT Abwasser- und Abfalltechnik GmbH & Co[a]	Evalor[a]	N.N.[b]
Location	Italy	Netherlands	Austria	France	Germany
Type of feedstock	Cattle slurry and manure and silage	Cattle slurry and glycerin	Pig slurry, energy crops/ agricultural residues	Pig slurry, leftovers and grass silage	Cattle slurry, poultry dung and energy crops (silage of maize, whole plant cereal crops and grass)
Amount of biomass treated (t/year)	31 000 (22 000 t of slurry, 7000 t of manure and 1850 t of silage)	3100 (3000 t cattle slurry and 100 t glycerine)	18 300 (11 000 t energy crops/agricultural residues and 7300 t manure plus leachates)	4150 (4000 t pig slurry, 50 t leftovers and 100 t grass silage)	14 900 (3280 t cattle slurry, 1640 t poultry dung and 9980 t energy crops)
Amount of biogas produced (m^3/year)	173 500	150 000	4 020 000	100 000	n.a.
Electric power of CHP (kW)	150	35	1000	110 (boiler only)	600
Generated electric energy (kWh/year)	1 301 400	280	8030	-	4 370 000
Generated thermal energy (kWh/year)	123	45	8 223 000	-	5 032 000
Investment cost (k€)	n.a.	65	n.a.	n.a.	3670

n.a., no information available.
[a]Data were collected from available documents produced in the aim of EU project BioEnergyFarm2 (https://www.bioenergyfarm.eu).
[b]Biogas-Messprogramm III, Eg. Fachagentur Nachwachsende Rohstoffe e. V. (2021); https://mediathek.fnr.de/biogas-messprogramm-iii.html.

and the sale of thermal energy to the neighbouring village heat grid was 1600 MWh/year.

Another interesting example is the Zeewolde biogas plant (Netherlands) with two CHP units, one on the farm and another in a nearby residential area, with a biogas pipeline between the farm and the residential area. The plant treats 30000 m^3 of manure and organic material (e.g. corn, grass and agro-industrial waste) per year. The energy production per year includes 7500 MWh of electricity and 7100 MWh of thermal energy. Approximately one-quarter of the biogas generated is used in the CHP plant on the farm, with the resulting heat used for heating the digester and the farm buildings.

The Kalmari dairy farm (Finland) is an example of biogas upgrading to biomethane production. The AD plant digests cattle slurry (2000 m^3/year), locally generated confectionery by-products (200 m^3/year) and fat waste (600 m^3/year) for the production of electricity, heat and vehicle fuel. Occasionally small amounts of energy crops, mainly grass silage, are digested as well. Digestion takes place under mesophilic conditions in a CSTR. Raw biogas (62–64% methane) is converted into a CHP with 25 kW electric and 50 kW thermal capacity. The farm also has an 80 kW gas boiler. Thermal energy is used for heating, hot water and crop drying. Electricity is mainly used on the farm, and a small amount is sold to the grid. A biogas upgrading unit using high-pressure water scrubbing can process 50 Nm^3 of biogas/h to 95% methane content. The biomethane produced (1000 MWh/year) can fuel about 200 light-duty gas vehicles. Alternatively, biomethane can be used as tractor fuel, significantly reducing emissions associated with agricultural field operations.

7 Case studies: centralized anaerobic digestion plants

An alternative to small-scale AD plants is bigger centralized plants that integrate manure, agriculture residues and agro-industrial waste produced in the surrounding area. An example is the Solrød biogas plant in Denmark which brings together local agro-industry actors looking for better ways to manage their waste. The plant has a treatment capacity of 200000 tonnes of feedstock/year, including 7400 tonnes of seaweed, 53200 of manure and 139400 tonnes of agro-industrial waste (79400 tonnes from the production of citrus pectin and 60000 tonnes from the production of microbial cultures and enzymes). The CHP unit produces 51 GWh/year of renewable energy. Electricity is sold to the grid (23 GWh/year), and the heat is supplied to the local district heating system (28 GWh/year). This strategy saves 40100 tons CO_2 eq. per year.

Another example is the Sønderjysk Biogas Plant Bevtoft (in Denmark) with the capacity to co-digest up to 600000 tonnes of biomass per year. The feedstock includes locally-produced animal slurry (425000 t/year), straw (50000 t/year), organic wastes (55000 t/year) and animal bedding/deep litter

(10 000 t/year). The biogas upgrading unit uses amine scrubbing and has the capacity to process 6000 Nm^3 of biogas/h, producing about 21 million Nm^3 of biomethane per year which is sold to the gas grid. This amount of biomethane is enough to satisfy the annual energy demand of 15 000 households or to biofuel 10 000 cars. The plant results in a reduction of emissions from waste of an estimated 51 000 t CO_2/year.

Maabjerg BioEnergy (Denmark) is a very-large-scale plant processing up to 800 000 tonnes of biomass per year, of which 480 000 tonnes are liquid and solid manure from local farms, 85 000 tonnes of wastewater sludge and 160 000 tonnes of agro-industrial waste (e.g. from dairy and potato processing). The 18 million Nm^3 of biogas produced per year are used for electricity generation (49.1 GWh, to be consumed at the plant) and for district heating (58.2 GWh).

8 Conclusion and future trends

Biogas and biomethane are expected to play an important role in energy transition goals given: the storage capacity and demand-based (flexible) utilization of these gases as energy vectors, an already established role in a wide range of applications (CHP production and transport) and the very diverse biomasses (livestock manure, residues and waste) that can be used for their production (Paterson et al., 2016; Meyer et al., 2018).

The continuous recovery of biomass resources and the reduction of waste are two of the main pillars of the 'European Green Deal' and are key to creating a circular economy. Organizations such as the European Biogas Association have encouraged a cross-sector approach to exploit the potential of biogas and biomethane to reduce methane emissions in non-energy sectors.

The biogas solution, mainly considered as a cascade biowaste processing technology, provides a new opportunity to promote a bioeconomy with the generation of high-value products: biogas, biomethane as carriers of energy and biofertilizer (Awasthi et al., 2019). Implementing a manure-based biorefinery can increase the profitability of AD operations and therefore promote its adoption, contributing to closing nutrient loops and reducing environmental impacts.

The potential to efficiently produce biogas from manure and crop residues in the EU depends on overcoming the challenges associated with the nature and composition of these substrates, either by supplementing them with suitable co-substrates or through further development of biogas technology (e.g. substrate pre-treatment and effluent recirculation) (Bumbiere et al., 2020).

Several strategies have been adopted to improve AD process performance, for example the pre-treatment of feedstock and developing co-digestion.

Research on the pre-treatment of manure to break down the recalcitrant structure of lignocellulosic fibres and increase organic matter availability to microbes has considered different types of mechanical, thermochemical, microwave thermochemical and ultrasonic solutions (Khoshnevisan et al., 2021). Future challenges include developing a pre-treatment solution that both efficiently disrupts lignocellulosic fibres and has a positive energy balance.

Another research trend will be using nanotechnology by adding nano-structured materials into the digester to improve cellular uptake of trace elements and to promote metabolic intermediates and enzyme activities involved in the AD process (Baniamerian et al., 2019). The addition of trace elements (e.g. Co, Ni and Fe) as nano-structured materials into manure-based digesters reduced the lag phase reduction and enhanced yield.

Tailor-made microbiota may also help to overcome problems that manure-based digesters can face. Microbial management through the use of bioaugmentation, microbial enrichment and process inhibition control has been researched as a strategy to enhance methane production in manure-based digesters (Ozbayram et al., 2018; Fotidis et al., 2014) and will certainly be addressed in future studies.

Regarding societal challenges, the engagement of farmers in biogas production is crucial for manure-based AD and needs to be addressed. Economic aspects should also be addressed in developing circular business models at a regional level, including either on-farm or centralized AD, which contribute to food chain sustainability and to industrial symbiosis by bringing together different players to achieve common environmental and economic benefits. One option is the creation of 'green gas' hubs that use locally connected biogas to generate electricity or upgrade it to be injected into the grid. The success of manure-based AD will also require better policy design in coordinating incentives, regulations and information for end users.

9 Where to look for further information

Interesting links and related EU projects:

- BioEnergyFarm2 https://www.bioenergyfarm.eu/en/.
- BiogasAction http://www.biogasaction.eu/.
- Biogas3 http://www.biogas3.eu/eng/.
- Sustaingas http://sustaingas.eu/.
- Green gas grids http://www.greengasgrids.eu/.
- Biogas, Energizing The Countryside http://www.biogas-etc.eu.
- ARBOR Biomass for energy http://www.arbornwe.eu/.
- BiogasXpose http://www.biogasxpose.eu/.

The following articles provide a further overview of the subject:

- Møller, H. B., Sørensen, P., Olesen, J. E., Petersen, S. O., Nyord, T. and Sommer, S. G. (2022). Agricultural biogas production-climate and environmental impacts. *Sustainability*, 14(3), 1849.
- Suhartini, S., Nurika, I., Paul, R. and Melville, L. (2021). Estimation of biogas production and the emission savings from anaerobic digestion of fruit-based agro-industrial waste and agricultural crops residues. *BioEnergy Research*, 14(3), 844–859.
- Swarup, S., Sarangi, P. K. and Nanda, S. (2022). Utilization of Waste Biomass Resources for Biogas Production. In *Biomethane* (pp. 1–11). Apple Academic Press.
- Deublein, D. and Steinhauser, A. (2011). *Biogas from Waste and Renewable Resources: An Introduction*. Second Edition. Wiley-VCH Verlag GmbH.

10 References

Achinas, S. and Euverink, G. J. W. (2020). Rambling facets of manure-based biogas production in Europe: A briefing. *Renew. Sustain. Energ. Rev.* 119, 109566.

Awasthi, M. K., Sarsaiya, S., Wainaina, S., Rajendran, K., Kumar, S., Quan, W., Duan, Y., Awasthi, S. K., Chen, H., Pandey, A. and Zhang, Z. (2019). A critical review of organic manure biorefinery models toward sustainable circular bioeconomy: Technological challenges, advancements, innovations, and future perspectives. *Renew. Sust. Energ. Rev.*, 111, 115–131.

Baniamerian, H., Isfahani, P. G., Tsapekos, P., Alvarado-Morales, M., Shahrokhi, M., Vossoughi, M. and Angelidaki, I. (2019). Application of nano-structured materials in anaerobic digestion: current status and perspectives. *Chemosphere* 229, 188.

Bumbiere, K., Gancone, A., Pubule, J., Kirsanovs, V., Vasarevicius, S. and Blumberga, D. (2020). Ranking of bioresources for biogas production. *Rigas Tehniskas Universitates Zinatniskie Raksti* 24(1), 368.

Caposciutti, G., Baccioli, A., Ferrari, L. and Desideri, U. (2020). Biogas from anaerobic digestion: power generation or biomethane production? *Energies* 13(3), 743.

Duarte, E., Fragoso, R., Smozinski, N. and Tavares, J. (2021). Enhancing bioenergy recovery from agro-food biowastes as a strategy to promote circular bioeconomy. *J. Sustain. Dev. Energy Water Environ. Syst.* 9(1). doi: 10.13044/j.sdewes.d8.0320.

EC (2020a). Communication from the Commission to the European Parliament, the Council, the European Economic and Social Committee and the Committee of the Regions on an EU strategy to reduce methane emissions. *Com* 663. Available at: https://ec.europa.eu/energy/sites/ener/files/eu_methane_strategy.pdf. Accessed February 24, 2022.

EC (2020b). Proposal for a decision of the European Parliament and of the Council on a General Union Environment Action Programme to 2030. Available at: https://ec.europa.eu/environment/pdf/8EAP/2020/10/8EAP-draft.pdf. Accessed February 25, 2022.

EC (2021a). A European Green Deal: striving to be the first climate-neutral continent. European Commission. Available at: https://ec.europa.eu/info/strategy/priorities-2019-2024/european-green-deal_en. Accessed February 8, 2022.

EC (2021b). Greenhouse gas emissions from agriculture in Europe. European Commission. Available at: https://www.eea.europa.eu/ims/greenhouse-gas-emissions-from-agriculture#footnote-SZNUYDDV. Accessed February 8, 2022.

EEA (2021). Trends and projections in Europe 2021. Available at: https://www.eea.europa.eu/publications/trends-and-projections-in-europe-2021. Accessed February 24, 2022.

Einarsson, R. and Persson, U. M. (2017). Analyzing key constraints to biogas production from crop residues and manure in the EU–A spatially explicit model. *PLoS ONE* 12(1), e0171001.

EurObserv'ER (2020). Biogas barometer. Accessed February 24, 2022. Available at: https://www.eurobserv-er.org/biogas-barometer-2020/.

Fangueiro, D., Alvarenga, P. and Fragoso, R. (2021). Horticulture and orchards as new markets for manure valorisation with less environmental impacts. *Sustainability* 13(3), 1436. doi: 10.3390/su13031436.

Fotidis, I. A., Wang, H., Fiedel, N. R., Luo, G., Karakashev, D. B. and Angelidaki, I. (2014). Bioaugmentation as a solution to increase methane production from an ammonia-rich substrate. *Environ. Sci. Technol.* 48(13), 7669.

Ghirardini, A., Grillini, V. and Verlicchi, P. (2020). A Review of the occurrence of selected micropollutants and microorganisms in different raw and treated manure-environmental risk due to antibiotics after application to soil. *Sci. Total Environ.* 707, 136118, doi: 10.1016/j.scitotenv.2019.136118.

Hidalgo, D., Sanz-Bedate, S., Martín-Marroquín, J. M., Castro, J. and Antolín, G. (2020). Selective separation of CH4 and CO2 using membrane contactors. *Renew. Energy* 150, 935, doi: 10.1016/j.renene.2019.12.073.

IEA (2020). Outlook for biogas and biomethane: prospects for organic growth. IEA, Paris. Available at: https://www.iea.org/reports/outlook-for-biogas-and-biomethane-prospects-for-organic-growth. Accessed July 7, 2022.

IEA (2021). Methane Tracker 2021. IEA, Paris. https://www.iea.org/reports/methane-tracker-2021. Accessed July 5, 2022.

Khoshnevisan, B., Duan, N., Tsapekos, P., Awasthi, M. K., Liu, Z., Mohammadi, A., Angelidaki, I., Tsang, D. C. W., Zhang, Z., Pan, J., Ma, L., Aghbashlo, M., Tabatabaei, M. and Liu, H. (2021). A critical review on livestock manure biorefinery technologies: Sustainability, challenges, and future perspectives. *Renew. Sustain. Energ. Rev.* 135, 110033. doi: 10.1016/j.rser.2020.110033.

Köninger, J., Lugato, E., Panagos, P., Kochupillai, M., Orgiazzi, A. and Briones, M. J. I. (2021). Manure management and soil biodiversity: towards more sustainable food systems in the EU. *Agric. Syst.* 194, 103251. doi: 10.1016/j.agsy.2021.103251.

Korbag, I., Omer, S. M. S., Boghazala, H. and Abusasiyah, M. A. A. (2020). Recent advances of biogas production and Future Perspective. In: *Biogas-Recent Advances and Integrated Approaches* (P. 3), IntechOpen.

KTBL (2013). Faustzahlen biogas (vol. 3). Technik, K. f.ü.r. (Ed.) und Bauwesen in der Landwirtschaft e.V. (KTBL), Darmstadt.

KTBL (2021). Gasausbeute in landwirtschaftlichen Biogasanlagen. In: Potenziale, Erträge and Einflussfaktoren (Eds). *Kuratorium für Technik und Bauwesen in der Landwirtschaft e.V. (KTBL)*, Darmstadt.

KTBL (2022). Wirtschaftlichkeitsrechner biogas (vol. 07) Technik, K. f.ü.r. (Ed.) und Bauwesen in der Landwirtschaft e. Available at: https:daten.ktbl.de/biogas/startseite.do.V.

Majer, S., Kornatz, P., Daniel-Gromke, J., Rensberg, N., Brosowski, A., Oehmichen, K. and Liebetrau, J. (2019). Stand und Perspektiven der. In: aus Gülle, B. (Ed.). *Deutsches Biomasseforschungszentrum Gemeinnützige GmbH (DBFZ)*, Leipzig.

Mata-Alvarez, J., Dosta, J., Romero-Güiza, M. S., Fonoll, X., Peces, M. and Astals, S. (2014). A critical review on anaerobic co-digestion achievements between 2010 and 2013. *Renew. Sustain. Energy Rev.* 36, 412.

Meyer, A. K. P., Ehimen, E. A. and Holm-Nielsen, J. B. (2018). Future European biogas: animal manure, straw and grass potentials for a sustainable European biogas production. *Biomass Bioenergy* 111, 154.

Ozbayram, E. G., Akyol, Ç., Ince, B., Karakoç, C. and Ince, O. (2018). Rumen bacteria at work: bioaugmentation strategies to enhance biogas production from cow manure. *J. Appl. Microbiol.* 124(2), 491. doi: 10.1111/jam.13668.

Paterson, M., Kayser, K., Bonhomme, S., Majewsk, E., Amrozy, M., Berruto, R., Franco Parola, F., Jan Willem Bijnagte, J. W. and Gysen, M. (2016). Implementation Guide for small-scale biogas plants - a farmers handbook. BioEnergy Farm 2 publication, vol. 1.2 Technik, K. f.ü.r. (Ed.) und Bauwesen in der Landwirtschaft e.V. (KTBL), Darmstadt.

Rösemann, C., Haenel, H.-D., Vos, C., Dämmgen, U., Döring, U., Wulf, S., Eurich-Menden, B., Freibauer, A., Döhler, H., Schreiner, C., Osterburg, B. and Fuß, R. (2021). Calculations of Gaseous and Particulate Emissions from German Agriculture 1990–2019 – Report on Methods and Data (RMD) Submission 2021. Thünen Report 84 84, Johann Heinrich von Thünen-Institut (VTI), Braunschweig. Available at: https://www.doi.org/10.3220/REP1616572444000.

Scarlat, N., Dallemand, J. F. and Fahl, F. (2018). Biogas: developments and perspectives in Europe. *Renew. Energy* 129, 457.

Tamburini, E., Gaglio, M., Castaldelli, G. and Fano, E. A. (2020). Is bioenergy truly sustainable when land-use-change (LUC) emissions are accounted for? The case-study of biogas from agricultural biomass in Emilia-Romagna region, Italy. *Sustainability* 12(8), 3260; doi: 10.3390/su12083260.

Zhou, J., Zhang, Y., Khoshnevisan, B. and Duan, N. (2021). Meta-analysis of anaerobic co-digestion of livestock manure in last decade: identification of synergistic effect and optimization synergy range. *Appl. Energy* 282(A), 116128, doi: 10.1016/j.apenergy.2020.116128.

Chapter 10

Multi-feedstock biorefineries for converting agricultural wastes and microalgae into co-products

Butch Bataller, University of the Philippines at Los Baños (UPLB), The Philippines; and Sergio Capareda, Texas A&M University, USA*

1 Introduction

Biomass provides the foundation for a more circular economy (Fig. 1). It has a traditional role as a source of food and feed and, more recently, as a feedstock for biofuel. In addition, there has been a growing interest in utilizing biomass residues and waste to address the challenges of more efficient resource use, reducing waste and mitigation of climate change (Kumar et al., 2018). The US government has conducted research on the potential of biorefining of waste materials both to meet future needs for energy and manufacture a range of specialty chemicals (Holladay et al., 2007; Werpy and Petersen, 2007; USDOE, 2016). Biomass waste is increasingly being exploited to produce a much wider range of high-value co-products, including food additives, nutraceuticals, pharmaceuticals and biobased plastics as well as materials such as graphene and activated carbon (Fig. 2).

* Corresponding author: bgbataller@up.edu.ph.

http://dx.doi.org/10.19103/AS.2023.0120.19

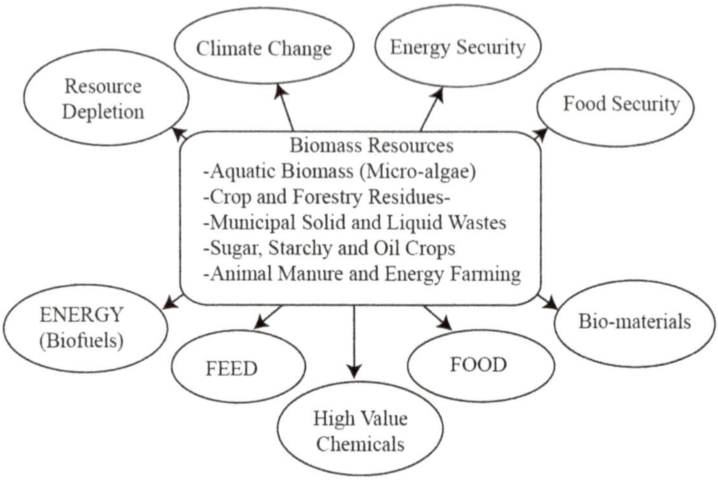

Figure 1 The role of biomass in the circular economy.

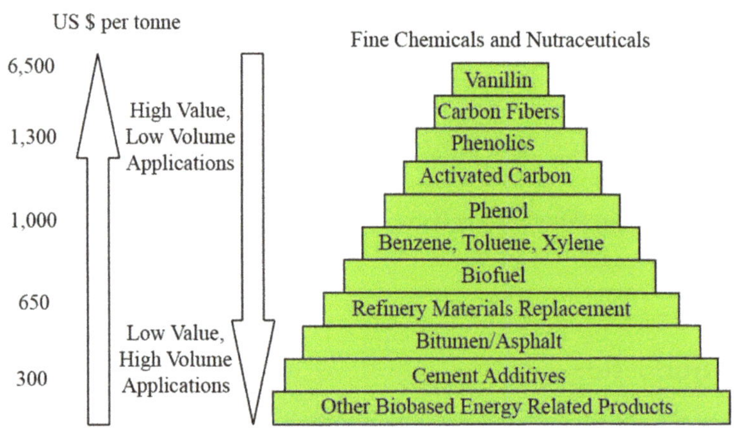

Figure 2 Valorization of biomass into different products by volume and value.

An example of this developing range of co-products is the synthesis of sugars from disaccharides and polysaccharides derived from lignocellulosic biomass (Murzin and Simiken, 2016). Examples of platform chemicals derived from sugars include methanol, ethanol, acetic acid, lactic acid, succinic acid, glycerol, aromatic hydrocarbons, furfurals and furanics, phenols, sorbitol, xylitol and levulinic acid, levoglucosan and levoglucosenone, hydroxypropionic acid and 2,5-furandicarboxylic acid. Ethanol is also a bulk chemical that can be dehydrated to ethylene, which can be polymerized to polyethylene. Lactic acid is another bulk chemical that can be converted to lactic acid esters or polylactic acid (PLA, polyethylene furanoate (PEF) and poly-γ-glutamic acid (γ-PGA); Yu et al., 2020). Succinic acid is a precursor for 1,4 butanediol, polymers and esters, polybutylene succinate, solvents and coatings, polyester polyols and plasticizers (Kumar et al., 2019). New research challenges are emerging in converting agricultural biomass resources into this growing range of co-products. This involves characterizing the range of biomass resources available, the various conversion pathways available to create co-products and the range of potential high-value co-products that can be produced from these resources.

2 Biomass feedstock sources available for conversion

The United States has an estimated biomass yield of 1 billion tons (US), equivalent to about 0.909 billion metric tonnes. Biomass is still the largest single source for renewable energy production in the United States, representing about 3.9 quadrillion (almost 41%) of 9.6 quadrillion British thermal units (Btu) generated from renewable energy sources in 2015 (EIA, 2016). These biomass resources include agricultural and forestry resources as well as municipal solid wastes (MSW) and microalgae.

Production of key crops has continued to grow, which has led to an increase in agricultural residues as a potential source of biomass. For example, the 2022 corn yields in the United States were estimated at a record high of 11.13 tonnes/ha, 0.35 tonnes above the 2020 yield of 10.78 (NAS, 2022). Capareda (2014) reported an average corn yield of only 7.8 tonnes/ha/year in 2011, indicating an increase of around 43% in over a decade. Some yields have reached 37.4 tonnes/ha/year (Meki et al., 2017). Based on these increases in production, residue from corn available as a biomass feedstock is estimated to have increased to around 28 tonnes/ha/year (as shown in Example 1). Sorghum waste has also grown by an average of 28.5% over similar timeframe, yielding an average of around 10.85 tonnes/ha/year. Texas A&M University, e.g. has developed a high yield sorghum biomass with a reported a yield of approximately 23.8 tonnes/ha/year (Rooney et al., 2007). Capareda (2014) has estimated the yield potential of different biomass feedstocks, showing that trees have a particularly high potential yield with, e.g. a leguminous tree (*Leucaena leucocephala*) having the highest reported yield of 30 dry tonnes equivalent/ha/year. This was followed by herbaceous crops such as hybrid Napier grass (also called elephant grass), a short rotation grass with a reported average yield of 27 dry tonnes equivalent/ha/year. Aquatic plants have even higher potential yields with, e.g. paper reed (*Cyperus papyrus* L.) having a reported yield of 50 dry tonnes/ha/year. Example 1 shows the calculations used to estimate potentially available biomass for corn and soybean.

Example 1. The average corn, sorghum and soybean production yield in 2021 was 177, 69 and 51.4 bushels/acre/year, respectively. Determine the yields in units of metric tonnes/ha/year. Note that a bushel of corn and sorghum weighs 56 lbs while that of soybean is 60 lbs per bushel. Estimate the residue production assuming a crop/residue ratio of 1:2.5 for all crops.

Solution:

1 The calculations to convert to SI units are as follows:

$$\text{Corn yield}\left(\frac{\text{tonnes}}{\text{ha}-\text{year}}\right)=\frac{177\text{ bu}}{\text{acre}}\times\frac{56\text{ lbs}}{\text{bu}}\times\frac{1\text{ kg}}{2.2\text{ lbs}}\times\frac{1\text{ tonne}}{1000\text{ kg}}\times\frac{2.47\text{ acre}}{\text{ha}}=11.13\frac{\text{tonnes}}{\text{ha}-\text{year}}$$

$$\text{Sorghum yield}\left(\frac{\text{tonnes}}{\text{ha}-\text{year}}\right)=\frac{69\text{ bu}}{\text{acre}}\times\frac{56\text{ lbs}}{\text{bu}}\times\frac{1\text{ kg}}{2.2\text{ lbs}}\times\frac{1\text{ tonne}}{1000\text{ kg}}\times\frac{2.47\text{ acre}}{\text{ha}}=4.34\frac{\text{tonnes}}{\text{ha}-\text{year}}$$

$$\text{Soybean yield}\left(\frac{\text{tonnes}}{\text{ha}-\text{year}}\right)=\frac{51.4\text{ bu}}{\text{acre}}\times\frac{60\text{ lbs}}{\text{bu}}\times\frac{1\text{ kg}}{2.2\text{ lbs}}\times\frac{1\text{ tonne}}{1000\text{ kg}}\times\frac{2.47\text{ acre}}{\text{ha}}=3.46\frac{\text{tonnes}}{\text{ha}-\text{year}}.$$

2 The equivalent waste residue generations are calculated as follows:

$$\text{Corn waste yield}\left(\frac{\text{tonnes}}{\text{ha}-\text{year}}\right)=11.13\frac{\text{tonnes}}{\text{ha}-\text{year}}\times 2.5=27.82\frac{\text{tonnes}}{\text{ha}-\text{year}}.$$

$$\text{Sorghum yield}\left(\frac{\text{tonnes}}{\text{ha}-\text{year}}\right)=4.34\frac{\text{tonnes}}{\text{ha}-\text{year}}\times 2.5=10.85\frac{\text{tonnes}}{\text{ha}-\text{year}}$$

$$\text{Soybean yield}\left(\frac{\text{tonnes}}{\text{ha}-\text{year}}\right)=3.46\frac{\text{tonnes}}{\text{ha}-\text{year}}\times 2.5=8.65\frac{\text{tonnes}}{\text{ha}-\text{year}}.$$

While agricultural residues remain the major source of biomass for conversion into energy or co-products, further growth is limited by factors such as the availability of arable land. Aquatic biomass, particularly microalgae, is therefore emerging as an important additional biomass resource (Lee, 2014). One advantage of this source is its ability to fix carbon. Photosynthetic organisms fix around 100 Gt (1×10^9 tonne) of carbon into biomass annually (Field et al., 1998). They play a central role in the global carbon cycle (Sayre, 2010). Between 1.6 g and 2 g of CO_2 is captured for every gram of algal biomass produced (Herzog and Golomb, 2004). This means that a potential billion tonnes of CO_2 could be captured for every billion ton of biomass produced, as estimated in Example 2.

Example 2. Convert the sequestered carbon dioxide rate of 2 g CO_2 for every gram of biomass produced in units of billion tons CO_2 captured by this yearly biomass production rate.

Solution:

1 The CO_2 utilization rate is calculated as follows:

$$CO_2 \text{ utilization}\left(\frac{\text{g } CO_2}{\text{tonne biomass}}\right) = \frac{2 \text{ g } CO_2}{\text{g biomass}} \times \frac{1000 \text{ g}}{1 \text{ kg}} \times \frac{1000 \text{ kg}}{\text{tonne}}$$
$$= 2 \times 10^6 \frac{\text{g } CO_2}{\text{tonne biomass}}$$

$$CO_2 \text{ utilization}\left(\frac{\text{tonne } CO_2}{\text{tonne biomass}}\right) = 2 \times 10^6 \frac{\text{g } CO_2}{\text{tonne biomass}}$$
$$\times \frac{\text{tonne}}{1 \times 10^6 \text{ g}} = 2 \frac{\text{tonne } CO_2}{\text{tonne biomass}}$$

$$CO_2 \text{ use}\left(\frac{\text{tonne } CO_2}{\text{ton biomass}}\right) = 2 \frac{\text{tonne } CO_2}{\text{tonne biomass}}$$
$$\times \frac{0.909 \text{ tonne}}{\text{ton biomass}} = 1.82 \frac{\text{tonne } CO_2}{\text{ton biomass}}.$$

2 Hence, close to a billion tonne of CO_2 is captured for every billion ton of biomass produced.

3 Note the conversion of ton to tonnes in the following equation.

$$1 \times 10^9 \text{ tons} \times \frac{2000 \text{ lbs}}{\text{ton}} \times \frac{\text{kg}}{2.2 \text{ lbs}} \times \frac{\text{tonne}}{1000 \text{ kg}} = 0.909 \times 10^9 \text{ tonnes}$$

Simply,

1 ton = 0.909 tonnes

3 Biological and thermochemical biomass conversion processes

Most plant-based biomass resources are primarily composed of cellulose with hemicellulose and lignin. Sugar and starchy crops as well as lignocellulosic biomass are ideal for bioethanol, while oil crops, containing triglycerides, are primarily used for biodiesel. As noted, biomass with a high cellulosic component includes lignin, a complex organic polymer deposited in the cell wall of many plants which gives them their structure. The complex structure of lignin makes conversion of cellulosic materials into monomeric components challenging (Singhvi and Gokhale, 2019). This requires breaking down long-chain chemical compounds into short-chain compounds.

There are two main pathways available for biomass conversion:

- Biological pathways; and
- Thermochemical pathways.

The former utilizes microorganisms to convert biomass, while the latter break down biomass via use of elevated temperature. An overview of the two different pathways is provided by Capareda (2014).

Biological pathways include anaerobic digestion and fermentation (Sun et al., 2013). They have high product selectivity but rates of conversion are slow. Since lignin is hard to break down, biological pathways also require pretreatment of the biomass to break down its components into simpler compounds which can be broken down further by microorganisms (Mauyra et al., 2015). Pretreatment adds to the capital and processing cost requirement (representing about one-third of the processing cost). Ethanol from lignocellulosic biomass, e.g. is produced by pretreatment to release enzymes, enzymatic hydrolysis to convert polysaccharides into monomer sugars followed by fermentation of sugars to produce ethanol. The cost of enzymes, typically used in enzymatic hydrolysis, also contributes substantially to the processing cost.

Since they do not require a pretreatment stage, thermal conversion pathways are often better suited to biomass resources with high cellulose and lignin components. Thermal biomass conversion technologies include the following: hydrothermal carbonization, torrefaction, slow and fast pyrolysis, supercritical and high-temperature gasification and combustion (Lewandowski et al., 2020). Torrefaction, pyrolysis and gasification are, e.g. reviewed in Saffron (2020). The main advantage of thermal conversion systems is the quicker breakdown of materials into simpler compounds and further conversion into high-value products with the use of appropriate catalysts and additives. Perhaps the main disadvantage of the thermochemical pathway is the formation of side products due to the complexity of the biomass structure which must then be disposed of.

Figure 3 shows an integrated thermal conversion process using a combination of gasification and pyrolysis to produce a wide range of products. Lignocellulosic materials are ideal for this conversion pathway. In gasification systems, e.g. the primary product is synthesis gas (CO and H_2) and some low-molecular-weight hydrocarbon gas. These combustible syngases can be used to generate electrical power. In some pathways, synthesis gases may also be converted into biofuels via biological conversion. For example, syngas may be fermented to bioethanol using *Clostridium autoethanogenum*. Pyrolysis is a versatile thermal conversion pathway in which biomass is converted into solid biochar, liquid bio-oil and combustible syngas. The resulting bio-oil liquid may be upgraded into high-value fine chemicals including biofuels. Bio-oil upgrading can be achieved via physical, chemical or catalytic means. Liquid fuel products include diesel, gasoline, kerosene, methane and LPG. Biochar

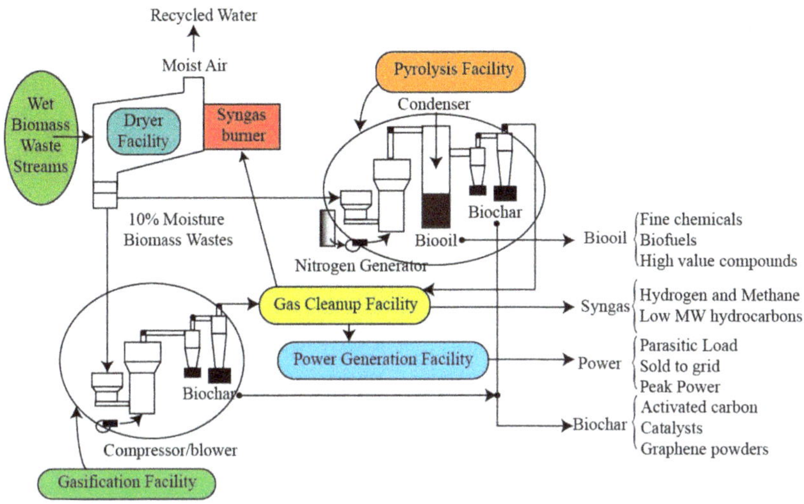

Figure 3 Integrated biomass thermal conversion for high-value biofuel, biopower and bioproduct development.

may be purified to carbon black, activated carbon and used to make high-value graphene materials.

4 Microalgae as a source of biomass

A fundamental process underpinning all biomass production is photosynthesis which is summarized in Equation 1:

$$6CO_2 + 6H_2O \rightarrow C_6H_{12}O_6 + 6O_2 \quad (1)$$

The equation requires 6 moles of CO_2 and H_2O, converted by light energy captured by the plant chlorophyll (and some bacteria) into a representative sugar molecule (in this case glucose sugar) and six moles of oxygen. As noted earlier, an emerging source of biomass for conversion is microalgae, primarily due to their high photosynthetic conversion efficiency (Shen, 2021).

Microalgae are rapidly growing photosynthetic organisms with the potential to transform 9–10% of solar energy (average sunlight irradiance) into biomass with a theoretical yield of about 77 g/biomass/m^2/day which is about 280 tonnes/ha/year (Formighieri et al., 2012; Melis, 2009); 1 kg of algal biomass can also fix 1.83 kg of CO_2 as well as absorbing SOx and NOx as nutrients along with CO_2 (Khan et al., 2018). These characteristics mean that microalgae are a promising feedstock for biomass production because they are potentially 10–20 times more productive than any other biofuel crop and large-scale cultivation does not need to compete for arable land (Narala et al., 2016).

Algae are comprised of significant amounts of protein as well as carbohydrates and lipids. They also contain high-value nutrients such as bio-pigments, amino acids, vitamins and antioxidants that have found use in the nutraceutical, pharmaceutical, food and feed sectors (Lei, 2021). Algae containing high amounts of lipids can also be used to produce biodiesel and oleochemicals. Carbohydrates can be converted to bioethanol while proteins can be extracted for food and feed. The residual biomass can be subjected to thermochemical processing for syngas, bio-oil and biochar production or it can be digested anaerobically for the production of methane. Algal biomass conversion has been widely researched and studied but has yet to become commercially competitive. Scalability of algal cultivation is still a major challenge.

Microalgae can be cultivated in controlled conditions in open ponds, e.g. in raceways or in photobioreactors (Khan et al., 2018). Figure 4 is an example of a simple laboratory raceway used to evaluate biomass growth rates and productivity. In these systems, artificial light is used coupled with controlled fertilization and aeration. The design uses paddle wheels for aeration, pumps for water circulation and flow and automatic feeders for fertigation. Illumination is provided by fluorescent or incandescent bulbs with different wattage ratings. Figure 5 shows a photobioreactor, an enclosed chamber which allows more precise process control.

Figure 4 Laboratory raceway facilities to evaluate controlled biomass production.

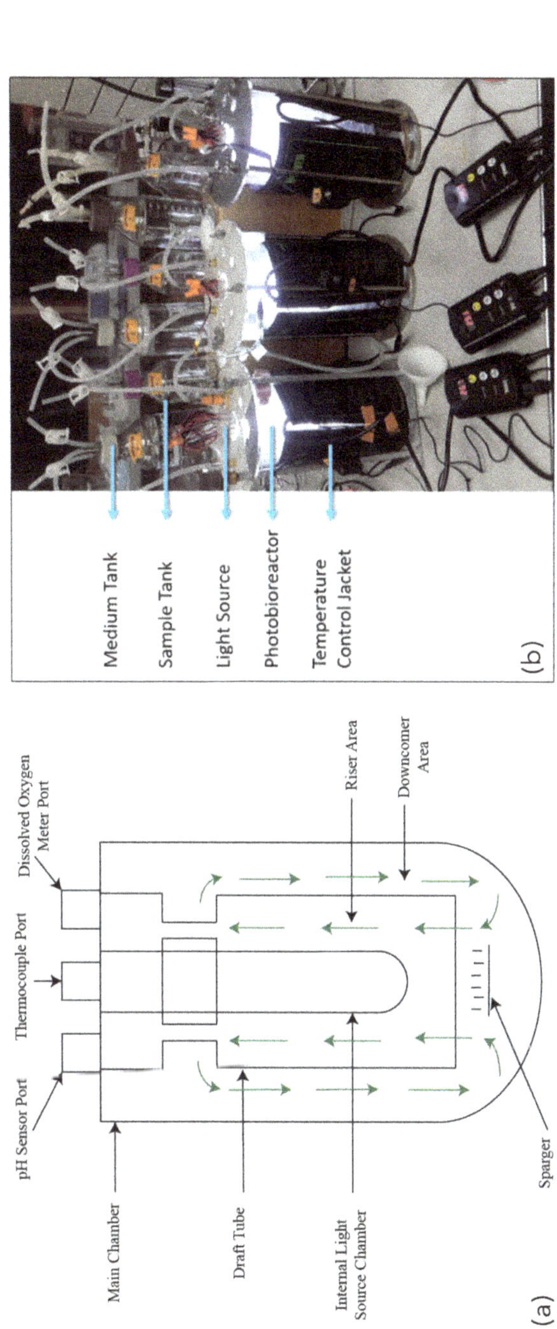

Figure 5 Design of a concentric tube internally illuminated photobioreactor: (a) schematic diagram and (b) actual bench-scale photobioreactor fabricated at the Bioenergy Testing and Analysis Laboratory (BETALab), TAMU, Texas, USA.

Table 1 Comparison between open pond raceways and photobioreactors

Culture systems	Photobioreactors	Open ponds
Contamination control	Easy	Difficult
Contamination risk	Reduced	High
Sterility	Achievable	None
Process control	Easy	Difficult
Species control	Easy	Difficult
Mixing	Uniform	Very poor
Operation regime	Batch or continuous	Batch or continuous
Area/volume	High (10–200/m)	Low (5–10/m)
Population density	High	Low
Investment	High	Low
Operating cost	3–10×higher	3–10× lower
Light utilization efficiency	High	Poor
Evaporation of medium	Low	High
Gas transfer control	High	Low
Productivity	3–5× more productive	3–5× less productive

There are advantages and disadvantages in each system. These are summarized in Table 1 (Khan et al., 2018). Open cultivation systems have up to 11 times lower capital and operating costs, including lower energy requirements, than photobioreactors (Banerjee and Ramaswamy, 2019). However, open systems such as raceways require large areas to scale up and are very susceptible to contamination and adverse weather conditions. Raceways are easier to clean and with minimal biofilm accumulation.

The primary advantage of a photobioreactor over an open pond is sterility and protection from contamination. If one is growing specialty algae species, good process control improves product purity. Given better process control, photobioreactors are three to five times more productive than open ponds. If costs can be managed, there is greater potential for scalability. Light utilization efficiency is also improved through use of light emitting diodes (LEDs). The main disadvantages are high set-up and operating costs. Where possible, the power needed to operate a photobioreactor should come from renewable sources.

Photobioreactors make it possible to control variables such as pH, temperature, mixing rates, CO_2 and O concentration. All these controls improve biomass production rates. To optimize microalgae production, the photobioreactor should be designed to illuminate cultures uniformly without interfering with the flow pattern required to keep the cells in suspension. Light intensities may be controlled for a given volume (e.g. 1.7 mW/cm^2 over specific surface area of say 3.2 cm^2/cm^3). High densities are achieved of up to 10^9 cells/mL or 3% w/v. A potential level of production is shown in Example 3.

Example 3. A pilot scale microalgae production system has a biomass yield of 5 g/L dry weight per unit volume. Translate this output into units of tonnes per year. The reactor has a volume of 11 355 L (3000 gallons) and the unit is operated 350 days in a year with weekly harvest (~50 weeks). There are four reactors used for this project and the harvest is done daily. For simplicity, assume density of growth media is close to that of water at 1 g/mL or 1 kg/L.

Solution:

1 The weekly production rate is calculated as follows:

$$\text{Weekly yield}\left(\frac{\text{kg}}{\text{day}}\right)=\frac{11355\ \text{L}}{\text{week}}\times\frac{5\text{g}}{\text{L}}\times\frac{1\ \text{kg}}{1000\ \text{g}}=56.8\frac{\text{kg}}{\text{week}}.$$

2 The weekly production for the four photobioreactors is calculated as follows:

$$\text{Total weekly yield}\left(\frac{\text{kg}}{\text{day}}\right)=56.8\frac{\text{kg}}{\text{week}}\times 4=227.2\frac{\text{kg dry algae}}{\text{week}}.$$

3 Hence, the yearly microalgae yield in units of tonnes/year is calculated as follows:

$$\text{Yearly}\left(\frac{\text{tonnes}}{\text{year}}\right)=227.2\frac{\text{kg}}{\text{week}}\times 50\frac{\text{weeks}}{\text{year}}\times\frac{1\ \text{tonne}}{1000\ \text{kg}}=11.4\frac{\text{tonnes}}{\text{year}}.$$

5 Growing microalgae

Figure 6 illustrates a complete pilot laboratory fermenter to investigate and optimize parameters for microalgae growth. Key features include an air compressor for efficient provision of oxygen, power for LED lighting system, raw materials input and mixing, a continuous sterilization process and provision of feed while controlling system pH, light and system temperature.

There are four basic phases of growth as shown in Fig. 7:

- The lag phase;
- The exponential phase, also sometimes referred to as log phase;
- The stationary phase; and
- The death or decline phase

This chapter focuses on the example of growing *Spirulina platensis*. Generation times, which vary among various microalgae species, is controlled by many environmental factors. The lag phase is the first major step in microbial growth in a batch growth process. This is the period of adaptation of the cells to their new environment. There is minimal increase in cell density which may be absent in some species.

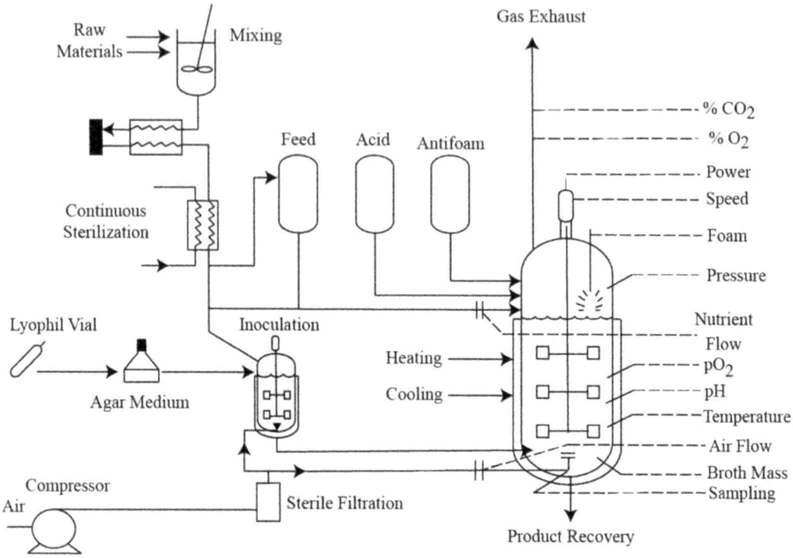

Figure 6 Schematic of laboratory fermenter for microalgae growth experiments.

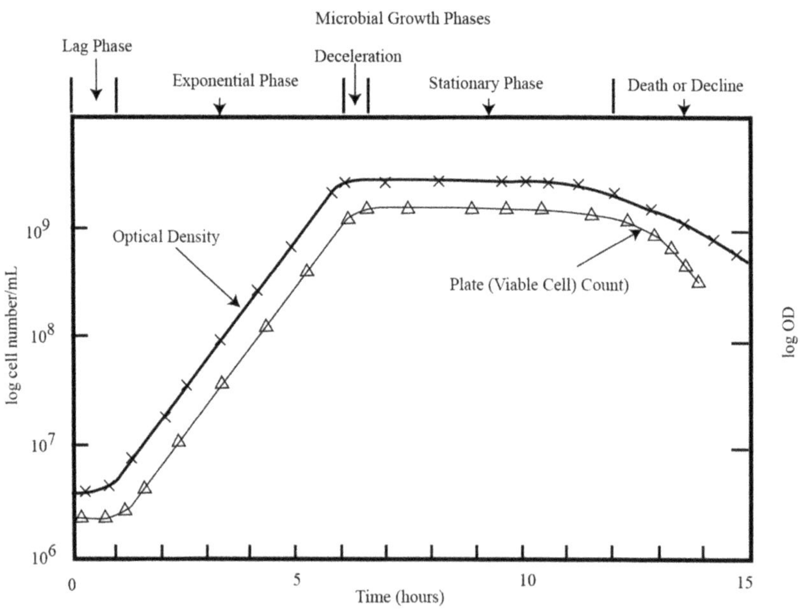

Figure 7 Microbial growth phases (Griffiths, 1972).

The second major step is the exponential phase also known as the logarithmic growth phase. Having adjusted to their new environment, cells start to divide at a constant rate, resulting in an exponential increase in the number of cells present. This is known as the specific growth rate. This chapter provides laboratory data to help researchers determine doubling time of growth (called t_d) and specific growth rate (μ). The specific growth rate can be represented mathematically by first-order kinetics as shown in Equation 2 (Doran, 2013) and shown in Fig. 8.

$$\frac{dX}{dt} = (\mu - k_d)X \quad (2)$$

where

X = cell concentration, g/L

μ = cell growth rate (day^{-1}) and

k_d = cell death rate (day^{-1}).

The term $(\mu - k_d)$ can be referred to as μ_{net}. The cell death rate is sometimes not used if it is considerably smaller than the cell growth rate. Cell growth rate is often substrate limited as shown in Fig. 8.

The growth curve is represented by classical Monod kinetics as shown in Equation 3 (Doran, 2013) that relates the relationship between the specific growth rate, μ, and the growth-limiting substrate, S.

$$\mu = \frac{\mu_{max} S}{K_s + S} \quad (3)$$

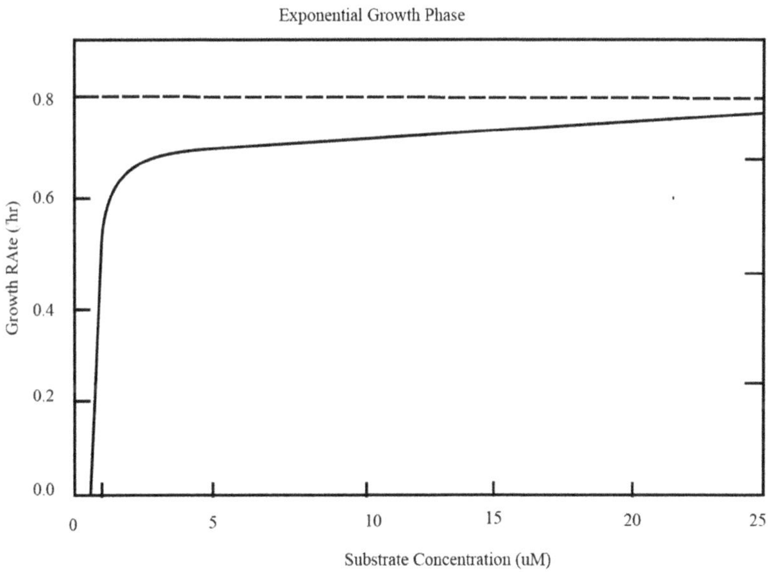

Figure 8 Exponential growth rate of biomass.

where

μ = specific growth rate (day^{-1}),

μ_{max} = maximum specific growth rate (day^{-1}),

S = growth limiting substrate concentration (g/L) and

K_s = saturation constant (g/L).

The doubling time may be established by simply determining the cell growth as a function of time if the number of cells expressed in terms of biomass concentration X (g/L) is doubled after some time $t = t_d$. Doubling time is shown in Equation 4, which was derived from the linearization of the exponential equation, $X_t = X_o 2^n = X_o 2^{t/t_d}$, representing the exponential growth phase:

$$\ln\left(\frac{X_t}{X_o}\right) = \frac{(\ln 2)t}{t_d} = 0.693 \times \frac{t}{t_d} \tag{4}$$

where

t = growth time, days

t_d = doubling time, days

X_t = biomass concentration at any time t (days)

If one needs to relate doubling time with specific growth rate, one may use Equation 2 to arrive at relationships presented in Equation 5.

$$\mu = \frac{1}{X}\frac{dX}{dt} = \frac{1}{X_t}\frac{X_t - X_o}{t_t - t_o} \tag{5}$$

Equation 6 shows the final relationship between specific growth rate and doubling time:

$$\mu = \frac{0.693}{t_d} \tag{6}$$

Substituting Equation 6 to Equation 4 gives the linearized equation as shown in Equation 7:

$$\ln(X_t / X_o) = \mu \Delta t \tag{7}$$

Table 2 shows experimental data for a *Spirulina platensis* growth curve. The first column shows time elapsed and the second column is the biomass concentration in units of g/L. The plot of biomass concentration versus substrate concentration is shown in Fig. 9. As substrate concentration is utilized, the *Spirulina platensis* follows a classical growth pattern. The calculation for doubling and specific growth rate is shown in Example 4.

Example 4. Determine the doubling time (t_d) and specific growth rate for *Spirulina platensis* samples using the data from Table 2. Note that these two parameters are taken during the linear portion of the data at around 142 h.

Table 2 Experimental growth data for *Spirulina platensis*

Elapsed time	Biomass concentration	Substrate concentration	μ	1/*S*	1/*u*	ln (*X*/*X*o)
t (h)	*X* (g/L)	*S* (g/L)				
0	0.6051	2.402				0.0000
4.233	0.6501	2.389	0.0164	0.419	61.15	0.0717
23.217	0.9667	2.194	0.0161	0.456	62.07	0.4685
42.083	1.5014	1.398	0.0142	0.715	70.49	0.9088
67.883	1.8288	0.602	0.0099	1.661	101.45	1.1060
93.85	2.5354	0.421	0.0081	2.375	123.27	1.4327
116.367	3.1457	0.367	0.0069	2.725	144.08	1.6484
142.15	3.8233	0.262	0.0059	3.817	168.88	1.8435
165.433	4.4077	0.205	0.0052	4.878	191.76	1.9857
189.083	5.0294	0.209	0.0047	4.785	214.94	2.1177
211.867	5.2719	0.189	0.0042	5.291	239.34	2.1648
237.833	6.2707	0.152	0.0038	6.579	263.23	2.3382
264.567	6.6437	0.145	0.0034	6.897	291.08	2.3960
306.883	7.8456	0.121	0.0030	8.264	332.53	2.5623
377.717	8.3844	0.102	0.0025	9.804	407.10	2.6287

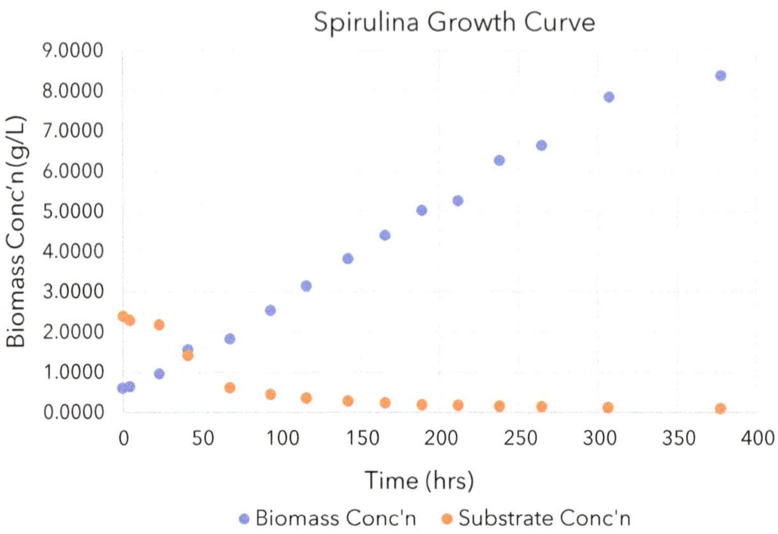

Figure 9 *Spirulina platensis* growth curve data.

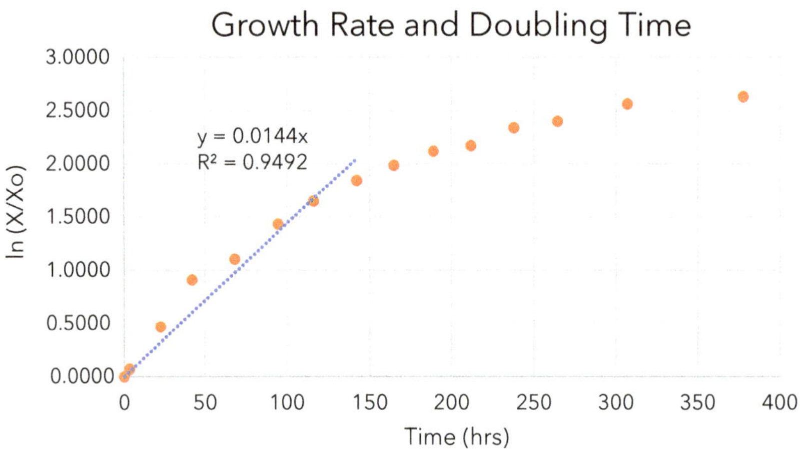

Figure 10 *Spirulina platensis* growth rate and doubling time.

This is done by doing regression analysis of columns 1 and 6 using the first eight rows of data. The linear regression line is shown in Fig. 10 with regression equation shown in the following.

Solution:

1 The equation of the line is $y = 0.0144x$. Thus, this will be equated to the following equation:

$\ln(X_t / X_o) = \mu t$.

2 The slope is simply equal to the specific growth rate of $\mu = 0.0144\text{h}^{-1}$.
3 The value of doubling time is calculated as follows (Fig. 10):

$$\mu = \frac{0.693}{t_d} \quad t_d = \frac{0.693}{\mu} = \frac{0.693}{0.0144h^{-1}} = 48.14\text{h} \cdot$$

4 Hence, the doubling time is around 2 days.

Having established the growth rate of *Spirulina platensis*, the classical first-order kinetic Monod model may also be derived using data from Table 2. This is shown in Example 5. In this example, the biomass yield coefficient is also calculated. This is given by Equation 8. This is defined as the weight of biomass produced per unit weight of the substrate consumed.

$$Y_{X/S} = \frac{\Delta X}{-\Delta S} = \frac{X_t - X_o}{-(S_t - S_o)} \tag{8}$$

Example 5. Determine the kinetic parameters of the classical Monod model using data from Table 2. Determine also the biomass yield coefficient to determine the weight of biomass produced per unit weight of substrate consumed.

Solution:

1 The Monod model is rewritten here and will be transformed in linear form as follows:

$$\mu = \frac{\mu_{\max} S}{K_s + S} \text{ or } \mu = \mu_{\max} \frac{S}{K_s + S} .$$

2 Simply take the inverse of the equation as follows:

$$\frac{1}{\mu} = \frac{K_s}{\mu_{\max}} \frac{1}{S} + \frac{1}{\mu_{\max}} .$$

3 The first term is the ordinate (y), the second term is the product of slope (m) and abscissa (x) and the third term is the intercept. The goal is then to perform linear regression analysis using the ordinate and the abscissa. This is done using the calculation for the value of μ, which is the fourth column in Table 2 and rewritten in the following equation.

$$\mu = \frac{1}{X_t} \frac{X_t - X_o}{t_t - t_o} .$$

4 The first and last values are calculated as follows:

$$\mu = \frac{1}{0.6501} \times \frac{(0.6501 - 0.6051)}{(4.233 - 0)} = 0.0164$$

$$\mu = \frac{1}{8.3844} \times \frac{(8.3844 - 0.6051)}{(377.717 - 0)} = 0.0025 \cdot$$

5 Then, the reciprocal of the values of all μ and that of *S* is calculated. The values are shown in columns 5 and 6 of Table 2.

6 When these values are plotted and regression line is determined, the plot is shown in Fig. 11. The linearized equation is also shown and rewritten here.

$$y = 35.758x + 41.552\,.$$

7 The μ_{max} is the reciprocal of the *y*-intercept and K_s is the slope multiplied by μ_{max} and calculated as follows:

$$\mu_{\max} = \frac{1}{41.552} = 0.0241\text{h}^{-1}$$

$$K_s = m \times \mu_{\max} = 35.758 \times 0.0241 = 0.861\frac{g}{L}.$$

8 The biomass yield coefficient is calculated in the following equation using Equation 8. The values used are the first-row data set and the last row data set to provide an estimated biomass yield coefficient.

$$Y_{X/S} = \frac{8.384 - 0.6051}{-(0.102 - 2.402)}$$

$$Y_{X/S} = 3.382\frac{\text{g biomass produced}}{\text{g substrate consumed}}\cdot$$

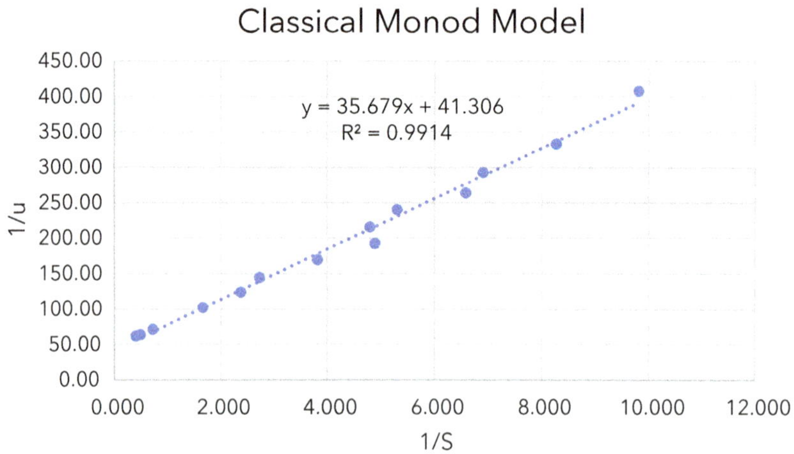

Figure 11 The plot of linearized classical Monod model using data from Table 2.

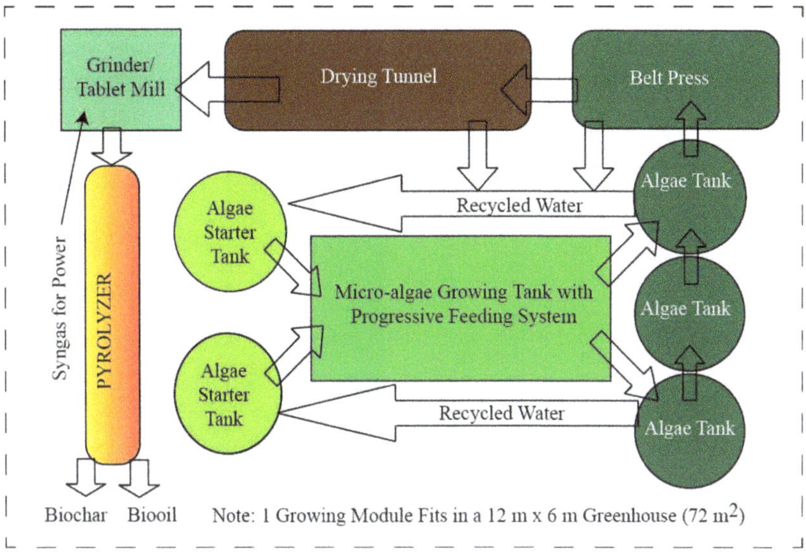

Figure 12 Pilot scale photobioreactor design for high-value product development.

These calculations show that biomass may be produced in a controlled manner to achieve its optimum potential yield given a proper environment for growth. With the laboratory experimental data presented here, process parameters may be established to design a larger scale system. An example is shown in Fig. 12 for one module of a microalgae growing facility. The components of such system might include

- an algae starter tank;
- a growing tank with a progressive feeding system; and
- algae concentrating tanks.

The concentrating or finishing tank contains concentrated microalgae biomass ready for harvesting. The harvested algae are then taken to a belt press to remove moisture and then to a drying tunnel. The dried microalgae may be further purified if necessary. It may then be milled/ground and pelleted for further processing into co-products. In the case of *Spirulina platensis*, the ground microalgae can be molded directly into tablets for sale as a high-protein supplement. All waste products may be pyrolyzed to generate syngas for heat and power as well as additional bio-oil and biochar co-products. Module systems like this may fit in a 12 m × 6 m greenhouse with about 50 units fitting into an acre (4049 m^2). Example 6 estimates yearly production on an acre basis.

Example 6. A medium scale (1 acre) microalgae production system has a biomass yield of 3.382 g biomass per g substrate with a doubling time of 2 days meaning harvest every two days. Assume that the growth rate is also per unit liter of sample. Translate this output into units of tonnes per year. Each reactor has a volume of 11 355 L (3000 gallons) and the unit will be operated year-round for around 350 days with harvest every 2 days (175 days in a year). There are around 50 reactors of this size used for this project for every acre (4049 m^2) of land.

Solution:

1 The monthly production rate (15 days) for every reactor is calculated as follows:

$$\text{Monthly yield}\left(\frac{\text{kg}}{\text{day}}\right)=\frac{11355\ \text{L}}{2\ \text{days}}\times\frac{3.382\ \text{g}}{\text{L}}\times\frac{\text{kg}}{1000\ \text{g}}\times 30\ \text{days}=576\frac{\text{kg}}{\text{month}}.$$

2 The monthly production for the 50 photobioreactors in an acre of land is calculated as follows:

$$\text{Monthly yield}\left(\frac{\text{tonnes}}{\text{month}-\text{acre}}\right)=576\frac{\text{kg}}{\text{month}}\times 50\frac{\text{units}}{\text{acre}}\times\frac{\text{tonne}}{1000\ \text{kg}}=88.8\frac{\text{tonnes dry algae}}{\text{month}-\text{acre}}.$$

3 Hence, the yearly microalgae yield in units of tonnes/year is calculated as follows:

$$\text{Yearly}\left(\frac{\text{tonnes}}{\text{acre}-\text{year}}\right)=88.8\frac{\text{tonnes}}{\text{acre}-\text{month}}\times 12\frac{\text{months}}{\text{year}}=345.6\frac{\text{tonnes}}{\text{acre}-\text{year}}.$$

Figure 13 Commercial-sized raceways for microalgae biomass production at Texas A&M University AgriLife Research Facility at Pecos, Texas.

This basic calculation of microalgae productivity shows a yield per unit of area that is more than 10 times that of traditional agricultural biomass. Progress in high-value biomass production from microalgae has been reviewed in Sun et al. (2018). Texas A&M University, e.g. has developed a commercial-sized raceway for microalgae production using species such as *Nannochloropsis oculata* and *Scenedesmus obliquus* (as shown in Fig. 13).

6 Rapid biomass characterization techniques

Operating a photobioreactor requires the development of rapid and non-destructive methods for quantifying microalgae and growth parameters. More than 90% of algal dry weight is composed of proteins, carbohydrates and lipids (Wagner et al., 2010). They are typically quantified by time-consuming conventional biochemical methods. Fourier-transform infrared (FTIR) spectroscopy coupled with chemometric techniques has the potential to determine single-cell biomass composition in a way that is reliable, fast and sensitive to species-specific changes in molecular composition. It is relatively simple to perform without the need for an extraction step.

FTIR characterizes biomass and measures the vibrational motion of functional groups of molecules at different wavelengths upon absorption of infrared rays. Interpretation of the infrared spectrum by identifying the presence of these functional groups indicates the identity of a molecule. The presence of proteins, carbohydrates and lipids in algal biomass can be determined by identifying band intensities related to their characteristic bonds such as the amide bond in proteins, C–C and C–O bonds in carbohydrates and C=O ester bond in lipids. Quantification can be achieved by relating the absorbance and the spectral structure of these functional groups to standard concentrations.

The simplest model that linearly relates absorbance to concentration is the Beer–Lambert's law. Linear regression can be done on the spectral peak absorbance or peak area against concentration. However, linear regression on spectral data may not be very reliable since such data often contain numerous variables. Partial least squares (PLS) regression is good in this situation especially where there are many variables but not necessarily many samples or observations. It also gives the maximal reduction in covariance of the data in order to give the minimum number of variables necessary to build a reliable prediction model.

Bataller and Capareda (2018) have demonstrated that FTIR can also be used to quantify the biochemical composition of algal samples by coupling it with PLS-regression techniques. The PLS models derived for proteins, carbohydrates and lipids in *Spirulina* samples have higher coefficient of determination (R^2) and residual prediction deviation (RDP) while having lower root mean square error of prediction (RMSEP) and relative error than the multipoint regression models. The predicted algal composition using the PLS models also obtained statistically similar results using conventional analytical methods, which are the Bradford Assay for proteins, phenol-sulfuric acid method for carbohydrates and the proposed method by Bigogno et al. (2002) for lipids. Bataller and Capareda (2020) have also successfully demonstrated the use of this novel method in monitoring the biochemical composition during the growth of *Spirulina* in a photobioreactor since compositional analysis results are made available within a few minutes. It must be noted, however, that this method is species-specific and that a different PLS model must be established for every algal species. The future direction for this novel method is to automate the procedure with the design and an FTIR module that can be installed directly to the photobioreactor coupled with a computer to analyze the spectra and report algal composition in real time.

The procedure for correlation of biomass properties such as concentration using the Beer–Lambert's law is as follows:

1 Algae samples are prepared with different cell densities or concentrations and are usually placed in a pre-weighed centrifuge tubes as shown in Fig. 14.
2 A sample is taken from each tube or absorbance reading.
3 The absorbance is read (in the experiment described earlier) at 680 nm in a UV-VIS spectrophotometer.
4 Centrifuge the tubes (e.g. 400 g for 10 min). Wash several times with deionized and distilled water to dissolve soluble salts. Then oven dry at temperature of around <90°C to prevent loss if volatile cell components.

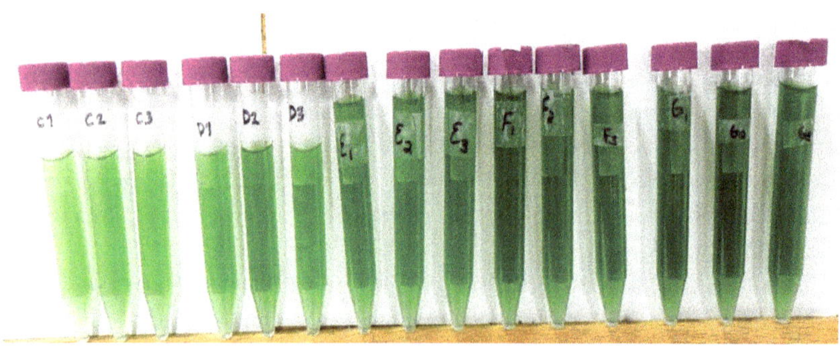

Figure 14 Photo of algae sampled with varying cell densities and concentrations.

5 The dried sample is then weighed and express the cell density/concentration as follows (Equation 9):

$$\text{Cell density} = \frac{\text{Weight of algae (mg)}}{\text{Volume of sample (mL)}}. \tag{9}$$

6 The cell density is plotted versus absorbance and the relationship should be about linear whereby regression analysis will be made. The regression equation is set up and hence, cell density is easily correlated with the sample absorbance for quick measurement of biomass concentration.

Example 7 shows the procedure for correlating microalgae cell density with absorbance.

Example 7. Establish the relationship between the cell density in units of g/L and absorbance to develop a predictive equation for reporting biomass concentration. Data are shown in the accompanying table.

Cell density (g/L)	Absorbance (nm)	Cell density (g/L)	Absorbance (nm)
0.152	0.141	0.582	0.700
0.162	0.161	0.462	0.785
0.183	0.182	0.475	0.790
0.165	0.162	0.595	0.810
0.212	0.325	0.755	0.900
0.292	0.338	0.705	0.945
0.305	0.500	0.629	0.950
0.323	0.550	0.810	1.150
0.425	0.710	0.920	1.180
0.505	0.690	0.790	1.200

Solution:

1 A regression analysis was performed and the X variable is found (meaning the slope) of 0.708 as shown in the figure. From the figure the predictive equation is as follows:

$$y = 0.708x$$

The $R^2 = 0.986$ showing high linearity. Hence, the biomass concentration in cell sample is easily determined by simply measuring its absorbance.

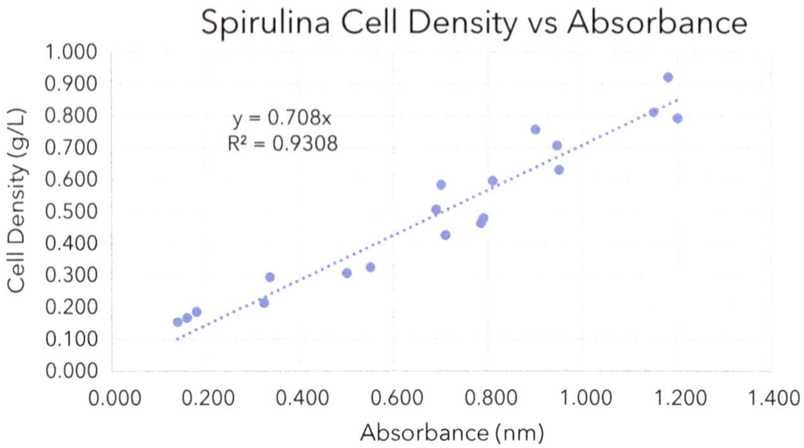

2 Hence, whenever a sample is taken from the photobioreactor its biomass concentration is easily determined.

The general procedure in developing a prediction model using FTIR spectroscopy and a PLS model is as follows:

1 Prepare different concentrations of standard grade bovine serum albumin (BSA) as protein standard, D(+)-glucose as carbohydrate standard and glycerol tripalmitate as lipid standard. The range of concentrations of these biomolecules shall be within the limit of quantification. In the study by Bataller and Capareda (2018), the limit of quantification was determined by multipoint regression.
2 Perform spectral scans in an FTIR. In the study by Bataller and Capareda (2018), Shimadzu IRAffinity-1 equipped with MIRacle™ Single Reflectance Attenuated Total Reflectance (ATR) sample cell and deuterated triglycine sulfate (DTGS) detector was used. The sample size pipetted to the ATR sample cell was 5 μL and dried with a nitrogen purge gas so that a solid sample is obtained. It was found that spectral scans of solid samples have higher resolution than liquid samples. The spectra of the standards were used to model the spectra of a *Spirulina* cell.
3 Build cell spectra models with different concentrations of protein, carbohydrates and lipids by iterative spectral subtraction and addition. These spectra models of the *Spirulina* cell will become the test spectra in creating the PLS regression model. Smoothing and baseline correction may be performed prior to building the cell spectra models. But make sure that the same parameters for spectral smoothing and baseline correction were used across all spectra.

4 Create the PLS regression models for protein, carbohydrates and lipids by subjecting the cell spectra models to PLS regression. This was done via the nonlinear iterative partial least squares (NIPALS) algorithm and leave-one-out cross-validation technique. The spectral data were centered, scaled and standardized prior to model development.
5 Confirm the robustness and accuracy of the PLS regression models by testing them against validation spectral data and by comparing the predicted composition with conventional analytical methods.

7 The range of microalgae species and their co-products

Numerous microalgae species have compounds of interest. Some examples are shown in Table 3. In this chapter, we have focused primarily on the production of *Spirulina platensis* for use as potential protein feed supplement. Various other types of fermentation products may be cultured for their components such as enzymes, metabolites or steroids. These microbes may be grown in a photobioreactor much like *Spirulina platensis*.

While *Spirulina platensis* is grown for its proteins, vitamins and minerals, other microalgae species are grown for materials such as antioxidants. *Haematococcus pluvialis* (HP), e.g. is specifically grown for the antioxidant astaxanthin. HP cells are radiated with bright light, subjected to high salinity and low nutrient availability (starvation), prompting cells to start producing astaxanthin. Under the right conditions in a photobioreactor, commercially cultivated HP can achieve more than 40 g of astaxanthin per kg of the HP biomass. Astaxanthin from HP is also reported to be more stable than

Table 3 Examples of natural microbes and their applications

Product	Application	Organism
1. Penicillin	Antibiotic	*Penicillium notatum*
2. Citric acid	Food flavoring	*Aspergillus niger*
3. Invertase	Candy	*Saccharomyces cerevisiae*
4. Lactase	Digestive aid	*Escherichia coli*
5. Pectinase	Fruit juice	*Aspergillus niger*
6. Riboflavin	Vitamin	*Ashbya gossypii*
7. Subtilisin	Laundry detergent	*Bacillus subtilis*
8. Tetracycline	Antibiotics	*Streptomyces aureofaciens*
9. Xanthan gum	Rheology modifier	*Xanthomonas campestris*
10. Acetone/butanol	Solvents	*Clostridium acetobutylicum*
11. α-amylase	Starch hydrolysis	*Bacillus subtilis*
12. Glycerol	Explosives production	*Saccharomyces cerevisiae*

synthetic compounds and have a longer shelf life without becoming oxidized. Astaxanthin from HP also has better bioavailability in humans who cannot produce carotenoids for themselves and must acquire them through an appropriate diet. There are two basic stages in the production of astaxanthin from HP. The first stage is called the green stage. This stage uses a single colony of HP microalgae. Cells are produced in a relaxed state. The goal is to create as many cells as possible using normal cell division. The second stage is called the stressed stage whereby algae cells are stressed to produce the maximum amount of astaxanthin. The stress stage occurs in a photobioreactor through high or controlled light levels and salt contents while nutrients are limited.

Other microalgae species are grown for their oil content which can be used to produce biofuel. For example, *N. oculata* accumulates high levels of polyunsaturated fatty acids with oil contents of at least 30% of its dry weight and as much as 60–70% of overall biomass weight. *Botryococcus braunii* (BB) produces high amounts of hydrocarbons (or triterpenes oil) representing about 65% of its dry weight. BBs can be genetically modified to generate oils as much as 86% of their dry weight. The oil has a chain length of C34 which, when split in half (C17), allows production of green diesel. Oils from BBs are particularly useful given their similarity to petroleum oils, being chemically almost identical to gasoline, kerosene (JP-8) and conventional green diesel. The primary issue with cultivating BB is that grows rather slowly, doubling its weight in about 4 days as opposed to 6–12 h for most high-oil-producing microalgae. Genetic modification will solve this issue.

Other commercially important microalgae include *Chlorella sp.*, *Dunaliella salina* and *Aphanizomenon*. Components in chlorella are reportedly able to decrease blood pressure, lower cholesterol levels and enhance the immune system. *Dunaliella salina* contains significant contents of beta-carotene, glycerol and other special antioxidants including lutein and lycopene. *Aphanizomenon* is a major source of numerous food and pharmaceutical ingredients. It is a prokaryotic cyanobacteria and commonly found in freshwater systems.

Figure 15 shows the wide range of products that may be derived from microalgae. Potential applications are also listed as follows:

Food industry:

1 Functional foods (e.g. essential amino acids, leucine, isoleucine and valine);
2 Food additives (e.g. beta carotene, DHA (docosahexaenoic acid), EPA (eicosapentaenoic acid) and alpha-linolenic acid (ALA);
3 Beverages (e.g. carrageenan, sweeteners and protein enriched smoothies); and

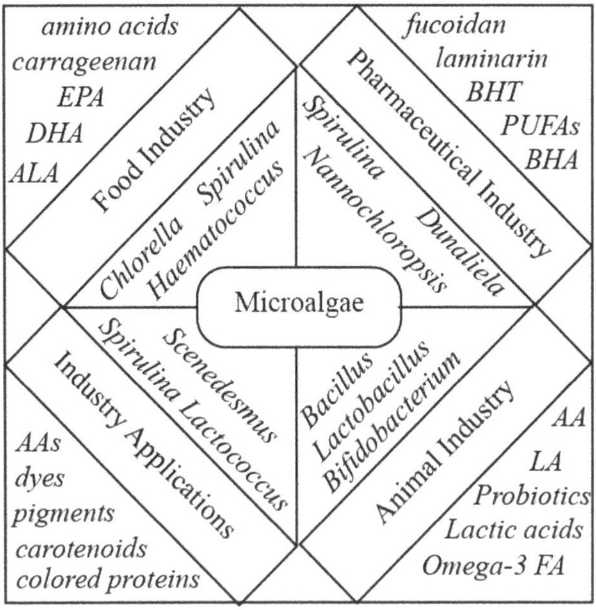

Figure 15 Range of products that may be produced from microalgae.

4 Food supplements (e.g. tocopherols, vitamins and minerals and high-protein powders).

Pharmaceutical industry:

1 Antibiotics (e.g. fucoidan and laminarin);
2 Antibacterial substances (e.g. fatty acids, glycolipids, phenolics, terpenes, beta-diketone and indole alkaloids);
3 Antioxidants [e.g. astaxanthin, butylated hydroxytoluene (BHT) and butylated hydroxyanisole (BHA)]; and
4 Cosmetics [e.g. polyunsaturated fatty acids (PUFAs), pigments (chlorophylls, carotenoids and phycobiliproteins), phytosterols and polyphenols].

Livestock production:

1 Feed supplements (e.g. proteins carbohydrates and fats);
2 Animal nutrition [e.g. EPA, DHA, ALA and arachidonic acid (AA), linolenic acid (LA) and omega fatty acids];
3 Fodder (e.g. omega-3 fatty acids, whole chlorella as dietary fiber and arthrospira);
4 Probiotics [e.g. polysaccharides (PS), *Bifidobacterium*, *Lactobacillus, Bacillus, Streptococcus* and *Lactococcus*]; and

5 Specialty rations (e.g. carrageenans, exopolysaccharides, fucoidans, alginates and lactic acid bacteria).

Other industrial applications:

1 Natural dyes [e.g. carotenoids and phycobiliproteins (the colored proteins)];
2 Functional pigments (e.g. immune enhancers, chlorophylls, canthaxanthin, violaxanthin and xanthophylls);
3 Feed enhancer for fishponds (e.g. essential amino acids, healthy triglycerides, vitamins and pigments); and
4 Enzymes for crude oil spills and degradations (e.g. lipases, enzymes from *Scenedesmus abundans*, sorption ability of *Spirulina* sp.).

8 Conclusion

Future biorefineries will not be used just to produce biofuels. In addition, these biorefineries will not just use agricultural wastes but will also exploit microalgae to generate higher value products. Depending on fluctuations in prices for fossil fuels (particularly the oil and gas markets), there is likely to be further diversification into high-value products for the food and pharmaceutical industries as well as for livestock production. It is possible to grow biomass in a controlled manner with the potential to achieve production rates much greater than land-based agriculture. There are numerous microalgae species that can produce specialty products with very high production rates through the use of technologies such as open ponds, raceways or photobioreactors. Whilst photobioreactors provide the greatest potential in terms of productivity and range of products, there are significant cost and resource issues to be addressed. More research is needed on the different parameters needed to optimize production of particular specialty compounds. Photobioreactors may need better mixing, regular sterilization, cyclic heating and cooling with proper control of pH and oxygen levels, as well as measures to prevent foaming and biofilm formation. Some algae species may require stress cycles such as high salt contents and restriction of nutrients. Energy requirements could be generated from waste biomass to provide the required power (primarily mixing and circulation) in addition to heating and cooling needs.

The last requirement in the operation of a photobioreactor is the final processing and isolation of the primary product. Moisture removal is particularly challenging due to the different viscosities of microalgae samples. If not done properly, it may lead to degradation of product in a short period of time. Possible solutions include solid-state cultivation of algae such as in a rotating algal biofilm reactor (RABR) or in a twin-layer photobioreactor (TL-PBR) to obtain an algal biomass with low moisture content upon harvest. This eliminates the need to remove large quantities of water during harvest. In the end, economics will dictate the feasibility of such projects.

9 Where to look for further information

Thermochemical biomass conversion processes are discussed in detail in the following:

- Capareda, S. C. (2014), 'Introduction to Biomass Energy Conversions,' CRC Press, Taylor and Francis Group, Boca Raton., FL.

Details of development of a novel method in algal biomass characterization can be found in:

- Bataller, B. G. and Capareda, S. C. (2018), 'A rapid and non-destructive method for quantifying biomolecules in *Spirulina platensis* via the Fourier transform infrared - Attenuated total reflectance spectroscopy,' Algal Research, 32, pp. 341-352, https://doi.org/10.1016/j.algal.2018.04.023.

Key research and programs related to biomass can be found at the following organization:

- US DOE Bioenergy Technologies Office (https://www.energy.gov/eere/bioenergy/bioenergy-technologies-office).
- USDA Regional Biomass Research Centers (https://www.ars.usda.gov/research/usda-regional-biomass-research-centers/).

10 References

Bataller, B. G. and Capareda, S. C. (2018). A rapid and non-destructive method for quantifying biomolecules in *Spirulina platensis* via the Fourier transform infrared: attenuated total reflectance spectroscopy, *Algal Research* 32, 341–352, https://doi.org/10.1016/j.algal.2018.04.023.

Bataller, B. G. and Capareda, S. C. (2020). Use of FTIR spectroscopy and PLS regression in monitoring biomolecules in *Spirulina platensis* during its growth in an internally-illuminated photobioreactor, *Chemical Engineering Transactions* 80, 103–108, https://doi.org/10.3303/CET2080018.

Banerjee, S. and Ramaswamy, S. (2019). Comparison of productivity and economic analysis of microalgae cultivation in open raceways and flat panel photobioreactor, *Bioresource Technology Reports* 8, https://doi.org/10.1016/j.biteb.2019.100328.

Bigogno, C., Khozin-Goldberg, I., Boussiba, S., Vonshak, A. and Cohen, Z. (2002). Lipid and fatty acid composition of the green oleaginous alga *Parietochloris incisa*, the riches plant source of arachidonic acid, *Phytochemistry* 60(5), 497–503, http://doi.org/10.1016/S0031-9422(02)00100-0.

Capareda, S. C. (2014). *Introduction to Biomass Energy Conversions*. CRC Press, Taylor and Francis Group, Boca Raton.

Doran, P. M. (2013). *Bioprocess Engineering Principles* (2nd edn.). Elsevier Ltd, Oxford, UK.

Field, C. B., Behrenfeld, M. J., Randerson, J. T. and Falkowski, P. (1998). Primary production of the biosphere: integrating terrestrial and oceanic components, *Science* 281(5374), 237–240.

Formighieri, C., Franck, F. and Bassi, R. (2012). Regulation of the pigment optical density of an algal cell: filling the gap between photosynthetic productivity in the laboratory and in mass culture, *Journal of Biotechnology* 162(1), 115–123, https://doi.org/10.1016/j.jbiotec.2012.02.021.

Griffiths, J. B. (1972). Scaling up in animal cell culture, *Journal of Cell Science* 1986, 33–69.

Herzog, H. and Golomb, D. (2004). 'Carbon capture and storage from fossil fuel use,' Encyclopedia of Energy 1, 1–11. Available at: http://sequestration.mit.edu/pdf/enclyclopedia_of_energy_PROOF.pdf.

Holladay, J. E., Bozell, J. J., White, J. F. and Johnson, D. (2007). Top value-added chemicals from biomass volume II: results of screening for potential candidates from biorefinery lignin report Prepared for the US Department of Energy under Contract DE-AC05-76RLL01830. Pacific Northwest National Laboratory, PNNL-16983.

Khan, M. I., Shin, J. H. and Kim, J. D. (2018). The promising future of microalgae: current status, challenges, and optimization of a sustainable and renewable industry for biofuels, feed, and other products, *Microbial Cell Factories* 17(1), 36, https://doi.org/10.1186/s12934-018-0879-x.

Kumar, M., Sundaram, S. and Gnansounou, E. (2018). Carbon capture, storage and production of biofuel and biomaterials by bacteria: a review, *Bioresource Technology* 247, 1059–1068, https://doi.org/10.1016/j.biortech.2017.09.050.

Kumar, G., Dharmaraja, J., Arvindnarayan, S., Shoban, S., Bakonyi, P., Saratale, G. D., Nemestóthy, N., Bélafi-Bakó, K., Yoon, J. and Kim, S. (2019). A comprehensive review on thermochemical, biological, biochemical and hybrid conversion methods of bio-derived lignocellulosic molecules into renewable fuels, *Fuel* 251, 352–367, https://doi.org/10.1016/j.fuel.2019.04.049.

Lee, J. J. (2014). Driven by climate change, algae blooms behind Ohio water scare are new normal, *National Geographic* Issue Published. Available at: https://www.nationalgeographic.com/science/article/140804-harmful-algal-bloom-lake-erie-climate-change-science. Accessed 6 August 2012.

Lei, X. G. (Eds.) (2021). *Seaweed and microalgae as alternative sources of protein*. Burleigh Dodds Science Publishing, Cambridge, UK.

Lewandowski, W. M., Ryms, M. and Kosakowski, W. (2020). Thermal biomass conversion: a review, *Processes* 8(5), 516.

Mauyra, D. P., Singla, A. and Negi, S. (2015). An overview of key pretreatment processes for biological conversion of lignocellulosic biomass to ethanol, *3 Biotech* 5(5), 597–609.

Meki, M. N., Ogoshi, R. M., Kiniry, J. R., Crow, S. E., Youkhana, A. H., Nakahata, M. H. and Littlejohn, K. (2017). Performance evaluation of biomass sorghum in Hawaii and Texas, *Industrial Crops and Products* 103, 257–266, http://doi.org/10.1016/j.indcrop.2017.04.014.

Melis, A. (2009). Solar energy conversion efficiencies in photosynthesis: minimizing the chlorophyll antennae to maximize efficiency, *Plant Science* 177(4), 272–280, https://doi.org/10.1016/j.plantsci.2009.06.005.

Murzin, D. and Simiken, O. (Eds.) (2016). *Biomass Sugars for Non-fuel Applications*. Royal Society of Chemistry, Cambridge, UK.

Narala, R. R., Garg, S., Sharma, K. K., Thomas-Hall, S. R., Deme, M., Li, Y. and Schenk, P. M. (2016). Comparison of microalgae cultivation in photobioreactor, open raceway pond, and a two-stage hybrid system, *Frontiers in Energy Research* 4(29), 1–10. https://doi.org/10.3389/fenrg.2016.00029.

National Agricultural Statistics (NAS) (2022). *Crop Production 2021 Summary*. United States Department of Agriculture (United States Department of Agriculture), Washington, DC. Available at: https://downloads.usda.library.cornell.edu/usda-esmis/files/k3569432s/sn00c1252/g158cj98r/cropan22.pdf. Accessed 17 January 2022.

Rooney, W. L., Blumenthal, J., Bean, B. and Mullet, J. E. (2007). Designing sorghum as a dedicated bioenergy feedstock, *Biofuels, Bioproducts and Biorefining* 1(2), 147–157.

Saffron, C. (Eds.) (2020). *Achieving Carbon-Negative Bioenergy Systems from Plant Materials*. Burleigh Dodds Science Publishing, Cambridge, UK.

Sayre, R. (2010). Microalgae: the potential for carbon capture, *BioScience* 60(9), 722–727, https://doi.org/10.1525/bio.2010.60.9.9.

Shen, G. (2021). Solar energy conversion, oxygen evolution and carbon assimilation in cyanobacteria and eukaryotic microalgae. In: Lei, X. G. (Ed.). *Seaweed and Microalgae as Alternative Sources of Protein*. Burleigh Dodds Science Publishing, Cambridge, UK, 17–50.

Singhvi, M. S. and Gokhale, D. V. (2019). Lignocellulosic biomass: hurdles and challenges in its valorisation, *Applied Microbiology and Biotechnology* 103(23–24), 9305–9320.

Sun, J., Ding, S. Y. and Peterson, J. D. (Eds.) (2013). *Biological Conversion of Biomass for Fuels and Chemicals: Explorations from Natural Utilization Systems*. Royal Society of Chemistry, Cambridge, UK.

Sun, H., Zhao, W., Mao, X., Li, Y., Wu, T. and Chen, F. (2018). High-value biomass from microalgae production platforms: strategies and progress based on carbon metabolism and energy conversion, *Biotechnology for Biofuels* 11(227), 227. https://doi.org/10.1186/s13068-018-1225-6.

U.S. Department of Energy (2016). *2016 Billion-Ton Report: Advancing Domestic Resources for a Thriving Bioeconomy, Volume 1: Economic Availability of Feedstocks,' M. H. Langholtz, B. J. Stokes, and L. M. Eaton (Leads), ORNL/TM-2016/160*. Oak Ridge National Laboratory, Oak Ridge. 448p. doi: 10.2172/1271651. Available at: http://energy.gov/eere/bioenergy/2016-billion-ton-report

USEIA (U.S. Energy Information Administration) (2016). Short-term energy outlook. Available at: https://www.eia.gov/forecasts/steo/report/renew_co2.cfm.

Wagner, H., Liu, Z., Langner, U., Stehfest, K. and Wilhelm, C. (2010). The use of FTIR spectroscopy to assess quantitative changes in the biochemical composition of microalgae, *Journal of Biophotonics* 3(8–9), 557–566, https://doi.org/10.1002/jbio.201000019.

Werpy, T. and Petersen, G. (Eds.) (2007). *Top Value Added Chemicals from Biomass Volume I - Results of Screening for Potential Candidates from Sugars and Synthesis*

Gas Produced by the Staff at the Pacific Northwest National Laboratory (PNNL). National renewable Energy Laboratory (NREL), Office of Biomass Program (EERE) for the Office of the Biomass Program.

Yu, I. K. M., Chen, H., Abeln, F., Auta, H., Fan, J., Budarin, V. L., Clark, J. H., Parsons, S., Chuck, C. J., Zhang, S., Luo, G. and Tsang, D. C. W. (2020). Chemicals from lignocellulosic biomass: A critical comparison between biochemical, microwave and thermochemical conversion methods, *Critical Reviews in Environmental Science and Technology* 41(14), 1479–1532, https://doi.org/10.1080/10643389.2020.1753632.

Chapter 11

Developing bioplastics from agro-industrial wastes for applications in food packaging

Angela Marchetti, Gaia Salvatori and Lionel Tayou Nguemna, Sapienza University of Rome, Italy; Mattia Grumi, Ahmet Ozan Basar, María Pardo-Figuerez, Jose Maria Lagaron and Cristina Prieto, Institute of Agrochemistry and Food Technology (IATA), Spain; Chiara Marcoaldi, Sapienza University of Rome, Italy and Institute of Agrochemistry and Food Technology (IATA), Spain; Marianna Villano and Mauro Majone, Research Center for Applied Sciences to the Safeguard of Environment and Cultural Heritage (CIABC) and Sapienza University of Rome, Italy; and David Bolzonella, University of Verona, Italy*

1 Introduction: the problem of plastic waste

Waste is one of the biggest challenges faced by both developing and developed countries. It is the result of unsustainable use of natural resources, leading to resource depletion and environmental degradation (Kundariya et al., 2021). According to the latest data, the global weight of municipal solid waste (MSW) in 2016 was 2.01 billion tons and is expected to increase to 3.40 billion tons in 2050. About 30–35% of MSW is made up of packaging waste and

* Correspondence author: cprieto@iata.csic.es.

http://dx.doi.org/10.19103/AS.2023.0120.20

with 60% of packaging waste used for food (Nemat et al., 2020). The highest contributions (by weight) to packaging waste were paper and paperboard (50.8%) and plastic (18%). In terms of landfill, the highest rates were observed for plastic (70%), wood (59.2%), aluminum (53.4%), and glass (53.1%). Materials with the highest rate of recycling were paper/paperboard (73.3%) and steel (73.1%), whilst glass totaled just 33.9%. The European Union (EU) generates 77.5 million tons (Mt) of packaging waste, comprising paper/cardboard (41%), plastic (19%), glass (18%), wood (17%), and metal (5%) (De Pilli et al., 2021; Eurostat, 2020).

Plastics represent a huge environmental problem due to their fossil-based origin, their key role in packaging, the resulting high percentage of overall waste that this material represents, and the persistence of waste plastics in the environment. In general, plastics are synthetic or semi-synthetic organic materials comprised of large molecules including many repeated sub-units known as monomers. During the polymerization stage, the monomers are linked together to form larger molecules called polymers. These polymers are then mixed with various chemical additives (at the compounding stage) that give the plastic its desired properties. The final products are lightweight, strong, durable, and low cost (van Eygen et al., 2017). These characteristics make their use ubiquitous around the globe.

Plastic are particularly useful in sectors such as packaging, construction, automotive, electrical and electronic, household, leisure, sports, furniture, medical technology, and textiles, among others. As a result, production has seen continuous growth over the past 50 years and is expected to double in the next 20 years (Gu et al., 2017). Approximately 367 Mt of plastics are produced annually worldwide (PlasticsEurope, 2020), the majority of which are not biodegradable. This creates serious concerns related to their disposal. Although plastic waste management is improving in the EU, around 70% of plastic waste is still sent to landfill or accumulates in the natural environment (De Pilli et al., 2021) with serious long-term adverse effects (Ellen MacArthur Foundation, 2016; Guillard et al., 2018; Jambeck et al., 2015; Magalhães et al., 2020). Incorrect disposal of plastic wastes can lead to the accumulation in the environment, causing harm to marine life (Besseling et al., 2017a; Jahnke et al., 2017a; Lasut et al., 2018), freshwater ecosystems (Besseling et al., 2017; Blettler et al., 2018; Rodrigues et al., 2018), soil (Bläsing and Amelung, 2018; Chae and An, 2018; Scheurer and Bigalke, 2018; Zubris and Richards, 2005), and the food chain (FAO United Nations, 2017; Romeo et al., 2015).

The greatest environmental impact comes from single-use plastics, the production and incorrect disposal of which have greatly increased environmental pollution, due to the emission of greenhouse gases (GHGs) during manufacture, waste generation, and the risk of littering. This critical situation has worsened due to the COVID-19 crisis. During the pandemic, there was a dramatic surge in global demand for plastic-based material including personal protective equipment such as masks (7.6%), take-away food packaging (4.7%), and plastic medical equipment (87.4%), which added further to the problem of waste (Uddin et al., 2022).

Figure 1 shows how the improper disposal of plastic negatively affects the environment and human health. Plastics can have detrimental effects on the environment, affecting air quality and the quality of the soil and oceans, and on animals due to the risk of entanglement, injury, or ingestion. A further problem is the fragmentation of spilled plastics into the environment, due to sunlight, high temperatures, wind, and waves, resulting in the diffusion of macro- and microplastics. The generation of macro and microplastics is a particular problem for the health of marine and terrestrial environments. This specific type of pollutant rapidly spreads in water, widely affecting the marine environment, including several hundreds of marine species (Wright et al., 2013). Moreover, they can compromise human health by threatening food security (through contamination of soil which can then transfer to crop plants and livestock) and, once ingested, acting as endocrine disrupters and causing reproductive and behavioral disorders, diabetes and obesity, asthma and cancers, among others (van Vliet and Jensen, 2016).

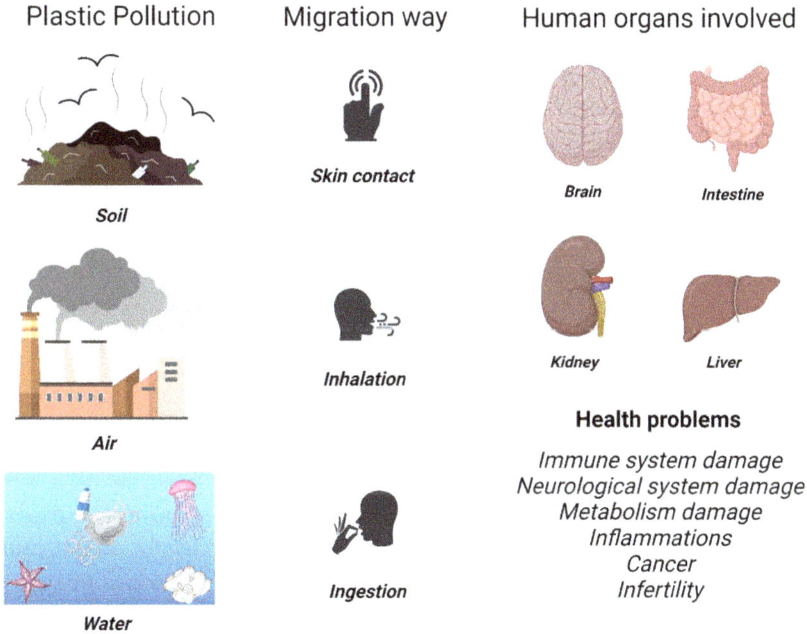

Figure 1 Impact of plastic waste on human health through different exposure routes: skin contact, inhalation, and ingestion. Source: Adapted from: Created with BioRender.com.

The problem of waste is bound up with the current linear economy principle of 'make, use, and dispose' in which raw materials are extracted, processed into value-added products, and, finally, sold to the end consumer, who must dispose of them once they reach the end of their useful life (Yadav et al., 2020). This supply chain model must be substituted by an alternative approach known as the circular economy, based on the principle of 'reuse, reduce, and recycle' (Borrelle et al., 2020). From this point of view, different solutions can be adopted in order to reduce waste, such as improving design, adopting higher quality plastics, encouraging and enabling reuse, repair, remanufacturing, and recycling. It is increasingly recognized that there is no single solution to this problem and that combined actions are required (Morseletto, 2020).

Recycling is a viable solution to reduce the need for new raw materials, decreasing the amount of waste. However, only around 30% of plastic waste is currently recycled (European Bioplastics Organization, 2019). Although plastic materials can be recycled in a variety of ways, the ease of recycling varies according to the polymer type and product design. For this reason, the overall recycling rate of plastic waste is still low, compared with other materials such as paper, glass, or metals (Statista, 2021). In addition, the low demand for recycled plastic, low prices, and uncertainties around market outlets have hampered the development of the EU plastic recycling sector (Milios et al., 2018).

A more lasting solution is to replace conventional fossil-based plastics with materials more in line with the concept of the circular economy such as biodegradable polymers from renewable resources (Rai et al., 2021). These have a potential double benefit. First, they can potentially be sourced either from renewable sources (e.g. from growing crops) or, even better, from wastes and by-products from other sectors (e.g. crop residues or food processing wastes, agro-industrial waste, and wastewaters), creating a closed-loop production system. Secondly, biodegradable materials can be more safely disposed of where they are not suitable for re-use in some other form. The following sections summarize the most important initiatives.

2 Bioplastics as a promising alternative

The lack of biodegradability, together with the scarcity of oil sources, has prompted global research into biodegradable and renewable alternatives to petroleum-based plastics for food packaging and other applications (Castro Mayorga, 2017; Torres-Giner et al., 2021). Viable alternatives include bioplastics obtained from biopolymers. These constitute a whole family of polymers with different properties and applications that include 'bio-based' polymers, 'biodegradable' polymers, and polymers combining both properties (Torres-Giner et al., 2021). Figure 2 summarizes polymer classifications according to their origin and biodegradability characteristics.

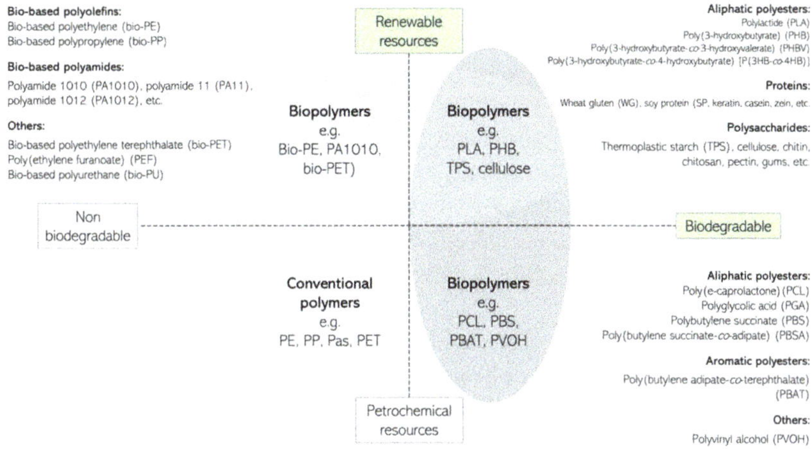

Figure 2 Classification of polymers widely used in packaging, according to their biodegradability and bio-based origin. Source: Adapted from (European Bioplastics Organization, 2018a; Torres-Giner et al., 2021).

Depending on their origin, polymers can be obtained from petrochemical resources or from renewable resources. The term 'bio-based' refers to all polymers produced from renewable resources, which include both synthetic polymers produced by means of monomers obtained from biological sources and naturally occurring polymers (Ashter, 2016). The latter are bio-macromolecules, i.e. molecules having large molecular weights (M_W) produced in nature by living organisms or by plants (Torres-Giner et al., 2021).

With respect to their degradation profile, polymers can be both biodegradable and non-biodegradable. The term 'biodegradable' describes those polymer materials whose physical and chemical properties undergo deterioration and complete degradation when exposed to the enzymatic action of microorganisms, breaking down into carbon dioxide (following the so-called aerobic process), methane (following the so-called anaerobic process), water (both aerobic and anaerobic processes), inorganic compounds, and biomass (Laycock et al., 2017).

Particular attention should be paid to the difference between the terms 'biodegradable' and 'compostable', as suggested by the European Bioplastics Organization (European Bioplastics Organization, 2018b). The term 'biodegradable' does not specify either the time frame or the environmental conditions necessary for the natural degradation of the material, the most important process being microbiological biodegradation. 'Compostable', instead, describes the material that allows it to biodegrade under specific conditions (e.g. a certain temperature and timeframe) described in standards such as the European Standard on industrial composting EN 13432, which is specific to packaging, or EN 14995, which is related to plastic materials in general

(Döhler et al., 2022). Materials and products complying with these standards can be certified and labeled as compostable, differentiating them from other materials (e.g. those marked to be oxo-biodegradable which, moreover, do not fulfill the requirements of EN 13432 on industrial compostability) that follow a more accurate end-life pathway (European Bioplastics Organization, 2018a). Furthermore, according to the EU Standard EN 13432, a product can be considered compostable if, under the conditions of an industrial composting plant, at least 90% biodegradation into CO_2 occurs within 6 months and no more than 1% of additives (which must be harmless) have been utilized during the entire process.

It is also important to note that bio-based polymers can be biodegradable but not all biodegradable polymers are bio-based (Castro Mayorga, 2017; Martínez Sanz, 2013; RameshKumar et al., 2020). Indeed, as reported by the European Bioplastics Organization (European Bioplastics Organization, 2020), slightly less than half (41%) of all biopolymers sold on the European market in 2019 were bio-based but not biodegradable. As an example, polyvinyl alcohol (PVOH) and polycaprolactone (PCL), two aliphatic polyesters, are biodegradable but, since they come from petrochemical resources, they are not bio-based. However, some synthetic biodegradable polymers could, in the near future, be partially or fully developed from bio-based monomers (Thakur et al., 2018; Tsang et al., 2019). The conservation of fossil resources by using renewable biomass is an extremely important advantage offered by bio-based polymers, together with the unique potential of carbon neutrality (Williams and Hillmyer, 2008).

On the other hand, biodegradability is an add-on property of certain types of polymers that offers additional means of recovery at the end-life of a product (Sirohi et al., 2020; Williams and Hillmyer, 2008). For instance, biodegradable polymers could require high temperatures in order to be organically recycled or biodegraded - conditions that are not achieved naturally but during industrial composting. Unfortunately, there is a lack of industrial composting facilities, and existing composters often reject biodegradable polymers because they require extended decomposition times in comparison with typical composting processes, resulting in lower quality composting products and an unreliable market for selling (Mastrolia et al., 2022; Rosenboom et al., 2022).

3 Types of bio-based polymers

Bio-based polymers, depending on their origin, can be classified into two macro-groups - synthetic and natural. Synthetic bio-based polymers are produced by means of polymerization from renewable monomers (Martínez Sanz, 2013) and include biodegradable polymers such as polylactic acid (PLA) and bio-based but not biodegradable polymers, such as bio-based polyethylene (bio-PE) and bio-based polypropylene (bio-PP).

PLA can be classified as a synthetic bio-based polymer because it is obtained from the bacterial fermentation of carbohydrates that can come from renewable starch and cellulosic resources (Coppola et al., 2021; Yadav et al., 2021). Fermentation turns the sugar into lactic acid, which is the building block for PLA production, even if most industrial applications make use of its dimer lactide (Torres-Giner et al., 2021). PLA is one of the most studied and commercialized biopolymers and is a suitable candidate to replace conventional polymers since it is bio-based and compostable and features similar properties to polymers such as polyethylene terephthalate (PET), polystyrene (PS), and polycarbonate (PC) (Cui et al., 2019; Höfer and Selig, 2012). PLA is considered safe for multiple applications since it is classified as 'generally recognized as safe' (GRAS) by the United States Food and Drug Administration (FDA).

Natural bio-based polymers are directly extracted from biomass, such as polysaccharides and gums, proteins, and lipids, or produced by microorganisms such as polyhydroxyalkanoates (PHAs) (Jeremic et al., 2020). Some of the most significant natural bio-based and biodegradable polymers are described below:

- *Starch* is a natural polysaccharide that can be obtained from a great variety of crops, including cassava and corn. Starch is one of the most promising biopolymers due to its availability, biodegradability, and low price (Chan et al., 2018). Native dry starch has a limited range of applications due to its high brittleness, high viscosity, retrogradation, insolubility in cold water, and poor melt processability because of its glass transition temperature (T_g) that is generally above decomposition point, making softening and flowing impossible (Khan et al., 2017; Li et al., 2019). However, starch can be modified by processing with a plasticizer, by grafting with vinyl monomers, and by blending with other polymers (Torres-Giner et al., 2021). When plasticizers are used, both the flexibility and processability of starch are improved. Modified starch that exhibits thermoplastic behavior is known as thermoplastic starch (TPS) (Tang et al., 2012). TPS can also be blended with other polymers and fillers to improve its mechanical properties and to achieve water resistance (Kaseem et al., 2012; Wang et al., 2009). Furthermore, the implementation of TPS in biopolymer blends appears particularly relevant in order to obtain materials with high elongation at break properties (Liu et al., 2009).
- *Proteins* stand out due to their film-forming capacity and cohesiveness, low cost, and biodegradability (Nazmi et al., 2020; Torres-Giner et al., 2021). Proteins show good barrier properties against oxygen and aroma, among other gases. Nevertheless, proteins present high water vapor permeability

due to their hydrophilic nature (Andrade et al., 2015; Azevedo et al., 2015). Protein-based films have been developed from different protein sources such as gelatin (Andreuccetti et al., 2011; Martucci and Ruseckaite, 2010; Said et al., 2021; Tongnuanchan et al., 2014), corn zein (Shalom et al., 2021; Spasojević et al., 2019), wheat gluten (Chavoshizadeh et al., 2020; Guna et al., 2023), soy protein (Koshy et al., 2015; Rani and Kumar, 2019), casein (Namratha et al., 2020; Pellá et al., 2020), and whey protein, among others.

- *PHAs* represent a promising example of natural bio-based polymers produced by microorganisms. PHAs are one of the most important alternatives to fossil-derived polymers in the frame of the circular economy (Torres-Giner et al., 2018c), and a potential replacement for polyolefins thanks to their biocompatibility, biodegradability, renewable origin, and physical properties (Mutlu et al., 2018). PHAs feature characteristics such as biodegradability, thermoplasticity, and similar mechanical strength and water resistance to other conventional polymers such as polypropylene and polystyrene (Reddy et al., 2003). PHAs represent a family of polyesters synthesized, as carbon storage material, by more than 300 different types of microorganisms (Lee et al., 1999; Tang et al., 2012). Among PHAs, the most widely studied is the poly(3-hydroxybutyrate) (PHB). This homopolyester is biodegradable not only in composting and controlled conditions but also in other environments such as marine water (Torres-Giner et al., 2021). However, its use has been limited due to its poor impact-strength resistance and a narrow processing temperature window (Yeo et al., 2018). To improve these drawbacks, its copolymers such as poly(3-hydroxybutyrate-*co*-3-hydroxyvalerate) [P(3HB-*co*-3HV)] and poly(3-hydroxybutyrate-*co*-4-hydroxybutyrate) [P(3HB-*co*-4HB)] have been widely studied (Melendez-Rodriguez et al., 2018). The [P(3HB-*co*-3HV)] appears to be a good candidate since it has a much lower crystallinity and melting temperature, decreased stiffness and brittleness, and higher ductility (McChalicher and Srienc, 2007). More recently, a terpolymer enriched in 3-hydroxyhexanoate (HHx) [P(3HB-*co*-3HV-*co*-3HH)] is receiving particular interest, since it presents improved elastomeric characteristics with a broader range of applications (Silva et al., 2022).

Currently, these biopolymers have a wide range of applications; however, bioplastics represent only about 1% of the approximately 367 Mt of annual global plastic production (PlasticsEurope, 2020), used mainly in packaging materials (Payne et al., 2019; Torres-Giner et al., 2021; Williams and Hillmyer, 2008). Thanks to increased environmental concern during the last decade, the implementation of biopolymers has increased considerably due to their sustainable feedstock, biodegradability, their similar processing characteristics as existing thermoplastics (Tharanathan, 2003), and the fact that they have similar features to conventional plastics (Coppola et al., 2021).

4 Bioplastics from renewable feedstocks: the example of polyhydroxyalkanoates from agricultural waste

The replacement of fossil-based materials with biomass as a raw material for biopolymers production is an interesting option and the driving force for the development of biorefinery platforms. Biorefineries can convert almost all biomass feedstocks into different classes of biochemicals (Cherubini, 2010) and biopolymers.

The term 'feedstock' refers to raw materials used in biorefinery. Renewable carbon-based raw materials for biorefinery are provided from five main different sectors:

- Agriculture;
- Forestry;
- Industries (secondary products, residues, and leftovers);
- Aquaculture; and
- Urban wastes

In this sense, bio-based plastics can be fully or partly derived from biological feedstock, typically oils, starches, sugars, cellulose, and even CO_2 (Menon and Lyng, 2021). In addition, the benefits of using renewable feedstock include reduced dependency on imports, reduced dependency on fossil resources, reduced GHG emissions, and, if locally sourced, increased rural development.

The main challenges faced by biorefinery development are achieving efficient and cost-effective production and efficient use of the land. The production of feedstock for bio-based plastics requires land, which is closely linked to environmental impacts on soil, biodiversity, GHG emissions, and water (Spierling et al., 2018), and competes with the food and energy industries. Some of these drawbacks can be partially overcome by using organic waste materials as feedstock for bioplastics production or as a filler for bioplastics. An example is represented by PHAs, which can be produced from different types of waste streams and reinforced via the use of agro-waste by-products as filler.

PHAs are promising candidates to replace petroleum-based plastics as they have similar physicochemical and mechanical characteristics (Foschi and Bonoli, 2019; Kourmentza and Kornaros, 2016) and can be processed in the same injection molding, film deposition and blowing, sheet extrusion, and thermoforming machines as conventional plastic products. In addition, PHAs are non-toxic, hydrophobic, insoluble in water, inert, and stable even when exposed to air (dos Santos et al., 2018). They also have a wide temperature range, have been shown to be resistant to water and moisture, and present low oxygen and water vapor permeability (Kim et al., 2020). All these factors differentiate them from most other bioplastics currently available on the market, which are generally sensitive to moisture and tolerate lower temperature ranges.

PHA polymerization is carried out naturally by bacteria, which are capable of performing the polymerization process intracellularly. This peculiarity makes PHAs different from the classic polymers of vegetable origin obtained from starch and its mixtures. PHAs are synthesized by more than 300 bacterial species, both Gram-positive and -negative (such as *Pseudomonas, Ralstonia, Azobacter*, and *Rhizobium*), photosynthetic aerobic bacteria (*Cyanobacteria*), anaerobic bacteria, and *Archaea*. Among those bacteria responsible for PHA accumulation are polyphosphate-accumulating organisms and glycogen-accumulating organisms. In both groups, this polymer is accumulated as intracellular carbon reserves via different biosynthetic pathways (Sudesh et al., 2000).

In particular, PHA is stored in the cytoplasm as an internal source of carbon and energy (up to 80–90% of microbial cell mass), in the form of intracellular granules (approximately 8–13 granules per cell with a diameter between 0.2 μm and 0.7 μm) (Braunegg et al., 1998). However, their monomeric composition depends both on the type of organic substrate used by the bacteria and on the bacterial family itself capable of carrying out this internal polymerization. More than 150 monomer types have been identified for hydroxyalkanoates (HAs) and even greater variability in the length and composition of the side chains. This structural variety gives them multiple properties, making them attractive for applications in various fields such as packaging, agriculture, textiles, chemicals, food, pharmaceuticals, and biomedicine (Chen, 2009). In fact, the incorporation of short-chain monomers other than 3-hydroxybutyrate (HB), such as 3-hydroxyvalerate (HV), or of medium-chain monomers, such as hydroxyhexanoate (HH), affects polymer crystallinity as well as thermal and mechanical properties (Reis et al., 2011; Silva et al., 2022).

Currently, the industrial production of PHAs is based on processes using pure microbial cultures. A necessary condition for this type of process is the specificity of the culture conditions required for the storage of the polymer, in relation to the type of microorganism used. Most of these bacteria require the restriction of an essential nutrient and sterile conditions. Initially, the culture is fed with a nutrient-rich medium and synthetic carbon substrates formulated specifically for the microbial strain. Only when a high cell density is reached, growth-limiting conditions (due to the depletion of an essential nutrient such as nitrogen) induce PHA storage and the subsequent accumulation phase is established. It is possible to achieve up to 90% of PHA on a dry weight of biomass, and the properties of the final polymer depend on the mixture of carbonaceous raw materials supplied to the microbial culture, the metabolic pathways, and the substrate specificities of the enzymes involved (Mobley, 1994). However, this process requires specific conditions that make it economically unsuitable (Mannina et al., 2020).

In order to reduce the problems and costs associated with the use of pure cultures, alternative processes have been tested to promote PHA production (Kourmentza et al., 2017). Specifically, these involve the use of mixed microbial cultures (MMCs), which allow the efficient exploitation of cheap or no-cost substrates consisting of waste feedstocks, without requiring sterile working conditions. MMC-PHA production requires multi-stage processes, typically composed of three main steps (Valentino et al., 2017):

- Acidogenic fermentation to convert waste feedstock into a volatile fatty acid (VFAs)-rich stream;
- The selection stage to enrich an activated sludge in PHA-storing microorganisms through the establishment of so-called Feast and Famine conditions; and
- The accumulation stage to maximize the intracellular polymer content.

Factors that regulate the polymer storage response versus the growth response in the culture selection reactors have been studied using either synthetic feedstocks or waste-based feedstocks. Indeed, in recent years, PHA production has spread to a greater extent, thanks to the potential for utilizing different organic fermentable wastes.

As an example, the potential to integrate MMC-PHA production with wastewater treatment plants (WWTPs) has attracted considerable attention over the last few years, also in the context of developing urban biorefineries (Valentino et al., 2021). Several feedstocks have been tested for PHA production, at a pilot scale, such as the organic fraction of municipal solid waste (OFMSW) (Valentino et al., 2018), mixture of OFMSW and waste activated sludge (WAS) (Moretto et al., 2020), municipal wastewater (Morgan-Sagastume et al., 2015), and pre-treated excess thickened WAS (Lorini et al., 2022).

Moreover, around 998 Mt of agricultural waste is generated every year (Obi et al., 2016), all of which can be potentially valorized through PHA production. However, agricultural wastes typically contain lignocellulosic biomass made of structural sugars (cellulose and hemicellulose) and lignin, together with smaller amounts of ash and extractives (Requiso et al., 2018). Lignocellulosic crop residues are widely abundant, but this feedstock is difficult to be fermented. Enhancing volatile fatty acid production from the fermentation process of organic residues is challenging. When the lignocellulosic component is high, pre-treatments capable of breaking the structure of lignin and improving the efficiency of saccharification are the most suitable options.

They should promote the solubilization of organic matter, thus increasing the amount of substrate that can be converted into acids. Different pre-treatment approaches can be used, such as microwave and steam (Anjum et al., 2016), ultrasonic pre-treatment (Cesaro et al., 2014), thermic procedure, and chemical pre-treatment based on acid or alkali to promote the decomposition of the lignocellulosic matrix and enhance the solubilization of hemicellulose (Wang et al., 2021). A complete fermentation of lignocellulose wastes makes them a promising feedstock for PHA production from both pure and mixed microbial cultures. In a recent study (Li et al., 2021), an acid pre-treatment on rice straw feedstock was applied, and the waste sugar was converted into PHB by *Bacillus megaterium* B-10 with a maximum content of 32.56% of dry cell weight.

There are many other available and plentiful feedstocks as corn stover and sugar cane bagasse that can also be converted into PHA. As an example, it was found that corn stover-derived sugars can be converted into PHA by using an isolated strain of *Paracoccus* sp. LL1, with maximal PHA concentration equal to 9.71 ± 0.27 g/L (corresponding to an intracellular polymer content of 72.4 ± 1.3%, w/w) observed at the 72nd hour (96 h in total) of the test (Sawant et al., 2015). In another study (Yin et al., 2020), lignocellulosic hydrolysate and acetate were used as feedstock for PHA synthesis from MMC, and a [P(3HB-*co*-3HV)] copolymer was produced up to about 2308.45 mg/L. The MMC–PHA production from crop wastes, such as sugar cane molasses, has also been investigated. As an example, it was demonstrated that the fermentation of sugar cane molasses allowed the conversion of all carbohydrates into VFA (about 75 ± 4% of the total) and lactate (Bengtsson et al., 2010). Moreover, the abovementioned multi-stage process that includes molasses acidogenic fermentation, as well as the use of the fermented molasses to feed reactors for the selection of PHA-accumulating microorganisms and PHA accumulation, has been successfully adopted (Albuquerque et al., 2007).

Another substrate that could be used as feedstock for PHA production is apple pomace. Apple juice industry produces huge amounts of waste as side streams (Chen and Lahaye, 2021). Typically, only 65–70% of apple mass is used while the remainder of the biomass, known as pomace, is considered waste, especially due to its low nutritional values that make it unsuitable for animal feed (Liu et al., 2021). This residue is composed of cellulose, hemicellulose, lignin, pectin, and fermentable sugars and has low protein content (Cargnin and Gnoatto, 2017; González-García et al., 2018). Despite this waste not having many economic uses, the fact it has high amounts of fermentable sugars such as galactose, glucose, sucrose, arabinose, and fructose could make it a cheap agricultural waste carbon feedstock to produce bioplastics. Apple pomace has suitable characteristics for PHA accumulation, as the carbon content is around

50% (on weight basis) and nitrogen around 2%, obtaining a desirable C:N ratio for the accumulation of the polymer (Liu et al., 2021).

In a recent study, apple waste was utilized by *Pseudomonas chlororaphis* sub sp. *Aurantiaca* to produce PHA monomers including 17.9 ± 1.0 (mol%) 3-hydroxyoctanoate, 42.7 ± 0.1 (mol%) 3-hydroxydecanoate, 14.5 ± 1.1 (mol%) 3-hydroxybutyrate, and 11.1 ± 0.6 (mol%) 3-hydroxytetradecanoate, with a polymer content of 49.25 ± 4.08% (mol/mol) (Pereira et al., 2021). In another study, *Pseudomonas citronellolis* utilized fructose- and glucose-rich apple pomace to produce different PHA monomers, namely 3-hydroxydecanoate (68%, mol), 3-hydroxyoctanoate (22%, mol), 3-hydroxydodecanoate (5%, mol), 3-hydroxytetradecanoate (4%, mol), and 3-hydroxyhexanoate (1%, mol) with a polymer concentration of 1.20 ± 0.05 (g/L) obtained at the end of the experiment, corresponding to an overall volumetric productivity of 0.025 g/L·h (Rebocho et al., 2019). Bagasse, orange peel, and banana peel make up other agro-industrial wastes with a high sugar content. In a study carried out by Kulkarni et al. (Kulkarni et al., 2015), significant amounts of total sugars were utilized and converted into cell mass and PHA, namely 62% sugar utilized from bagasse extract, 84% from orange peel extract, and 71% from banana peel extract, as compared to 51% in the case of maltose, on a weight basis. These feedstocks were directly converted into substrates for PHA accumulation through a selected culture of *Halomonas campisalis*, and 0.365 g PHA/L was formed from 1% (v/v) aqueous extract of bagasse, 0.329 g PHA/L from orange peel extract, and 0.11 g PHA/L from banana peel extract.

These studies highlight a potential application for agricultural residues for PHA production. The next challenge could be to implement these substrates in MMC-PHA production in order to further reduce overall process costs.

Another agricultural waste that can be valorized through polymer production is olive oil mill wastewaters (OMW). This waste generates serious environmental problems due to its high production and its chemical composition, which is characterized by a high concentration of organic matter and phenolic compounds, which in turn make its disposal difficult. Nevertheless, through biopolymer production, it is possible to treat and simultaneously valorize OMW, which can be used as a no-cost substrate for the MMC-PHA production process. Indeed, it was shown that, by using OMW as a feedstock for the multi-stage MMC-PHA process, good results in terms of both PHA production (maximum concentration reached in the accumulation stage up to about 2470 mg COD/L) and reduction of the chemical oxygen demand (COD) content in the wastewater liquid fraction (around 70% on COD basis) can be achieved (Campanari et al., 2017). Therefore, a relevant COD conversion into methane (up to 100%) through anaerobic digestion tests on the OMW solid fraction was also obtained. Likewise, in another study (Campanari et al., 2014), a good performance in terms of storage yield (0.56 ± 0.05 COD/COD) and polymer volumetric productivity (calculated from the combination of both the microbial selection and the accumulation reactors), accounting for 1.50 g PHA/L·d, were obtained. By taking into account all the process stages (including phenols removal and wastewater fermentation), an abatement of about 85% of the OMW initial COD was estimated.

Finally, wastewaters resulting from the production of yeast (WWY) can also be counted as agricultural waste to produce compounds of higher commercial value. The organic matter and nitrogen content present in these wastewaters is very high. WWY contains mainly organic matter with a COD content up to 130 g/L, non-biodegradable organic pollutants, and nitrogen (250–1450 mg of total nitrogen/L) (Pirsaheb et al., 2018) and, due to these specific characteristics, it can be potentially harmful to the environment. Therefore, converting WWY into a fermented feedstock to obtain an eco-friendly biopolymer could lead to the dual benefit of improving its management and turning it into resources. In a recent study of MMC–PHA production (Ospina-Betancourth et al., 2022), feeding the microbial selection reactor (i.e. a sequencing batch reactor, SBR) with treated yeast wastewaters, a maximum PHA accumulation of 17% (w/w, based on cell dry weight) with a concentration of 1.2 g PHB/L was achieved. In this study, the WWY, before being used as feedstock for PHA production, underwent a process of acidogenic fermentation and distillation. By eliminating the methanogenic bacteria in the acidogenic reactor, together with the distillation of its effluent, it favored the availability of the produced VFAs as a substrate for PHA production in the SBR.

A common problem faced in the PHA production process from organic wastes is the seasonal availability of the feedstocks since the polymer composition is strongly linked to the type of fermented waste, which might not be compatible with market requirements. In this context, the potential to combine the fermentation of different renewable agricultural residues as cheap feedstocks for MMC-PHA production could be exploited. This could significantly improve the whole process, resulting in both increased waste recovery and operative cost reduction.

5 Cost evaluation of biopolymers

According to the estimates from European Bioplastics Organization (European Bioplastics Organization, 2022), currently only around 1% of the more than 367 Mt of plastic produced worldwide is bio-based and/or biodegradable (Wellenreuther et al., 2022). One reason for the limited market penetration of bio-based plastics, despite their potential to replace fossil-based polymers, is that biopolymers and, concretely, PHAs have higher production costs than petrochemical plastics (1.18–6.12 €/kg vs. <1 €/kg) (Pagliano et al., 2021), a situation made worse by low oil prices. However, prices have continuously decreased over the past decade. As more companies and brands switch to bio-based plastics, and as production capacities rise, supply chains and processes are becoming more efficient, which has brought prices down significantly. With increasing volumes of bioplastics on the market, taking into account that oil prices are expected to rise, the costs of bioplastics will soon be comparable with those of conventional plastics (European Bioplastics, 2016).

The large-scale industrialization and commercialization of PHAs have been restricted by production costs (Chen et al., 2020), which result from the combination of upstream and post-processing costs. The upstream variable takes into account substrate cost, polymer yield, growth rate, PHAs productivity, and quality and quantity of accumulated biopolymer (Leong et al., 2016). Efforts have been made to reduce upstream costs by using cheap and environmentally friendly raw materials (as mentioned earlier), together with alternative microbial consortia for fermenting unusual feedstocks. However, additional effort is required in downstream processes, such as PHA recovery and purification (Pagliano et al., 2021). The efficiency of these downstream processes could be enhanced by the use of appropriate organic solvents to extract PHA granules from inside the bacterial cells, or additives to disrupt the cellular matrix and release intracellular PHA, which could be combined with pre-treatments to enhance the permeability of cellular membranes (Pagliano et al., 2021). All these measurements will contribute to make this polymer more competitive.

Based on these considerations, the following sections focus on the use of bioplastics for the development of novel food packaging materials.

6 Applications of bioplastics in food packaging

Plastics began to replace materials traditionally used in packaging, such as metal, glass, and cardboard, as they were more flexible, lighter, and habitually more cost effective (Torres-Giner et al., 2018b). However, plastic packaging presents a huge environmental issue, worsened by the increase of single-use packaging items. Packaging represents the last stage of the production chain before products are transported to the market for consumers. In this sense, food packaging can be defined as a protection system to increase food shelf-life, maintaining product quality and safety during transportation, distribution, storage, retail, and end-use for the satisfaction of the consumer at an affordable cost (Gnanasekaran, 2019). Therefore, the requirements of food packaging materials include thermal, mechanical, optical, barrier, and antimicrobial properties to prevent or inhibit microbial growth as well as food decomposition by blocking light, oxygen, and moisture and to prevent spoilage caused by microorganisms or small insects (Pellicer et al., 2017). They must also have chemical resistance and be food contact approved (Sidwell, 1992).

The barrier properties of polymers are associated with their intrinsic ability to modulate the permeation of substances with low molecular weights, such as moisture, carbon dioxide, oxygen, and flavors (Pinto et al., 2021), and they directly influence the shelf-life of the packed product (Atta et al., 2022). The mechanical properties of polymers, which are influenced by both the polymer architecture and its molecular mass (Koide and Shi, 2007; Mangaraj et al., 2019), are of great importance in food packaging applications as they ensure an adequate physical barrier to the packed product (Mangaraj et al., 2009), as well as protection against mechanical stress during transportation and storage. Closely linked to mechanical properties are thermal properties, which relate to the polymer glass transition temperature and its melting temperature. They define the thermal stability of the material (Figueroa-Lopez et al., 2018) as well as its processability. These properties are a key aspect when designing packaging systems (Shaikh et al., 2021), since some of them need to be formed, filled, sealed, and sterilized in the same production line (Castro Mayorga, 2017) or resist low temperatures (Mangaraj et al., 2019). The thermal and mechanical properties of polymeric materials may be modified by varying polymeric composition, allowing the production of many different packaging materials, varying from highly flexible structures used to manufacture bags and wrapping to rigid structures used for bottles, lids, and caps (Ebnesajjad, 2013). Furthermore, the optical properties of the material need to be taken into account, which determine light protection as well as allow the consumer to see through the packaging. Finally, the material needs to be printable to provide all the required information about the packed product (Mangaraj et al., 2019; Porta et al., 2020).

Social and environmental demands, as well as compliance with new circular economy politics and the United Nations Sustainable Development Goals (United Nations, 2021), are forcing the development of a new generation of food packaging. In recent years, the development of alternative food packaging materials has become the focus of scientific research for two main reasons: (1) environmental pollution caused by the fabrication and disposal of synthetic plastic-based packaging and (2) the large amount of food waste produced annually. Researchers are now focused on the development of innovative biodegradable packaging based on bioplastics for its application in the food industry to replace plastic packaging and to develop more effective food packaging strategies to extend the shelf-life of the food (Debeaufort et al., 1998).

One of the main drawbacks of using bioplastics as biopolymers for food packaging applications is their lack of adequate thermal, mechanical, and barrier properties (Rosenboom et al., 2022; Sergio Torres-Giner et al., 2021). To tackle this challenge, diverse strategies may be applied, using, for example, a blend of biopolymers, fillers, nanotechnology, or multilayer structures. The efficacy of these strategies to reinforce biopolymers for food packaging applications has already been proven in the context of the Ypack (https://www.ypack.eu/) and Usable Packaging (https://www.usable-packaging.eu/) European projects.

Fillers are defined as particulate materials that are added to polymers to improve their mechanical and/or barrier properties as well as to reduce costs (Baird, 2003). Generally, such fillers comprise finely ground inorganic materials, such as chalk, silica, or clay (Gradin et al., 1989). However, the use of natural organic fillers in the place of mineral inorganic fillers allows a considerable reduction in polymer requirement, a reduction in environmental impact, and cost reduction since these fillers can be obtained from plants or agro-based by-products. Moreover, agro-based by-products, in combination with biodegradable biopolymers, provide fully biobased and biodegradable solutions (Cecchi and de Carolis, 2021). The most widely known and used natural organic fillers are wood flour and fibers, as well as agro-based by-products, which represent an attractive and sustainable method of valorization (la Mantia and Morreale, 2011; Sánchez-Safont et al., 2018). Several examples can be found in the literature to date - wheat straw (Avella et al., 2000, 1993; Berthet et al., 2015), flax fibers (Lee et al., 2003; Shanks et al., 2004; Wong et al., 2002), jute (Ma and Joo, 2011), coconut (Macedo et al., 2010), kenaf (Kuciel and Liber-Knec, 2011), olive pomace (Berthet et al., 2015), and almond shell (Quiles-Carrillo et al., 2018), among others, have been explored.

However, the implementation of biocomposite materials based on biopolymers with agro-based by-products is relatively slow because of various intrinsic limitations due to the inherent characteristics of their constituents (Cecchi and de Carolis, 2021). There is a great variability in the obtained results due to filler characteristics, also depending on the matrix type, surface treatment of the materials, and the use of coupling agents, specifically when it comes to mechanical performance (Sánchez-Safont et al., 2018). For that reason, more research work is required. For packaging, multilayer structures are often used when various functions and properties are required, such as high barrier, mechanical strength, and heat sealability, which no single material possesses (Özdemir and Floros, 2004).

Finally, nanotechnology is an emerging discipline for fabricating, manipulating, characterizing, and producing materials at the nanoscale (1–100 nm). It is used in the fabrication of packaging solutions with improved physical and mechanical properties, as well as improved functionalities, by the incorporation of nanostructures, in the form of particles or fibers (Ashfaq et al., 2022).

7 Innovative food packaging

Innovations in food packaging are not only based on the use of biopolymers. Traditional food packaging, also known as passive packaging, was initially

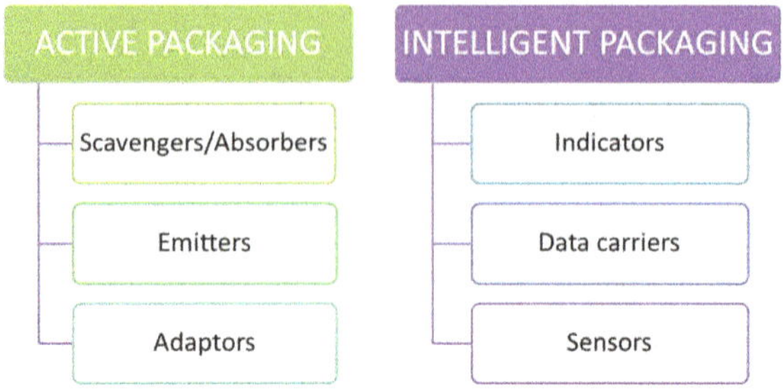

Figure 3 Classification of currently available packaging solutions.

developed to contain food products and protect them from external conditions (e.g. moisture, oxygen, dust, light, and microorganisms) (Zhang et al., 2020). However, in the past decade, thanks to technological advancements in the food industry, passive food packaging has gradually changed to innovative packaging systems – a combination of both active and intelligent systems that can increase shelf-life and prolong food quality throughout the distribution and supply chain to the end-consumer (Kuswandi and Jumina, 2020). Figure 3 shows the classification of currently available food packaging solutions.

Intelligent packaging includes materials and articles that monitor the condition of packaged food or the environment surrounding the food and provide information about its quality during transport and distribution (Kuswandi et al., 2011). It can include sensors for gases, chemical compounds or microorganism spoilage, indicators for freshness, ripeness, thawing, humidity, leakage or time–temperature variations, or data carriers in the form of bar or QR codes, among others (Alp-Erbay, 2022; Müller and Schmid, 2019).

Active packaging is defined by the European Regulation of the European Community No. 450/2009 as a system that deliberately incorporates active substances that change the environmental conditions of the packaged food environment to enhance the performance of the packaging system and to preserve safety and sensory properties, as well as prevent food quality loss (European Food Safety Authority (EFSA), 2009). These active substances can be categorized as (1) scavengers/absorbers; (2) emitters; and (3) adaptors (Cherpinski Correa et al., 2016).

7.1 Scavengers

Active scavenging systems are substances used to remove gases like CO_2, O_2, ethylene, or moisture from food packaging. The presence of oxygen in packaged food can negatively influence food quality and its shelf-life by allowing the growth of microorganisms and/or oxidation of the product (Vilela et al., 2018). Most commercially applied oxygen scavengers are iron-based solutions, whose mechanism of action is based on the principle of iron oxidation (Cooksey, 2010). However, the disadvantage of this system is the risk of involuntary content consumption (Vilela et al., 2018). In addition, iron-based systems need moisture to perform properly. Because of this, several alternatives have been developed, such as unsaturated hydrocarbon-based systems (e.g. α-tocopherol or ascorbic acid-based scavenging systems), enzyme-based scavenging systems (e.g. glucose oxidase and catalase), and microorganisms-based scavengers (e.g. dried yeast immobilized in solid wax). All the techniques mentioned earlier do not work fast enough to ensure product stability during food storage, especially under refrigerated conditions. An alternative is represented by the use of metal nanoparticles, such as palladium nanoparticles and titanium dioxide nanotubes, among others (Close et al., 2015). Some, such as titanium dioxide, change color in the presence of oxygen, from dark blue to brown, making them suitable as oxygen indicators to tamper-proof highly oxygen-sensitive products (Tulsyan et al., 2017).

Another compound responsible for the reduced shelf-life of fresh products is ethylene, being the main cause of post-harvest spoilage of fruit and vegetables. Strategies for removing exogenous ethylene mainly rely on inhibitors such as 1-methylcyclopropene, absorbents such as zeolites, and catalytic oxidants such as $KMnO_4$, ozone, and TiO_2 (Wei et al., 2021).

Excess moisture inside packaging promotes bacterial and mold growth, which in turn leads to food quality loss and reduced shelf-life. This is produced as a result of the respiration of fresh food, temperature fluctuations, and tissue fluids from cut meats and poultry. Several absorbing substances have been found to act as scavenging systems, including silica gel, calcium oxide, and modified starch (Ozdemir and Floros, 2004).

7.2 Emitters

Emitters, also known as release agents, include compounds such as antioxidants, antimicrobials, flavors, or CO_2, which promote the extended shelf-life of food by slowing down the growth of fungi and bacteria, improving organoleptic characteristics of the food, and preventing oxidation of food or by controlling ripening (De Pilli, 2020). The role of antimicrobial packaging is to reduce, inhibit, or delay foodborne bacterial growth. Antimicrobial substances are classified as either organic or inorganic materials, both of which are highly active against Gram-negative and Gram-positive bacteria. However, most organic substances (e.g. essential oils, enzymes, and organic acids) are sensitive to high temperatures and pressures, while inorganic materials (e.g. metals, metal oxides, and functionalized nanoclays) are more resistant (Huang et al., 2019; Joerger, 2007; Sharma et al., 2020).

Antioxidant packaging materials have been developed to avoid oxidative reactions that decrease nutritional foodstuff value caused by essential fatty

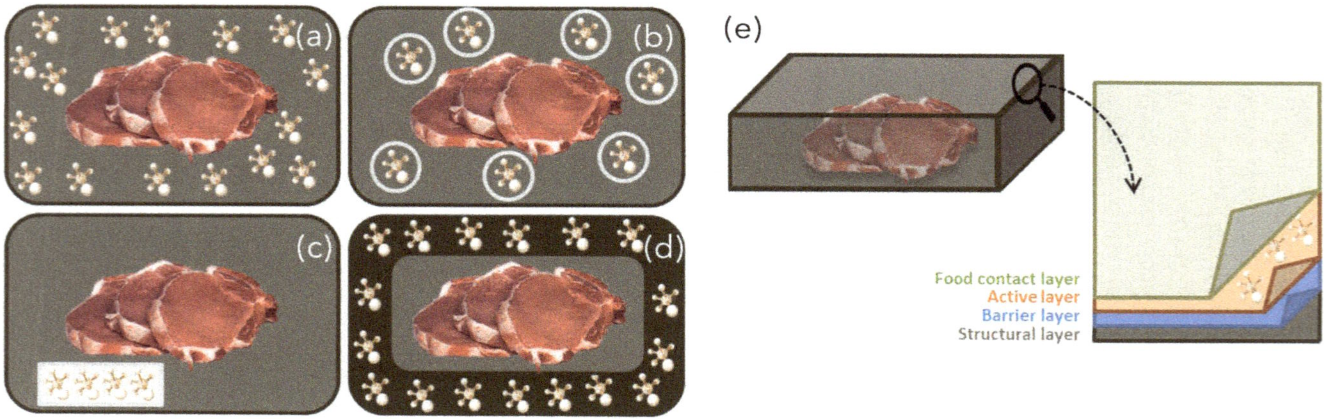

Figure 4 Classification of the different methods to incorporate the active agent into packaging items: (a) direct contact of the active agent with the food product, (b) direct contact of the microencapsulated active agent with the food product, (c) the active agent embedded into sachets or pads, d) the active agent coating the packaging surface or embedded into the interior of the polymeric film, and (e) multilayer film approach. Source: Adapted from Figueroa-Lopez (2021).

acids, proteins, and lipid-soluble vitamin degradation, which generate off-flavors, odors, and color change. Antioxidant packaging includes substances such as metal chelators, UV absorbers, singlet oxygen quenchers, polyphenols, tocopherols, carotenoids, caffeic acid, ferulic acid, ascorbic acid, plant and fruit extracts, as well as essential oils from herbs and spices, which prevent oxidation by inhibiting the oxidative chain reaction or forming stable radicals that are unreactive (Perumalla and Hettiarachchy, 2011).

7.3 Adaptors

Adaptors represent solutions that do not absorb or release substances but cause desirable physical, chemical, or biological changes in the packed product or the packaging's inner atmosphere (Cherpinski Correa et al., 2016). These substances include insulating materials, temperature-controlling materials (e.g. self-heating and self-cooling), temperature-sensitive films, microwave susceptors, modifiers, and UV and surface-treated packaging materials (Day, 2008; Kerry et al., 2006).

All these agents can be incorporated in packaging using the various alternatives summarized in Fig. 4. These can be classified as (Rai et al., 2019):

- Direct contact of the active agent with the food product;
- Direct contact of the microencapsulated active agent with the food product;
- The active agent embedded into sachets or pads; and
- The active agent coating the packaging surface or embedded into the interior of the polymeric film.

These agents are typically encapsulated or included in the polymeric film in order to achieve a controlled release of substances, promoting extended shelf-life of food by slowing down the growth of fungi and bacteria, improving the organoleptic characteristics of the food, preventing oxidation of food, or controlling ripening (De Pilli, 2020). Additionally, the multilayer film structure of packaging is often studied because the different layers have various functions and desirable properties, such as high barrier, mechanical strength, or heat sealability that no single material possesses (Ozdemir and Floros, 2004). In this sense, multilayer active films can be composed of several layers - the structural layer, the barrier layer, the active layer, and the food contact layer, as shown in Fig. 4e.

8 Packaging production techniques

Currently, multiple processes are available for packaging manufacture, including casting, blow molding, extrusion, thermoforming, injection molding, welding, and foaming (Market Prospects, 2021). Among these conventional techniques, extrusion, film blowing, and casting are the most implemented for biopolymers (Mangaraj et al., 2019; Nesic et al., 2020).

Extrusion is a versatile and continuous technology that allows a wide range of materials to be processed and often represents the first key step in plastic processing, thanks to its ability to convert resins from solid to molten phase (Emin and Schuchmann, 2017; Nesic et al., 2020). The high temperature and pressure during processing, and high-pressure drop in the extruder die, are responsible for physical and chemical transformations of material, generating changes in their properties and features (La Fuente Arias et al., 2021). Moreover, multilayer systems can be produced by using a co-extrusion process (Shaikh et al., 2021). To promote strong interfacial adhesion, additives or special adhesion-improving processes are needed (La Fuente Arias et al., 2021).

Film blowing, also known as blown extrusion, is the most common industrial film-producing method (Liu et al., 2020; Nesic et al., 2020). Films are produced by an extrusion process with a circle-shaped extrusion die, with air pressure used to further expand the film. Once the film reaches the desired dimensions and characteristics, it undergoes a cooling step to promote the solidification of the material (McKeen, 2012). This technique requires materials with certain melt viscosity resistance to be able to withstand the process, as well as relatively high operator skill to produce films with the desired size and thickness for specific applications (Biron, 2013; Nesic et al., 2020). However, when it comes to bioplastic materials packaging, several disadvantages have been found, including thermal instability, low heat sealability, high brittleness, and poor impact resistance (Cabedo et al., 2006; Cyras et al., 2009; Modi, 2010; Nesic et al., 2020).

Solvent casting is the most popular lab method used to produce bio-based monolayer films. Normally, water is used as a solvent, but the addition of a plasticizer is required (do Val Siqueira et al., 2021; La Fuente Arias et al., 2021). To obtain the polymeric film, the film-forming solution is poured in an adequate support (teflon plates, petri dishes, or acrylic plates), followed by a drying step, which can be conducted at room or controlled temperature. This latter step, however, is time-consuming, which makes scaling it up for industrial applications difficult (do Val Siqueira et al., 2021).

It is important to note that these conventional processes have several disadvantages, including the use of high temperatures, the production of thick layers, the lack of homogeneity of the active agent, the lack of stability of the active compound during processing, and being difficult to scale up (Das et al., 2020; Grothe et al., 2012; Hemmati et al., 2021; Sanchez et al., 2009; Soltani Firouz et al., 2021; Ur Rahim et al., 2021). Emerging technologies such as electrohydrodynamic processing could be an alternative to produce this new generation of sustainable packaging items (La Fuente Arias et al., 2021).

Novel electrohydrodynamic processing has emerged as a versatile and simple approach for the manufacture of ultrathin (both submicron and nano-sized) polymer structures with high specific surface areas and porosities for a wide variety of industrial applications, including food packaging production (Sameen et al., 2021; Subbiah et al., 2005; Torres-Giner et al., 2018a). Electrohydrodynamic processing is a non-mechanical, cost-effective technique that uses a homogeneous electrohydrodynamic force to break up liquids into thin jets (Charles et al., 2021; Wu and Clark, 2008). Using this technique, the polymeric solution is fed to the nozzle with the assistance of a syringe pump (Echegoyen et al., 2017). The atomizer injector, usually shaped as a metal capillary, is polarized by a high voltage. Due to the applied electric field, the liquid surface is subjected to a shear stress, which causes the formation of the Taylor cone and the elongation of a jet and production of micro- or nanostructures, whose morphology and size can be adjusted by changing the process conditions (Echegoyen et al., 2017). The basic setup consists of three components - a high-voltage power supply, a spinneret, which is a metallic needle, and a collecting surface, which is a single-piece conductive substrate (Echegoyen et al., 2017). A schematic representation of the laboratory scale of the electrohydrodynamic process, including electrospray and electrospinning processes, can be seen in Fig. 5.

Electrohydrodynamic processing can be subdivided into two main processing techniques - electrospraying and electrospinning - which produce particles or fibers, respectively (Anu Bhushani and Anandharamakrishnan, 2014). Fine particles or fibers are produced depending upon the operating

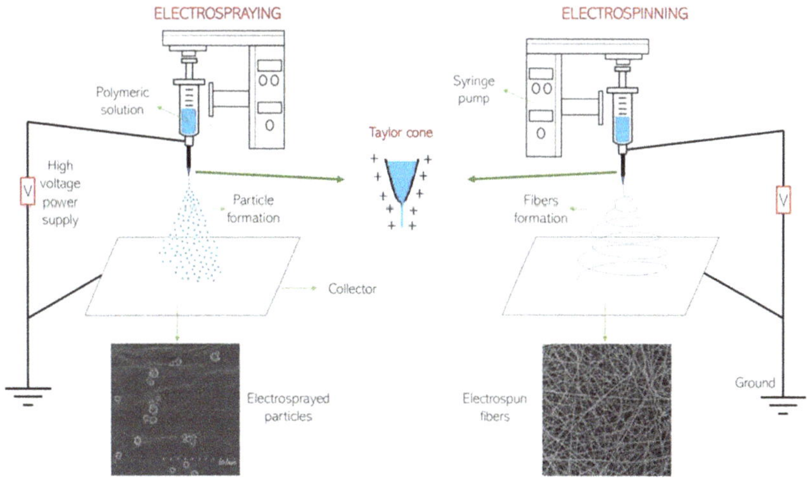

Figure 5 Schematic diagram of the two main electrohydrodynamic processes. On the left-hand side, a vertical setup of electrospraying is shown. On the right-hand side, a typical electrospinning vertical set-up is shown. Source: Adapted from Charles et al. (2021).

conditions, the nature of the polymers, and the solution dynamics (Drosou et al., 2017). Electrospraying and electrospinning follow the same physics; however, if the level of molecular cohesion, which is inferred from the solution's viscosity and polymer concentration (Chakraborty et al., 2009), is lower than the critical droplet value, particles are formed instead of fibers and the process is called electrospraying (Bock et al., 2011; Chen et al., 1995; Yurteri et al., 2010). Electrospinning is a process that obtains continuous polymer fibers in the submicrometer range (Doshi and Reneker, 1995).

Electrospinning appears to be very promising in the valorization of the processing of renewable polymers leading to the creation of mono- or interlayers or coatings with enhanced flexibility and transparency, while potentially retaining the barrier performance to gases and vapors of conventionally produced films (Vázquez-González et al., 2021). The process takes place at room temperature, so the active properties are maintained, and the use of adhesive layers is not required (Figueroa-Lopez et al., 2020b). Additionally, this technique is highly versatile in terms of the materials that can be processed.

To date, this technology has demonstrated its potential in the development of innovative packaging solutions based on biopolymeric systems. Figueroa et al. have developed antimicrobial and antioxidant packaging solutions based on poly(3-hydroxybutyrate-co-3-hydorxyvalerate) biopolymers and essential oils such as oregano and eugenol (Figueroa-Lopez et al., 2020a). Cherpinski et al. have developed oxygen-scavenging films made of poly(3-hydroxybutyrate) containing palladium nanoparticles by electrospinning technology (Cherpinski et al., 2018). Castro Mayorga et al. studied the antimicrobial properties of electrospun poly(3-hydroxybutyrate-co-3-hydroxyvalerate) films incorporating silver, zinc oxide, or copper oxide nanoparticles for food active packaging applications (Castro Mayorga, 2017; Castro Mayorga et al., 2018, 2017). Finally, Perez-Masiá et al. have developed coating materials with energy storage capacity for thermal insulation applications based on biodegradable polyesters and phase change materials processed by electrospinning (Perez-Masia et al., 2013).

9 Case study: the Horizon 2020 Usable Packaging project

Despite efforts to develop novel packaging solutions based on sustainable alternatives to plastics, to date, most existing bioplastics display a lack of adequate functional properties for packaging requirements. The main objective of the Usable Packaging Project (RIA BBI-JU H2020 ID: 836884) was the definition of a new vertically integrated and circular value chain in the plastic sector for the production of recyclable/biodegradable plastic items from bio-based materials that did not compete with the food chain or cause land-use problems. Indeed, the development of biodegradable and bio-based plastics is key to implementing circular economy models in the packaging sector. The project focused on the development of food, drinking, and pharma packaging solutions with high barrier properties, following a reverse engineering approach, in which the needs of the end-users of the packaging solutions were described to guide the specific functionalization of PHAs obtained from food-processing by-products.

The overall objectives of this project were:

- to develop high-performance bio-based packaging items that are environmentally sustainable and suitable for organic recycling;
- to improve circularity by eco-design over existing fossil-based and bio-based plastics across the full range of applications, including complex multilayer structures and functionalities to achieve high barrier and high mechanical performances, along with versatility of design;
- to ensure the end-of-life phase is wholly sustainable by creating a fully circular, vertically integrated supply chain;
- to create genuinely new and innovative designs for packaging and production, by developing market back applications, to meet specifications of large enterprise partners in the first instance; and
- to deliver packaging materials capable of biodegrading in different environments, including marine and soil settings.

To achieve this, the Usable Packaging project involved different segments of the value chain, as summarized in Fig. 6:

1. **The feedstock.** The Usable Packaging project focused on using both residues and non-waste by-products of the food-processing industry, which are cost effective and widely available, as well as the CO_2 produced by food processing (e.g. wine or beer production) or within the process itself via anaerobic digestion of residual organic material from the biorefinery.
2. **The process for raw material production.** The project converted feedstocks into PHAs, including the poly(3-hydroxybutyrate-co-3-hy dorxyvalerate) copolymer, which were selected for specific packaging needs and specifications. The [P(3HB-*co*-3HV)] was then functionalized by an advanced technologies platform to obtain bio-based sustainable materials. These were aimed at replacing the currently used fossil-based materials to prepare high-performance packaging items.
3. **The process for packaging production to the product.** At the packaging processing level, several techniques were used, including compounding to achieve target functionality, extrusion, coextrusion,

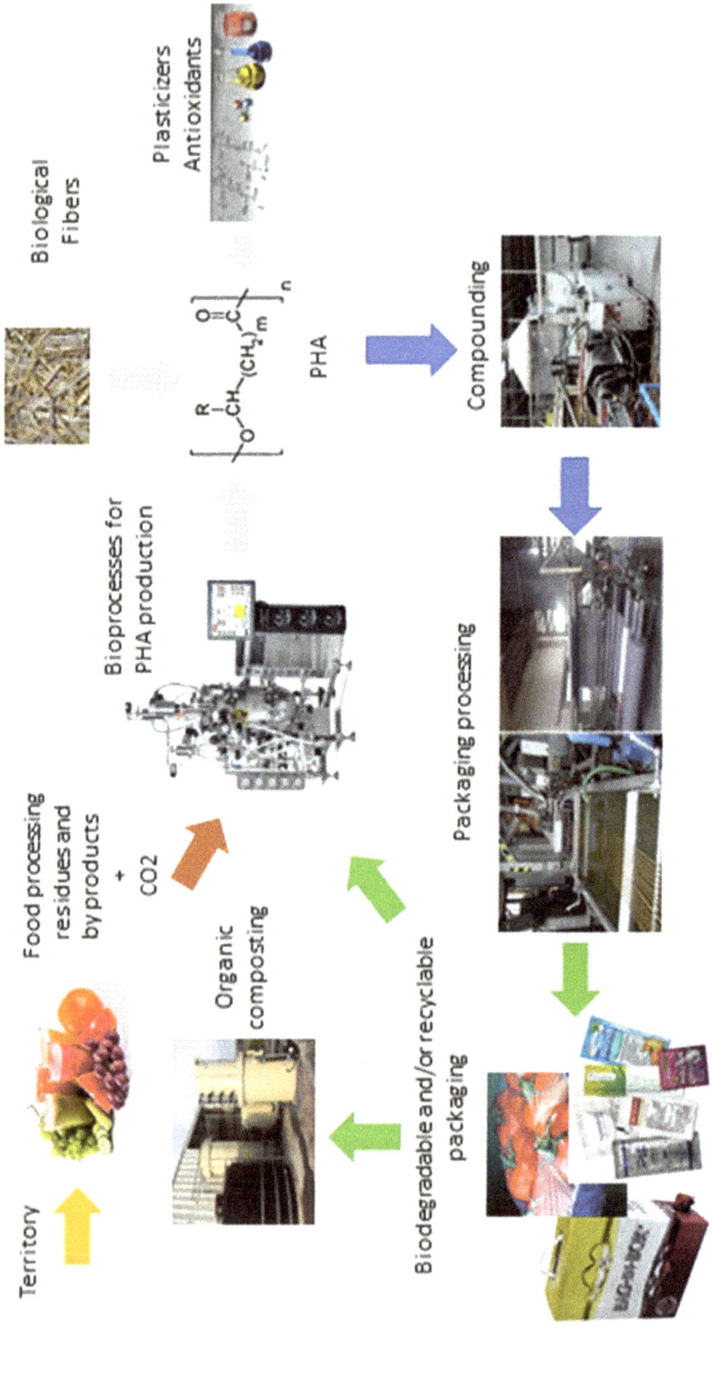

Figure 6 Project concept throughout the different value chains.

lamination, or electrospinning, which was an attractive application for obtaining fibers and monolayers with high performances, including those of hot tack, which were enablers of bio-based laminated and coated structures. The generated bio-based polymers were processed by these technologies to prepare high-performance packaging systems such as hot tack adhesives, multilayer packaging systems for food and drinks sectors, pharmaceutical packaging, and agriculture packaging, among others, while retaining biodegradable characteristics. The use of these technologies, along with the generated biodegradable material, addressed the problem of plastic packaging waste, providing a new and more sustainable end-of-life solution with higher material performance.

Overall, the activities within the Usable Packaging project focused on supporting new market opportunities and company growth while addressing environmental, technical, and societal issues.

10 Conclusion and future research needs

This chapter has looked at potential solutions to the environmental problem of fossil-based plastic. Recent research has shown that it is possible to obtain biodegradable and bio-based polymers (i.e. PHAs) that adhere to circular economy principles, by using mixed microbial cultures and renewable feedstocks, such as food, agro-industrial waste, and wastewaters. All this allows the recovery of waste, which is incorporated into the production chain, thus reducing its negative impact on the environment and human health, as well as the production of a plastic material that does not compete with the food chain or with ground use. Creating new packaging items with improved properties and based on these biopolymers represents another way to reduce the amount of waste, in terms of both food and plastic materials. The development of new processing technologies such as electrospinning favors this new generation of packaging materials with enhanced properties. The European Usable Packaging project evidenced the development of a circular and vertically integrated value chain within the plastics sector, which could be transferred to the industry as an alternative to current materials. However, future social and environmental challenges require a European policy to promote the use and production of biopolymers, with incentives for the industry, consumer awareness campaigns, and, where appropriate, taxes and prohibitions on the use of certain plastics of fossil origin. In addition, more research will be required to validate the use of PHA from a technological point of view as an alternative to commonly used plastics for multiple applications in different sectors such as packaging, agriculture, textiles, chemicals, food, biomedicine, and pharmaceuticals.

11 References

Albuquerque, M. G. E., Eiroa, M., Torres, C., Nunes, B. R. and Reis, M. A. M. 2007. Strategies for the development of a side stream process for polyhydroxyalkanoate (PHA) production from sugar cane molasses. *J. Biotechnol.* 130(4), 411–421. https://doi.org/10.1016/j.jbiotec.2007.05.011.

Alp-Erbay, E. 2022. Nanomaterials utilized in food packaging: State-of-the-art. *Food Eng. Rev.* 14(4), 629–654. https://doi.org/10.1007/s12393-022-09318-y.

Andrade, R., Skurtys, O. and Osorio, F. 2015. Drop impact of gelatin coating formulated with cellulose nanofibers on banana and eggplant epicarps. *LWT Food Sci. Technol.* 61(2), 422–429. https://doi.org/10.1016/j.lwt.2014.12.035.

Andreuccetti, C., Carvalho, R. A., Galicia-García, T., Martínez-Bustos, F. and Grosso, C. R. F. 2011. Effect of surfactants on the functional properties of gelatin-based edible films. *J. Food Eng.* 103(2), 129–136. https://doi.org/10.1016/j.jfoodeng.2010.10.007.

Anjum, M., Khalid, A., Mahmood, T. and Aziz, I. 2016. Anaerobic co-digestion of catering waste with partially pretreated lignocellulosic crop residues. *J. Clean. Prod.* 117, 56–63. https://doi.org/10.1016/j.jclepro.2015.11.061.

Anu Bhushani, J. and Anandharamakrishnan, C. 2014. Electrospinning and electrospraying techniques: Potential food based applications. *Trends Food Sci. Technol.* 38(1), 21–33. https://doi.org/10.1016/j.tifs.2014.03.004.

Ashfaq, A., Khursheed, N., Fatima, S., Anjum, Z. and Younis, K. 2022. Application of nanotechnology in food packaging: Pros and cons. *J. Agric. Food Res.* 7, 100270. https://doi.org/10.1016/j.jafr.2022.100270.

Ashter, S. A. 2016. 5-types of biodegradable polymers. In: Ashter, S. A. (Ed.), *Introduction to Bioplastics Engineering*. Elsevier, Oxford, pp. 81–151. https://doi.org/10.1016/B978-0-323-39396-6.00005-1.

Atta, O. M., Manan, S., Shahzad, A., Ul-Islam, M., Ullah, M. W. and Yang, G. 2022. Biobased materials for active food packaging: A review. *Food Hydrocoll.* 125, 107419. https://doi.org/10.1016/j.foodhyd.2021.107419.

Avella, M., Martuscelli, E., Pascucci, B., Raimo, M., Focher, B. and Marzetti, A. 1993. A new class of biodegradable materials: Poly-3-hydroxy-butyrate/steam exploded straw fiber composites. I. Thermal and impact behavior. *J. Appl. Polym. Sci.* 49(12), 2091–2103. https://doi.org/10.1002/app.1993.070491205.

Avella, M., La Rota, G., Martuscelli, E., Raimo, M., Sadocco, P., Elegir, G. and Riva, R. 2000. Poly(3-hydroxybutyrate-co-3-hydroxyvalerate) and wheat straw fibre composites: Thermal, mechanical properties and biodegradation behaviour. *J. Mater. Sci.* 35(4), 829–836. https://doi.org/10.1023/A:1004773603516.

Azevedo, V. M., Silva, E. K., Gonçalves Pereira, C. F., da Costa, J. M. G. and Borges, S. V. 2015. Whey protein isolate biodegradable films: Influence of the citric acid and montmorillonite clay nanoparticles on the physical properties. *Food Hydrocoll.* 43, 252–258. https://doi.org/10.1016/j.foodhyd.2014.05.027.

Baird, D. G. 2003. Polymer processing. In: *Encyclopedia of Physical Science and Technology*. Elsevier, pp. 611–643. https://doi.org/10.1016/B0-12-227410-5/00593-7.

Bengtsson, S., Pisco, A. R., Reis, M. A. M. and Lemos, P. C. 2010. Production of polyhydroxyalkanoates from fermented sugar cane molasses by a mixed culture enriched in glycogen accumulating organisms. *J. Biotechnol.* 145(3), 253–263. https://doi.org/10.1016/j.jbiotec.2009.11.016.

Berthet, M.-A., Angellier-Coussy, H., Machado, D., Hilliou, L., Staebler, A., Vicente, A. and Gontard, N. 2015. Exploring the potentialities of using lignocellulosic fibres derived from three food by-products as constituents of biocomposites for food packaging. *Ind. Crops Prod.* 69, 110–122. https://doi.org/10.1016/j.indcrop.2015.01.028.

Besseling, E., Quik, J. T. K., Sun, M. and Koelmans, A. A. 2017. Fate of Nano- and microplastic in freshwater systems: A modeling study. *Environ. Pollut.* 220(A), 540–548. https://doi.org/10.1016/j.envpol.2016.10.001.

Biron, M. 2013. 5-thermoplastic processing. In: Biron, M. (Ed.), *Thermoplastics and Thermoplastic Composites*. Elsevier, pp. 715–768. https://doi.org/10.1016/B978-1-4557-7898-0.00005-6.

Bläsing, M. and Amelung, W. 2018. Plastics in soil: Analytical methods and possible sources. *Sci. Total Environ.* 612, 422–435. https://doi.org/10.1016/j.scitotenv.2017.08.086.

Blettler, M. C. M., Abrial, E., Khan, F. R., Sivri, N. and Espinola, L. A. 2018. Freshwater plastic pollution: Recognizing research biases and identifying knowledge gaps. *Water Res.* 143, 416–424. https://doi.org/10.1016/j.watres.2018.06.015.

Bock, N., Woodruff, M. A., Hutmacher, D. W. and Dargaville, T. R. 2011. Electrospraying, a reproducible method for production of polymeric microspheres for biomedical applications. *Polymers* 3(1), 131–149. https://doi.org/10.3390/polym3010131.

Borrelle, S. B., Ringma, J., Law, K. L., Monnahan, C. C., Lebreton, L., McGivern, A., Murphy, E., Jambeck, J., Leonard, G. H., Hilleary, M. A., Eriksen, M., Possingham, H. P., De Frond, H., Gerber, L. R., Polidoro, B., Tahir, A., Bernard, M., Mallos, N., Barnes, M. and Rochman, C. M. 2020. Predicted growth in plastic waste exceeds efforts to mitigate plastic pollution. *Science* 369(6510), 1515–1518. https://doi.org/10.1126/science.aba3656.

Braunegg, G., Lefebvre, G. and Genser, K. F. 1998. Polyhydroxyalkanoates, biopolyesters from renewable resources: Physiological and engineering aspects. *J. Biotechnol.* 65(2–3), 127–161. https://doi.org/10.1016/S0168-1656(98)00126-6.

Cabedo, L., Luis Feijoo, J., Pilar Villanueva, M., Lagarón, J. M. and Giménez, E. 2006. Optimization of biodegradable nanocomposites based on aPLA/PCL blends for food packaging applications. *Macromol. Symp.* 233(1), 191–197. https://doi.org/10.1002/masy.200690017.

Campanari, S., Augelletti, F., Rossetti, S., Sciubba, F., Villano, M. and Majone, M. 2017. Enhancing a multi-stage process for olive oil mill wastewater valorization towards polyhydroxyalkanoates and biogas production. *Chem. Eng. J.* 317, 280–289. https://doi.org/10.1016/j.cej.2017.02.094.

Campanari, S., Silva, F. A., Bertin, L., Villano, M. and Majone, M. 2014. Effect of the organic loading rate on the production of polyhydroxyalkanoates in a multi-stage process aimed at the valorization of olive oil mill wastewater. *Int. J. Biol. Macromol.* 71, 34–41. https://doi.org/10.1016/j.ijbiomac.2014.06.006.

Cargnin, S. T. and Gnoatto, S. B. 2017. Ursolic acid from apple pomace and traditional plants: A valuable triterpenoid with functional properties. *Food Chem.* 220, 477–489. https://doi.org/10.1016/j.foodchem.2016.10.029.

Castro Mayorga, J. L. 2017. Biotechnological routes for the development of antimicrobial nano-metal based polyhydroxyalkanoates for active food packaging applications (Tesis doctoral). Universitat Politècnica de València. https://doi.org/10.4995/Thesis/10251/85678.

Castro Mayorga, J. L., Fabra, M. J., Pourrahimi, A. M., Olsson, R. T. and Lagaron, J. M. 2017. The impact of zinc oxide particle morphology as an antimicrobial and when incorporated in poly(3-hydroxybutyrate-co-3-hydroxyvalerate) films for food packaging and food contact surfaces applications. *Food Bioprod. Process.* 101, 32–44. https://doi.org/10.1016/j.fbp.2016.10.007.

Castro Mayorga, J. L., Fabra Rovira, M. J., Cabedo Mas, L., Sánchez Moragas, G. and Lagarón Cabello, J. M. 2018. Antimicrobial nanocomposites and electrospun coatings based on poly(3-hydroxybutyrate- *co* -3-hydroxyvalerate) and copper oxide nanoparticles for active packaging and coating applications. *J. Appl. Polym. Sci.* 135(2), 45673. https://doi.org/10.1002/app.45673.

Cecchi, T. and de Carolis, C. 2021. *Biobased Products from Food Sector Waste*. Springer International Publishing, Cham. https://doi.org/10.1007/978-3-030-63436-0.

Cesaro, A., Velten, S., Belgiorno, V. and Kuchta, K. 2014. Enhanced anaerobic digestion by ultrasonic pretreatment of organic residues for energy production. *J. Clean. Prod.* 74, 119–124. https://doi.org/10.1016/j.jclepro.2014.03.030.

Chae, Y. and An, Y. J. 2018. Current research trends on plastic pollution and ecological impacts on the soil ecosystem: A review. *Environ. Pollut.* 240, 387–395. https://doi.org/10.1016/j.envpol.2018.05.008.

Chakraborty, S., Liao, I. C., Adler, A. and Leong, K. W. 2009. Electrohydrodynamics: A facile technique to fabricate drug delivery systems. *Adv. Drug Deliv. Rev.* 61(12), 1043–1054. https://doi.org/10.1016/j.addr.2009.07.013.

Chan, C. M., Vandi, L.-J., Pratt, S., Halley, P., Richardson, D., Werker, A. and Laycock, B. 2018. Composites of wood and biodegradable thermoplastics: A review. *Polym. Rev.* 58(3), 444–494. https://doi.org/10.1080/15583724.2017.1380039.

Charles, A. P. R., Jin, T. Z., Mu, R. and Wu, Y. 2021. Electrohydrodynamic processing of natural polymers for active food packaging: A comprehensive review. *Compr. Rev. Food Sci. Food Saf.* 20(6), 6027–6056. https://doi.org/10.1111/1541-4337.12827.

Chavoshizadeh, S., Pirsa, S. and Mohtarami, F. 2020. Conducting/smart color film based on wheat gluten/chlorophyll/polypyrrole nanocomposite. *Food Packag. Shelf Life* 24, 100501. https://doi.org/10.1016/j.fpsl.2020.100501.

Chen, D.-R., Pui, D. Y. H. and Kaufman, S. L. 1995. Electrospraying of conducting liquids for monodisperse aerosol generation in the 4 nm to 1.8 μm diameter range. *J. Aerosol Sci.* 26(6), 963–977. https://doi.org/10.1016/0021-8502(95)00027-A.

Chen, G. Q. 2009. A microbial polyhydroxyalkanoates (PHA) based bio- and materials industry. *Chem. Soc. Rev.* 38(8), 2434–2446. https://doi.org/10.1039/b812677c.

Chen, G.-Q., Chen, X.-Y., Wu, F.-Q. and Chen, J.-C. 2020. Polyhydroxyalkanoates (PHA) toward cost competitiveness and functionality. *Adv. Ind. Eng. Polym. Res.* 3(1), 1–7. https://doi.org/10.1016/j.aiepr.2019.11.001.

Chen, M. and Lahaye, M. 2021. Natural deep eutectic solvents pretreatment as an aid for pectin extraction from apple pomace. *Food Hydrocoll.* 115, 106601. https://doi.org/10.1016/j.foodhyd.2021.106601.

Cherpinski, A., Torres-Giner, S., Vartiainen, J., Peresin, M. S., Lahtinen, P. and Lagaron, J. M. 2018. Improving the water resistance of nanocellulose-based films with poly hydroxyalkanoates processed by the electrospinning coating technique. *Cellulose* 25(2), 1291–1307. https://doi.org/10.1007/s10570-018-1648-z.

Cherpinski Correa, A., Dantuma, A., Silva, F., Gonçalves, I., Vecbiskena, L., Delgado-Aguilar, M., Kvalvåg Pettersen, M., Lavoine, N., Sumińska, P. and Brabbs, S. 2016. Active packaging - ActinPak. https://www.actinpak.eu/wp-content/uploads/2018/06/Active_packaging_English.pdf (accessed 20 November 2023).

Cherubini, F. 2010. The biorefinery concept: Using biomass instead of oil for producing energy and chemicals. *Energy Convers. Manag.* 51(7), 1412–1421. https://doi.org/10.1016/j.enconman.2010.01.015.

Close, T., Tulsyan, G., Diaz, C. A., Weinstein, S. J. and Richter, C. 2015. Reversible oxygen scavenging at room temperature using electrochemically reduced titanium oxide nanotubes. *Nat. Nanotechnol.* 10(5), 418–422. https://doi.org/10.1038/nnano.2015.51.

Cooksey, K. 2010. Oxygen scavenging packaging systems. In: *Encyclopedia of Polymer Science and Technology*. John Wiley & Sons, Inc., Hoboken, NJ. https://doi.org/10.1002/0471440264.pst570.

Coppola, G., Gaudio, M. T., Lopresto, C. G., Calabro, V., Curcio, S. and Chakraborty, S. 2021. Bioplastic from renewable biomass: A facile solution for a greener environment. *Earth Syst. Environ.* 5(2), 231–251. https://doi.org/10.1007/s41748-021-00208-7.

Cui, S., Borgemenke, J., Liu, Z. and Li, Y. 2019. Recent advances of "soft" bio-polycarbonate plastics from carbon dioxide and renewable bio-feedstocks via straightforward and innovative routes. *J. CO2 Util.* 34, 40–52. https://doi.org/10.1016/j.jcou.2019.05.027.

Cyras, V. P., Soledad, C. M. and Analía, V. 2009. Biocomposites based on renewable resource: Acetylated and non acetylated cellulose cardboard coated with polyhydroxybutyrate. *Polymer* 50(26), 6274–6280. https://doi.org/10.1016/j.polymer.2009.10.065.

Das, A., Chatham, C. A., Fallon, J. J., Zawaski, C. E., Gilmer, E. L., Williams, C. B. and Bortner, M. J. 2020. Current understanding and challenges in high temperature additive manufacturing of engineering thermoplastic polymers. *Addit. Manuf.* 34, 101218. https://doi.org/10.1016/j.addma.2020.101218.

Day, B. P. F. 2008. Active packaging of food. In: *Smart Packaging Technologies for Fast Moving Consumer Goods*. John Wiley & Sons, Ltd, Chichester, pp. 1–18. https://doi.org/10.1002/9780470753699.ch1.

De Pilli, T. 2020. Development of a vegetable oil and egg proteins edible film to replace preservatives and primary packaging of sweet baked goods. *Food Control* 114, 107273. https://doi.org/10.1016/j.foodcont.2020.107273.

De Pilli, T., Baiano, A., Lopriore, G., Russo, C. and Cappelletti, G. M. 2021. An overview on the environmental impact of food packaging. In: *Sustainable Innovations in Food Packaging*. Springer International Publishing, Cham, pp. 1–14. https://doi.org/10.1007/978-3-030-80936-2_1.

Debeaufort, F., Quezada-Gallo, J. A. and Voilley, A. 1998. Edible films and coatings: Tomorrow's packagings: A review. *Crit. Rev. Food Sci. Nutr.* 38(4), 299–313. https://doi.org/10.1080/10408699891274219.

do Val Siqueira, L., Arias, C. I. L. F., Maniglia, B. C. and Tadini, C. C. 2021. Starch-based biodegradable plastics: Methods of production, challenges and future perspectives. *Curr. Opin. Food Sci.* 38, 122–130. https://doi.org/10.1016/j.cofs.2020.10.020.

Döhler, N., Wellenreuther, C. and Wolf, A. 2022. Market dynamics of biodegradable bio-based plastics: Projections and linkages to European policies. *EFB Bioeconomy Journal* 2, 100028.

dos Santos, A. J., Oliveira Dalla Valentina, L. V., Hidalgo Schulz, A. A. and Tomaz Duarte, M. A. 2018. From obtaining to degradation of PHB: A literature review. Part II. *Ing. Cienc.* 14(27), 207–228. https://doi.org/10.17230/ingciencia.14.27.9.

Doshi, J. and Reneker, D. H. 1995. Electrospinning process and applications of electrospun fibers. *J. Electrostat.* 35(2–3), 151–160. https://doi.org/10.1016/0304-3886(95)00041-8.

Drosou, C. G., Krokida, M. K. and Biliaderis, C. G. 2017. Encapsulation of bioactive compounds through electrospinning/electrospraying and spray drying: A comparative assessment of food-related applications. *Drying Technol.* 35(2), 139–162. https://doi.org/10.1080/07373937.2016.1162797.

Ebnesajjad, S. 2013. *Plastic Films in Food Packaging*. Elsevier. https://doi.org/10.1016/C2012-0-00246-3.

Echegoyen, Y., Fabra, M. J., Castro-Mayorga, J. L., Cherpinski, A. and Lagaron, J. M. 2017. High throughput electro-hydrodynamic processing in food encapsulation and food packaging applications: Viewpoint. *Trends Food Sci. Technol.* 60, 71–79. https://doi.org/10.1016/j.tifs.2016.10.019.

Ellen MacArthur Foundation. 2016. *The New Plastics Economy: Rethinking the Future of Plastics*. Ellen MacArthur Foundation, United Kingdom.

Emin, M. A. and Schuchmann, H. P. 2017. A mechanistic approach to analyze extrusion processing of biopolymers by numerical, rheological, and optical methods. *Trends Food Sci. Technol.* 60, 88–95. https://doi.org/10.1016/j.tifs.2016.10.003.

European Bioplastics. 2016. Are bioplastics more expensive then conventional plastics? [WWW Document]. © European Bioplastics e.V.

European Bioplastics Organization. 2022. Bioplastics market development update 2022, European Bioplastics Org.

European Bioplastics Organization. 2020. Bioplastics market development update 2020, European Bioplastics Org.

European Bioplastics Organization. 2018a. What are bioplastics? Material types, terminology, and labels - an introduction, European Bioplastics.

European Bioplastics Organization. 2018b. What are bioplastics? Material types, terminology, and labels - an introduction, European Bioplastics.

European Food Safety Authority (EFSA) 2009. Guidelines on submission of a dossier for safety evaluation by the EFSA of active or intelligent substances present in active and intelligent materials and articles intended to come into contact with food. *EFSA J.* 7(8), 1208.

Eurostat. 2020. Packing waste statistics [WWW Document]. https://ec.europa.eu/eurostat/statisticsexplained/index.php/Packaging_waste_statistics#Time_series_of_packaging_waste_generation_ and_treatment (accessed 8.19.20).

FAO United Nations. 2017. Microplastics in fisheries and aquaculture. Status of knowledge on their occurrence and implications for aquatic organisms and food safety. https://www.fao.org/3/I7677E/I7677E.pdf (accessed 20 November 2023).

Figueroa-Lopez, K. J. 2021. Biodegradable mono and multilayer materials with antimicrobial capacity based on circular bioeconomy of application interest in food packaging. Doctoral Thesis, Universidad Politécnica de Valencia.

Figueroa-Lopez, K. J., Cabedo, L., Lagaron, J. M. and Torres-Giner, S. 2020a. Development of electrospun poly (3-hydroxybutyrate-co-3-hydroxyvalerate) monolayers containing eugenol and their application in multilayer antimicrobial food packaging. *Front. Nutr.* 7, 140. https://doi.org/10.3389/fnut.2020.00140.

Figueroa-Lopez, K. J., Torres-Giner, S., Angulo, I., Pardo-Figuerez, M., Escuin, J. M., Bourbon, A. I., Cabedo, L., Nevo, Y., Cerqueira, M. A. and Lagaron, J. M. 2020b. Development of active barrier multilayer films based on electrospun antimicrobial hot-tack food waste derived poly(3-hydroxybutyrate-co-3-hydroxyvalerate) and cellulose nanocrystal interlayers. *Nanomaterials (Basel)* 10(12), 2356. https://doi.org/10.3390/nano10122356.

Figueroa-Lopez, K. J., Castro-Mayorga, J. L., Andrade-Mahecha, M. M., Cabedo, L. and Lagaron, J. M. 2018. Antibacterial and barrier properties of gelatin coated by electrospun polycaprolactone ultrathin fibers containing black pepper oleoresin of interest in active food biopackaging applications. *Nanomaterials (Basel)* 8(4), 199. https://doi.org/10.3390/nano8040199.

Foschi, E. and Bonoli, A. 2019. The commitment of packaging industry in the framework of the European strategy for plastics in a circular economy. *ADM Sci.* 9(1), 18. https://doi.org/10.3390/admsci9010018.

Gnanasekaran, D. 2019. *Green Biopolymers and Their Nanocomposites, Materials Horizons: From Nature to Nanomaterials*. Springer, Singapore. https://doi.org/10.1007/978-981-13-8063-1.

González-García, S., Argiz, L., Míguez, P. and Gullón, B. 2018. Exploring the production of bio-succinic acid from apple pomace using an environmental approach. *Chem. Eng. J.* 350, 982–991. https://doi.org/10.1016/j.cej.2018.06.052.

Gradin, P., Howgate, P. G., Seldén, R. and Brown, R. A. 1989. 16 - Dynamic-mechanical properties. In: Allen, G. and Bevington, J. C. (Eds.), *Comprehensive Polymer Science and Supplements* (Vol. 16). Pergamon, Amsterdam, pp. 533–569. https://doi.org/10.1016/B978-0-08-096701-1.00053-7.

Grothe, J., Kaskel, S. and Leuteritz, A. 2012. Nanocomposites and hybrid materials. In: Matyjaszewski, K. and Möller, M. (Eds.), *Polymer Science: A Comprehensive Reference*. Elsevier, Amsterdam, pp. 177–209. https://doi.org/10.1016/B978-0-444-53349-4.00206-5.

Gu, F., Guo, J., Zhang, W., Summers, P. A. and Hall, P. 2017. From waste plastics to industrial raw materials: A life cycle assessment of mechanical plastic recycling practice based on a real-world case study. *Sci. Total Environ.* 601–602, 1192–1207. https://doi.org/10.1016/j.scitotenv.2017.05.278.

Guillard, V., Gaucel, S., Fornaciari, C., Angellier-Coussy, H., Buche, P. and Gontard, N. 2018. The next generation of sustainable food packaging to preserve our environment in a circular economy context. *Front. Nutr.* 5, 121. https://doi.org/10.3389/fnut.2018.00121.

Guna, V., Touchaleaume, F., Saulnier, B., Grohens, Y. and Reddy, N. 2023. Sustainable bioproducts through thermoplastic processing of wheat gluten and its blends. *J. Thermoplast. Compos. Mater.* 36(4), 1775–1806. https://doi.org/10.1177/08927057211048540.

Hemmati, F., Bahrami, A., Esfanjani, A. F., Hosseini, H., McClements, D. J. and Williams, L. 2021. Electrospun antimicrobial materials: Advanced packaging materials for food applications. *Trends Food Sci. Technol.* 111, 520–533. https://doi.org/10.1016/j.tifs.2021.03.014.

Höfer, R. and Selig, M. 2012. Green chemistry and green polymer chemistry. In: Matyjaszewski, K. and Möller, M. (Eds.), *Polymer Science: A Comprehensive Reference*. Elsevier, Amsterdam, pp. 5–14. https://doi.org/10.1016/B978-0-444-53349-4.00252-1.

Huang, T., Qian, Y., Wei, J. and Zhou, C. 2019. Polymeric antimicrobial food packaging and its applications. *Polymers (Basel)* 11(3). https://doi.org/10.3390/polym11030560.

Jahnke, A., Arp, H. P. H., Escher, B. I., Gewert, B., Gorokhova, E., Kühnel, D., Ogonowski, M., Potthoff, A., Rummel, C., Schmitt-Jansen, M., Toorman, E. and MacLeod, M. 2017. Reducing uncertainty and confronting ignorance about the possible impacts of weathering plastic in the marine environment. *Environ. Sci. Technol. Lett.* 4(3), 85–90. https://doi.org/10.1021/acs.estlett.7b00008.

Jambeck, J. R., Geyer, R., Wilcox, C., Siegler, T. R., Perryman, M., Andrady, A., Narayan, R. and Law, K. L. 2015. Plastic waste inputs from land into the ocean. *Science* 347, 768–771. https://doi.org/10.1126/science.1260352.

Jeremic, S., Milovanovic, J., Mojicevic, M., Skaro-Bogojevic, S. and Nikodinovic-Runic, J. 2020. Understanding bioplastic materials - current state and trends. *J. Serb. Chem. Soc.* 85(12), 1507–1538. https://doi.org/10.2298/JSC200720051J.

Joerger, R. D. 2007. Antimicrobial films for food applications: A quantitative analysis of their effectiveness. *Packaging Technol. Sci.* 20(4), 231–273. https://doi.org/10.1002/pts.774.

Kaseem, M., Hamad, K. and Deri, F. 2012. Thermoplastic starch blends: A review of recent works. *Polym. Sci. Ser. A* 54(2), 165–176. https://doi.org/10.1134/S0965545X1202006X.

Kerry, J. P., O'Grady, M. N. and Hogan, S. A. 2006. Past, current and potential utilisation of active and intelligent packaging systems for meat and muscle-based products: A review. *Meat Sci.* 74(1), 113–130. https://doi.org/10.1016/j.meatsci.2006.04.024.

Khan, B., Niazi, M. B. K., Samin, G. and Jahan, Z. 2017. Thermoplastic starch: A possible biodegradable food packaging material-a review. *J. Food Process Eng.* 40, e12447. https://doi.org/10.1111/jfpe.12447.

Kim, J., Oh, S., Cho, S. M., Jun, J. and Kwak, S. 2020. Oxygen barrier properties of polyketone/EVOH blend films and their resistance to moisture. *J. Appl. Polym. Sci.* 137(47). https://doi.org/10.1002/app.49537.

Koide, S. and Shi, J. 2007. Microbial and quality evaluation of green peppers stored in biodegradable film packaging. *Food Control* 18(9), 1121–1125. https://doi.org/10.1016/j.foodcont.2006.07.013.

Koshy, R. R., Mary, S. K., Thomas, S. and Pothan, L. A. 2015. Environment friendly green composites based on soy protein isolate – a review. *Food Hydrocoll.* 50, 174–192. https://doi.org/10.1016/j.foodhyd.2015.04.023.

Kourmentza, C. and Kornaros, M. 2016. Biotransformation of volatile fatty acids to polyhydroxyalkanoates by employing mixed microbial consortia: The effect of pH and carbon source. *Bioresour. Technol.* 222, 388–398. https://doi.org/10.1016/j.biortech.2016.10.014.

Kourmentza, C., Plácido, J., Venetsaneas, N., Burniol-Figols, A., Varrone, C., Gavala, H. N. and Reis, M. A. M. 2017. Recent advances and challenges towards sustainable polyhydroxyalkanoate (PHA) production. *Bioengineering (Basel)* 4(2), 55. https://doi.org/10.3390/bioengineering4020055.

Kuciel, S. and Liber-Knec, A. 2011. Biocomposites based on PHB filled with wood or kenaf fibers. *Polimery* 56(3), 218–223. https://doi.org/10.14314/polimery.2011.218.

Kulkarni, S. O., Kanekar, P. P., Jog, J. P., Sarnaik, S. S. and Nilegaonkar, S. S. 2015. Production of copolymer, poly (hydroxybutyrate-co-hydroxyvalerate) by Halomonas campisalis MCM B-1027 using agro-wastes. *Int. J. Biol. Macromol.* 72, 784–789. https://doi.org/10.1016/j.ijbiomac.2014.09.028.

Kundariya, N., Mohanty, S. S., Varjani, S., Hao Ngo, H., W C Wong, J., Taherzadeh, M. J., Chang, J. S., Yong Ng, H., Kim, S. H. and Bui, X. T. 2021. A review on integrated approaches for municipal solid waste for environmental and economical relevance: Monitoring tools, technologies, and strategic innovations. *Bioresour. Technol.* 342, 125982. https://doi.org/10.1016/J.BIORTECH.2021.125982.

Kuswandi, B. and Jumina. 2020. 12-active and intelligent packaging, safety, and quality controls. In: Siddiqui, M. W. (Ed.), *Fresh-Cut Fruits and Vegetables*. Elsevier, pp. 243-294. https://doi.org/10.1016/B978-0-12-816184-5.00012-4.

Kuswandi, B., Wicaksono, Y., Jayus, J., Abdullah, A., Heng, L. Y. and Ahmad, M. 2011. Smart packaging: Sensors for monitoring of food quality and safety. *Sens. Instrum. Food Qual. Saf.* 5, 137–146. https://doi.org/10.1007/s11694-011-9120-x.

La Fuente Arias, C. I., Kubo, M. T. K., Tadini, C. C. and Augusto, P. E. D. 2021. Bio-based multilayer films: A review of the principal methods of production and challenges. *Crit. Rev. Food Sci. Nutr.* 1-17. https://doi.org/10.1080/10408398.2021.1973955.

La Mantia, F. P. and Morreale, M. 2011. Green composites: A brief review. *Compos. A* 42(6), 579–588. https://doi.org/10.1016/j.compositesa.2011.01.017.

Lasut, M. T., Weber, M., Pangalila, F., Rumampuk, N. D. C., Rimper, J. R. T. S. L., Warouw, V., Kaunang, S. T. and Lott, C. 2018. From coral triangle to trash triangle—How the hot spot of global marine biodiversity is threatened by plastic waste. In: Mariacristina, C., Di Pace, E., Errico, M. E., Gentile, G., Montarsolo, A., Mossotti, R. and Avella, M. (Eds.), *Proceedings of the International Conference on Microplastic Pollution in the Mediterranean Sea*. Springer International Publishing, Cham, pp. 107–113.

Laycock, B., Nikolić, M., Colwell, J. M., Gauthier, E., Halley, P., Bottle, S. and George, G. 2017. Lifetime prediction of biodegradable polymers. *Prog. Polym. Sci.* 71, 144–189. https://doi.org/10.1016/j.progpolymsci.2017.02.004.

Lee, S. G., Choi, S.-S., Park, W. H. and Cho, D. 2003. Characterization of surface modified flax fibers and their biocomposites with PHB. *Macromol. Symp.* 197, 089-100. https://doi.org/10.1002/masy.200350709.

Lee, S. Y., Choi, J. and Wong, H. H. 1999. Recent advances in polyhydroxyalkanoate production by bacterial fermentation: Mini-review. *Int. J. Biol. Macromol.* 25(1-3), 31–36. https://doi.org/10.1016/S0141-8130(99)00012-4.

Leong, Y. K., Show, P. L., Lin, H. C., Chang, C. K., Loh, H.-S., Lan, J. C.-W. and Ling, T. C. 2016. Preliminary integrated economic and environmental analysis of polyhydroxyalkanoates (PHAs) biosynthesis. *Bioresour. Bioprocess.* 3(1), 41. https://doi.org/10.1186/s40643-016-0120-x.

Li, H., Qi, Y., Zhao, Y., Chi, J. and Cheng, S. 2019. Starch and its derivatives for paper coatings: A review. *Prog. Org. Coat.* 135, 213–227. https://doi.org/10.1016/j.porgcoat.2019.05.015.

Li, J., Yang, Z., Zhang, K., Liu, M., Liu, D., Yan, X., Si, M. and Shi, Y. 2021. Valorizing waste liquor from dilute acid pretreatment of lignocellulosic biomass by Bacillus megaterium B-10. *Ind. Crops Prod.* 161, 113160. https://doi.org/10.1016/j.indcrop.2020.113160.

Liu, H., Kumar, V., Jia, L., Sarsaiya, S., Kumar, D., Juneja, A., Zhang, Z., Sindhu, R., Binod, P., Bhatia, S. K. and Awasthi, M. K. 2021. Biopolymer poly-hydroxyalkanoates (PHA) production from apple industrial waste residues: A review. *Chemosphere* 284, 131427. https://doi.org/10.1016/j.chemosphere.2021.131427.

Liu, H., Xie, F., Yu, L., Chen, L. and Li, L. 2009. Thermal processing of starch-based polymers. *Prog. Polym. Sci.* 34(12), 1348–1368. https://doi.org/10.1016/j.progpolymsci.2009.07.001.

Liu, W., Wang, Z., Liu, J., Dai, B., Hu, S., Hong, R., Xie, H., Li, Z., Chen, Y. and Zeng, G. 2020. Preparation, reinforcement and properties of thermoplastic starch film by film blowing. *Food Hydrocoll.* 108, 106006. https://doi.org/10.1016/j.foodhyd.2020.106006.

Lorini, L., Munarin, G., Salvatori, G., Alfano, S., Pavan, P., Majone, M. and Valentino, F. 2022. Sewage sludge as carbon source for polyhydroxyalkanoates: A holistic approach at pilot scale level. *J. Clean. Prod.* 354, 131728. https://doi.org/10.1016/j.jclepro.2022.131728.

Ma, H. and Joo, C. W. 2011. Investigation of jute-lignin-poly (3-hydroxybutyrate) hybrid biodegradable composites with low water absorption. *Fibers Polym.* 12(3), 310–315. https://doi.org/10.1007/s12221-011-0310-2.

Macedo, J. de S., Costa, M. F., Tavares, M. I. B. and Thiré, R. M. S. M. 2010. Preparation and characterization of composites based on polyhydroxybutyrate and waste powder from coconut fibers processing. *Polym. Eng. Sci.* 50(7), 1466–1475. https://doi.org/10.1002/pen.21669.

Magalhães, S., Alves, L., Medronho, B., Romano, A. and Rasteiro, M. D. G. 2020. Microplastics in ecosystems: From current trends to bio-based removal strategies. *Molecules* 25(17), 3954. https://doi.org/10.3390/molecules25173954.

Mangaraj, S., Goswami, T. K. and Mahajan, P. V. 2009. Applications of plastic films for modified atmosphere packaging of fruits and vegetables: A review. *Food Eng. Rev.* 1(2), 133–158. https://doi.org/10.1007/s12393-009-9007-3.

Mangaraj, S., Yadav, A., Bal, L. M., Dash, S. K. and Mahanti, N. K. 2019. Application of biodegradable polymers in food packaging industry: A comprehensive review. *J. Packag. Technol. Res.* 3(1), 77–96. https://doi.org/10.1007/s41783-018-0049-y.

Mannina, G., Presti, D., Montiel-Jarillo, G., Carrera, J. and Suárez-Ojeda, M. E. 2020. Recovery of polyhydroxyalkanoates (PHAs) from wastewater: A review. *Bioresour. Technol.* 297, 122478. https://doi.org/10.1016/j.biortech.2019.122478.

Market Prospects. 2021. What are the common plastics manufacturing processes? [WWW Document]. https://www.market-prospects.com/articles/plastic-product-manufacturing-processes.

Martínez Sanz, M. 2013. Bacterial cellulose nanowhiskers to enhance the properties of plastics and bioplastics of interest in food packaging (Thesis doctoral). Universitat Politècnica de València. https://doi.org/10.4995/Thesis/10251/30312.

Martucci, J. F. and Ruseckaite, R. A. 2010. Biodegradable three-layer film derived from bovine gelatin. *J. Food Eng.* 99(3), 377–383. https://doi.org/10.1016/j.jfoodeng.2010.02.023.

Mastrolia, C., Giaquinto, D., Gatz, C., Pervez, Md., Hasan, S., Zarra, T., Li, C.-W., Belgiorno, V. and Naddeo, V. 2022. Plastic pollution: Are Bioplastics the right solution? *Water* 14(22), 3596. https://doi.org/10.3390/w14223596.

McChalicher, C. W. J. and Srienc, F. 2007. Investigating the structure-property relationship of bacterial PHA block copolymers. *J. Biotechnol.* 132(3), 296–302. https://doi.org/10.1016/j.jbiotec.2007.04.014.

McKeen, L. W. 2012. Production of films, containers, and membranes. In: McKeen, L. W. (Ed.), *Permeability Properties of Plastics and Elastomers*. Elsevier, Oxford, pp. 39–58. https://doi.org/10.1016/B978-1-4377-3469-0.10003-7.

Melendez-Rodriguez, B., Castro-Mayorga, J. L., Reis, M. A. M., Sammon, C., Cabedo, L., Torres-Giner, S. and Lagaron, J. M. 2018. Preparation and characterization of electrospun food biopackaging films of poly(3-hydroxybutyrate-co-3-hydroxyvaler ate) derived from fruit pulp biowaste. *Front. Sustain. Food Syst.* 2, 1–16. https://doi .org/10.3389/fsufs.2018.00038.

Menon, A. and Lyng, J. G. 2021. Circular bioeconomy solutions: Driving anaerobic digestion of waste streams towards production of high value medium chain fatty acids. *Rev. Environ. Sci. Biotechnol.* 20(1), 189–208. https://doi.org/10.1007/s11157 -020-09559-5.

Milios, L., Holm Christensen, L., McKinnon, D., Christensen, C., Rasch, M. K. and Hallstrøm Eriksen, M. 2018. Plastic recycling in the Nordics: A value chain market analysis. *Waste Manag.* 76, 180–189. https://doi.org/10.1016/j.wasman.2018.03.034.

Mobley, D. P. 1994. *Plastics from Microbes: Microbial Synthesis of Polymers and Polymer Precursors*. Hanser Publishers, Germany.

Modi, S. J. 2010. *Assessing the Feasibility of Poly-(3-Hydroxybutyrate-co-3-Hydroxyvale rate) (PHBV) and Poly-(Lactic Acid) for Potential Food Packaging Applications*. Master thesis from the Ohio State University. https://etd.ohiolink.edu/acprod/odb_etd/ws/send_file/send?accession=osu1268921056&disposition=inline (accessed 20 November 2023).

Moretto, G., Russo, I., Bolzonella, D., Pavan, P., Majone, M. and Valentino, F. 2020. An urban biorefinery for food waste and biological sludge conversion into polyhydroxyalkanoates and biogas. *Water Res.* 170, 115371. https://doi.org/10 .1016/j.watres.2019.115371.

Morgan-Sagastume, F., Hjort, M., Cirne, D., Gérardin, F., Lacroix, S., Gaval, G., Karabegovic, L., Alexandersson, T., Johansson, P., Karlsson, A., Bengtsson, S., Arcos-Hernández, M. V., Magnusson, P. and Werker, A. 2015. Integrated production of polyhydroxyalkanoates (PHAs) with municipal wastewater and sludge treatment at pilot scale. *Bioresour. Technol.* 181, 78–89. https://doi.org/10.1016/j.biortech.2015 .01.046.

Morseletto, P. 2020. Targets for a circular economy. *Resour. Conserv. Recycl.* 153, 104553. https://doi.org/10.1016/j.resconrec.2019.104553.

Müller, P. and Schmid, M. 2019. Intelligent packaging in the food sector: A brief overview. *Foods* 8(1), 16. https://doi.org/10.3390/foods8010016.

Mutlu, G., Calamak, S., Ulubayram, K. and Guven, E. 2018. Curcumin-loaded electrospun PHBV nanofibers as potential wound-dressing material. *J. Drug Deliv. Sci. Technol.* 43, 185–193. https://doi.org/10.1016/j.jddst.2017.09.017.

Namratha, S., Sreejit, V. and Preetha, R. 2020. Fabrication and evaluation of physicochemical properties of probiotic edible film based on pectin-alginate-casein composite. *Int. J. Food Sci. Technol.* 55(4), 1497–1505. https://doi.org/10 .1111/ijfs.14550.

Nazmi, N. N. M., Isa, M. I. N. and Sarbon, N. M. 2020. Characterisation of biodegradable protein based films from gelatin alternative: A review. *Int. Food Res. J.* 27, 971–987.

Nemat, B., Razzaghi, M., Bolton, K. and Rousta, K. 2020. The potential of food packaging attributes to influence consumers' decisions to sort waste. *Sustainability* 12(6), 2234. https://doi.org/10.3390/su12062234.

Nesic, A., Castillo, C., Castaño, P., Cabrera-Barjas, G. and Serrano, J. 2020. Bio-based packaging materials. In: *Biobased Products and Industries*. Elsevier, pp. 279–309. https://doi.org/10.1016/B978-0-12-818493-6.00008-7.

Obi, F., Ugwuishiwu, B. and Nwakaire, J. 2016. Agricultural waste concept, generation, utilization and management. *Niger. J. Technol.* 35(4), 957. https://doi.org/10.4314/njt.v35i4.34.

Ospina-Betancourth, C., Echeverri, S., Rodriguez-Gonzalez, C., Wist, J., Combariza, M. Y. and Sanabria, J. 2022. Enhancement of PHA production by a mixed microbial culture using VFA obtained from the fermentation of wastewater from yeast industry. *Fermentation* 8(4), 180. https://doi.org/10.3390/fermentation8040180.

Ozdemir, M. and Floros, J. D. 2004. Active food packaging technologies. *Crit. Rev. Food Sci. Nutr.* 44(3), 185–193. https://doi.org/10.1080/10408690490441578.

Pagliano, G., Galletti, P., Samorì, C., Zaghini, A. and Torri, C. 2021. Recovery of polyhydroxyalkanoates from single and mixed microbial cultures: A review. *Front. Bioeng. Biotechnol.* 9, 624021. https://doi.org/10.3389/fbioe.2021.624021.

Payne, J., McKeown, P. and Jones, M. D. 2019. A circular economy approach to plastic waste. *Polym. Degrad. Stab.* 165, 170–181. https://doi.org/10.1016/j.polymdegradstab.2019.05.014.

Pellá, M. C. G., Silva, O. A., Pellá, M. G., Beneton, A. G., Caetano, J., Simões, M. R. and Dragunski, D. C. 2020. Effect of gelatin and casein additions on starch edible biodegradable films for fruit surface coating. *Food Chem.* 309, 125764. https://doi.org/10.1016/j.foodchem.2019.125764.

Pellicer, E., Nikolic, D., Sort, J., Baró, M. D., Zivic, F., Grujovic, N., Grujic, R. and Pelemis, S. 2017. *Advances in Applications of Industrial Biomaterials, Advances in Applications of Industrial Biomaterials*. Springer International Publishing, Cham. https://doi.org/10.1007/978-3-319-62767-0.

Pereira, J. R., Araújo, D., Freitas, P., Marques, A. C., Alves, V. D., Sevrin, C., Grandfils, C., Fortunato, E., Reis, M. A. M. and Freitas, F. 2021. Production of medium-chain-length polyhydroxyalkanoates by Pseudomonas chlororaphis subsp. aurantiaca: Cultivation on fruit pulp waste and polymer characterization. *Int. J. Biol. Macromol.* 167, 85–92. https://doi.org/10.1016/j.ijbiomac.2020.11.162.

Perez-Masia, R., Lopez-Rubio, A., Fabra, M. J. and Lagaron, J. M. 2013. Biodegradable polyester-based heat management materials of interest in refrigeration and smart packaging coatings. *J. Appl. Polym. Sci.* 130(5), 3251–3262. https://doi.org/10.1002/app.39555.

Perumalla, A. V. S. and Hettiarachchy, N. S. 2011. Green tea and grape seed extracts – potential applications in food safety and quality. *Food Res. Int.* 44(4), 827–839. https://doi.org/10.1016/j.foodres.2011.01.022.

Pinto, L., Bonifacio, M. A., De Giglio, E., Santovito, E., Cometa, S., Bevilacqua, A. and Baruzzi, F. 2021. Biopolymer hybrid materials: Development, characterization, and food packaging applications. *Food Packag. Shelf Life* 28, 100676. https://doi.org/10.1016/j.fpsl.2021.100676.

Pirsaheb, M., Mohamadi, S., Rahmatabadi, S., Hossini, H. and Motteran, F. 2018. Simultaneous wastewater treatment and biogas production using integrated anaerobic baffled reactor granular activated carbon from baker's yeast wastewater. *Environ. Technol.* 39(21), 2724–2735. https://doi.org/10.1080/09593330.2017.1365939.

PlasticsEurope, A. of P.M. 2020. Plastics – the Facts 2020. An analysis of European plastics production, demand and waste data. PlasticEurope, 1–64.

Porta, R., Sabbah, M. and Di Pierro, P. 2020. Biopolymers as food packaging materials. *Int. J. Mol. Sci.* 21(14), 1–3. https://doi.org/10.3390/ijms21144942.

Quiles-Carrillo, L., Montanes, N., Sammon, C., Balart, R. and Torres-Giner, S. 2018. Compatibilization of highly sustainable polylactide/almond shell flour composites by reactive extrusion with maleinized linseed oil. *Ind. Crops Prod.* 111, 878–888. https://doi.org/10.1016/j.indcrop.2017.10.062.

Rai, M., Ingle, A. P., Gupta, I., Pandit, R., Paralikar, P., Gade, A., Chaud, M. V. and dos Santos, C. A. 2019. Smart nanopackaging for the enhancement of food shelf life. *Environ. Chem. Lett.* 17(1), 277–290. https://doi.org/10.1007/s10311-018-0794-8.

Rai, P., Mehrotra, S., Priya, S., Gnansounou, E. and Sharma, S. K. 2021. Recent advances in the sustainable design and applications of biodegradable polymers. *Bioresour. Technol.* 325, 124739. https://doi.org/10.1016/j.biortech.2021.124739.

RameshKumar, S., Shaiju, P., O'Connor, K. E. and Babu P., R. 2020. Bio-based and biodegradable polymers - State-of-the-art, challenges and emerging trends. *Curr. Opin. Green Sustain. Chem.* 21, 75–81. https://doi.org/10.1016/j.cogsc.2019.12.005.

Rani, S. and Kumar, R. 2019. A review on material and antimicrobial properties of soy protein isolate film. *J. Polym. Environ.* 27(8), 1613–1628. https://doi.org/10.1007/s10924-019-01456-5.

Rebocho, A. T., Pereira, J. R., Freitas, F., Neves, L. A., Alves, V. D., Sevrin, C., Grandfils, C. and Reis, M. A. M. 2019. Production of medium-chain length polyhydroxyalkanoates by Pseudomonas citronellolis grown in apple pulp waste. *Appl. Food Biotechnol.* 6, 71–82. https://doi.org/10.22037/afb.v6i1.21793.

Reddy, C. S. K., Ghai, R., Rashmi and Kalia, V. C. 2003. Polyhydroxyalkanoates: An overview. *Bioresour. Technol.* 87(2), 137–146. https://doi.org/10.1016/S0960-8524(02)00212-2.

Reis, M., Albuquerque, M., Villano, M. and Majone, M. 2011. Mixed culture processes for polyhydroxyalkanoate production from agro-industrial surplus/wastes as feedstocks. In: Moo-Young, M. (Ed.), *Comprehensive Biotechnology*. Elsevier, Burlington, pp. 669–683. https://doi.org/10.1016/B978-0-08-088504-9.00464-5.

Requiso, P. J., Nayve, F. R. P., Alfafara, C. G., Ventura, R. L. G., Escobar, E. C. and Ventura, J.-R. S. 2018. Agricultural residue feedstock selection for polyhydroxyalkanoates production using AHP-GRA. *Philipp. J. Sci.* 147, 693–704.

Rodrigues, M. O., Abrantes, N., Gonçalves, F. J. M., Nogueira, H., Marques, J. C. and Gonçalves, A. M. M. 2018. Spatial and temporal distribution of microplastics in water and sediments of a freshwater system (Antuã River, Portugal). *Sci. Total Environ.* 633, 1549–1559. https://doi.org/10.1016/j.scitotenv.2018.03.233.

Romeo, T., Pietro, B., Pedà, C., Consoli, P., Andaloro, F. and Fossi, M. C. 2015. First evidence of presence of plastic debris in stomach of large pelagic fish in the Mediterranean Sea. *Mar. Pollut. Bull.* 95(1), 358–361. https://doi.org/10.1016/j.marpolbul.2015.04.048.

Rosenboom, J. G., Langer, R. and Traverso, G. 2022. Bioplastics for a circular economy. *Nat. Rev. Mater.* 7(2), 117–137. https://doi.org/10.1038/s41578-021-00407-8.

Said, N. S., Howell, N. K. and Sarbon, N. M. 2021. A review on potential use of gelatin-based film as active and smart biodegradable films for food packaging application. *Food Rev. Int.* 0, 1–23. https://doi.org/10.1080/87559129.2021.1929298.

Sameen, D. E., Ahmed, S., Lu, R., Li, R., Dai, J., Qin, W., Zhang, Q., Li, S. and Liu, Y. 2021. Electrospun nanofibers food packaging: Trends and applications in food systems. *Crit. Rev. Food Sci. Nutr.* 0, 1–14. https://doi.org/10.1080/10408398.2021.1899128.

Sanchez, J.-Y., Iojoiu, C., Alloin, F., Guindet, J. and Leprêtre, J.-C. 2009. Fuel cells – proton-exchange membrane fuel cells | Membranes: Non-fluorinated. In: Garche, J. (Ed.),

Encyclopedia of Electrochemical Power Sources. Elsevier, Amsterdam, pp. 700–715. https://doi.org/10.1016/B978-044452745-5.00887-X.

Sánchez-Safont, E. L., Aldureid, A., Lagarón, J. M., Gámez-Pérez, J. and Cabedo, L. 2018. Biocomposites of different lignocellulosic wastes for sustainable food packaging applications. *Compos. B Eng.* 145, 215–225. https://doi.org/10.1016/j.compositesb.2018.03.037.

Sawant, S. S., Salunke, B. K. and Kim, B. S. 2015. Degradation of corn stover by fungal cellulase cocktail for production of polyhydroxyalkanoates by moderate halophile Paracoccus sp. LL1. *Bioresour. Technol.* 194, 247–255. https://doi.org/10.1016/j.biortech.2015.07.019.

Scheurer, M. and Bigalke, M. 2018. Microplastics in Swiss floodplain soils. *Environ. Sci. Technol.* 52(6), 3591–3598. https://doi.org/10.1021/acs.est.7b06003.

Shaikh, S., Yaqoob, M. and Aggarwal, P. 2021. An overview of biodegradable packaging in food industry. *Curr. Res Food Sci.* 4, 503–520. https://doi.org/10.1016/j.crfs.2021.07.005.

Shalom, T. Ben, Belsey, S., Chasnitsky, M. and Shoseyov, O. 2021. Cellulose nanocrystals and corn zein oxygen and water vapor barrier biocomposite films. *Nanomaterials (Basel)* 11(1), 1–16. https://doi.org/10.3390/nano11010247.

Shanks, R. A., Hodzic, A. and Wong, S. 2004. Thermoplastic biopolyester natural fiber composites. *J. Appl. Polym. Sci.* 91(4), 2114–2121. https://doi.org/10.1002/app.13289.

Sharma, R., Jafari, S. M. and Sharma, S. 2020. Antimicrobial bio-nanocomposites and their potential applications in food packaging. *Food Control* 112, 107086. https://doi.org/10.1016/j.foodcont.2020.107086.

Sidwell, J. A. 1992. *Food Contact Polymeric Materials*. iSmithers Rapra Publishing, United Kingdom.

Silva, F., Matos, M., Pereira, B., Ralo, C., Pequito, D., Marques, N., Carvalho, G. and Reis, M. A. M. 2022. An integrated process for mixed culture production of 3-hydroxyhexanoate-rich polyhydroxyalkanoates from fruit waste. *Chem. Eng. J.* 427, 131908. https://doi.org/10.1016/j.cej.2021.131908.

Sirohi, R., Prakash Pandey, J., Kumar Gaur, V., Gnansounou, E. and Sindhu, R. 2020. Critical overview of biomass feedstocks as sustainable substrates for the production of polyhydroxybutyrate (PHB). *Bioresour. Technol.* 311, 123536. https://doi.org/10.1016/j.biortech.2020.123536.

Soltani Firouz, M., Mohi-Alden, K. and Omid, M. 2021. A critical review on intelligent and active packaging in the food industry: Research and development. *Food Res. Int.* 141, 110113. https://doi.org/10.1016/j.foodres.2021.110113.

Spasojević, L., Katona, J., Bučko, S., Savić, S. M., Petrović, L., Milinković Budinčić, J., Tasić, N., Aidarova, S. and Sharipova, A. 2019. Edible water barrier films prepared from aqueous dispersions of zein nanoparticles. *LWT* 109, 350–358. https://doi.org/10.1016/j.lwt.2019.04.038.

Spierling, S., Knüpffer, E., Behnsen, H., Mudersbach, M., Krieg, H., Springer, S., Albrecht, S., Herrmann, C. and Endres, H.-J. 2018. Bio-based plastics - a review of environmental, social and economic impact assessments. *J. Clean. Prod.* 185, 476–491. https://doi.org/10.1016/j.jclepro.2018.03.014.

Statista. 2021. Plastic waste in Europe - Statistics & facts [WWW Document]. Available at: https://www.statista.com/topics/5141/plastic-waste-in-europe/#topicHeader__wrapper (accessed 7.13.22).

Subbiah, T., Bhat, G. S., Tock, R. W., Parameswaran, S. and Ramkumar, S. S. 2005. Electrospinning of nanofibers. *J. Appl. Polym. Sci.* 96(2), 557–569. https://doi.org/10.1002/app.21481.

Sudesh, K., Abe, H. and Doi, Y. 2000. Synthesis, structure and properties of polyhydroxyalkanoates: Biological polyesters. *Prog. Polym. Sci.* 25(10), 1503-1555. https://doi.org/10.1016/S0079-6700(00)00035-6.

Tang, X. Z., Kumar, P., Alavi, S. and Sandeep, K. P. 2012. Recent advances in biopolymers and biopolymer-based nanocomposites for food packaging materials. *Crit. Rev. Food Sci. Nutr.* 52(5), 426–442. https://doi.org/10.1080/10408398.2010.500508.

Thakur, S., Chaudhary, J., Sharma, B., Verma, A., Tamulevicius, S. and Thakur, V. K. 2018. Sustainability of bioplastics: Opportunities and challenges. *Curr. Opin. Green Sustain. Chem.* 13, 68–75. https://doi.org/10.1016/j.cogsc.2018.04.013.

Tharanathan, R. N. 2003. Biodegradable films and composite coatings: Past, present and future. *Trends Food Sci. Technol.* 14(3), 71–78. https://doi.org/10.1016/S0924-2244(02)00280-7.

Tongnuanchan, P., Benjakul, S. and Prodpran, T. 2014. Structural, morphological and thermal behaviour characterisations of fish gelatin film incorporated with basil and citronella essential oils as affected by surfactants. *Food Hydrocoll.* 41, 33–43. https://doi.org/10.1016/j.foodhyd.2014.03.015.

Torres-Giner, S., Busolo, M., Cherpinski, A. and Lagaron, J. M. 2018a. Chapter 10 electrospinning in the packaging industry. In: *Electrospinning: From Basic Research to Commercialization*. The Royal Society of Chemistry, pp. 238–260. https://doi.org/10.1039/9781788012942-00238.

Torres-Giner, S., Gil, L., Pascual-Ramírez, L. and Garde-Belza, J. 2018b. *Packaging Food Waste Reduction, 1990–2009*. https://doi.org/10.4324/9781351019422-120054093.

Torres-Giner, S., Montanes, N., Fombuena, V., Boronat, T. and Sanchez-Nacher, L. 2018c. Preparation and characterization of compression-molded green composite sheets made of poly(3-hydroxybutyrate) reinforced with long pita fibers. *Adv. Polym. Technol.* 37(5), 1305–1315. https://doi.org/10.1002/adv.21789.

Torres-Giner, S., Figueroa-Lopez, K. J., Melendez-Rodriguez, B., Prieto, C., Pardo-Figuerez, M. and Lagaron, J. M. 2021. Emerging trends in Biopolymers for food packaging. In: Athanassiou, A. (Ed.), *Sustainable Food Packaging Technology*. John Wiley & Sons, Ltd, pp. 1–33. https://doi.org/10.1002/9783527820078.ch1.

Tsang, Y. F., Kumar, V., Samadar, P., Yang, Y., Lee, J., Ok, Y. S., Song, H., Kim, K. H., Kwon, E. E. and Jeon, Y. J. 2019. Production of bioplastic through food waste valorization. *Environ. Int.* 127, 625–644. https://doi.org/10.1016/j.envint.2019.03.076.

Tulsyan, G., Richter, C. and Diaz, C. A. 2017. Oxygen scavengers based on titanium oxide nanotubes for packaging applications. *Packaging Technol. Sci.* 30(6), 251–256.

Uddin, M. A., Afroj, S., Hasan, T., Carr, C., Novoselov, K. S. and Karim, N. 2022. Environmental impacts of personal protective clothing used to combat COVID-19. *Adv. Sustain. Syst.* 6(1), 2100176. https://doi.org/10.1002/adsu.202100176.

United Nations. 2021. The sustainable development goals report, United Nations publication issued by the department of economic and social affairs. https://unstats.un.org/sdgs/report/2021/The-Sustainable-Development-Goals-Report-2021.pdf (accessed 20 November 2023).

Ur Rahim, H., Qaswar, M., Uddin, M., Giannini, C., Herrera, M. L. and Rea, G. 2021. Nano-enable materials promoting sustainability and resilience in modern agriculture. *Nanomaterials (Basel)* 11(8), 2068. https://doi.org/10.3390/nano11082068.

Valentino, F., Gottardo, M., Micolucci, F., Pavan, P., Bolzonella, D., Rossetti, S. and Majone, M. 2018. Organic fraction of municipal solid waste recovery by conversion into added-value polyhydroxyalkanoates and biogas. *ACS Sustainable Chem. Eng.* 6(12), 16375–16385. https://doi.org/10.1021/acssuschemeng.8b03454.

Valentino, F., Lorini, L., Pavan, P. and Majone, M. 2021. Development of a biorefinery platform for urban waste valorisation into biogas and added-value products. *Chem. Eng. Trans.* 86, 13–18. https://doi.org/10.3303/CET2186003.

Valentino, F., Morgan-Sagastume, F., Campanari, S., Villano, M., Werker, A. and Majone, M. 2017. Carbon recovery from wastewater through bioconversion into biodegradable polymers. *N. Biotechnol.* 37(A), 9–23. https://doi.org/10.1016/j.nbt.2016.05.007.

van Eygen, E., Feketitsch, J., Laner, D., Rechberger, H. and Fellner, J. 2017. Comprehensive analysis and quantification of national plastic flows: The case of Austria. *Resour. Conserv. Recycl.* 117, 183–194. https://doi.org/10.1016/j.resconrec.2016.10.017.

van Vliet, L. and Jensen, G. 2016. Food contact materials and chemical contamination. https://www.env-health.org/IMG/pdf/15022016_-_heal_briefing_fcm_final.pdf (accessed 20 November 2023).

Vázquez-González, Y., Prieto, C., Filizoglu, M. F., Ragazzo-Sánchez, J. A., Calderón-Santoyo, M., Furtado, R. F., Cheng, H. N., Biswas, A. and Lagaron, J. M. 2021. Electrosprayed cashew gum microparticles for the encapsulation of highly sensitive bioactive materials. *Carbohydr. Polym.* 264, 118060. https://doi.org/10.1016/j.carbpol.2021.118060.

Vilela, C., Kurek, M., Hayouka, Z., Röcker, B., Yildirim, S., Antunes, M. D. C., Nilsen-Nygaard, J., Pettersen, M. K. and Freire, C. S. R. 2018. A concise guide to active agents for active food packaging. *Trends Food Sci. Technol.* 80, 212–222. https://doi.org/10.1016/j.tifs.2018.08.006.

Wang, J., Liu, S., Huang, J. and Qu, Z. 2021. A review on polyhydroxyalkanoate production from agricultural waste biomass: Development, advances, circular approach, and challenges. *Bioresour. Technol.* 342, 126008 https://doi.org/10.1016/j.biortech.2021.126008.

Wang, Z. F., Peng, Z., Li, S. D., Lin, H., Zhang, K. X., She, X. D. and Fu, X. 2009. The impact of esterification on the properties of starch/natural rubber composite. *Compos. Sci. Technol.* 69(11–12), 1797–1803. https://doi.org/10.1016/j.compscitech.2009.04.018.

Wei, H., Seidi, F., Zhang, T., Jin, Y. and Xiao, H. 2021. Ethylene scavengers for the preservation of fruits and vegetables: A review. *Food Chem.* 337, 127750.

Wellenreuther, C., Wolf, A. and Zander, N. 2022. Cost competitiveness of sustainable bioplastic feedstocks – A Monte Carlo analysis for polylactic acid. *Clean Eng. Technol.* 6, 100411. https://doi.org/10.1016/j.clet.2022.100411.

Williams, C. K. and Hillmyer, M. A. 2008. Polymers from renewable resources: A perspective for a special issue of polymer reviews. *Polym. Rev.* 48(1), 1–10. https://doi.org/10.1080/15583720701834133.

Wong, S., Shanks, R. and Hodzic, A. 2002. Properties of poly(3-hydroxybutyric acid) composites with flax fibres modified by plasticiser absorption. *Macromol. Mater. Eng.* 287(10), 647–655. https://doi.org/10.1002/1439-2054(200210)287:10<647::AID-MAME647>3.0.CO;2-7.

Wright, S. L., Thompson, R. C. and Galloway, T. S. 2013. The physical impacts of microplastics on marine organisms: A review. *Environ. Pollut.* 178, 483–492 https://doi.org/10.1016/j.envpol.2013.02.031.

Wu, Y. and Clark, R. L. 2008. Electrohydrodynamic atomization: A versatile process for preparing materials for biomedical applications. *J. Biomater. Sci. Polym. Ed.* 19(5), 573–601. https://doi.org/10.1163/156856208784089616.

Yadav, B., Chavan, S., Atmakuri, A., Tyagi, R. D. and Drogui, P. 2020. A review on recovery of proteins from industrial wastewaters with special emphasis on PHA production process: Sustainable circular bioeconomy process development. *Bioresour. Technol.* 317, 124006. https://doi.org/10.1016/j.biortech.2020.124006.

Yadav, N., Nain, L. and Khare, S. K. 2021. Studies on the degradation and characterization of a novel metal-free polylactic acid synthesized via lipase-catalyzed polymerization: A step towards curing the environmental plastic issue. *Environ. Technol. Innov.* 24, 101845. https://doi.org/10.1016/j.eti.2021.101845.

Yeo, J. C. C., Muiruri, J. K., Thitsartarn, W., Li, Z. and He, C. 2018. Recent advances in the development of biodegradable PHB-based toughening materials: Approaches, advantages and applications. *Mater. Sci. Eng. C Mater. Biol. Appl.* 92, 1092–1116. https://doi.org/10.1016/j.msec.2017.11.006.

Yin, F., Li, D., Ma, X., Li, J. and Qiu, Y. 2020. Poly(3-hydroxybutyrate-3-hydroxyvalerate) production from pretreated waste lignocellulosic hydrolysates and acetate co-substrate. *Bioresour. Technol.* 316, 123911. https://doi.org/10.1016/j.biortech.2020.123911.

Yurteri, C. U., Hartman, R. P. A. and Marijnissen, J. C. M. 2010. Producing pharmaceutical particles via electrospraying with an emphasis on nano and nano structured particles - a review. *KONA Powder Part. J.* 28, 91–115. https://doi.org/10.14356/kona.2010010.

Zhang, C., Li, Y., Wang, P. and Zhang, H. 2020. Electrospinning of nanofibers: Potentials and perspectives for active food packaging. *Compr. Rev. Food Sci. Food Saf.* 19(2), 479–502.

Zubris, K. A. V. and Richards, B. K. 2005. Synthetic fibers as an indicator of land application of sludge. *Environ. Pollut.* 138(2), 201–211. https://doi.org/10.1016/j.envpol.2005.04.013.

Chapter 12

Developing polyphenols from agricultural wastes

J. Echave[1], M. Fraga-Corral[1,2], S. Seyyedi-Mansour[1], L. Cassani[1], P. Garcia-Oliveira[1], Hui Cao[1], J. Simal-Gandara[1,2] and M. A. Prieto[1,2], University of Vigo, Spain; and Jianbo Xiao[1,3], University of Vigo, Spain and Jiangsu University, China

1 Introduction

Phenolic compounds (PCs) are a large and diverse group of plant secondary metabolites characterized as containing at least a benzene ring and one or more hydroxyl radicals (Albuquerque et al., 2021). PCs are synthesized as a response to abiotic (e.g. drought and ultraviolet radiation) or biotic (e.g. competition, predation and infection) stress factors (Kessler and Heil, 2011). PC may display various degrees of polymerization and/or glycosylation and are found in fruits, herbs and vegetables. PCs accumulate particularly in roots, leaves and fruits though they appear in differing forms and amounts in all parts of plants, even in the same species (Jiménez et al., 2018; Liu et al., 2011). More

[1] Universidade de Vigo, Nutrition and Bromatology Group, Department of Analytical Chemistry and Food Science, Faculty of Science, 32004 Ourense, Spain.
[2] Correspondence: M. Fraga-Corral (mfraga@uvigo.es), J. Simal-Gandara (jsimal@uvigo.es) and M. A. Prieto (mpriet@uvigo.es).
[3] International Joint Research Laboratory of Intelligent Agriculture and Agri-products Processing, Jiangsu University, Zhenjiang, China.

http://dx.doi.org/10.19103/AS.2023.0120.21

than 10 000 different structures of PCs have been described to date, indicating the great structural diversity of this chemical group (Kennedy and Wightman, 2011). Depending on their structure, PCs are classified into different groups such as phenolic acids (PAs), flavonoids, tannins, stilbenes or lignans.

The use of PC for functional applications has a long history, starting with the use of plants in traditional medicine (Garcia-Oliveira et al., 2021). From the 1990s, interest has focused particularly on the antioxidant properties of PC which have been linked to a range of health benefits such as antioxidant, antidiabetic, anti-inflammatory or antitumoral properties (Oliveira et al., 2019; Pinto et al., 2016; Pires et al., 2018). They have also been linked to improving animal health (Maria Fraga-Corral et al., 2020).

The numerous properties associated with PCs mean they have a wide range of potential applications, e.g. as natural food preservatives or cosmetic ingredients (thanks to their antioxidant and antimicrobial activities) (Moein and Moein, 2010; Wu et al., 2019), as well as flavorings or colorants, especially in the case of anthocyanins (López et al., 2019). They are used in a wide range of sectors such as the food, animal feed, pharmaceutical, cosmetic and nutraceutical product industries (Burri et al., 2017).

Current global agricultural production and food processing generates a diverse number of by-products (seeds, peel, leaves, pomace etc.) that result in large amounts of organic waste (OW). These volumes are only likely to increase as the global population increases (to an estimated 9 billion in 2050). It has been estimated by the FAO-led Global Initiative on Food Loss and Waste Reduction (known as SAVE FOOD) that at least one-third of all food produced globally (an equivalent to at least 1300 million metric tons) is wasted every year with fresh horticulture commodities, the biggest category as measured by weight (Ishangulyyev et al., 2019). It has been estimated that 800 000 tons of fruit and vegetable residues are produced annually (Baiano, 2014). This has led to growing legislative intervention to tackle the problem of food waste, e.g. in the European Union (European Commission, 2005; Directive 2006/12/EC of the European Parliament and of the Council of 5 April 2006 on Waste, 2006). In order to implement sustainable management of biomass, the European Commission has developed a legal framework mostly based on Directive 2008/98/EC on waste (European Commission, 2008). This makes the development of OW management strategies mandatory in EU member countries. Similar legal frameworks have also been developed in other regions.

Potential solutions to manage this OW include using it as raw materials for energy production (bioethanol, biogas) or as fertilizers (compost) (Jimenez-Lopez et al., 2020). OW can also be an affordable and effective source of PCs, reducing waste by recycling OW as valuable yet affordable raw materials for a range of applications (Fig. 1). Many of these crop residues have been found to contain high amounts of PCs and, in many cases, at higher levels than the

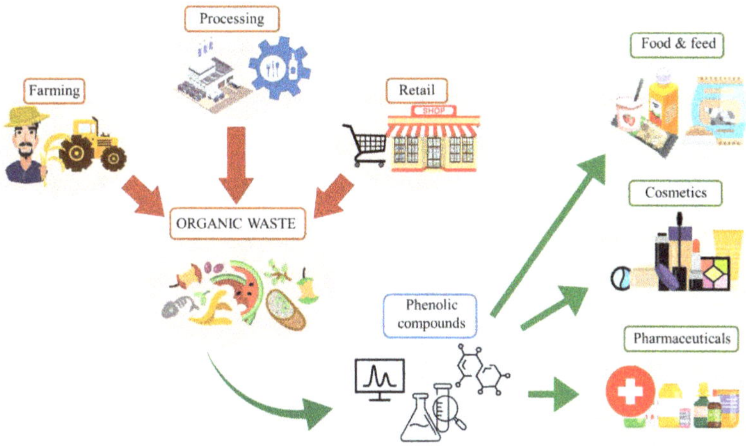

Figure 1 Depiction of potential obtention of phenolic compounds from diverse agricultural waste and their applications.

plants from which they are derived (Mayanga-Torres et al., 2017; Munekata et al., 2016). They are a rich potential source of bioactive compounds for the production of high-value-added products (Kazemi et al., 2019).

It has been suggested that, from a functional point of view, PCs can be classified into two main groups (De la Rosa et al., 2009):

- Flavonoids; and
- Non-flavonoids.

This chapter discusses the chemistry of PCs, their recovery from agri-food waste biomass and potential applications.

2 Phenolic compounds: flavonoids

Flavonoids are the more abundant and common group of PCs (>8000 compounds). Flavonoids have a basic structure consisting of 15 carbon atoms, distributed in C6-C3-C6, with two benzene rings (a and b) and one heterocyclic pyran ring (c) (Fig. 2) (Albuquerque et al., 2021). Depending on the bond between B and C rings and the substitution patterns of the C ring, there are six subclasses (Durazzo et al., 2019):

- Isoflavonoids;
- Anthocyanins;
- Flavanols;
- Flavanones;
- Flavonols; and
- Flavones.

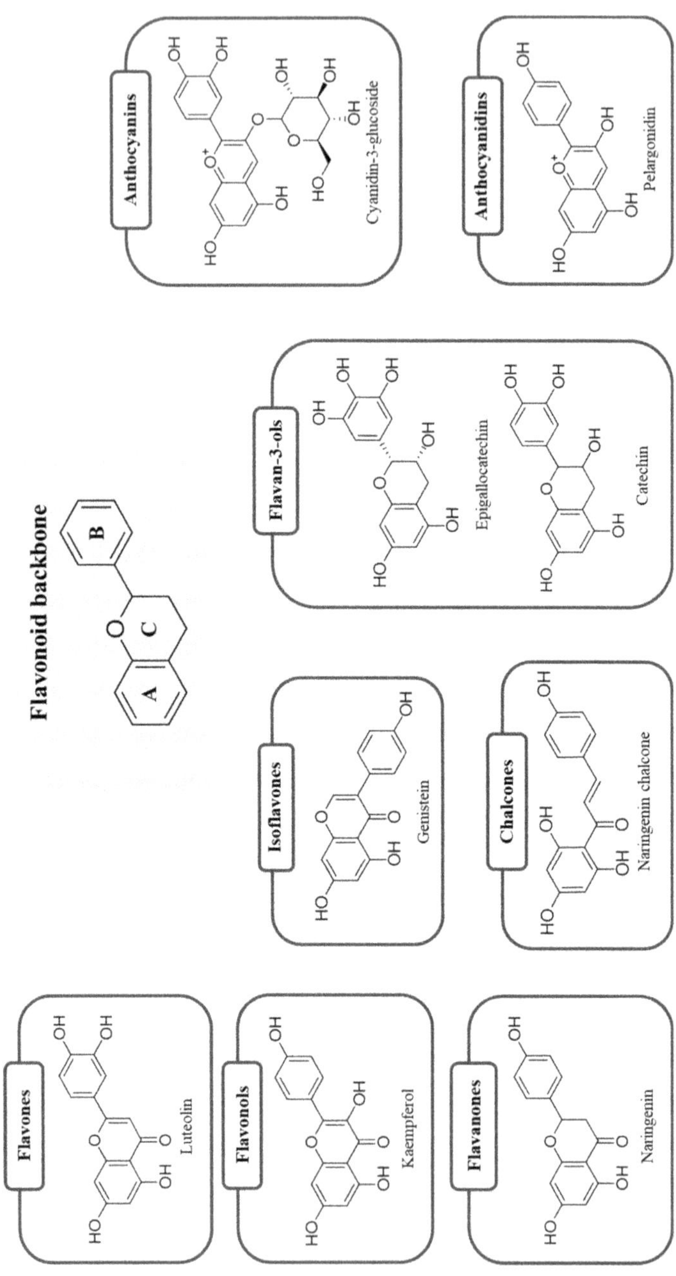

Figure 2 Representative chemical structures of the major flavonoid groups.

This group of compounds is known to possess antioxidant, anti-inflammatory, anti-cancer and anti-obesity properties.

Isoflavonoids possess an isoflavan (3-phenylchroman) backbone but are structurally diverse de- spite having a low distribution in the plant kingdom. Genistein and daidzein, which are found in soy, are important isoflavonoids with antioxidant and estrogenic activities (Papuc et al., 2017; Thangavel et al., 2019). Chalcones are α,β-unsaturated ketones (trans-1,3-diaryl-2-propen-1-ones), consisting of two aromatic rings (A and B) attached by α,β-unsaturated carbonyl system with a variety of substituents. In these flavonoids, the C ring is missing (Raut et al., 2016). As precursors of isoflavonoids and other flavonoids, they are considered as open-chain flavonoids found abundantly in edible plants. Many heterocycles of biological importance, such as 1,4-diketones, benzothiazepine, flavones and pyrazolines, can be synthesized using chalcones as key precursors. The two rings A and B linked by α,β-unsaturated carbonyl system are the unique features of chalcones (Raut et al., 2016).

The most important subclasses for the food industry are anthocyanins and flavanols (Fig. 2). Anthocyanins are responsible for the color of fruits and flowers. The most common anthocyanins are cyanidin, delphinidin, malvidin and peonidin glucosides (Durazzo et al., 2019). This flavonoid subclass can be found abundantly in black currants, red raspberry, elderberries, chokeberries or strawberries, plums, pomegranates, blood oranges, beans, black olives or red onions, i.e. from fruits or vegetables that display vivid red, blue or black colors (Rajauria and Tiwari, 2018). They are widely used as a natural coloring additive in foodstuffs. Anthocyanins possess different biological properties, including antioxidant, anti-inflammatory and antimicrobial properties, as well as the ability to inhibit tumor cell growth and reduce the glucose levels in the blood (Albuquerque et al., 2021; Khoo et al., 2017).

Flavanols (also known as flavan-3-ols) are one of the most complex subclasses of flavonoids, with a saturated C-3 element and folded structure. The main food sources of flavanols are fruits (e.g. grapes, blackberries, apples, nuts and almonds), cocoa-based products such as chocolate (rich in catechin and epicatechin) and green and oolong tea (Cassino et al., 2019; Durazzo et al., 2019). Flavanols are widely used in the food industry either directly in food or encapsulated to extend their bioactive effects (Carocho et al., 2015; Kaewprachu et al., 2015). Catechin, e.g. is used as a food additive (E300) in the US and EU because of its high antioxidant activity in extending shelf life (Carocho et al., 2015).

3 Phenolic compounds: non-flavonoids

The basic structure of these compounds contains at least an aromatic ring. Phenolic acids, xanthones, stilbenes, lignans and tannins are examples of non-flavonoid PCs (Durazzo et al., 2018). Two aromatic rings are found in the structures of xanthones, stilbenes and lignans, whereas tannins have more than two aromatic rings (Durazzo et al., 2019). Stilbenes and tannins are widely used in the food industry. This section discusses

- phenolic acids;
- stilbenes; and
- tannins.

These are discussed in the following sections.

3.1 Phenolic acids

PAs are monomeric PCs characterized by a carboxyl group linked to a benzene ring (Lafay and Gil-Izquierdo, 2008). PA may be again subdivided into two main groups based on their carbon skeleton (Durazzo et al., 2019):

- Benzoic acids; and
- Hydroxycinnamic acids.

The basic structure of hydroxybenzoic acids consists of seven carbon atoms distributed in C6-C1 (Albuquerque et al., 2021), (Fig. 3). Some examples are as follows (Amarowicz et al., 2009):

- Gallic acid;
- Vanillic acid;
- *p*-Hydroxybenzoic acid; and
- Syringic acid.

They possess antioxidant and antimicrobial properties with a role in preventing chronic diseases (Durazzo et al., 2019). They are also used in the food industry, particularly in food preservation (Sun et al., 2014).

Hydroxycinnamic acids are the most common type of PA, containing nine carbon atoms distributed in C6-C3 (Albuquerque et al., 2021), (Fig. 3). Examples include the following (Amarowicz et al., 2009):

- Caffeic acid;
- Ferulic acid;
- Sinapic acid; and
- *p*-Coumaric acid.

These acids are present in fruits, particularly red-colored fruits, vegetables, tea, cocoa, wine, tea leaves, coffee and whole grains (El Gharras, 2009). Hydroxycinnamic acids have also been used for food preservation (Kumar and Pruthi, 2014).

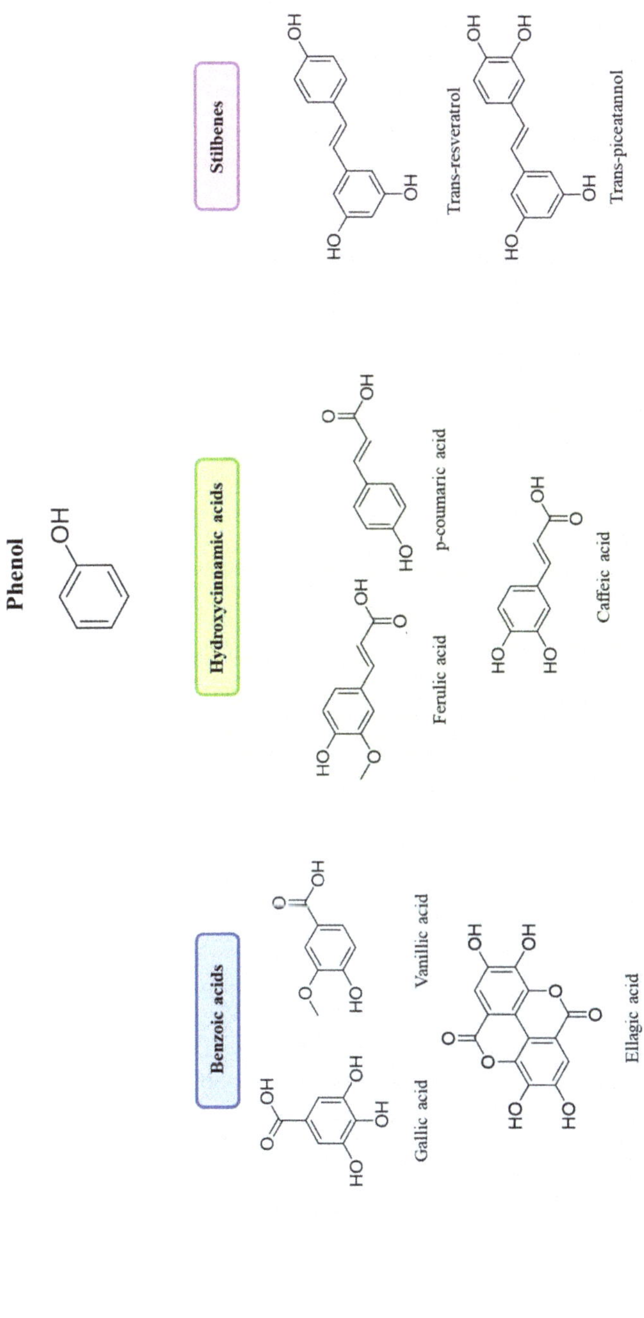

Figure 3 Representative chemical structures of major phenolic acids (blue and green) and stilbenes (purple).

3.2 Stilbenes

This non-flavonoid class has 14 carbon atoms distributed in C6-C2-C6 with 2 benzene rings connected by a double bond (Albuquerque et al., 2021), (Fig. 3). Stilbenes are derived from grapes, almond, bean, blueberries, bilberries, peanuts, grapevine, cranberries, mulberries, plum and wine (Błaszczyk et al., 2019). Trans-resveratrol is the best-known stilbene extracted from roots, stems, leaves, flowers, fruits and seeds, e.g. from grape OW (Tian and Liu, 2020). Resveratrol is well-known to possess antioxidant, anti-inflammatory, anti-obesity and anticancer properties (Murcia and Martínez-Tomé, 2001). It has potential applications as a food preservative (Pando et al., 2015; Shimamura et al., 2017). Stilbenes such as resveratrol, quercetin and hydroxycinnamic acids have also been tested as effective photoprotective agents in sunscreen (de Lima Cherubim et al., 2020).

3.3 Tannins

This group of non-flavonoid PC has the highest molecular weight in the group, due to their generally high degree of polymerization and may be divided into two main subclasses:

- Hydrolyzable tannins; and
- Non-hydrolyzable or condensed tannins (also known as proanthocyanins).

Hydrolyzable tannins can be subdivided into gallotannins and ellagitannins (depending on whether the polymer consists of gallic or ellagic acids) (Durazzo and Lucarini, 2019) (Fig. 4). Some common examples of hydrolyzable tannins are trigalloyl, tetragalloyl or pentagalloyl glucose. Ellagitannins are more variable, with some major examples being castalagin, castalin or punicalagin (María Fraga-Corral et al., 2021). Sources of ellagitannins include chestnuts (leaves and barks) and pomegranate skins.

The basic structure of non-hydrolyzable or condensed tannins consists of two or more condensing monomers of flavan-3-ol units, linked together mostly by bonds between the A and B rings of the flavanol units and pyran rings of other flavanols (Albuquerque et al., 2021). Catechin, epicatechin or epigallocatechin are some examples of flavan-3-ols that build up condensed tannins by polymerization, such as procyanidins A2, or B1 (Serrano et al., 2009). Due to their structure, hydrolysable tannins tend to be more easily degraded than condensed tannins, which in contrast, are very persistent molecules (María Fraga-Corral et al., 2021). A third, rarer class of tannins are complex tannins, which are formed through the union of gallo- or ellagitannin groups with flavan-3-ol subunits (Fig. 4). These PC are usually part of seeds, bark and fruit skins and have photoprotective and antioxidant properties. Tannins are abundant in fruits such as berries and nuts, bean legumes and especially grape skin and seeds (Boudjou et al., 2013; Lucarini et al., 2019).

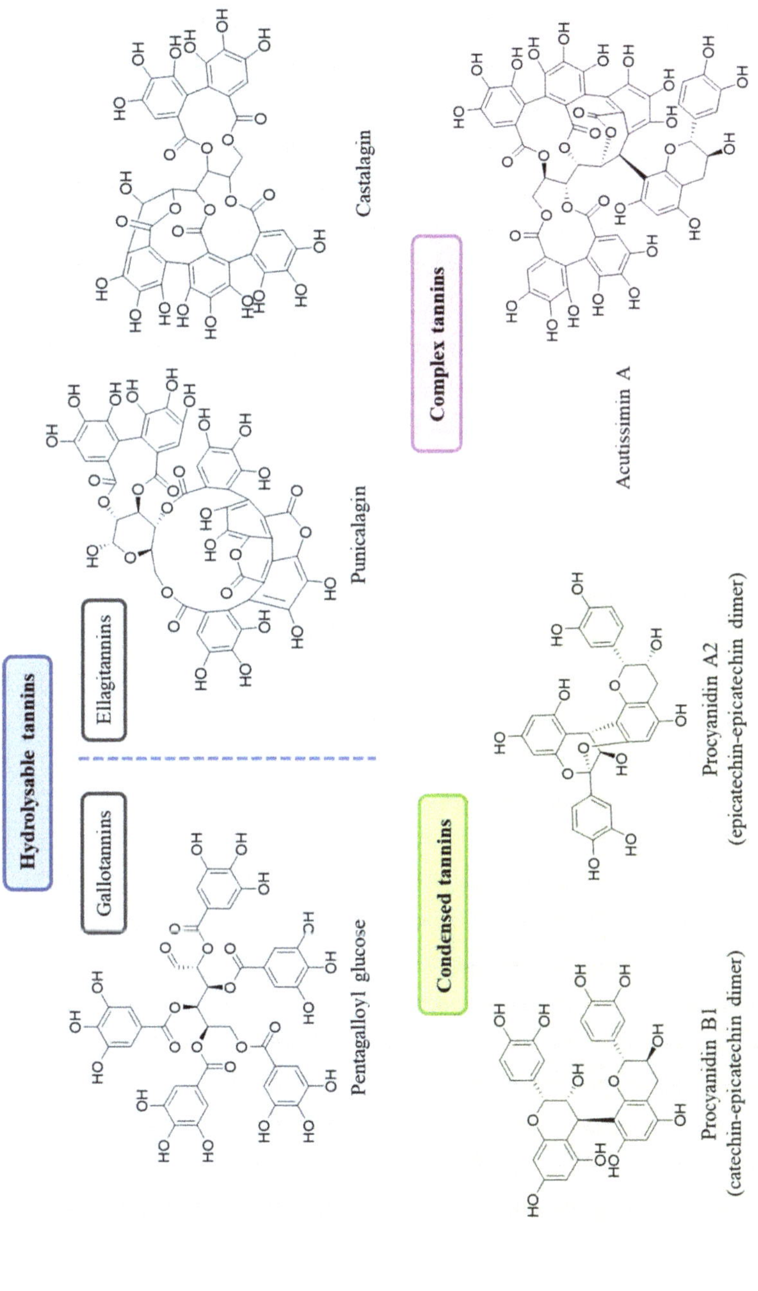

Figure 4 Representative chemical structures of some common and relevant tannin groups found in terrestrial plants.

Table 1 Biological activities and potential applications of PC recovered from agri-food waste

Biological properties with potential therapeutic applications			
Bioactivity	**Mechanism of action**	**Effect**	**Reference**
Antioxidant	Strong scavenging action of free radicals by donating a proton of a hydroxyl group attached to the aromatic ring	ROS levels decrease and sensitive biomolecules oxidation prevention	(Badhani et al., 2015; Yahfoufi et al., 2018)
Anti-inflammatory	Inhibitory action of toll-like membrane receptors. Scavenging of NO radicals	Proinflammatory genes expression and proinflammatory mediator's synthesis reduction (ILs, NF-κB, mTOR etc.)	(Takeda et al., 2014; Yahfoufi et al., 2018)
Cardioprotection	Scavenging of ROS. Vascular smooth muscle tissue proliferation inhibition	Atherosclerosis prevention. Platelet aggregation reduction	(Steinberg, 1997; Testai et al., 2013; Uhrin et al., 2018)
Anticancer	Scavenging of ROS. Cell-membrane receptors interaction	DNA damage prevention Angiogenesis reduction. Cell life-cycle mediation	(Cory et al., 2018; Grivennikov et al., 2010; Khalid et al., 2019)
Food and feed applications			
Application	**Effects**	**Compounds**	**Ref.**
Food additives	Synthetic additives substitution. Microorganism growth inhibition. Organoleptic properties preservation. Food shelf-life prolongation	Flavonoids, catechins, curcumin, luteolin, quercetin, ferulic acid, gallic acid	(Borowski et al., 2020; Brewer, 2011; Medina et al., 2007)
Animal feed	Animal welfare and productivity increase. Animal infection incidence reduction. Antibiotic resistance prevention	Resveratrol, tannins, quercetin flavonoids	(de Camargo et al., 2018; Yahfoufi et al., 2018; C. Zhang et al., 2019)
Other applications			
Application	**Effects**	**Compounds**	**Ref.**
Cosmetic ingredients	Dermis cellular density conservation. Skin oxidation prevention. UV radiation protection. Skin aging slowdown	Quercetin, rutin, luteolin, epigallocatechin gallate, resveratrol	(Herrling and Jung, 2012; Ratz-Łyko and Arct, 2019; Scapagnini et al., 2011)

(Continued)

Table 1 (*Continued*)

Natural pesticides	Less persistence in soils and produce. Plant biostimulant action. Safety for users. Bacterial and fungal control	Hydrolysable tannins, flavonoids	(Bargiacchi et al., 2012; Goławska et al., 2014; Oleszek et al., 2019; Patzke and Schieber, 2018)

ROS, reactive oxygen species; NO, nitric oxide; IL, interleukins; NF-κB, nuclear factor kappa-light-chain-enhancer of activated B cells; mTOR, mammalian target of rapamycin; DNA, deoxyribonucleic acid.

4 Types of organic waste containing phenolic compounds

Table 2 summarizes data on PC found in OW from different crops. As an example, fruit seeds are typical by-products of processing fruit. These seeds contain bioactive substances that remain in the residue obtained after fruit pressing (Fazio et al., 2013). An example of the potential of these by-products is the extraction of polyphenols from olive pruning materials (green leaves, dried leaves and olive pulp). The sustainability of the process on an industrial scale has been demonstrated from an economic and environmental point of view, with green chemistry technologies used to extract PC for subsequent re-use (Romani et al., 2016). Some particular examples of OW sources of PC are discussed further.

Grape OW includes pomace (the pulpy residue after pressing to extract juice) containing important bioactive compounds such as PCs. About 9 million tons of grape OW are produced globally each year representing 20% of the total weight of grapes harvested for winemaking (Goula et al., 2016). Grape pomace is the main by-product of the wine industry. It is rich in grape seeds (20 to 26%) that can be potentially used as a source of PCs such as quercetin, catechin, epicatechin, rutin, myricetin, kaempferol, ellagic acid, gallic acid, syringic acid, caffeic acid and trans-resveratrol (Chamorro et al., 2012; Drosou et al., 2015). Anthocyanins can also be recovered from this pomace for use as natural colorants and preservatives (Drosou et al., 2015). Grapevine leaves or stalks also contain anthocyanins and flavonoids (Dresch et al., 2014).

Tomato is a widely produced vegetable with an annual production of 106 million tons worldwide. However, approximately 40% of the tomato plant is discarded during production, harvesting and processing. This OW is even more significant in greenhouse tomato production which produces up to 49 tons per hectare per year of vegetative biomass (e.g. lower leaves and stems that are regularly removed from the vines to improve fruit production). Both leaves, stalks and tomato processing by-products (pomace, skin, seeds) contain PCs (Arab et al., 2019; Palomo et al., 2019).

Table 2 Main phenolic compounds extracted from some major agri-food byproducts following conventional extraction methods

Crop	Waste	Phenolic compounds determined	Extraction	Solvents	PC (mg/g)	Reference
Apple	Seed	Phlorizin, PA, epicatechin, catechin	HAE	MeOH	2-5	(Gunes et al., 2019)
	Pomace	Chlorogenic acid, catechin, epicatechin, rutin, quercetin-3-O-glucoside, quercetin-3-O-galactoside, quercetin-3-O-rhamnoside, phlorizin, phloretin	HAE	MeOH/W (80:20 v/v) + 1% FA	5	(Li et al., 2020)
	Peel	5-*O*-caffeoylquinic acid, catechin, procyanidins B1 and B2, various quercetin glycosides	HAE	MeOH/W (80:20 v/v)	25	(Kosińska et al., 2012)
Avocado	Seed	Catechin hydrate, epicatechin	HAE	MeOH	18	(Morais et al., 2015)
		3-*O*-caffeoylquinic acid, 3-*O*-*p*-coumaroylquinic acid, procyanidins A1 and A2, catechin/epicatechin gallate	HAE	MeOH/W (80:20 v/v)	10	(Kosińska et al., 2012)
		Catechin, epigallocatechin	HAE	MeOH	15	(Morais et al., 2015)
	Peel	Epicatechin	HAE	MeOH	42	(Morais et al., 2015)
Banana		-	Maceration	W	18	(Šeremet et al., 2020)
			HAE		25	
Blueberry	Peel	Anthocyanins	HAE	EtOH/W (60:40 v/v), 0.8% HCl	20	(Grillo et al., 2020)
Broccoli	Stems and leaves	Neochlorogenic, chlorogenic, sinapic and ferulic acids, catechin, quercetin-3-rutinoside, kaempferol-3-rutinoside	HAE	MeOH/W (80:20 v/v), 1M HCl	4	(Aires et al., 2017)
Carrot	Peel	Gallic, *p*-coumaric and protocatechuic acids, rutin, apigenin, naringenin	Maceration	W, MeOH, EtOH	9	(Nguyen and Scarlett, 2016)
Cashew	Peel	Epicatechin	HAE	EtOH	109	(Kamath and Rajini, 2007)

Coffee	Cherry pulp	Gallic, protocatechuic and chlorogenic acids, rutin	HAE	W	9	(Cantergiani et al., 2017)
	Spent grounds	Dihydroxybenzoic, caffeic, *p*-coumaric and ferulic acids	Soxhlet	EtOH	3	(Araújo et al., 2019)
Garlic	Husk	Trans-coumaric acid, guaiacylglycerol-β-caffeic acid ether	HAE	EtOH/W (80:20 v/v)	0.05	(Ichikawa et al., 2003)
		Ferulic, gallic, hydroxybenzoic, caffeic, *p*-coumaric acid, chlorogenic and coumaroylquinic acids	Soxhlet	MeOH	14	(Chhouk et al., 2017)
Red grape	Seeds	Gallic acid, catechin, epicatechin, several catechin/epicatechin polymers	HAE	EtOH/W (70:30 v/v)	86	(Lucarini et al., 2019)
	Leaves	Anthocyanins, flavonols	Maceration	W	52	(Dresch et al., 2014)
Kiwifruit	Peel	Several PA, vanillin, hyperoside, rutin, quercetin, kaempfero , catechin, epicatechin	Maceration	EtOH/W (70:30 v/v) + 0.2% HCl	28	(Y. Wang et al., 2018)
	Seeds	Several PA, catechin, epicatechin			11	
	Pomace	Caffeic, chlorogenic and ferulic acids, catechin, epicatechin, rutin, quercitrin	HAE	EtOH/W (70:30 v/v)	76	(Aires and Carvalho, 2020)
Lemon	Pomace	Several hydroxybenzoic and hydroxycinnamic acids, eriocitrin, narirutin, naringenin, tangeritin, flavonols, hesperidin, furanocoumarins	Cold maceration	MeOH/W (80:20 v/v)	8	(Multari et al., 2020)
Mandarin	Peel	Hesperidin, hesperitin, neohesperidin, naringin, naringenin tangeritin, nobiletin, several PA	HAE	Ace/W (80:20 v/v)	20	(Wang et al., 2016)
	Seed	Naringin, hesperidin, nobiletin, caffeic, *p*-coumaric and ferulic acids			3	

(*Continued*)

Table 2 (*Continued*)

Crop	Waste	Phenolic compounds determined	Extraction	Solvents	PC (mg/g)	Reference
Mango	Peel	Gallic, digallic and trigallic acids, methyl gallate, tetra-O-galloyl, mangiferin gallate	HAE	W	127	(Rojas et al., 2020)
	Seed	Theogallin, mangiferin, quercetin-3-O-galactoside, quercetin pentoside, ethyl *p*-trigallate, ellagic acid derivatives, other flavonols	HAE	Ace/W (50:50 v/v)	447	(Dorta et al., 2014)
Melon	Peel	Gallic, protocatechuic, isovanillic, chlorogenic and hydroxybenzoic acids, tyrosol, hydroxytyrosol, oleuropein, luteolin-7-glucoside	Maceration	EtOH/W (95:5 v/v)	3	(Mallek-Ayadi et al., 2017)
Onion	Peel	Quercetin, quercetin 3-glucoside, quercetin 4′-glucoside, isorhamnetin 3,4′-diglucoside	Maceration	MeOH/W (80:20 v/v) 0.5% HCl	52	(Benítez et al., 2011)
Olive	Pomace Oil wastewater	Gallic, caffeic, syringic, ferulic and vanillic acids, tyrosol, hydroxytyrosol, oleuropein, apigenin, luteolin-7-*O*-glucoside, pinoresinol	HAE	MeOH	4 7	(De Bruno et al., 2021)
Sweet orange	Pomace	Several hydroxybenzoic and hydroxycinnamic acids, eriocitrin, narirutin, naringenin, tangeritin, flavonols, hesperidin, furanocoumarins	Cold maceration	MeOH/W (80:20 v/v)	8	(Multari et al., 2020)
	Peel	Hesperidin vanillic acids, *p*-hydroxybenzoic acid, rutin, naringin, quercitrin, quercetin, kaempferol, luteolin	Decoction	W	92	(Omoba et al., 2015)
Pecan	Shell	Gallic, ellagic, protocatechuic and *p*-hydroxybenzoic acids, epigallocatechin gallate	HAE	EtOH/W (50:50 v/v)	69	(Villasante et al., 2019)
Pistachio	Shell skin	Gallic, protocatechuic, caffeic, *p*-hydroxybenzoic, *p*-coumaric and syringic acids, catechin, epicatechin, hesperidin, quercetin, apigenin	Maceration	MeOH	19	(Kilic et al., 2016)

Pomegranate	Peel	Punicalagin, cyanidin-3,5-*O*-diglucoside, delphinidin-3-*O*-glucoside, cyanidin-3-*O*- glucoside, pelargonidin-3-*O*-glucoside	Maceration	EtOH/W (80:20 v/v)	100	(Galaz et al., 2017)
		Catechin, epicatechin, hesperidin, rutin, PA	Maceration	MeOH	42	(Kupnik et al., 2022)
			Soxhlet		40	
Pumpkin	Seed	-	Soxhlet	EtOH/W (60:40 v/v)	42	(Ferreira et al., 2019)
Raspberry	Extrudate	Pelargonidin-3-glucoside, cyanidin-3-glucoside, quercetin-3-glucoside, ellagic acid derivatives, kaempferol	Cold maceration	Ace	19	(Vázquez-González et al., 2020)
Strawberry	Extrudate				7	

Ace, acetone; EtOH, ethanol; FA, formic acid; HAE, heat-assisted extraction; HCl, chlorohydric acid; MeOH, methanol; PA, phenolic acids; PC, phenolic compounds; W, water.

Coffee cherry is the main by-product of coffee production together with other by-products such as coffee pulp and husk (Mourtzinos and Goula, 2019). All of these contain polyphenols such as chlorogenic acid (Esquivel and Jiménez, 2012). Other PCs include protocatechuic acid, gallic acid and rutin (Cantergiani et al., 2017). The processing of avocado into guacamole produces waste such as peel and seeds with potential antioxidant compounds (Fidelis et al., 2019; Rosero et al., 2019). This waste includes PCs such as catechins, procyanidins or various flavonoids (Kosińska et al., 2012; Tremocoldi et al., 2018).

5 Nutraceutical properties of polyphenols and potential applications

PCs have shown a range of health-promoting, nutraceutical antimicrobial, anticancer, anti-inflammatory and cardioprotective properties related to their antioxidant content (Cory et al., 2018; de Camargo et al., 2018; Panche et al., 2016). PCs are generally regarded as safe with a low risk of potential toxicity (Lage et al., 2020; Saad et al., 2019). Table 1 summarizes these properties which are also discussed in the following sections.

5.1 Antioxidant properties

Oxidative stress is the result of excessive levels of reactive oxygen species (ROS) which are produced by normal oxidative metabolism and exposure to the environment (Moloney and Cotter, 2018). The term ROS refers to molecules like superoxide radical (O_2^{2-}), hydroxyl radicals (OH^-) or peroxide radical (HOO^-), which are reactive and may disrupt the chemical structure of labile molecules. They are produced by the sequential reduction of molecular oxygen in metabolic reactions (Metcalfe and Alonso-Alvarez, 2010). Other important free radicals responsible for oxidative stress are reactive nitrogen species (NOS) (Shankar and Mehendale, 2014). Oxidative stress may ultimately cause damage to biomolecules and can lead to an increased rate of DNA mutations and cell mitosis (Shankar and Mehendale, 2014). They are also one of the major markers of cell death which, in turn, triggers inflammation via the pro-inflammatory NOS nitric oxide (NO), which is released by macrophages (Aguirre et al., 2005; Metcalfe and Alonso-Alvarez, 2010). Oxidative stress is known to be a factor leading to inflammatory, cardiovascular, neurodegenerative or metabolic disorders and can result in cancer development (de Camargo et al., 2018; Metcalfe and Alonso-Alvarez, 2010; Yahfoufi et al., 2018).

All living aerobic organisms have developed defensive mechanisms against oxidative stress through the synthesis of reductive biochemicals (like glutathione) or enzymes such as superoxide dismutase, catalase or glutathione-*S*-transferase

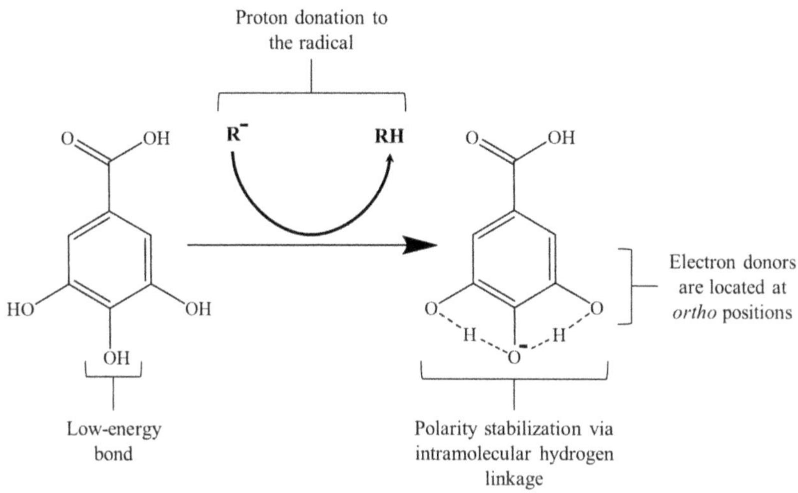

Figure 5 Possible main antioxidant mechanism of phenolic compounds, depicting gallic acid as an example. Phenolic compounds with hydroxyl groups in ortho positions exert the highest antioxidant effects.

(Kaulmann and Bohn, 2014). PCs are the major antioxidants in nature and generally show greater antioxidant capacities than other antioxidants like carotenoids or tocopherols (Ariza et al., 2018). It has been suggested that the antioxidant effect of PC consists of scavenging-free superoxide and hydroxyl radicals by donating a proton of a hydroxyl group attached to the aromatic ring (Fig. 5). This donated proton stabilizes the free radicals and avoids accumulation of high levels of ROS, NOS and subsequent oxidation of sensitive biomolecules like proteins or lipids (Badhani et al., 2015; Yahfoufi et al., 2018). Olive-derived polyphenols, e.g. are legally recognized in the EU as effective antioxidants that support cardiovascular health (European Commission, 2017).

These antioxidant properties, besides their potential beneficial effects on health, make PCs useful for applications such as cosmetic ingredients, e.g. sunscreen ingredients (de Lima Cherubim et al., 2020). PCs have also shown great potential as food preservatives, sometimes proving more effective than synthetic antioxidants such as dibutylhydroxytoluene (BHT) (Papuc et al., 2017). Various polyphenols have been reported to exert these antioxidant effects.

5.2 Anti-inflammatory properties

The inflammatory process is a natural response to the presence of invasive agents in the body (de Camargo et al., 2018). However, the inflammatory response may be deregulated in many metabolic diseases like diabetes mellitus or obesity, as well as in chronic autoimmune diseases like arthritis (Grivennikov et al., 2010). Likewise, in the later stages of life, almost all tissues show signs of low-grade chronic inflammation which could lead to the development of other severe chronic inflammatory diseases (Fortin et al., 2006). Chronic inflammation, whether as a result of infection or other health impairment, is also a well-recognized factor for cancer development or neurodegenerative diseases (Grivennikov et al., 2010; Laplante and Sabatini, 2012).

The anti-inflammatory effect of polyphenols primarily relates to immune-cell regulation involving proinflammatory cytokine synthesis as well as scavenging of NO radicals produced by macrophages (Mukherjee et al., 2010). Some of the most well-known proinflammatory mediators are interleukins (IL6 and IL8) produced by cells in the immune system such as dendritic cells and macrophages as well as cell proliferation and apoptosis modulators like mTOR, NF-κB, MAPk and PI3K (de Camargo et al., 2018; Pandey et al., 2017; Qiu et al., 2019; Sanches-Silva et al., 2020). Some PCs such as hydroxytyrosol can suppress toll-like receptors (a crucial cell membrane receptor for inducing inflammatory response) activation, which partially inhibits the expression of pro-inflammatory cytokines (Takeda et al., 2014; Yahfoufi et al., 2018). PCs such as resveratrol, tannins and luteolin have been proposed as a natural alternative to prevent inflammatory-related ailments (Yahfoufi et al., 2018).

5.3 Anti-diabetic and anti-obesity properties

Obesity involves genetic and environmental factors that result in an excessive increase in body fat and a decrease in energy expenditure. Obesity is associated with a whole host of metabolic diseases known as metabolic syndrome (Caimari et al., 2013). At present, there is a growing interest in the potential of PCs as inhibitors of the carbohydrate digestion enzymes α-glucosidase and α-amylase (digestive enzymes involved in diabetes) to prevent increased postprandial glycemia (Cardinali et al., 2010). Fruit skin and pomace (containing flavanones, dihydroflavonols and stilbenes) have been targeted for their anti-diabetic and anti-obesity properties (Shahidi and Ambigaipalan, 2015). Pomegranate peel extracts, e.g. were found to exert inhibitory properties against α-glucosidase and pancreatic lipase in reducing blood glucose levels and lipid cholesterol (Ambigaipalan et al., 2016; Durazzo et al., 2019; El-Hadary and Ramadan, 2019) The inhibitory potential of α-glucosidase from PC-rich extracts of mango, jackfruit, pineapple, papaya or banana peel, among many other sources, has been extensively reported (Y. Zhang et al., 2016).

5.4 Other biological properties

The antioxidant and anti-inflammatory properties of PC have been linked with cancer-preventive effects. Various studies have shown that PC may act as chemopreventive agents against tumor development by inducing an apoptosis response, e.g. by increasing caspase expression and leading to tumoral cell death (Qiu et al., 2019). Various PCs are also reported to inhibit tumor growth, as observed in animal models (Garcia-Oliveira et al., 2021). For instance, luteolin (a type of flavone) is used in the treatment of tumor cell lines, since it contains a synergetic effect when used with tumor necrosis factor-related apoptosis-inducing ligand (TRIAL) in the treatment of lung cancer (Yan et al., 2012). Another study showed that a moderate consumption of epigallocatechin-3-gallate, present in green tea, reduced the risk of both chronic cardiovascular disease and cancer (Peluso and Serafini, 2017). Another possibility is that PCs may further act as cardioprotective agents, not only by lowering inflammation or obesity-related symptoms, but also as anticoagulant agents. In a clinical study carried out with PC-rich extracts of tomato pomace, participants showed lower platelet aggregation than the control group (Palomo et al., 2019).

6 Extraction methods: conventional solid-liquid extraction technologies

One of the main challenges in obtaining PCs from OW is the choice of extraction method. This requires assessing both the nature and composition of the matrix and the target compounds. As an example, several phytochemicals can be recovered from grape by-products including anthocyanins from pomace and tannins and flavonoids from peels and polyphenol-rich oils from seeds (Varzakas et al., 2016). Targeting specific compounds requires assessing the extraction method together with extraction parameters, such as solvent, solvent gradient, time, temperature or other specific parameters (e.g. power, pressure and charge).

Solid-liquid extraction (S-L) techniques are chemically based extraction processes. They require pre-treatment steps such as milling, grinding, drying and homogenization (Carreira-Casais et al., 2021; Pinela et al., 2019). Different S-L techniques include

- maceration;
- heat-assisted extraction (HAE); and
- Soxhlet extraction.

Maceration involves soaking and agitating a milled sample in a solvent to separate out target compounds. Conventional extraction is typically carried out at 20-50°C. By speeding up the reaction, HAE requires less solvent and extraction time. Soxhlet extraction both applies heat and refluxes the solvent in a continuous process that requires less time and solvent, improving extraction time and yield, though heat can degrade PCs and reduce extract yield. Temperatures exceeding 70°C can lead to degradation or oxidation of some PCs (Brglez Mojzer et al., 2016).

As highly polar compounds, PCs are usually extracted with solvents such as methanol, ethanol, acetone, diethyl ether and ethyl acetate, typically mixed with water. Methanol, e.g. is more appropriate for lower molecular weight polyphenols (e.g. resveratrol), whilst acetone is better for higher molecular weight polyphenols (e.g. flavonols). Other solvents may prove effective depending on the type of matrix. For example, addition of acids such as formic acid, chlorohydric acid or acetic acid may enhance extraction yield, especially for lignocellulosic biomass such as seeds or stalks (Gullón et al., 2020).

6.1 Maceration and heat-assisted extraction

These conventional extraction techniques possess a number of advantages that justify their continuing use. They require low-cost and simple equipment compared with some other innovative techniques. A large range of molecules can also be extracted in large volumes by changing the protocols and adapting variables such as solvent, temperature, agitation rate and time in order to optimize extraction of the desired compounds (Carreira-Casais et al., 2021).

However, both maceration and HAE also have some major drawbacks. They often require long extraction times when compared to newer, alternative methods, large volumes of solvent (which can be hazardous to use and environmentally damaging as a waste by-product of the process), high temperatures which increase energy-related costs (and can reduce yields), several filtration or centrifugation steps in order to separate the extract from the biomass and produce less pure, lower-quality PCs (Carreira-Casais et al., 2021). The use of large volumes of solvent also requires purification steps to produce a safe, food-grade product and this in turn creates a significant waste problem, particularly given growing legal restrictions on disposal of solvent residues (Torres-Valenzuela et al., 2020). This has resulted in the development of alternative, more efficient and environmentally friendly extraction techniques which are discussed later.

6.2 Soxhlet extraction

Soxhlet is a reference extraction method to evaluate the performance of other liquid-solid extraction methodologies (Baiano, 2014). This technique, first developed in 1879, uses a particular type of condenser known as the Soxhlet apparatus. The traditional Soxhlet extractor is composed of a thimble-holder. The sample is placed inside the thimble linked to a distillation flask where fresh solvent is added. When the solvent reaches boiling point, it vaporizes and enters the matrix, solubilizing compatible compounds. The solvent makes contact with the cooling tubes of the condenser and refluxes back into the initial flask with the extracted compounds. The process is repeated until full extraction is completed (Garcia-Vaquero et al., 2020). This is one of the main variations in comparison with percolation, which also passes a heated solvent solution through the matrix but without solvent condensation (Botineştean et al., 2015).

This type of extraction has several advantages. The constant renovation and flux of the solvent in contact with the matrix allows for a constant unsaturation of the latter, thus allowing a constant transfer of target compounds to the solvent. Secondly, the temperature of the system is maintained throughout the process. A main advantage of Soxhlet extraction is that the extract usually requires no filtration or centrifugation, being well separated from the biomass matrix. Additionally, it allows for the treatment of several samples in parallel at a relatively low cost with relatively simple equipment (Luque de Castro and Priego-Capote, 2010). However, Soxhlet extraction also has some disadvantages. Large amounts of solvents and long extraction periods are required until the final number of cycles is completed. In many cases, as the solvent must be boiled to allow its reflux, the set temperature must be high and may degrade the target compounds or require the use of organic solvents, which may not be suitable for food-grade extractions or ideal for PCs given their polarity (Dang et al., 2019). Additionally, Soxhlet extraction cannot be accelerated by agitation.

Nevertheless, the Soxhlet extraction has continued to evolve to try to compensate for some of these disadvantages, e.g. by automating the process to shorten extraction times. Soxhlet extraction has recently been coupled to innovative technologies, e.g. through the development of high-pressure Soxhlet extraction, supercritical fluid–Soxhlet extraction or by applying ultrasound or microwaves to improve efficiency compared to conventional Soxhlet extraction.

7 Extraction methods: green technologies

There is now a range of new 'greener' extraction techniques in development or commercial use (Carreira-Casais et al., 2021). These new techniques are less reliant on solvents, reduce extraction times and can improve both yields and quality (Poveda et al., 2018). Milder processing conditions can also potentially reduce oxidation and/or degradation of labile PCs, though process parameters need to be carefully established, monitored and maintained to avoid or minimize PC degradation (Panja, 2018). Table 3 summarizes studies of PC extraction from OW using green extraction technologies. A great number of these techniques rely on physical processes to achieve extraction. These include

- ultrasound-assisted extraction (UAE);
- microwave assisted extraction (MAE); and
- pulsed-electric fields (PEFs).

Table 3 Main phenolic compounds extracted from some major agri-food byproducts following green extraction methods

Crop	Waste	Phenolic compounds determined	Extraction	Solvents	PC (mg/g)	Reference
Apple	Peel	Chlorogenic, homovanilic and rosmarinic acids, rutin, isoquercetin, quercitrin, phlorizin	UAE	EtOH/W (80:20 v/v)	13	(Kalinowska et al., 2020)
Avocado	Peel	Epicatechin, procyanidin B2	UAE	EtOH/W (80:20 v/v)	120	(Tremocoldi et al., 2018)
	Seed	Trans-5-O-caffeoyl-D-quinic acid, catechin, epicatechin, procyanidin B1			59	
Banana	Peel	-	UAE	W	18	(Šeremet et al., 2020)
			MAE		3	
Beetroot	Pomace	Ferulic, vanillic, caffeic, protocatechuic and *p*-hydroxybenzoic acids, catechin, epicatechin, rutin	UAE	EtOH/W (50:50 v/v), 0.5% AcOH	2	(Vulić et al., 2014)
Blueberry	Peel	Anthocyanins	MAE	NADES/W (78:22 v/v)	25	(Grillo et al., 2020)
			UAE		21	
Broccoli	Leaves and stems	Neochlorogenic, chlorogenic, sinapic and ferulic acids, catechin, quercetin-3-rutinoside, kaempferol-3-rutinoside	UAE	MeOH/W (80:20 v/v), 1M HCl	3	(Aires et al., 2017)
Carrot	Peel	Gallic, *p*-coumaric and protocatechuic acids, apigenin, rutin, naringenin	MAE	MeOH	28	(Nguyen and Le, 2018)
Cocoa	Bean shell	Epicatechin, 5-caffeoylquinic acid, tannins	PEF	EtOH/W (5:95 v/v)	55	(Barbosa-Pereira et al., 2018)
Coffee	Wet pulp		UAE	EtOH/W (50:50 v/v)	14	(Kieu Tran et al., 2020)
	Spent grounds	Dihydroxybenzoic, caffeic, *p*-coumaric and ferulic acids	PLE	EtOH	4	(Araújo et al., 2019)
			SFE	$scCO_2$/EtOH (33:67, w/w)	4	

		-	MAE	EtOH/W (20:80 v/v)	8	(Ranic et al., 2014)
	Silverskin	5-caffeoylquinic acid, caffeine	PEF	EtOH/W (5:95 v/v)	13	(Barbosa-Pereira et al., 2018)
	Spent grounds	Flavonoids and phenolic acids	EAE	Cellulase or hemicellulase	0.6	(Scully et al., 2016)
Corn	Cob	Gallic, protocatechuic, vanillic, chlorogenic, caffeic, *p*-coumaric and ferulic acids, catechin, quercetin, rutin	UAE	Ace/W (50:50 v/v)	6	(Saha et al., 2019)
Garlic	Husk	Gallic, hydroxybenzoic, caffeic, *p*-coumaric and ferulic acids	PLE	EtOH	28	(Chhouk et al., 2017)
			SFE	$scCO_2$/EtOH (33:67, w/w)	30	
Red grape	Peel	-	SFE	$scCO_2$/EtOH (75:25, w/w)	12	(Manna et al., 2015)
	Seeds				8	
	Pomace	Epicatechin, catechin, gallocatechin, epigallocatechin, epicatechin 3-O-gallate, procyanidins B1 and B2	EAE	Pectinase, cellulase and tannase	240	(Chamorro et al., 2012)
		Catechin, epicatechin, procyanidins B1 and C1	PEF	EtOH/W (50:50 v/v)	113	(Brianceau et al., 2015)
		Several anthocyanins, flavonols, procyanidins and aglycone derivatives	UAE	EtOH/W (50:50 v/v)	438	(Drosou et al., 2015)
				W	51	
			MAE	EtOH/W (50:50 v/v)	200	
				W	52	

(Continued)

Table 3 (*Continued*)

Crop	Waste	Phenolic compounds determined	Extraction	Solvents	PC (mg/g)	Reference
Kiwifruit	Peel	-	PLE	W	51	(Guthrie et al., 2020)
	Pomace	Chlorogenic, *p*-coumaric, protocatechuic and caffeic acids, catechin			60	(Kheirkhah et al., 2019)
Mandarin	Peel	Narirutin, hesperidin, naringin, naringenin	PLE	W	113	(Ko et al., 2016)
Mango	Peel	Mangiferin	MAE	Na-acetate/Gly (33:67 v/v)	155	(Pal and Jadeja, 2022)
		-	PLE	EtOH	23	(Garcia-Mendoza et al., 2015)
Melon	Peel	Gallic, hydroxybenzoic, caffeic acid, syringic, *p*-coumaric, ferulic and cinnamic acids, catechin, quercetin, kaempferol, resveratrol, rutin	MAE	MeOH/W (80:20 v/v)	150	(Ahmed et al., 2021)
Onion	Peel	Flavonols	MAE	MeOH/W (74:26 v/v)	12	(González-de-Peredo et al., 2022)
Olive	Pomace	Hydroxytyrosol	MAE	EtOH/W (90:10 v/v)	53	(Xie et al., 2019)
			UAE		55	
	Leaves	Hydroxytyrosol, lignans, verbascoside, flavonoids and secoiridoid derivatives	PLE	W	289	(Romani et al., 2016)
Sweet orange	Peel	Gallic and ellagic acids, narirutin, naringin, hesperidin, hesperitin, diosmetin, tangeritin	PLE	EtOH/W (75:25, v/v)	19	(Barrales et al., 2018)
Pomegranate	Peel	-	SFE	CO_2	389	(Cavalcanti et al., 2012)
		Several PA, flavonoids, hydrolyzable tannins and proanthocyanidins	UAE	Ace/W (70:30 v/v)	21	(Ambigaipalan et al., 2016)

Potato	Peel	Chlorogenic acid, caffeic acid and derivatives	UAE	EtOH/W (55:45 v/v)	7	(Riciputi et al., 2018)
Pear	Peel	Gallic, chlorogenic, vanillic, ferulic and *p*-coumaric acids, arbutin, rutin	UAE	MeOH/W (60:40 v/v)	8	(Wang et al., 2015)
Pumpkin	Seed	-	UAE	EtOH/W (60:40 v/v)	34	(Ferreira et al., 2019)
			MAE		35	
Strawberry	Leaves	Caffeic, chlorogenic, *p*-coumaric and synapic acids, catechin, epicatechin, rutin	MAE	EtOH/W (50:50 v/v)	89	(Lin et al., 2020)
Tomato	Leaves	Flavonoids	SFE	$scCO_2$	240	(Arab et al., 2019)
	Pomace	Coumaric acid, phlorizin, phloretin, kaempferol, quercitrin, procyanidin B2, kaempferol-3-glucoside	UAE	W	46	(Palomo et al., 2019)
	Peel and seeds	Lycopene	EAE	Pectinase	11.5	(Catalkaya and Kahveci, 2019)

Ace, acetone; AcOH, acetic acid; EAE, enzyme-assisted extraction; EtOH, ethanol; FA, formic acid; HAE, heat-assisted extraction; HCl, chlorohydric acid; MAE, microwave-assisted extraction; MeOH, methanol; NADES, natural deep eutectic solvents; PA, phenolic acids; PEF, pulsed-electric fields; PLE, pressurized-liquid extraction; $scCO_2$, supercritical carbon dioxide; SFE, supercritical fluid extraction; UAE, ultrasound-assisted extraction; W, water.

Other techniques include

- supercritical fluid extraction (SFE);
- enzymatic-assisted extraction (EAE); and
- pressurized liquid-assisted extraction (PLE).

Some of the main techniques are discussed in the following sections.

7.1 Enzymatic-assisted extraction

Bound or dispersed phytochemicals inside cells or attached to cell walls can be difficult to extract by conventional solvent extraction. Enzymes help to release those components in a unique way by digesting the surrounding molecules, thus releasing the target molecules to be extracted. EAE can be very effective for the extraction of polyphenols bound to proteins or polysaccharides. Enzyme classes such as cellulases, pectinases, proteases etc. are commonly used, in order to permeate and break down plant cell walls, as well as free polyphenols from wall-bound polysaccharides and proteins (Nadar et al., 2018). The particle size and the ratio of enzyme and sample are the key controlling factors to maximize polyphenol yield. The main step in extraction is enzymatic hydrolysis, where a mixture of sample, enzyme and solvent (buffered water in most cases) is incubated at low temperatures, around 35–50°C, with adjusted pH. In general terms, hydrolytic enzymes work best at mildly acidic pH (5.5–6.5), but others, especially proteases, may work at strong acidic pH (e.g. pepsin) or at very basic pH (e.g. Alcalase®) (Xiao and Niu, 2015). Hydrolysis is stopped by denaturing the enzymes at an elevated temperature of 80–90°C for a short time (~15 min) (Panja, 2018).

The main advantage of EAE is that it is an environment-friendly process, as water is used as solvent and enzymes, being proteins, can be non-hazardous once denatured. Low-temperature extraction also prevents degradation and the process has lower energy-related costs. EAE can also achieve high extraction yields given its targeted approach (Sowbhagya and Chitra, 2010). The main drawback of EAE is its generally prolonged extraction time, which varies from 1 h to 48 h, as well as its higher reagent costs, as enzymes are generally more expensive than conventional solvents (Gligor et al., 2019).

7.2 Ultrasound-assisted extraction

UAE involves the application of high-intensity ultrasonic waves which promotes the breakdown of cell walls and solvent diffusion (Ameer et al., 2017). This reduces solvent requirement and process time. The technique is regarded as versatile, safe and relatively cost-effective (Savic and Savic Gajic, 2020). The main parameters of this method are the ultrasonic frequency (>20 kHz) and the power (W) applied (Panzella et al., 2020). Low frequency and high-power conditions produce cell disruption and the subsequent release of compounds present in the matrix, without applying excessive shear force that could disrupt the structure of extractable compounds (Chemat et al., 2017). Bubbles produced by ultrasound agitation of the medium grow in such a way that they result in micro explosions, causing the rupture of cell walls and consequent release of stored molecules (Lavilla and Bendicho, 2017). Another important characteristic of UAE is that the hot spots created by cavitation bubbles during the extraction process generate little heating and cool down rapidly while the process is taking place. As UAE extractions take place usually below 30°C, it is considered suitable for the extraction of thermolabile compounds (Chemat et al., 2017). However, when UAE times are required to be longer (e.g. >5 min) at high power, it is necessary to use a refrigeration system to keep the temperature constant and avoid possible overheating.

Factors affecting UAE include reactor type, the matrix to be extracted and the optimized parameters to be applied. Greater ultrasonic amplitude is not directly related to better process efficiency but can create problems such as erosion of the probe and reduce cavitation which can even promote the degradation of the extracted compounds (Chemat et al., 2017). Frequencies must be selected together with ultrasonic intensity to obtain the desired cavitation. Higher extraction yields are reported in the low-frequency range (20–40 kHz) (Kadam et al., 2015). High temperatures help to interrupt the interaction of the solvent and the matrix and improve the diffusion rates of the solvent, while low temperatures improve cavitation. Longer extraction time may improve extraction yields at lower frequencies; however, it may also result in excessive cavitation and degradation of target compounds (Wen et al., 2020). The use of a viscous solvent reduces cavitation while a volatile solvent can evaporate if extraction is carried out at a higher temperature for a long period. Finally, the size of the matrix, its interaction with the solvent and the ratio of solvent to matrix are also parameters to consider when UAE is used (Tiwari, 2015) Among the advantages of this technology are reduction of solvent requirement, temperature and time, low equipment investment and ease of implementation which make it a widely used commercial technique (Chemat et al., 2017). Drawbacks include a potential reduction of extraction yield if power is too low or degradation of target compounds if too much power is used.

7.3 Microwave-assisted extraction

MAE uses microwave energy to disrupt the plant material matrix (Wen et al., 2020). Microwaves are non-ionizing electromagnetic waves with a spectral frequency ranging from 300 to 300 000 MHz involving dipole rotation and/or the ionic conduction through a medium (Chaves et al., 2020). MAE results in lower extraction time, less solvent needed, high extract purity and yield with relatively low energy consumption and cost. There are two MAE systems using either a closed or open vessel. The former allows application of high temperatures (~100°C), while an open system allows solvent reflux and agitation, improving extraction mass transfer (Ameer et al., 2017).

One of the potential disadvantages of MAE is the degradation of some compounds due to the heat produced by irradiation. MAE efficiency depends on factors such as the nature of the extractant, amount of microwave irradiation, temperature, extraction time as well as characteristics of matrices and the solvent-food relationship. Extraction efficiency is usually directly proportional to the microwave power applied up to a limit. With power between 400 W and 500 W, PC yield is generally greater than when power exceeds 500 W (Alara et al., 2018). This is because overexposure to microwaves and/or high temperatures causes overheating and degradation of PC. MAE is not commonly used for the extraction of polymeric PCs, e.g. tannins or anthocyanins (Alara et al., 2021).

The process requires the use of solvents of at least partial polarity with an elevated dielectric point that allows an effective absorption of microwaves. Water, e.g. is considered a good solvent, as it has an elevated dielectric constant ($\varepsilon = 78$) (Chaves et al., 2020). Water/ethanol mixtures with or without organic solvents generally result in greater penetration of the extractant in the matrix, allowing for higher yields than with organic solvents alone. This is due to the good microwave absorption capacity of these polar solvents (Montenegro-Landívar et al., 2021). MAE extraction time may be less than 30 min (V González-de-Peredo et al., 2022).

7.4 Pressurized-liquid extraction

PLE is based on the application of high temperature (~80–200°C) and pressure (~100–200 bar) to S–L extraction using solvents such as water and ethanol (Chaves et al., 2020). The main solvent used for PC extraction is water, hence the alternative name for the technique: subcritical water extraction. At elevated pressure and temperature, solvents such as water reach subcritical states, altering their ambient physical-chemical properties and improving their penetration into the matrix (Plaza and Turner, 2015). This allows a solvent to remain a liquid after its vapor pressure and boiling point have been exceeded (Makris, 2018). Solvent polarity may also be altered. For example, at 300°C and 230 bar, the dielectric constant of water is reduced from 78 to 21, with similar values to that of ethanol ($\varepsilon = 24$ at 25°C) or acetone ($\varepsilon = 20.7$ at 25°C), thus greatly altering extraction selectivity without altering the solvent (Chaves et al.,

2020). The increase in temperature also results in increased solvent solubility and permeation of the solvent into the matrix.

There are three PLE techniques - static, dynamic and mixed - though only static PLE appears to be extensively used (Tena, 2018). Static PLE is carried out by mixing the matrix with an inert desiccant agent (e.g. sand and diatomaceous earth) in order to prevent sample aggregation and improve the exposure area to the solvent. The solvent is then introduced and the cell is sealed, applying the set pressure for a given time (5–50 min). The extract is obtained by releasing the pressure (Tena, 2018). The operation can be repeated for 2 or 3 cycles for maximum extraction (Carreira-Casais et al., 2021). Advantages of PLE include a significant reduction in solvent volume, together with greater solvent flexibility since its polarity can be altered. PLE extracts usually do not require filtration or centrifugation and extraction cycle times are typically very short (5–15 min) (Chaves et al., 2020). Drawbacks include the relatively high cost in terms of equipment investment and challenges in the optimization of operational parameters. The latter can become an issue since both temperature and pressure must be carefully adjusted in accordance with both matrix and solvent, as well as to avoid potential PC degradation (Carreira-Casais et al., 2021).

7.5 Supercritical fluid extraction

SFE relies on the usage of supercritical fluids as solvents. These are fluids in which gaseous and liquid phases coexist with the solvent behaving as both. This is achieved by applying high pressures (100–400 bars) and moderately high temperatures (40–80°C) to liquify gaseous solvents (Ameer et al., 2017; Brglez Mojzer et al., 2016). This allows the use of gaseous and more inert solvents. CO_2 is the most widely used supercritical solvent given its ability to penetrate plant materials and high solvent capacity due to its low density and low supercritical point (31°C, 73 bar) (Osorio-Tobón, 2020). CO_2 is usually combined with ethanol or in hydroethanolic solutions as a co-solvent which greatly increases the solubility of PC in SFE (Tyśkiewicz et al., 2018). Advantages include relative simplicity, low extract time and solvent requirement, high extract quality and stability as well as reduced environmental impact (Chaves et al., 2020).

The extraction process involves several stages. The plant matrix initially absorbs the supercritical solvent, producing a swelling of the cell structure with consequent dilation of intercellular spaces. This results in a decrease in resistance to mass transfer and penetration of the solvent. In addition, there is a simultaneous transfer of matter from the internal matrix to the surface. These compounds are then transported from the surface to the supercritical solvent and finally removed from the solvent, the extraction cell is cooled down and the extract is obtained by pressure release (Osorio-Tobón, 2020). As in PLE, SFE operational parameters must be carefully considered. Temperature and pressure variations have important effects on fluid hydrodynamics, solubility and mass transfer. A low temperature is crucial to preserve PC to achieve higher yields or greater extract bioactivity. The best temperature range appears to be 40–50°C with an optimum pressure range between 200 bar and 400 bar (Carreira-Casais et al., 2021). Optimal pressure and temperature conditions can be influenced by the type of biomass and its particle morphology (Herrero et al., 2015).

Despite the large number of studies using this extraction technique with effective results, SFE requires significant investment in equipment as well as pure CO_2 which increases operating costs (Brglez Mojzer et al., 2016). This has limited its commercial application. Recent studies suggest that SFE is economically viable. SFE is used commercially in the decaffeination of coffee or tea, extraction of essential oils from spices for oleoresin production or extraction of hop compounds (Knez et al., 2013). CO_2 can also be considered a 'green' solvent with minimum environmental impact. In addition, it generally requires short extraction times (<1 h) with greater extraction efficiency than other methods (Ameer et al., 2017).

7.6 Pulsed-electric fields

PEF is one of the most innovative and recent extraction techniques in the field of non-thermal food processing and chemical extraction. Its mechanism of action is based on electroporation. This involves breaking cell walls and membranes by applying short-duration pulses to matrices (μs to ms) of high-intensity electric voltage (20–80 kV/cm, but typically 0.5–20 kV/cm in the case of plant materials), with the matrix placed between two electrodes and submerged in a conductive solvent (Carreira-Casais et al., 2021; Zeng and Zhang, 2019). These characteristics have led to an increased interest in the method to improve the extraction performance of bioactive compounds from plant tissues and their by-products, such as PC, carotenoids and oils, as well as soluble intracellular polysaccharides or proteins. Its main application, however, is as a pre-treatment prior to extraction, which results in a significant improvement in extraction yields (Brianceau et al., 2015). Its effectiveness is greatly dependent on the permeability of the matrix, which will consequently affect setting appropriate parameters and optimizing extraction yield. Critical parameters include electric field intensity, treatment time (number of pulses × pulse duration), pulse waveform, matrix conductivity, pH and ionic strength of the medium (Panja, 2018).

The technology, either as PEF or high voltage electrical discharges (HVED) is still being explored as an effective tool for PC extraction from OW. It has been applied to winemaking OW with promising results but further research is required (Puértolas and Barba, 2016). Some of the main advantages of this method are sustainability, as almost no solvent or salts are required, as well as its minimal effect on compound quality (Andreou et al., 2020). Although the method uses electrical discharges, overall energy consumption is low as the process time is extremely short (Puértolas and Barba, 2016). PEF is also used as a non-thermal sterilization/preservation process (Z. W. Liu et al., 2018).

8 Comparing extraction techniques

The development of more efficient extraction techniques is a key goal for industry to maximize the throughput of agricultural wastes for the recovery of high-value natural compounds. In most cases, several variables need to be optimized, requiring the use of response surface methodology as a method to determine optimal operating conditions (Pinela et al., 2019).

A number of reviews provide comparative reviews of various extraction techniques. Ameer et al. (2017), e.g. discussed and compared the principles, technologies and applications of MAE, PLE, pressurized hot water extraction, SFE and UAE. Alara et al. (2021) compared MAE, UAE, PLE, SFE and EAE.

On a more practical level, a number of studies provide tables showing how individual extraction techniques can be applied to particular plant materials. Mojzer et al. (2016), e.g. described polyphenol classes, typical plant sources and recommended extraction methods. Table 2 and Table 3 in this chapter summarize sources and types of PCs and recommends extraction methods and solvents required.

Traditional solid/liquid extractions are established techniques for recovering biocompounds, particularly maceration and HAE. The main parameters to optimize these techniques include biomass-solvent ratio, the chemical nature of the solvent or their combinations, time and temperature of incubation. Solvents used to recover biomolecules from agricultural wastes are mainly methanol, acetone and ethanol, sometimes diluted with water or acidified with formic or hydrochloric acid (Table 2). Optimum yields have been achieved, e.g. from the peel of cashews (109 mg/g using ethanol), peel and seeds of mango (127 mg/g and 447 mg/g extracted with water and a mixture 1:1 of acetone and water, respectively) and the peel of pomegranate (100 mg/g extracted with aqueous ethanol 80%) (Table 2). Other innovative techniques require the use of specific equipment or reagents, e.g. for EAE, UAE, MAE, PLE, SFE) or PEFs.

EAE is a key method for vegetal biomass since it hydrolyzes cell walls to allow the release of biocompounds from the cell. Different enzymes can be used such as pectinase, cellulase, hemicellulose or tannase. EAE has been demonstrated to achieve both good yields, e.g. in processing pomace from red grapes (240 mg/g of PC, Table 3) and purer extracts. However, it requires specific equipment and conditions to optimize the process which can be complex and laborious.

UAE has been widely applied given its relative simplicity and ease of optimization as well as lower solvent requirement. The application of ultrasounds has been proved to efficiently disrupt cell walls and provide high yields, e.g. in the case of avocado seeds and red grape pomace with yields of 120 mg/g and 438 mg/g of PC, respectively (Table 3). However, this technique implies individual treatment of each sample which increases the process time.

MAE is another extensively used technique given its efficiency and cost-effectiveness. Equipment is affordable and simple to use with less solvent needed and efficient cell disruption has achieved yields, e.g. of 155 mg/g PC for the peel of mango or 200 mg/g PC for red grape pomace (Table 3). MAE allows the simultaneous extraction of several samples depending on experimental conditions and available racks.

PLE has also been found to be feasible for recovering biocompounds from agricultural wastes since the process is cost-effective, allowing simultaneous extraction of samples with less solvent required. High recovery rates of PC can be achieved, e.g. peel of mandarin (113 mg/g) or olive leaves (289 mg/g, Table 3). However, optimal extractive conditions may require longer process times. However, equipment for the method (as well as for SFE and PEF) is relatively expensive with specific requirements for facilities and trained personnel. SFE, e.g. has achieved high recovery rates for tomato leaves (240 mg/g PCs) or sweet orange peels (389 mg/g PCs), whereas PEF applied on red grape pomace has achieved levels around 113 mg/g of PCs (Table 3).

In general terms, conventional extraction methods are advantageous from a short-term economic point of view since the required equipment is easy to set up on an industrial scale. However, these methods require large volumes of solvent with additional cost and a significant environmental impact. In addition, optimized extraction yields using conventional methods (either maceration, HAE or Soxhlet) also tend to require long process times, which may not be efficient from an operational point of view (Jimenez-Lopez et al., 2020). Alternative extraction technologies such as EAE, MAE and UAE provide shorter process times, minimal use of extractive solvents, and highly effective yields, making their industrial scale-up easier and more effective.

9 Conclusion

PCs have undoubtedly attracted a great deal of attention in recent decades due to their antioxidant and other properties, while also being preferred by consumers given their natural origin. OW is potentially one of the most accessible and cost-effective sources for obtaining PCs. Exploiting OW in this way reduces waste, adds value to crops (through the development of co-products) and promotes a more sustainable circular bioeconomy in which renewable natural resources are optimally exploited for a wide range of applications. Achieving these goals requires suitable and effective extraction methods which also need to be based on green and environment-friendly techniques. Considering that many of these methods are relatively novel and can require a considerable initial investment, S-L methods are still the most used. As a field of study that is increasingly gaining attention, further scientific research may provide novel insights on potential sources, biological effects and ways of obtaining PC from OW in the future.

10 Acknowledgements

The research leading to these results was supported by MICINN supporting the Ramón y Cajal grant for M. A. Prieto (RYC-2017-22891) and Jianbo Xiao (RYC-2020-030365-I), the Juan de la Cierva Incorporación grant for Hui Cao (IJC2020-046055-I); by Xunta de Galicia for supporting the program EXCELENCIA-ED431F 2020/12, the postdoctoral grants of M. Fraga-Corral (ED481B-2019/096) and L. Cassani (ED481B-2021/152), and the pre-doctoral grants of P. Garcia-Oliveira (ED481A-2019/295). The research leading to these results was supported by the European Union through the EcoChestnut Project (Erasmus + KA202) which supports the work of J. Echave. The authors are grateful to the Bio Based Industries Joint Undertaking (JU) under grant agreement No 888003 UP4HEALTH Project (H2020-BBI-JTI-2019). The JU receives support from the European Union's Horizon 2020 research and innovation program and the Bio Based Industries Consortium. The project SYSTEMIC Knowledge hub on Nutrition and Food Security has received funding from national research funding parties in Belgium (FWO), France (INRA), Germany (BLE), Italy (MIPAAF), Latvia (IZM), Norway (RCN), Portugal (FCT) and Spain (AEI) in a joint action of JPI HDHL, JPI-OCEANS and FACCE-JPI launched in 2019 under the ERA-NET ERA-HDHL (n° 696295).

11 Where to look for further information

A general overview of the European situation regarding the management of agricultural wastes can be found at several institutional websites. Indeed, the European Commission has prompted the implementation of alternative productive models aimed to promote more sustainable practices that reduce their negative environmental impact. Some of the adopted European strategies are as follows:

- Bio-economy: adopted by 2012 with the main objective of producing renewable bio-logical resources and their conversion into vital products and bioenergy (https://op.europa.eu/en/publication-detail/-/publication/1f0d8515-8dc0-4435-ba53-9570e47dbd51/language-en).

- Circular Economy Action Plan (CEAP): adopted by the end of 2015, this comprehensive policy package, comprise 54 actions and 4 legislative proposals on waste, regarding landfill, reuse, and recycling (https://eur-lex.europa.eu/legal-content/EN/TXT/?uri=CELEX:52015DC0614).
- European Green Deal: adopted by 2019 with the aim of reducing net green-house gas (GHG) emissions by >55% by 2030 (https://eur-lex.europa.eu/legal-content/EN/TXT/?uri=COM%3A2019%3A640%3AFIN). CEAP belongs and represents one building block of this policy package.
- 'Farm to fork': this strategy was presented by the European Commission by May 2020 to achieve climate neutrality by 2050 by shifting current EU food system towards more sustainable models (https://eur-lex.europa.eu/legal-content/EN/TXT/?uri=CELEX%3A52020DC0381). This strategy belongs to the Common Agricultural Policy and also represents a part of the Green Deal.

12 References

Aguirre, J., Ríos-Momberg, M., Hewitt, D. and Hansberg, W. (2005). Reactive oxygen species and development in microbial eukaryotes. *Trends in Microbiology* 13(3), 111–118. https://doi.org/10.1016/j.tim.2005.01.007.

Ahmed, I. A. M., Al Juhaimi, F., Özcan, M. M., Uslu, N., Babiker, E. E., Ghafoor, K., Osman, M. A. and Salih, H. A. A. (2021). A comparative study of bioactive compounds, antioxidant activity and phenolic compounds of melon (Cucumis melo L.) slices dehydrated by oven, microwave and infrared systems. *Journal of Food Processing and Preservation* 45(7), 1–8. https://doi.org/10.1111/jfpp.15605.

Aires, A. and Carvalho, R. (2020). Kiwi fruit residues from industry processing: study for a maximum phenolic recovery yield. *Journal of Food Science and Technology* 57(11), 4265–4276. https://doi.org/10.1007/s13197-020-04466-7.

Aires, A., Carvalho, R. and Saavedra, M. J. (2017). Reuse potential of vegetable wastes (broccoli, green bean and tomato) for the recovery of antioxidant phenolic acids and flavonoids. *International Journal of Food Science and Technology* 52(1), 98–107. https://doi.org/10.1111/ijfs.13256.

Alara, O. R., Abdurahman, N. H. and Olalere, O. A. (2018). Optimization of microwave-assisted extraction of flavonoids and antioxidants from Vernonia amygdalina leaf using response surface methodology. *Food and Bioproducts Processing* 107, 36–48. https://doi.org/10.1016/j.fbp.2017.10.007.

Alara, O. R., Abdurahman, N. H. and Ukaegbu, C. I. (2021). Extraction of phenolic compounds: a review. *Current Research in Food Science* 4(March), 200–214. https://doi.org/10.1016/j.crfs.2021.03.011.

Albuquerque, B. R. and Heleno, S. A., Oliveira, M. B. P. P., Barros, L. and Ferreira, I. C. F. R. (2021). Phenolic compounds: Current industrial applications, limitations and future challenges. *Food and Function*, *12*(1), 14–29. https://doi.org/10.1039/d0fo02324h.

Amarowicz, R., Carle, R., Dongowski, G., Durazzo, A., Galensa, R., Kammerer, D., Maiani, G. and Piskula, M. K. (2009). Influence of postharvest processing and storage on the content of phenolic acids and flavonoids in foods. *Molecular Nutrition and Food Research* 53 (Suppl. 2), S151–S183. https://doi.org/10.1002/mnfr.200700486.

Ambigaipalan, P., de Camargo, A. C. and Shahidi, F. (2016). Phenolic compounds of pomegranate byproducts (outer skin, mesocarp, divider membrane) and their antioxidant activities. *Journal of Agricultural and Food Chemistry* 64(34), 6584–6604. https://doi.org/10.1021/acs.jafc.6b02950.

Ameer, K., Shahbaz, H. M. and Kwon, J. H. (2017). Green extraction methods for polyphenols from plant matrices and their byproducts: a review. *Comprehensive Reviews in Food Science and Food Safety* 16(2), 295–315. https://doi.org/10.1111/1541-4337.12253.

Andreou, V., Dimopoulos, G., Dermesonlouoglou, E. and Taoukis, P. (2020). Application of pulsed electric fields to improve product yield and waste valorization in industrial tomato processing. *Journal of Food Engineering* 270, 109778. https://doi.org/10.1016/j.jfoodeng.2019.109778.

Arab, M., Bahramian, B., Schindeler, A., Valtchev, P., Dehghani, F. and McConchie, R. (2019). Extraction of phytochemicals from tomato leaf waste using subcritical carbon dioxide. *Innovative Food Science and Emerging Technologies* 57(May), 102204. https://doi.org/10.1016/j.ifset.2019.102204.

Araújo, M. N., Azevedo, A. Q. P. L., Hamerski, F., Voll, F. A. P. and Corazza, M. L. (2019). Enhanced extraction of spent coffee grounds oil using high-pressure CO2 plus ethanol solvents. *Industrial Crops and Products* 141(August), 111723. https://doi.org/10.1016/j.indcrop.2019.111723.

Ariza, M. T., Reboredo-Rodríguez, P., Cervantes, L., Soria, C., Martínez-Ferri, E., González-Barreiro, C., Cancho-Grande, B., Battino, M. and Simal-Gándara, J. (2018). Bioaccessibility and potential bioavailability of phenolic compounds from achenes as a new target for strawberry breeding programs. *Food Chemistry* 248(June 2017), 155–165. https://doi.org/10.1016/j.foodchem.2017.11.105.

Badhani, B., Sharma, N. and Kakkar, R. (2015). Gallic acid: A versatile antioxidant with promising therapeutic and industrial applications. *RSC Advances* 5(35), 27540–27557. https://doi.org/10.1039/C5RA01911G.

Baiano, A. (2014). Recovery of biomolecules from food wastes: a review. *Molecules* 19(9), 14821–14842. https://doi.org/10.3390/molecules190914821

Barbosa-Pereira, L., Guglielmetti, A. and Zeppa, G. (2018). Pulsed electric field assisted extraction of bioactive compounds from cocoa bean shell and coffee silverskin. *Food and Bioprocess Technology* 11(4), 818–835. https://doi.org/10.1007/s11947-017-2045-6.

Bargiacchi, E., Miele, S., Romani, A. and Campo, M. (2013). Biostimulant activity of hydrolyzable tannins from sweet chestnut (Castanea sativa Mill.). *Acta Horticulturae* 1009, 111–116. https://doi.org/10.17660/ActaHortic.2013.1009.13.

Barrales, F. M., Silveira, P., Barbosa, Pd. P. M., Ruviaro, A. R., Paulino, B. N., Pastore, G. M., Macedo, G. A. and Martinez, J. (2018). Recovery of phenolic compounds from citrus by-products using pressurized liquids: an application to orange peel. *Food and Bioproducts Processing* 112, 9–21. https://doi.org/10.1016/j.fbp.2018.08.006.

Benítez, V., Mollá, E., Martín-Cabrejas, M. A., Aguilera, Y., López-Andréu, F. J., Cools, K., Terry, L. A. and Esteban, R. M. (2011). Characterization of industrial onion wastes (Allium cepa L.): dietary fibre and bioactive compounds. *Plant Foods for Human Nutrition* 66(1), 48–57. https://doi.org/10.1007/s11130-011-0212-x.

Błaszczyk, A., Sady, S. and Sielicka, M. (2019). The stilbene profile in edible berries. *Phytochemistry Reviews* 18(1), 37–67. https://doi.org/10.1007/s11101-018-9580-2.

Borowski, R. G. V. Von, Barros, M. P., da Silva, D. B., Lopes, N. P., Zimmer, K. R., Staats, C. C., de Oliveira, C. B., Giudice, E., Gillet, R., Macedo, A. J., Gnoatto, S. C. B. and Zimmer, A. R. (2020). Red pepper peptide coatings control Staphylococcus epidermidis adhesion and biofilm formation. *International Journal of Pharmaceutics* 574(December 2019), 118872. https://doi.org/10.1016/j.ijpharm.2019.118872.

Botineştean, C., Gruia, A. T. and Jianu, I. (2015). Utilization of seeds from tomato processing wastes as raw material for oil production. *Journal of Material Cycles and Waste Management* 17(1), 118–124. https://doi.org/10.1007/s10163-014-0231-4.

Boudjou, S., Oomah, B. D., Zaidi, F. and Hosseinian, F. (2013). Phenolics content and antioxidant and anti-inflammatory activities of legume fractions. *Food Chemistry* 138(2–3), 1543–1550. https://doi.org/10.1016/j.foodchem.2012.11.108.

Brewer, M. S. (2011). Natural antioxidants: sources, compounds, mechanisms of action, and potential applications. *Comprehensive Reviews in Food Science and Food Safety* 10(4), 221–247. https://doi.org/10.1111/j.1541-4337.2011.00156.x.

Brglez Mojzer, E., Knez Hrnčič, M., Škerget, M., Knez, Ž. and Bren, U. (2016). Polyphenols: extraction methods, antioxidative action, bioavailability and anticarcinogenic effects. *Molecules* 21(7), 901. https://doi.org/10.3390/molecules21070901.

Brianceau, S., Turk, M., Vitrac, X. and Vorobiev, E. (2015). Combined densification and pulsed electric field treatment for selective polyphenols recovery from fermented grape pomace. *Innovative Food Science and Emerging Technologies* 29, 2–8. https://doi.org/10.1016/j.ifset.2014.07.010.

Burri, S. C. M., Ekholm, A., Håkansson, Å., Tornberg, E. and Rumpunen, K. (2017). Antioxidant capacity and major phenol compounds of horticultural plant materials not usually used. *Journal of Functional Foods* 38(A), 119–127.

Caimari, A., Del Bas, J. M., Crescenti, A. and Arola, L. (2013). Low doses of grape seed procyanidins reduce adiposity and improve the plasma lipid profile in hamsters. *International Journal of Obesity* 37(4), 576–583. https://doi.org/10.1038/ijo.2012.75.

Cardinali, A., Cicco, N., Linsalata, V., Minervini, F., Pati, S., Pieralice, M., Tursi, N. and Lattanzio, V. (2010). Biological activity of high molecular weight phenolics from olive mill wastewater. *Journal of Agricultural and Food Chemistry* 58(15), 8585–8590. https://doi.org/10.1021/jf101437c.

Carocho, M., Morales, P. and Ferreira, I. C. F. R. (2015). Natural food additives: quo vadis? *Trends in Food Science and Technology* 45(2), 284–295. https://doi.org/10.1016/j.tifs.2015.06.007.

Carreira-Casais, A., Lourenço-Lopes, C., Otero, P., Carpena, M., Gonzalez Pereira, A., Echave, J., Soria-Lopez, A., Chamorro, F., Prieto, M. A. and Simal-Gandara, J. (2021). Application of green extraction techniques for natural additives production. *Food Additives [Working Title]*, August. https://doi.org/10.5772/intechopen.100320.

Cassino, C., Tsolakis, C., Bonello, F., Gianotti, V. and Osella, D. (2019). Wine evolution during bottle aging, studied by 1H NMR spectroscopy and multivariate statistical analysis. *Food Research International* 116(April 2018), 566–577. https://doi.org/10.1016/j.foodres.2018.08.075.

Catalkaya, G. and Kahveci, D. (2019). Optimization of enzyme assisted extraction of lycopene from industrial tomato waste. *Separation and Purification Technology* 219, 55–63. Available at: https://doi.org/10.1016/j.seppur.2019.03.006.

Cavalcanti, R. N., Navarro-Díaz, H. J., Santos, D. T., Rostagno, M. A. and Meireles, M. A. A. (2012). Supercritical carbon dioxide extraction of polyphenols from pomegranate (Punica granatum L.) leaves: chemical composition, economic evaluation and chemometric approach. *Journal of Food Research* 1(3), 282. https://doi.org/10.5539/jfr.v1n3p282.

Chamorro, S., Viveros, A., Alvarez, I., Vega, E. and Brenes, A. (2012). Changes in polyphenol and polysaccharide content of grape seed extract and grape pomace after enzymatic treatment. *Food Chemistry* 133(2), 308–314. https://doi.org/10.1016/j.foodchem.2012.01.031.

Chaves, J. O., de Souza, M. C., da Silva, L. C., Lachos-Perez, D., Torres-Mayanga, P. C., Machado, A. P. D. F., Forster-Carneiro, T., Vázquez-Espinosa, M., González-de-Peredo, A. V., Barbero, G. F. and Rostagno, M. A. (2020). Extraction of flavonoids from natural sources using modern techniques. *Frontiers in Chemistry* 8(September), 507887. https://doi.org/10.3389/fchem.2020.507887.

Chemat, F., Rombaut, N., Sicaire, A. G., Meullemiestre, A., Fabiano-Tixier, A. S. and Abert-Vian, M. (2017). Ultrasound assisted extraction of food and natural products. Mechanisms, techniques, combinations, protocols and applications. A review. In: *Ultrasonics Sonochemistry*. Elsevier 34. https://doi.org/10.1016/j.ultsonch.2016.06.035.

Chhouk, K., Uemori, C., Wahyudiono, Kanda, H. and Goto, M. (2017) Extraction of phenolic compounds and antioxidant activity from garlic husk using carbon dioxide expanded ethanol. *Chemical Engineering and Processing: Process Intensification* 117, 113–119. https://doi.org/10.1016/j.cep.2017.03.023.

Cory, H., Passarelli, S., Szeto, J., Tamez, M. and Mattei, J. (2018). The role of polyphenols in human health and food systems: A mini-review. *Frontiers in Nutrition* 5(September), 87. https://doi.org/10.3389/fnut.2018.00087.

Dang, X. C., Hoan, N. X., Dong, D. H., Thuy, L. T. M., Thanh, N. Van, Ha, H. T., Tuyen, D. T. T. and Chinh, D. X. (2019). Tannins: extraction from plants. In: *Tannins: Structural Properties, Biological Properties and Current Knowledge* (pp. 1–20). IntechOpen. https://doi.org/10.5772/intechopen.86040.

De Bruno, A., Romeo, R., Piscopo, A. and Poiana, M. (2021). Antioxidant quantification in different portions obtained during olive oil extraction process in an olive oil press mill. *Journal of the Science of Food and Agriculture* 101(3), 1119–1126. https://doi.org/10.1002/jsfa.10722.

de Camargo, A. C., Schwember, A. R., Parada, R., Garcia, S., Maróstica Júnior, M. R., Franchin, M., Regitano-D'arce, M. A. B. and Shahidi, F. (2018). Opinion on the hurdles and potential health benefits in value-added use of plant food processing by-products as sources of phenolic compounds. *International Journal of Molecular Sciences*. MDPI AG 19(11). https://doi.org/10.3390/ijms19113498.

De la Rosa, L. A., Alvarez-Parrilla, E. and González-Aguilar, G. A. (2009). Fruit and vegetable phytochemicals: chemistry, nutritional value, and stability. In: Ames (Ed.). *Fruit and Vegetable Phytochemicals: Chemistry, Nutritional Value, and Stability* (1st ed., vol. I). W.-B. https://doi.org/10.1002/9780813809397.

de Lima Cherubim, D. J., Buzanello Martins, C. V., Oliveira Fariña, L. and da Silva de Lucca, R. A. (2020). Polyphenols as natural antioxidants in cosmetics applications. *Journal of Cosmetic Dermatology* 19(1), 33–37. https://doi.org/10.1111/jocd.13093.

Directive 2006/12/EC of the European Parliament and of the Council of 5 April 2006 on waste, April (2006). *Official Journal of the European Union L 114/9.*

Dorta, E., González, M., Lobo, M. G., Sánchez-Moreno, C. and de Ancos, B. (2014). Screening of phenolic compounds in by-product extracts from mangoes (Mangifera indica L.) by HPLC-ESI-QTOF-MS and multivariate analysis for use as a food ingredient. *Food Research International* 57, 51–60. https://doi.org/10.1016/j.foodres.2014.01.012.

Dresch, R. R., Dresch, M. K., Guerreiro, A. F., Biegelmeyer, R., Holzschuh, M. H., Rambo, D. F. and Henriques, A. T. (2014). Phenolic compounds from the leaves of Vitis labrusca and Vitis vinifera L. as a source of waste byproducts: development and validation of LC method and antichemotactic activity. *Food Analytical Methods* 7(3), 527–539. https://doi.org/10.1007/s12161-013-9650-4.

Drosou, C., Kyriakopoulou, K., Bimpilas, A., Tsimogiannis, D. and Krokida, M. (2015). A comparative study on different extraction techniques to recover red grape pomace polyphenols from vinification byproducts. *Industrial Crops and Products* 75, 141–149. https://doi.org/10.1016/j.indcrop.2015.05.063.

Durazzo, A. and Lucarini, M. (2019). Extractable and non-extractable antioxidants. *Molecules* 24(10). https://doi.org/10.3390/molecules24101933.

Durazzo, A., Lucarini, M., Camilli, E., Marconi, S., Gabrielli, P., Lisciani, S., Gambelli, L., Aguzzi, A., Novellino, E., Santini, A., Turrini, A. and Marletta, L. (2018). Dietary lignans: definition, description and research trends in databases development. *Molecules* 23(12), 1–14. https://doi.org/10.3390/molecules23123251.

Durazzo, A., Lucarini, M., Souto, E. B., Cicala, C., Caiazzo, E., Izzo, A. A., Novellino, E. and Santini, A. (2019). Polyphenols: A concise overview on the chemistry, occurrence, and human health. *Phytotherapy Research* 33(9), 2221–2243. https://doi.org/10.1002/ptr.6419.

El-Hadary, A. E. and Ramadan, M. F. (2019). Phenolic profiles, antihyperglycemic, antihyperlipidemic, and antioxidant properties of pomegranate (Punica granatum) peel extract. *Journal of Food Biochemistry* 43(4), e12803. https://doi.org/10.1111/jfbc.12803.

El Gharras, H. (2009). Polyphenols: food sources, properties and applications: a review. *International Journal of Food Science and Technology* 44(12), 2512–2518. https://doi.org/10.1111/j.1365-2621.2009.02077.x.

Esquivel, P. and Jiménez, V. M. (2012). Functional properties of coffee and coffee by-products. *Food Research International* 46(2), 488–495. https://doi.org/10.1016/j.foodres.2011.05.028.

European Commission (2005). Taking Sustainable use of resources forward: A Thematic Strategy on the prevention and recycling of waste. *Com 666 Final.*

European Commission (2008). Council Directive 2008/98/EC of 19 November 2008 on waste and repealing certain Directives. *Official Journal of the European Union* 312(3), 3–30. https://doi.org/2008/98/EC.; 32008L0098.

European Commission (2017). Commission Regulation (EU) No 432/2012 of 16 May 2012 establishing a list of permitted health claims made on foods, other than those referring to the reduction of disease risk and to children's development and health. Https://Eur-Lex.Europa.Eu/, 10(December 2016), 1–21. https://data.europa.eu/eli/reg/2012/432/2021-05-17.

Fazio, A., Plastina, P., Meijerink, J., Witkamp, R. F. and Gabriele, B. (2013). Comparative analyses of seeds of wild fruits of Rubus and Sambucus species from Southern Italy : fatty acid composition of the oil, total phenolic content , antioxidant and anti-inflammatory properties of the methanolic extracts. *Food Chemistry* 140(4), 817–824. https://doi.org/10.1016/j.foodchem.2012.11.010.

Ferreira, D. F., Barin, J. S., Binello, A., Veselov, V. V. and Cravotto, G. (2019). Highly efficient pumpkin-seed extraction with the simultaneous recovery of lipophilic and hydrophilic compounds. *Food and Bioproducts Processing* 117, 224–230. https://doi.org/10.1016/j.fbp.2019.07.014.

Fidelis, M., de Moura, C., Kabbas Junior, T., Pap, N., Mattila, P., Mäkinen, S., Putnik, P., Bursać Kovačević, D., Tian, Y., Yang, B. and Granato, D. (2019). Fruit seeds as sources of bioactive compounds: sustainable production of high value-added ingredients from by-products within circular economy. *Molecules* 24(21), 1–54. https://doi.org/10.3390/molecules24213854.

Fortin, C. F., Larbi, A., Lesur, O., Douziech, N. and Fulop, T. (2006). Impairment of SHP-1 down-regulation in the lipid rafts of human neutrophils under GM-CSF stimulation contributes to their age-related, altered functions. *Journal of Leukocyte Biology* 79(5), 1061–1072. https://doi.org/10.1189/jlb.0805481.

Fraga-Corral, M., García-Oliveira, P., Pereira, A. G., Lourenço-Lopes, C., Jimenez-Lopez, C., Prieto, M. A. and Simal-Gandara, J. (2020). Technological application of tannin-based extracts. *Molecules* 25(3), 1–27.

Fraga-Corral, M., Otero, P., Echave, J., Garcia-Oliveira, P., Carpena, M., Jarboui, A., Nuñez-Estevez, B., Simal-Gandara, J. and Prieto, M. A. (2021). By-products of Agri-food industry as tannin-rich sources: a review of tannins' biological activities and their potential for valorization. *Foods* 10(1). https://doi.org/10.3390/foods10010137.

Galaz, P., Valdenegro, M., Ramírez, C., Nuñez, H., Almonacid, S. and Simpson, R. (2017). Effect of drum drying temperature on drying kinetic and polyphenol contents in pomegranate peel. *Journal of Food Engineering* 208, 19–27. https://doi.org/10.1016/j.jfoodeng.2017.04.002.

Garcia-Mendoza, M. P., Paula, J. T., Paviani, L. C., Cabral, F. A. and Martinez-Correa, H. A. (2015). Extracts from mango peel by-product obtained by supercritical CO2 and pressurized solvent processes. *LWT: Food Science and Technology* 62(1), 131–137. https://doi.org/10.1016/j.lwt.2015.01.026.

Garcia-Oliveira, P., Barral, M., Carpena, M., Gullón, P., Fraga-Corral, M., Otero, P., Prieto, M. A. and Simal-Gandara, J. (2021). Traditional plants from Asteraceae family as potential candidates for functional food industry. *Food and Function* 12(7), 2850–2873. https://doi.org/10.1039/d0fo03433a.

Garcia-Oliveira, P., Otero, P., Pereira, A. G., Chamorro, F., Carpena, M., Echave, J., Fraga-Corral, M., Simal-Gandara, J. and Prieto, M. A. (2021). Status and challenges of plant-anticancer compounds in cancer treatment. In: *Pharmaceuticals* 14(2). https://doi.org/10.3390/ph14020157.

Garcia-Vaquero, M., Rajauria, G. and Tiwari, B. (2020). Conventional extraction techniques: solvent extraction. In: Torres, M. D., Kraan, S. and Dominguez, H. (Eds). *Sustainable Seaweed Technologies* (pp. 171–189). Elsevier. https://doi.org/10.1016/B978-0-12-817943-7.00006-8.

Gligor, O., Mocan, A., Moldovan, C., Locatelli, M., Crişan, G. and Ferreira, I. C. F. R. (2019). Enzyme-assisted extractions of polyphenols – A comprehensive review. *Trends in Food Science and Technology* 88(March), 302–315. https://doi.org/10.1016/j.tifs.2019.03.029.

Goławska, S., Sprawka, I., Łukasik, I. and Goławski, A. (2014). Are naringenin and quercetin useful chemicals in pest-management strategies? *Journal of Pest Science* 87(1), 173–180. https://doi.org/10.1007/s10340-013-0535-5.

Goula, A. M., Thymiatis, K. and Kaderides, K. (2016). Food and Bioproducts processing Valorization of grape pomace : drying behavior and ultrasound extraction of phenolics. *Food and Bioproducts Processing* 100, 132–144. https://doi.org/10.1016/j.fbp.2016.06.016.

Grillo, G., Gunjević, V., Radošević, K., Redovniković, I. R. and Cravotto, G. (2020). Deep eutectic solvents and nonconventional technologies for blueberry-peel extraction: kinetics, anthocyanin stability, and antiproliferative activity. *Antioxidants* 9(11), 1069. https://doi.org/10.3390/antiox9111069.

Grivennikov, S. I., Greten, F. R. and Karin, M. (2010). Immunity, inflammation, and cancer. *Cell* 140(6), 883–899. https://doi.org/10.1016/j.cell.2010.01.025.

Gullón, P., Gullón, B., Astray, G., Carpena, M., Fraga-Corral, M., Prieto, M. A. and Simal-Gandara, J. (2020). Valorization of by-products from olive oil industry and added-value applications for innovative functional foods. *Food Research International* 137, 109683. https://doi.org/10.1016/j.foodres.2020.109683.

Gunes, R., Palabiyik, I., Toker, O. S., Konar, N. and Kurultay, S. (2019). Incorporation of defatted apple seeds in chewing gum system and phloridzin dissolution kinetics. *Journal of Food Engineering* 255, 9–14. https://doi.org/10.1016/j.jfoodeng.2019.03.010.

Guthrie, F., Wang, Y., Neeve, N., Young, S. Y., Mohammadi, K. and Baroutian, S. (2020). Recovery of phenolic antioxidants from green kiwifruit peel using subcritical water extraction. *Food and Bioproducts Processing* 122, 136–144. https://doi.org/10.1016/j.fbp.2020.05.002.

Heeger, A., Kosińska-Cagnazzo, A., Cantergiani, E. and Andlauer, W. (2017). Bioactives of coffee cherry pulp and its utilisation for production of Cascara beverage. *Food Chemistry* 221, 969–975. https://doi.org/10.1016/j.foodchem.2016.11.067.

Herrero, M., Sánchez-Camargo, A. del P., Cifuentes, A. and Ibáñez, E. (2015). Plants, seaweeds, microalgae and food by-products as natural sources of functional ingredients obtained using pressurized liquid extraction and supercritical fluid extraction. *TrAC Trends in Analytical Chemistry* 71, 26–38. https://doi.org/10.1016/j.trac.2015.01.018.

Herrling, T. and Jung, K. (2012). The Radical Status Factor (RSF): A novel metric to characterize skin products. *International Journal of Cosmetic Science* 34(4), 285–290. https://doi.org/10.1111/j.1468-2494.2012.00718.x.

Ichikawa, M., Ryu, K., Yoshida, J., Ide, N., Kodera, Y., Sasaoka, T. and Rosen, R. T. (2003). Identification of six phenylpropanoids from garlic skin as major antioxidants. *Journal of Agricultural and Food Chemistry* 51(25), 7313–7317. https://doi.org/10.1021/jf034791a.

Ishangulyyev, R., Kim, S. and Lee, S. H. (2019). Understanding food loss and waste-why are we losing and wasting food? *Foods* 8(8). https://doi.org/10.3390/foods8080297.

Jimenez-Lopez, C., Fraga-Corral, M., Carpena, M., García-Oliveira, P., Echave, J., Pereira, A. G., Lourenço-Lopes, C., Prieto, M. A. and Simal-Gandara, J. (2020). Agriculture waste valorisation as a source of antioxidant phenolic compounds within a circular and sustainable bioeconomy. *Food and Function* 11(6), 4853–4877. https://doi.org/10.1039/D0FO00937G.

Kadam, S. U., Tiwari, B. K., Álvarez, C. and O'Donnell, C. P. (2015). Ultrasound applications for the extraction, identification and delivery of food proteins and bioactive peptides. *Trends in Food Science and Technology* 46(1), 60–67. https://doi.org/10.1016/j.tifs.2015.07.012.

Kaewprachu, P., Osako, K., Benjakul, S. and Rawdkuen, S. (2015). Quality attributes of minced pork wrapped with catechin-lysozyme incorporated gelatin film. *Food Packaging and Shelf Life* 3, 88–96. https://doi.org/10.1016/j.fpsl.2014.11.002.

Kalinowska, M., Gryko, K., Wróblewska, A. M., Jabłońska-Trypuć, A. and Karpowicz, D. (2020). Phenolic content, chemical composition and anti-/pro-oxidant activity of Gold Milenium and Papierowka apple peel extracts. *Scientific Reports* 10(1), 14951. https://doi.org/10.1038/s41598-020-71351-w.

Kamath, V. and Rajini, P. S. (2007). The efficacy of cashew nut (Anacardium occidentale L.) skin extract as a free radical scavenger. *Food Chemistry* 103(2), 428–433. https://doi.org/10.1016/j.foodchem.2006.07.031.

Kaulmann, A. and Bohn, T. (2014). Carotenoids, inflammation, and oxidative stress-implications of cellular signaling pathways and relation to chronic disease prevention. *Nutrition Research* 34(11), 907–929. https://doi.org/10.1016/j.nutres.2014.07.010.

Kazemi, M., Khodaiyan, F., Hosseini, S. S. and Najari, Z. (2019). An integrated valorization of industrial waste of eggplant: simultaneous recovery of pectin, phenolics and sequential production of pullulan. *Waste Management* 100, 101–111. https://doi.org/10.1016/j.wasman.2019.09.013.

Kennedy, D. O. and Wightman, E. L. (2011). Herbal Extracts and Phytochemicals: plant Secondary Metabolites and the Enhancement of Human Brain function. *Advances in Nutrition* 2(1), 32–50.

Kessler, A. and Heil, M. (2011). The multiple faces of indirect defences and their agents of natural selection. *Functional Ecology* 25(2), 348–357.

Khalid, M., Saeed-ur-Rahman, Bilal, M. and HUANG, D.-F. (2019). Role of flavonoids in plant interactions with the environment and against human pathogens – a review. *Journal of Integrative Agriculture* 18(1), 211–230. https://doi.org/10.1016/S2095-3119(19)62555-4.

Kheirkhah, H., Baroutian, S. and Young, S. Y. (2019). Evaluation of bioactive compounds extracted from Hayward kiwifruit pomace by subcritical water extraction. *Food and Bioproducts Processing* 115, 143–153. https://doi.org/10.1016/j.fbp.2019.03.007.

Khoo, H. E., Azlan, A., Tang, S. T. and Lim, S. M. (2017). Anthocyanidins and anthocyanins: colored pigments as food, pharmaceutical ingredients, and the potential health benefits. *Food and Nutrition Research* 61(1), 1361779. https://doi.org/10.1080/16546628.2017.1361779.

Kieu Tran, T. M., Kirkman, T., Nguyen, M. and Van Vuong, Q. (2020). Effects of drying on physical properties, phenolic compounds and antioxidant capacity of robusta wet coffee pulp (*Coffea canephora*). *Heliyon* 6(7), e04498. https://doi.org/10.1016/j.heliyon.2020.e04498.

Kilic, I. H., Sarikurkcu, C., Karagoz, I. D., Uren, M. C., Kocak, M. S., Cilkiz, M. and Tepe, B. (2016). A significant by-product of the industrial processing of pistachios: shell skin - RP-HPLC analysis, and antioxidant and enzyme inhibitory activities of the methanol extracts of Pistacia vera L. shell skins cultivated in Gaziantep, Turkey. *RSC Advances* 6(2), 1203–1209. https://doi.org/10.1039/C5RA24530C.

Knez, Ž., Škerget, M. and KnezHrnčič, M. (2013). Principles of supercritical fluid extraction and applications in the food, beverage and nutraceutical industries. In: *Separation, Extraction and Concentration Processes in the Food, Beverage and Nutraceutical Industries* (pp. 3–38). Elsevier. https://doi.org/10.1533/9780857090751.1.3.

Ko, M.-J., Kwon, H.-L. and Chung, M.-S. (2016). Pilot-scale subcritical water extraction of flavonoids from satsuma mandarin (Citrus unshiu Markovich) peel. *Innovative Food Science and Emerging Technologies* 38, 175–181. https://doi.org/10.1016/j.ifset.2016.10.008.

Kosińska, A., Karamać, M., Estrella, I., Hernández, T., Bartolomé, B. and Dykes, G. A. (2012). Phenolic compound profiles and antioxidant capacity of Persea americana Mill. Peels and Seeds of two varieties. *Journal of Agricultural and Food Chemistry* 60(18), 4613–4619. https://doi.org/10.1021/jf300090p.

Kumar, N. and Pruthi, V. (2014). Potential applications of ferulic acid from natural sources. *Biotechnology Reports* 4(1), 86–93. https://doi.org/10.1016/j.btre.2014.09.002.

Kupnik, K., Leitgeb, M., Primožič, M., Postružnik, V., Kotnik, P., Kučuk, N., Knez, Ž. and Marevci, M. K. (2022). Supercritical fluid and conventional extractions of high value-added compounds from pomegranate peels waste: production, quantification and antimicrobial activity of bioactive constituents. *Plants* 11(7). https://doi.org/10.3390/plants11070928.

Lafay, S. and Gil-Izquierdo, A. (2008). Bioavailability of phenolic acids. *Phytochemistry Reviews* 7(2), 301–311. https://doi.org/10.1007/s11101-007-9077-x.

Lage, N. N., Layosa, M. A. A., Arbizu, S., Chew, B. P., Pedrosa, M. L., Mertens-Talcott, S., Talcott, S. and Noratto, G. D. (2020). Dark sweet cherry (*Prunus aviumz*) phenolics enriched in anthocyanins exhibit enhanced activity against the most aggressive breast cancer subtypes without toxicity to normal breast cells. *Journal of Functional Foods* 64.

Laplante, M. and Sabatini, D. M. (2012). MTOR signaling in growth control and disease. *Cell* 149(2), 274–293. https://doi.org/10.1016/j.cell.2012.03.017.

Lavilla, I. and Bendicho, C. (2017). Fundamentals of ultrasound-assisted extraction. In: *Water Extraction of Bioactive Compounds: From Plants to Drug Development*, (291–316). https://doi.org/10.1016/B978-0-12-809380-1.00011-5.

Li, W., Yang, R., Ying, D., Yu, J., Sanguansri, L. and Augustin, M. A. (2020). Analysis of polyphenols in apple pomace: A comparative study of different extraction and hydrolysis procedures. *Industrial Crops and Products* 147, 112250. https://doi.org/10.1016/j.indcrop.2020.112250.

Lin, D., Ma, Q., Zhang, Y. and Peng, Z. (2020). Phenolic compounds with antioxidant activity from strawberry leaves: a study on microwave-assisted extraction optimization. *Preparative Biochemistry and Biotechnology* 50(9), 874–882. https://doi.org/10.1080/10826068.2020.1762213.

Liu, P., Kallio, H. and Yang, B. (2011). Phenolic compounds in hawthorn (Crataegus grayana) fruits and leaves and changes during fruit ripening. *Journal of Agricultural and Food Chemistry* 59(20), 11141–11149.

Liu, Z. W., Zeng, X. A. and Ngadi, M. (2018). Enhanced extraction of phenolic compounds from onion by pulsed electric field (PEF). *Journal of Food Processing and Preservation* 42(9), 4–11. https://doi.org/10.1111/jfpp.13755.

López, C. J., Caleja, C., Prieto, M. A., Barreiro, M. F., Barros, L. and Ferreira, I. C. F. R. (2018). Optimization and comparison of heat and ultrasound assisted extraction techniques to obtain anthocyanin compounds from Arbutus unedo L . Fruits. *Food Chemistry* 264(May), 81–91. https://doi.org/10.1016/j.foodchem.2018.04.103.

López, C. J., Caleja, C., Prieto, M. A., Sokovic, M., Calhelha, R. C., Barros, L. and Ferreira, I. C. F. R. (2019). Stability of a cyanidin-3-O-glucoside extract obtained from *Arbutus unedo* L. and incorporation into wafers for colouring purposes. *Food Chemistry* 275(June 2018), 426–438.

Lucarini, M., Durazzo, A., Kiefer, J., Santini, A., Lombardi-Boccia, G., Souto, E. B., Romani, A., Lampe, A., Ferrari Nicoli, S., Gabrielli, P., Bevilacqua, N., Campo, M., Morassut, M. and Cecchini, F. (2019). Grape seeds: chromatographic profile of fatty acids and phenolic compounds and qualitative analysis by FTIR-ATR spectroscopy. *Foods* 9(1), 10. https://doi.org/10.3390/foods9010010.

Luque de Castro, M. D. and Priego-Capote, F. (2010). Soxhlet extraction: past and present panacea. *Journal of Chromatography. A* 1217(16), 2383–2389. https://doi.org/10.1016/j.chroma.2009.11.027.

Makris, D. P. (2018). Green extraction processes for the efficient recovery of bioactive polyphenols from wine industry solid wastes – recent progress. *Current Opinion in Green and Sustainable Chemistry* 13, 50–55. https://doi.org/10.1016/j.cogsc.2018.03.013.

Mallek-Ayadi, S., Bahloul, N. and Kechaou, N. (2017). Characterization, phenolic compounds and functional properties of Cucumis melo L. peels. *Food Chemistry* 221, 1691–1697. https://doi.org/10.1016/j.foodchem.2016.10.117.

Manna, L., Bugnone, C. A. and Banchero, M. (2015). Valorization of hazelnut, coffee and grape wastes through supercritical fluid extraction of triglycerides and polyphenols. *Journal of Supercritical Fluids* 104, 204–211. https://doi.org/10.1016/j.supflu.2015.06.012.

Mayanga-Torres, P. C., Lachos-Perez, D., Rezende, C. A., Prado, J. M., Ma, Z., Tompsett, G. T., Timko, M. T. and Forster-Carneiro, T. (2017). Valorization of coffee industry residues by subcritical water hydrolysis: recovery of sugars and phenolic compounds. *Journal of Supercritical Fluids* 120, 75–85. https://doi.org/10.1016/j.supflu.2016.10.015.

Medina, I., Gallardo, J. M., González, M. J., Lois, S. and Hedges, N. (2007). Effect of molecular structure of phenolic families as hydroxycinnamic acids and catechins on their antioxidant effectiveness in minced fish muscle. *Journal of Agricultural and Food Chemistry* 55(10), 3889–3895. https://doi.org/10.1021/jf063498i.

Metcalfe, N. B. and Alonso-Alvarez, C. (2010). Oxidative stress as a life-history constraint: the role of reactive oxygen species in shaping phenotypes from conception to death. *Functional Ecology* 24(5), 984–996. https://doi.org/10.1111/j.1365-2435.2010.01750.x.

Moein, S. and Moein, M. R. (2010). Relationship between antioxidant properties and phenolics in Zhumeria majdae. *Journal of Medicinal Plants Research* 4(7), 517–521.

Moloney, J. N. and Cotter, T. G. (2018). ROS signalling in the biology of cancer. *Seminars in Cell and Developmental Biology* 80, 50–64. https://doi.org/10.1016/j.semcdb.2017.05.023.

Montenegro-Landívar, M. F., Tapia-Quirós, P., Vecino, X., Reig, M., Valderrama, C., Granados, M., Cortina, J. L. and Saurina, J. (2021). Fruit and vegetable processing wastes as natural sources of antioxidant-rich extracts: evaluation of advanced extraction technologies by surface response methodology. *Journal of Environmental Chemical Engineering* 9(4). https://doi.org/10.1016/j.jece.2021.105330.

Morais, D. R., Rotta, E. M., Sargi, S. C., Schmidt, E. M., Bonafe, E. G., Eberlin, M. N., Sawaya, A. C. H. F. and Visentainer, J. V. (2015). Antioxidant activity, phenolics and UPLC-ESI(-)-MS of extracts from different tropical fruits parts and processed peels. *Food Research International* 77, 392–399. https://doi.org/10.1016/j.foodres.2015.08.036.

Mourtzinos, I. and Goula, A. (2019). Polyphenols in agricultural byproducts and food waste. In: *Polyphenols in Plants* (2nd ed.). Elsevier Inc. https://doi.org/10.1016/b978-0-12-813768-0.00002-5.

Mukherjee, S., Dudley, J. I. and Das, D. K. (2010). Dose-dependency of resveratrol in providing health benefits. *Dose-Response* 8(4), 478–500. https://doi.org/10.2203/dose-response.09-015.Mukherjee.

Multari, S., Carlin, S., Sicari, V. and Martens, S. (2020). Differences in the composition of phenolic compounds, carotenoids, and volatiles between juice and pomace of four citrus fruits from Southern Italy. *European Food Research and Technology* 246(10), 1991–2005. https://doi.org/10.1007/s00217-020-03550-8.

Munekata, P. E. S., Franco, D., Trindade, M. A. and Lorenzo, J. M. (2016). Characterization of phenolic composition in chestnut leaves and beer residue by LC-DAD-ESI-MS. *LWT: Food Science and Technology* 68, 52–58. https://doi.org/10.1016/j.lwt.2015.11.017.

Murcia, M. A. and Martínez-Tomé, M. (2001). Antioxidant activity of resveratrol compared with common food additives. *Journal of Food Protection* 64(3), 379–384. https://doi.org/10.4315/0362-028X-64.3.379.

Nadar, S. S., Rao, P. and Rathod, V. K. (2018). Enzyme assisted extraction of biomolecules as an approach to novel extraction technology: a review. *Food Research International* 108(October 2017), 309–330. https://doi.org/10.1016/j.foodres.2018.03.006.

Nguyen, V. and Scarlett, C. (2016). Mass proportion, bioactive compounds and antioxidant capacity of carrot peel as affected by various solvents. *Technologies* 4(4), 36. https://doi.org/10.3390/technologies4040036.

Nguyen, V. T. and Le, M. D. (2018). Influence of various drying conditions on phytochemical compounds and antioxidant activity of carrot peel. *Beverages* 4(4). https://doi.org/10.3390/beverages4040080.

Oleszek, M., Pecio, Ł., Kozachok, S., Lachowska-Filipiuk, Ż., Oszust, K. and Frąc, M. (2019). Phytochemicals of apple pomace as prospect bio-fungicide agents against mycotoxigenic fungal species–in vitro experiments. *Toxins* 11(6). https://doi.org/10.3390/toxins11060361.

Oliveira, D. F., Costa, V. A., Terra, W. C., Campos, V. P., Paula, P. M. and Martins, S. J. (2019). Impact of phenolic compounds on *Meloidogyne incognita in vitro* and in tomato plants. *Experimental Parasitology* 199(December 2018), 17–23.

Omoba, O. S., Obafaye, R. O., Salawu, S. O., Boligon, A. A. and Athayde, M. L. (2015). HPLC-DAD phenolic characterization and antioxidant activities of ripe and unripe sweet orange peels. *Antioxidants* 4(3), 498–512. https://doi.org/10.3390/antiox4030498.

Osorio-Tobón, J. F. (2020). Recent advances and comparisons of conventional and alternative extraction techniques of phenolic compounds. *Journal of Food Science and Technology* 57(12), 4299–4315. https://doi.org/10.1007/s13197-020-04433-2.

Pal, C. B. T. and Jadeja, G. C. (2022). Optimization and kinetics of polyphenol recovery from raw mango (Mangifera indica L.) peel using a glycerol-sodium acetate deep eutectic solvent system. *Biomass Conversion and Biorefinery* 89, 01234567. https://doi.org/10.1007/s13399-022-02550-w.

Palomo, I., Concha-Meyer, A., Lutz, M., Said, M., Sáez, B., Vásquez, A. and Fuentes, E. (2019). Chemical characterization and antiplatelet potential of bioactive extract from tomato pomace (byproduct of tomato paste). *Nutrients* 11(2), 456. https://doi.org/10.3390/nu11020456.

Panche, A. N., Diwan, A. D. and Chandra, S. R. (2016). Flavonoids: an overview. *Journal of Nutritional Science* 5, e47. https://doi.org/10.1017/jns.2016.41.

Pandey, M. K., Gupta, S. C., Nabavizadeh, A. and Aggarwal, B. B. (2017). Regulation of cell signaling pathways by dietary agents for cancer prevention and treatment. In: *Seminars in Cancer Biology*. Academic Press 46. https://doi.org/10.1016/j.semcancer.2017.07.002.

Pando, D., Beltrán, M., Gerone, I., Matos, M. and Pazos, C. (2015). Resveratrol entrapped niosomes as yoghurt additive. *Food Chemistry* 170, 281–287. https://doi.org/10.1016/j.foodchem.2014.08.082.

Panja, P. (2018). Green extraction methods of food polyphenols from vegetable materials. *Current Opinion in Food Science* 23, 173–182. https://doi.org/10.1016/j.cofs.2017.11.012.

Panzella, L., Moccia, F., Nasti, R., Marzorati, S., Verotta, L. and Napolitano, A. (2020). Bioactive phenolic compounds from Agri-food wastes: an update on green and sustainable extraction methodologies. *Frontiers in Nutrition* 7, 60. https://doi.org/10.3389/fnut.2020.00060.

Papuc, C., Goran, G. V., Predescu, C. N., Nicorescu, V. and Stefan, G. (2017). Plant polyphenols as antioxidant and antibacterial agents for shelf-life extension of meat and meat products: classification, structures, sources, and action mechanisms. *Comprehensive Reviews in Food Science and Food Safety* 16(6), 1243–1268. https://doi.org/10.1111/1541-4337.12298.

Patzke, H. and Schieber, A. (2018). Growth-inhibitory activity of phenolic compounds applied in an emulsifiable concentrate: ferulic acid as a natural pesticide against Botrytis cinerea. *Food Research International* 113(March), 18–23. https://doi.org/10.1016/j.foodres.2018.06.062.

Peluso, I. and Serafini, M. (2017). Antioxidants from black and green tea: from dietary modulation of oxidative stress to pharmacological mechanisms. *British Journal of Pharmacology* 174(11), 1195–1208. https://doi.org/10.1111/bph.13649.

Pinela, J., Prieto, M. A., Pereira, E., Jabeur, I., Barreiro, M. F., Barros, L. and Ferreira, I. C. F. R. (2019). Optimization of heat- and ultrasound-assisted extraction of anthocyanins from Hibiscus sabdariffa calyces for natural food colorants. *Food Chemistry* 275, 309–321. Available at: https://doi.org/10.1016/j.foodchem.2018.09.118.

Pinto, E. P., Perin, E. C., Schott, I. B., da Silva Rodrigues, R., Lucchetta, L., Manfroi, V. and Rombaldi, C. V. (2016). The effect of postharvest application of UV-C radiation on the phenolic compounds of conventional and organic grapes (*Vitis labrusca* cv. 'Concord'). *Postharvest Biology and Technology* 120, 84–91.

Pires, T. C. S. P., Dias, M. I., Barros, L., Calhelha, R. C., Alves, M. J., Oliveira, M. B. P. P., Santos-Buelga, C. and Ferreira, I. C. F. R. (2018). Edible flowers as sources of phenolic compounds with bioactive potential. *Food Research International* 105, 580–588.

Plaza, M. and Turner, C. (2015). Pressurized hot water extraction of bioactives. In: *TrAC Trends in Analytical Chemistry* 71, 39–54. https://doi.org/10.1016/j.trac.2015.02.022.

Poveda, J. M., Loarce, L., Alarcón, M., Díaz-Maroto, M. C. and Alañón, M. E. (2018). Revalorization of winery by-products as source of natural preservatives obtained by means of green extraction techniques. *Industrial Crops and Products* 112, 617–625. https://doi.org/10.1016/j.indcrop.2017.12.063.

Puértolas, E. and Barba, F. J. (2016). Electrotechnologies applied to valorization of by-products from food industry: main findings, energy and economic cost of their industrialization. In: *Food and Bioproducts Processing* 100, 172–184. https://doi.org/10.1016/j.fbp.2016.06.020.

Qiu, J. G., Wang, L., Liu, W. J., Wang, J. F., Zhao, E. J., Zhou, F. M., Ji, X. B., Wang, L. H., Xia, Z. K., Wang, W., Chia-Mi Lin, M. C., Liu, L. Z., Huang, Y. X. and Jiang, B. H. (2019). Apigenin inhibits IL-6 transcription and suppresses esophageal carcinogenesis. *Frontiers in Pharmacology* 10(September), 1002. https://doi.org/10.3389/fphar.2019.01002.

Rajauria, G. and Tiwari, B. K. (2018). Extraction, composition, quality and analysis. In: Tiwari, B. K. and Rajauria, G. (Eds.). *Fruit Juices* (p. 910). Holland: A.

Ranic, M., Nikolic, M., Pavlovic, M., Buntic, A., Siler-Marinkovic, S. and Dimitrijevic-Brankovic, S. (2014). Optimization of microwave-assisted extraction of natural antioxidants from spent espresso coffee grounds by response surface methodology. *Journal of Cleaner Production* 80, 69–79. https://doi.org/10.1016/j.jclepro.2014.05.060.

Ratz-Łyko, A. and Arct, J. (2019). Resveratrol as an active ingredient for cosmetic and dermatological applications: a review. *Journal of Cosmetic and Laser Therapy* 21(2), 84–90. https://doi.org/10.1080/14764172.2018.1469767.

Raut, N. A., Dhore, P. W., Saoji, S. D. and Kokare, D. M. (2016). Selected bioactive natural products for diabetes mellitus. In: *Studies in Natural Products Chemistry* (1st ed., Vol. 48, Issue Ad 130, pp. 287–322). Elsevier B.V. https://doi.org/10.1016/B978-0-444-63602-7.00009-6.

Riciputi, Y., Diaz-de-Cerio, E., Akyol, H., Capanoglu, E., Cerretani, L., Caboni, M. F. and Verardo, V. (2018). Establishment of ultrasound-assisted extraction of phenolic compounds from industrial potato by-products using response surface methodology. *Food Chemistry* 269, 258–263. https://doi.org/10.1016/j.foodchem.2018.06.154.

Rojas, R., Alvarez-Pérez, O. B., Contreras-Esquivel, J. C., Vicente, A., Flores, A., Sandoval, J. and Aguilar, C. N. (2020). Valorisation of mango peels: extraction of pectin and antioxidant and antifungal polyphenols. *Waste and Biomass Valorization* 11(1), 89–98. https://doi.org/10.1007/s12649-018-0433-4.

Romani, A., Pinelli, P., Ieri, F. and Bernini, R. (2016). Sustainability, innovation, and green chemistry in the production and valorization of phenolic extracts from Olea europaea L. *Sustainability* 8(10). https://doi.org/10.3390/su8101002.

Rosero, J. C., Cruz, S., Osorio, C. and Hurtado, N. (2019). Analysis of phenolic composition of byproducts (seeds and peels) of avocado (Persea americana Mill.) cultivated in Colombia. *Molecules* 24(17), 3209. https://doi.org/10.3390/molecules24173209.

Saad, F., Mrabti, H. N., Sayah, K., Bouyahya, A., Salhi, N., Cherrah, Y. and My El Abbes, F. (2019). Phenolic content, acute toxicity of *Ajuga iva* extracts and assessment of their antioxidant and carbohydrate digestive enzyme inhibitory effects. *South African Journal of Botany*.

Saha, S. K., Dey, S. and Chakraborty, R. (2019). Effect of microwave power on drying kinetics, structure, color, and antioxidant activities of corncob. *Journal of Food Process Engineering* 42(4), 1–13. https://doi.org/10.1111/jfpe.13021.

Sanches-Silva, A., Testai, L., Nabavi, S. F., Battino, M., Pandima Devi, K., Tejada, S., Sureda, A., Xu, S., Yousefi, B., Majidinia, M., Russo, G. L., Efferth, T., Nabavi, S. M. and Farzaei, M. H. (2020). Therapeutic potential of polyphenols in cardiovascular diseases: regulation of mTOR signaling pathway. *Pharmacological Research* 152(July 2019), 104626. https://doi.org/10.1016/j.phrs.2019.104626.

Savic, I. M. and Savic Gajic, I. M. (2020). Optimization of ultrasound-assisted extraction of polyphenols from wheatgrass (Triticum aestivum L.). *Journal of Food Science and Technology* 57(8), 2809–2818. https://doi.org/10.1007/s13197-020-04312-w.

Scapagnini, G., Vasto, S., Abraham, N. G., Caruso, C., Zella, D. and Fabio, G. (2011). Modulation of Nrf2/ARE pathway by food polyphenols: A nutritional neuroprotective strategy for cognitive and neurodegenerative disorders. *Molecular Neurobiology* 44(2), 192–201. https://doi.org/10.1007/s12035-011-8181-5.

Scully, D. S., Jaiswal, A. K. and Abu-Ghannam, N. (2016). An investigation into spent coffee waste as a renewable source of bioactive compounds and industrially important sugars. *Bioengineering* 3(4). https://doi.org/10.3390/bioengineering3040033.

Šeremet, D., Durgo, K., Jokić, S., Hudek, A., Cebin, A. V., Mandura, A., Jurasović, J. and Komes, D. (2020). Valorization of banana and red beetroot peels: determination of basic macrocomponent composition, application of novel extraction methodology and assessment of biological activity in vitro. *Sustainability* 12(11). https://doi.org/10.3390/su12114539.

Serrano, J., Puupponen-Pimiä, R., Dauer, A., Aura, A. M. and Saura-Calixto, F. (2009). Tannins: current knowledge of food sources, intake, bioavailability and biological effects. *Molecular Nutrition and Food Research* 53(Suppl. 2), S310–S329. https://doi.org/10.1002/mnfr.200900039.

Shahidi, F. and Ambigaipalan, P. (2015). Phenolics and Polyphenolics in foods, beverages and spices: antioxidant activity and health effects: a review. *Journal of Functional Foods* 18, 820–897. https://doi.org/10.1016/j.jff.2015.06.018.

Shankar, K. and Mehendale, H. M. (2014). Oxidative stress. In: *Encyclopedia of Toxicology* (3rd edn.) (pp. 735–737). https://doi.org/10.1016/B978-0-12-386454-3.00345-6.

Shimamura, Y., Hirai, C., Sugiyama, Y., Shibata, M., Ozaki, J., Murata, M., Ohashi, N. and Masuda, S. (2017). Inhibitory effects of food additives derived from polyphenols on staphylococcal enterotoxin A production and biofilm formation by Staphylococcus aureus. *Bioscience, Biotechnology, and Biochemistry* 81(12), 2346–2352. https://doi.org/10.1080/09168451.2017.1395681.

Sowbhagya, H. B. and Chitra, V. N. (2010). Enzyme-assisted extraction of flavorings and colorants from plant materials. *Critical Reviews in Food Science and Nutrition* 50(2), 146–161. https://doi.org/10.1080/10408390802248775.

Steinberg, D. (1997). Oxidation and its pathobiological significance. *The Journal of Biological Chemistry* 272(34), 20963–20966.

Sun, X., Wang, Z., Kadouh, H. and Zhou, K. (2014). The antimicrobial, mechanical, physical and structural properties of chitosan-gallic acid films. *LWT: Food Science and Technology* 57(1), 83–89. https://doi.org/10.1016/j.lwt.2013.11.037.

Takeda, Y., Bui, V. N., Iwasaki, K., Kobayashi, T., Ogawa, H. and Imai, K. (2014). Influence of olive-derived hydroxytyrosol on the toll-like receptor 4-dependent inflammatory response of mouse peritoneal macrophages. *Biochemical and Biophysical Research Communications* 446(4), 1225–1230. https://doi.org/10.1016/j.bbrc.2014.03.094.

Tena, M. T. (2018). Pressurized liquid extraction. In: *Reference Module in Chemistry, Molecular Sciences and Chemical Engineering* (3rd ed., pp. 78–83). Elsevier. https://doi.org/10.1016/B978-0-12-409547-2.14407-5.

Testai, L., Martelli, A., Cristofaro, M., Breschi, M. C. and Calderone, V. (2013). Cardioprotective effects of different flavonoids against myocardial ischaemia/reperfusion injury in Langendorff-perfused rat hearts. *Journal of Pharmacy and Pharmacology* 65(5), 750–756. https://doi.org/10.1111/jphp.12032.

Thangavel, P., Puga-Olguín, A., Rodríguez-Landa, J. F. and Zepeda, R. C. (2019). Genistein as potential therapeutic candidate for menopausal symptoms and other related diseases. *Molecules* 24(21), 3892. https://doi.org/10.3390/molecules24213892.

Tian, B. and Liu, J. (2020). Resveratrol: a review of plant sources, synthesis, stability, modification and food application. *Journal of the Science of Food and Agriculture* 100(4), 1392–1404. https://doi.org/10.1002/jsfa.10152.

Tiwari, B. K. (2015). Ultrasound: A clean, green extraction technology. *TrAC Trends in Analytical Chemistry* 71, 100–109. https://doi.org/10.1016/j.trac.2015.04.013.

Torres-Valenzuela, L. S., Ballesteros-Gómez, A. and Rubio, S. (2020). Green solvents for the extraction of high added-value compounds from Agri-food waste. *Food Engineering Reviews*. Springer 12(1), 83–100. https://doi.org/10.1007/s12393-019-09206-y.

Tremocoldi, M. A., Rosalen, P. L., Franchin, M., Massarioli, A. P., Denny, C., Daiuto, É. R., Paschoal, J. A. R., Melo, P. S. and de Alencar, S. M. (2018). Exploration of avocado by-products as natural sources of bioactive compounds. *PLoS ONE* 13(2), e0192577. https://doi.org/10.1371/journal.pone.0192577.

Tyśkiewicz, K., Konkol, M. and Rój, E. (2018). The application of supercritical fluid extraction in phenolic compounds isolation from natural plant materials. *Molecules* 23(10), 2625. https://doi.org/10.3390/molecules23102625.

Uhrin, P., Wang, D., Mocan, A., Waltenberger, B., Breuss, J. M., Tewari, D., Mihaly-Bison, J., Huminiecki, Ł., Starzyński, R. R., Tzvetkov, N. T., Horbańczuk, J. and Atanasov, A. G. (2018). Vascular smooth muscle cell proliferation as a therapeutic target. Part 2: Natural products inhibiting proliferation. *Biotechnology Advances* 36(6), 1608–1621. https://doi.org/10.1016/j.biotechadv.2018.04.002.

V González-de-Peredo, A., Vázquez-Espinosa, M., Espada-Bellido, E., Ferreiro-González, M., F Barbero, G., Palma, M. and Carrera, C. (2022). Optimization of a microwave assisted extraction method for maximum flavonols and antioxidant activity of onion extracts. *Antioxidants* 11(12). https://doi.org/10.3390/antiox11122393.

Varzakas, T., Zakynthinos, G. and Verpoort, F. (2016). Plant food residues as a source of nutraceuticals and functional foods. *Foods* 5(4), 88. https://doi.org/10.3390/foods5040088.

Vázquez-González, M., Fernández-Prior, Á., Bermúdez Oria, A., Rodríguez-Juan, E. M., Pérez-Rubio, A. G., Fernández-Bolaños, J. and Rodríguez-Gutiérrez, G. (2020). Utilization of strawberry and raspberry waste for the extraction of bioactive compounds by deep eutectic solvents. *LWT* 130(May), 109645. https://doi.org/10.1016/j.lwt.2020.109645.

Villasante, J., Pérez-Carrillo, E., Heredia-Olea, E., Metón, I. and Almajano, M. P. (2019). In vitro antioxidant activity optimization of nut shell (Carya illinoinensis) by extrusion using response surface methods. *Biomolecules* 9(12), 883. https://doi.org/10.3390/biom9120883.

Vulić, J. J., Ćebović, T. N., Čanadanović-Brunet, J. M., Ćetković, G. S., Čanadanović, V. M., Djilas, S. M. and Tumbas Šaponjac, V. T. (2014). In vivo and in vitro antioxidant effects of beetroot pomace extracts. *Journal of Functional Foods* 6(1), 168–175. https://doi.org/10.1016/j.jff.2013.10.003.

Wang, H., Chen, G., Guo, X., Abbasi, A. M. and Liu, R. H. (2016). Influence of the stage of ripeness on the phytochemical profiles, antioxidant and antiproliferative activities in different parts of Citrus reticulata Blanco cv. Chachiensis. *LWT: Food Science and Technology* 69, 67–75. https://doi.org/10.1016/j.lwt.2016.01.021.

Wang, T., Li, X., Zhou, B., Li, H., Zeng, J. and Gao, W. (2015). Anti-diabetic activity in type 2 diabetic mice and α-glucosidase inhibitory, antioxidant and anti-inflammatory potential of chemically profiled pear peel and pulp extracts (Pyrus spp.). *Journal of Functional Foods* 13, 276–288. https://doi.org/10.1016/j.jff.2014.12.049.

Wang, Y., Li, L., Liu, H., Zhao, T., Meng, C., Liu, Z. and Liu, X. (2018). Bioactive compounds and in vitro antioxidant activities of peel, flesh and seed powder of kiwi fruit. *International Journal of Food Science and Technology* 53(9), 2239–2245. https://doi.org/10.1111/ijfs.13812.

Wen, L., Zhang, Z., Sun, D. W., Sivagnanam, S. P. and Tiwari, B. K. (2020). Combination of emerging technologies for the extraction of bioactive compounds. *Critical Reviews in Food Science and Nutrition* 60(11), 1826–1841. https://doi.org/10.1080/10408398.2019.1602823.

Wu, G., Chang, C., Hong, C., Zhang, H., Huang, J., Jin, Q. and Wang, X. (2019). Phenolic compounds as stabilizers of oils and antioxidative mechanisms under frying conditions: A comprehensive review. *Trends in Food Science and Technology* 92(July), 33–45.

Xiao, J. and Niu, L. (2015). Antilisterial peptides released by enzymatic hydrolysis from grass carp proteins and activity on controlling L. monocytogenes inoculated in surimi noodle. *Journal of Food Science* 80(11), M2564–M2569. https://doi.org/10.1111/1750-3841.13104.

Xie, P., Huang, L., Zhang, C., Deng, Y., Wang, X. and Cheng, J. (2019). Enhanced extraction of hydroxytyrosol, maslinic acid and oleanolic acid from olive pomace: process parameters, kinetics and thermodynamics, and greenness assessment. *Food Chemistry* 276, 662–674. https://doi.org/10.1016/j.foodchem.2018.10.079.

Yahfoufi, N., Alsadi, N., Jambi, M. and Matar, C. (2018). The immunomodulatory and anti-inflammatory role of polyphenols. In: *Nutrients*. MDPI AG 10(11). https://doi.org/10.3390/nu10111618.

Yan, J., Wang, Q., Zheng, X., Sun, H., Zhou, Y., Li, D., Lin, Y. and Wang, X. (2012). Luteolin enhances TNF-related apoptosis-inducing ligand's anticancer activity in a lung cancer xenograft mouse model. *Biochemical and Biophysical Research Communications* 417(2), 842–846. https://doi.org/10.1016/j.bbrc.2011.12.055.

Zeng, X. and Zhang, Z. (2019). Pulsed electric field assisted extraction of bioactive compounds. In: Jia, J., Liu, D. and Ma, H. (Eds). *Advances in Food Processing Technology* (pp. 125–135). Springer, Singapore. https://doi.org/10.1007/978-981-13-6451-8_5.

Zhang, C., Kang, X., Zhang, T. and Huang, J. (2019). Positive effects of resveratrol on egg-laying ability, egg quality, and antioxidant activity in hens. *Journal of Applied Poultry Research* 28(4), 1099–1105. https://doi.org/10.3382/japr/pfz072.

Zhang, Y., Zhu, K., He, S., Tan, L. and Kong, X. (2016). Characterizations of high purity starches isolated from five different jackfruit cultivars. *Food Hydrocolloids* 52, 785–794. https://doi.org/10.1016/j.foodhyd.2015.07.037.

Part 4

Case studies

Chapter 13

Developing closed-loop dairy value chains and tools to support decision-makers

Jack B. Hetherington, University of Adelaide/CSIRO Agriculture and Food/Fight Food Waste Cooperative Research Centre, Australia; Pablo Juliano, CSIRO Agriculture and Food, Australia; and Rodolfo García-Flores, CSIRO Data61, Australia

1 Introduction

The dairy sector is an important contributor to economies and diets globally, employing many people in the production of food that provides an important source of protein and micro-nutrients to consumers. However, dairy systems also contribute to environmental problems through emissions of greenhouse gases (GHG) (i.e. methane (CH_4) and nitrous oxide (N_2O)) and ammonia (NH_3) throughout the process of cattle digestion and manure management (Gac et al., 2007; Knapp et al., 2014). Additionally, losses and waste of dairy products along the value chain present a twofold problem. The first problem is the direct effects of decomposing organic material in landfill, which contributes to GHG emissions (FAO, 2013), and the discharge of pollutants into the environment (Smithers, 2008). The second problem is the inefficient use of resources such as water, fertiliser, electricity and fuel that have gone into producing, processing, cooling and transporting the food product until the point of disposal (Kummu et al., 2012).

http://dx.doi.org/10.19103/AS.2023.0120.22

Reduction of food loss and waste (FLW) has been recognised in the United Nation's (UN) Sustainable Development Goals as a global challenge for food systems, with significant economic, social and environmental ramifications. Kummu et al. (2012) estimated that approximately one-quarter of global water, fertiliser and cropland used to produce food are wasted. FLW also contributes significantly to GHG and nitrogen emissions (FAO, 2013) causing environmental issues such as contamination of freshwater, ocean acidification and damage to biosphere integrity (Lade et al., 2020). The social implications of FLW include lost meals that could feed food-insecure people (Reynolds et al., 2015). In the case of dairy products, this means loss of macro-nutrients (e.g. proteins) and micro-nutrients (e.g. calcium, choline, riboflavin, zinc and vitamin B12) (Chen et al., 2020; de Wit, 2001). FLW also adds significant costs to households, businesses and society more broadly. A study by Hanson and Mitchell (2017) of over 700 businesses across 17 countries found that, for every dollar invested in addressing FLW, there was a net return of $14. This is because properly managing FLW can reduce input, labour and overhead costs and even generate new revenue through developing new products or selling co-products as an input to other industries.

FLW is generated at different stages of the value chain and varies significantly across regions (Gustavsson et al., 2011). As shown in Fig. 1, the majority of dairy FLW in developing countries is generated during production, storage, distribution and processing, whereas in developed countries, most of the waste occurs at the consumption stage of the chain. There are many opportunities to utilise these waste streams and by-products in close-looped systems or the circular economy. This has been defined as 'an economic system that replaces the "end-of-life" concept with reducing, alternatively reusing, recycling and recovering materials in production/distribution and consumption processes' (Kirchherr et al., 2017, p. 229).

This chapter focuses on issues and opportunities for managing dairy FLW for countries with developed dairy value chains, i.e. economies with established

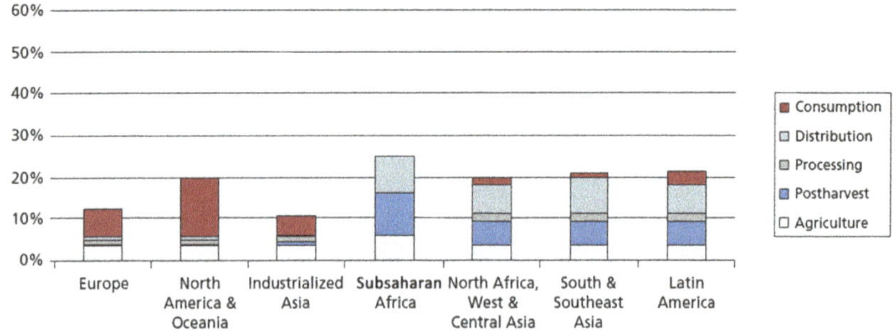

Figure 1 Loss and waste of dairy products for each region at different stages of the supply chain. Source: Reproduced with permission from: Gustavsson et al. (2011).

refrigeration infrastructure, traceability and quality control. While there will be some relevance to regions with less developed supply chains, this chapter does not intend to address the full spectrum of issues affecting these regions (Lore et al., 2005).

In countries with developed dairy chains, like Australia, dairy products generate different types of 'avoidable' and 'unavoidable' FLW, which refers to the components that are intended for human consumption (Papargyropoulou et al., 2014). Examples of avoidable dairy FLW include liquid milk or cheese products that have expired (i.e. reached their 'use by date') and are no longer safe to consume. A key example of often unavoidable dairy FLW is the generation of whey as a by-product of small-scale cheese production. As discussed at length in this book chapter, both avoidable and unavoidable FLW can be used as a valuable resource, such as an input to another business or transformed into new food products. Therefore, they do not always have to be considered waste.

In Australia, FIAL (2021) estimated that most of the waste from liquid milk products occurs within households, while most of the waste generated from cheese products occurs at the processing stage (see Fig. 2). The FIAL (2021) study did not consider any dairy by-product fed to animals as this was considered a desirable management option and therefore not measured as 'waste'. In the case of cheese whey, an estimated 3.1 million tonnes were produced in Australia in the 2018-9 financial year, with two-thirds being used as animal feed (personal communication, PA Bontinck, Lifecycles, 9 December 2021). This highlights there is significantly more whey that needs to be managed. Additionally, it underscores that the study did not consider the standard definition for FLW and its measurement, which is described in the FLW Protocol (Hanson et al., 2016). Furthermore, the whey accounted into animal feed ignores that there are high-value return options for its use. For example,

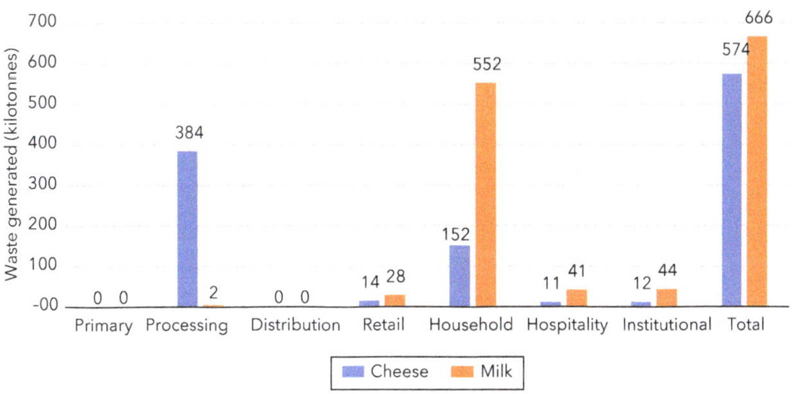

Figure 2 Quantity of dairy product waste along the value chain in Australia in 2018-9. Source: personal communication, PA Bontinck, Lifecycles, 9 December 2021.

instead of feeding whey to pigs, it may be more valuable to manufacture it into new food products or other high-value products (e.g. pharmaceutical uses).

2 Frameworks for identifying and managing food loss and waste in developed dairy chains

The Food Waste Hierarchy, proposed by Papargyropoulou et al. (2014) and described in detail in the following paragraphs, has been adopted by governments and industry as a framework for managing food waste. This is because it broadly outlines different management practices, with prevention of surplus food production as being the most desirable management practice, through to recovering soil nutrients and energy and, lastly, disposal (e.g. through landfill or wastewater).

There have been many iterations of the Food Waste Hierarchy (Jones et al., 2022; Redlingshöfer et al., 2020). The version proposed by Teigiserova et al. (2020) attempts to clarify some of the inconsistencies in the terminology and the order of preference (see Fig. 3). In Teigiserova et al.'s (2020) version of the hierarchy, the order of preference is as follows:

1. Prevention (remains the most desired management practice);
2. Reusing edible food waste in the human supply chain (e.g. through donations to food rescue organisations);
3. Reuse of inedible food wastes as animal feed;
4. Recycling of food waste streams to extract valuable materials (i.e. for non-food applications);
5. Recovering soil nutrients (e.g. composting which can be used again for agricultural production);

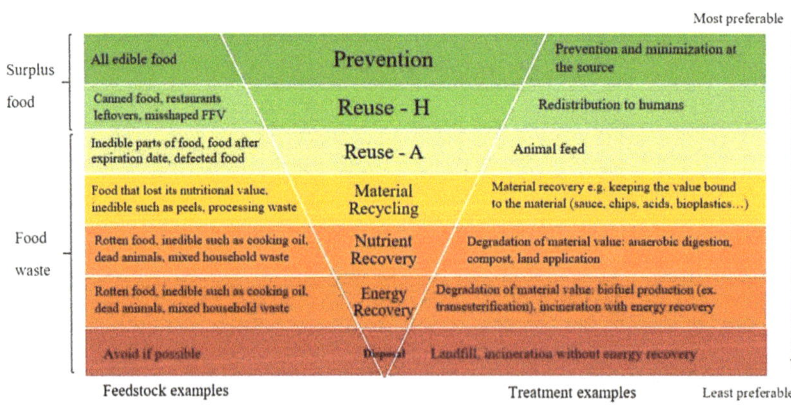

Figure 3 Food surplus and waste hierarchy. Source: Reproduced with permission from: Teigiserova et al. (2020).

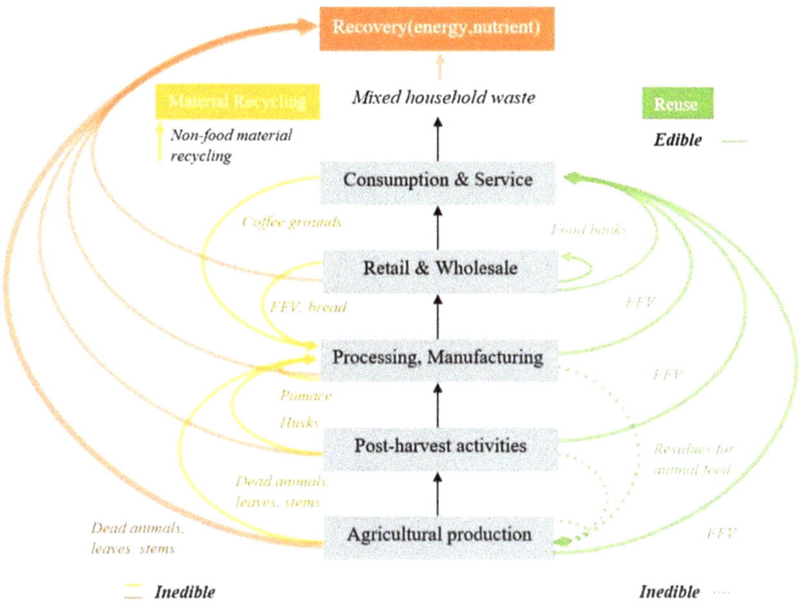

Figure 4 The circular economy framework for food surplus, waste and loss. Source: Reproduced with permission from: Teigiserova et al. (2020).

6 Energy recovery (e.g. production of biofuels); and
7 Disposal.

Teigiserova et al. (2020) apply the updated hierarchy to a typical food value chain, incorporating circular economy thinking (as shown in Fig. 4). The colours translate to possible management options for each stage of the chain. Green represents reuse in human (solid line) or animal feed (dotted line); yellow represents material recycling of non-food components; and orange represents recovery of energy or nutrients.

The framework by Teigiserova et al. (2020) provides a foundation that can be adapted to specific food sectors, including dairy. However, an issue with this framework is that recovered soil nutrients and energy are not directed back to agricultural production, meaning it is not truly 'circular', and it does not consider the recovery of water from FLW an important outcome. We propose the expanded framework in Fig. 5 to overcome these limitations and apply management practices available to developed dairy chains. In this framework, dairy products move from production to processing, retail or wholesale, consumption (e.g. in households, food service or institutions), recovery and then back to production. Additionally, there are multiple management practices that could be employed at each stage of the chain that would prevent the disposal of waste. These are summarised in Section 3.

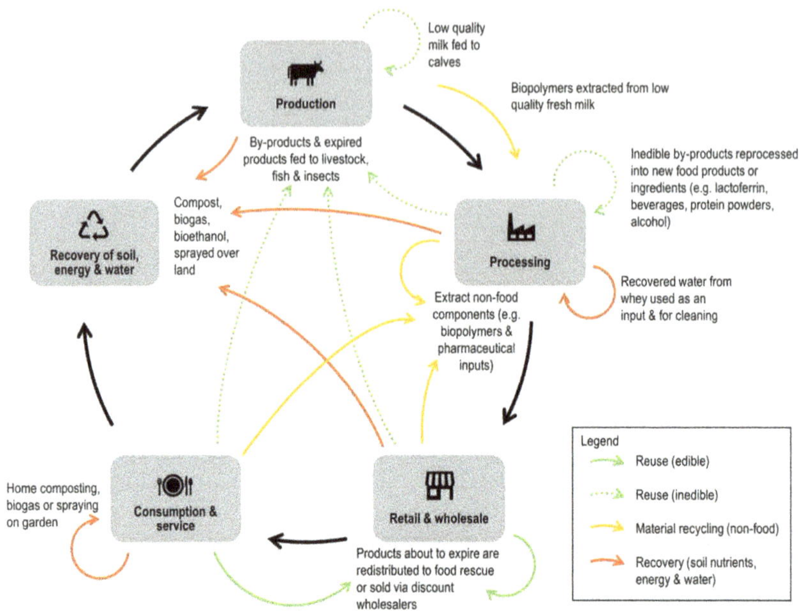

Figure 5 Proposed framework for utilising dairy food waste streams in a close-looped paradigm.

3 Key stages and management practices for reducing food loss and waste in dairy chains

3.1 Production

Production losses on farms in developed dairy chains are relatively low compared to regions with under-developed chains. The major source of on-farm loss is milk that does not meet minimum quality parameters due to issues such as bacterial contamination (Lore et al., 2005) or antibiotic residues (March et al., 2019). A common management practice is feeding this milk to calves or to other livestock species such as pigs.

According to the Food Waste Hierarchy, recycling of non-food components present in FLW is preferred when it cannot be reused as human or animal food. Biopolymers can be extracted from milk that does not meet minimum quality specifications, which can be used to manufacture bio-plastics (Jefferson et al., 2020). Although high production costs have been reported as a significant issue for bio-plastic production from dairy products (de Castro et al., 2022), there is some evidence that it can return a positive return on investment (Chalermthai et al., 2020). The negative environmental impacts of such bio-plastic production still remain because of the chemical by-products from the manufacturing process (Chalermthai et al., 2021; de Castro et al., 2022). This is discussed in the Section 3.2.

Other on-farm management practices for dealing with waste from milk production include recovering soil nutrients, energy and water from milk that does not meet the minimum specifications through anaerobic digestion (AD). These can be co-digested by mixing with dairy cattle effluent (i.e. manure) (Wu et al., 2011).

3.2 Processing

During the processing stage of the value chain, there can be significant waste due to the generation of by-products. Whey, which is the most significant by-product from the manufacturing of dairy products such as cheese and yoghurt, has historically been disposed of through wastewater treatment (Smithers, 2008). However, whey management options other than disposal include:

1 re-processing into new products;
2 as a feed source in livestock production;
3 recycling non-food components; and
4 recovering soil nutrients, energy and water.

Tsermoula et al. (2021) describe the emergence of more sustainable and profitable processing techniques to valorise whey. Fig. 6, which has been adapted from Juliano et al. (2017), illustrates different processing options and end products and how these align with the management practices outlined in Fig. 3:

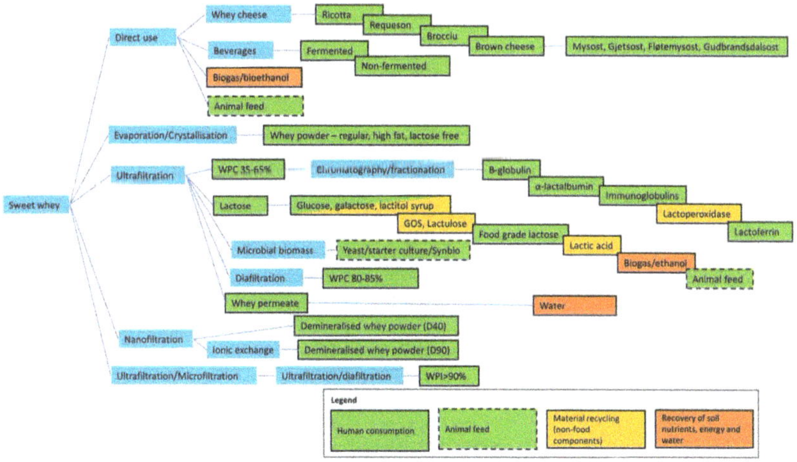

Figure 6 Processing options for cheese whey. Source: Adapted from: Juliano et al. (2017).

- Products that are intended for human consumption (green);
- Animal feed (green with a dotted outline);
- Material recycling of non-food components (yellow); and
- Recovery of soil nutrients, energy and water (orange).

While this is not an exhaustive list of processing techniques and end products, it highlights the many management options available to dairy processors.

The resulting products (e.g. powders or protein and lactose) can be used in a variety of industries such as nutraceuticals, infant formulas and confectionery products (Oliveira et al., 2020; Tsermoula et al., 2021). Lactoferrin is used as a nutritional supplement with anti-inflammatory properties and has been investigated as a potential treatment for severe coronavirus disease 2019 (COVID-19) (Chang et al., 2020). Lactose isolated from whey can be further fermented, or co-fermented with other inputs such as wheat, into ethanol. The ethanol can be used in alcoholic beverages, vinegars and biofuels (Ling, 2008; Murari et al., 2019; Parashar et al., 2016; Zotta et al., 2020). Some processing options, including forward osmosis/reverse osmosis and ultra-filtration/reverse osmosis, are cost-effective means of producing whey powder and recovering water (Aydiner et al., 2014; Meneses and Flores, 2016).

There are many processing options for whey, but as García-Flores et al. (2015) show, the optimal choice for processing whey depends on various socioeconomic and production factors, such as the quantity of whey being generated. Figure 7 illustrates the expected earnings before interest and tax (EBIT) from a range of processing options and the capacity ranges to make it financially viable. An additional consideration to make these financially viable is the capital and operating costs.

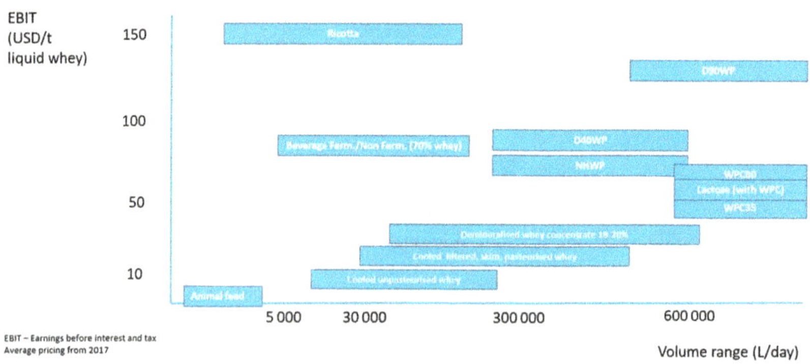

Figure 7 Options for value adding whey with consideration to the capacity range to make processing options viable and the expected earnings before interest and tax (EBIT). Source: Adapted from: Juliano (2021).

Other options that circular dairy systems can employ to minimise FLW during processing include reusing waste streams in animal industries. By-products, such as whey, are good sources of protein and energy and, to a lesser extent, micronutrients such as calcium and phosphorus (Schingoethe, 1976). This means they are an adequate feed source for ruminants, pigs, poultry, fish, horses, and rats (Sallam et al., 2022; Schingoethe, 1976). Ruminants and pigs can consume liquid and dry whey effectively, and dried whey has been shown to improve the quality of silage. Another option for diverting whey, and other dairy waste products, is by feeding it to *Hermetia illucens* larvae, commonly known as black soldier flies (BSF) (Hadj Saadoun et al., 2020). By feeding dairy FLW to BSF, the larvae can be fed to other livestock industries such as pigs, poultry and fish (Barragan-Fonseca et al., 2017). However, the feasibility of this will depend on the cost of transport and the price the processor can sell the whey to the animal industry.

Recycling non-food components from dairy FLW during processing includes extracting biopolymers and lactoperoxidase. Biopolymers extracted from whey, such as polylactic acid (PLA) and polyhydroxyalkanoates (PHA), can be used to manufacture bio-plastics, which can be a substitute for plastics derived from fossil fuels (Asunis et al., 2020). This can be achieved by processing liquid and powdered whey which has been shown to be economically feasible (Chalermthai et al., 2020). However, there are persistent environmental issues arising from processing, in particular, from the chemical input polyethylene glycol methyl ether methacrylate (PEGMA) (Chalermthai et al., 2021). Further research is needed to find alternative chemical processes that are environmentally sustainable. Lactoperoxidase has antimicrobial properties and has been recommended to reduce microbial levels in milk where pasteurisation or cold chain infrastructure is not available (Codex, 1991). It can also be combined with biopolymers to act as antimicrobial in edible packaging films (Zhang and Rhim, 2022). A study undertaken by Saravani et al. (2019) found that lactoperoxidase was an effective additive to biodegradable coatings for Gouda cheese products.

Recovery of soil nutrients, energy and water from dairy FLW is possible through composting and AD, e.g. for use in biogas or as compost (Hauser, 2017). Tait et al. (2021) estimate high energy recovery potential from dairy processing waste streams through AD (see Fig. 8), which can be coupled with effluent from dairy farms (Mostafa Imeni et al., 2019). This can contribute to offsetting the intensive energy requirements from processing and cooling milk (Mainardis et al., 2019). However, there are a number of factors related to the individual business that affect the viability of AD systems. This includes waste production volumes, space availability, infrastructure cost, energy costs, ability to utilise biogas energy, skilled labour, location and proximity to the processing plant (GHD, 2017). Utilising whey as a feedstock for bio-hydrogen has been explored

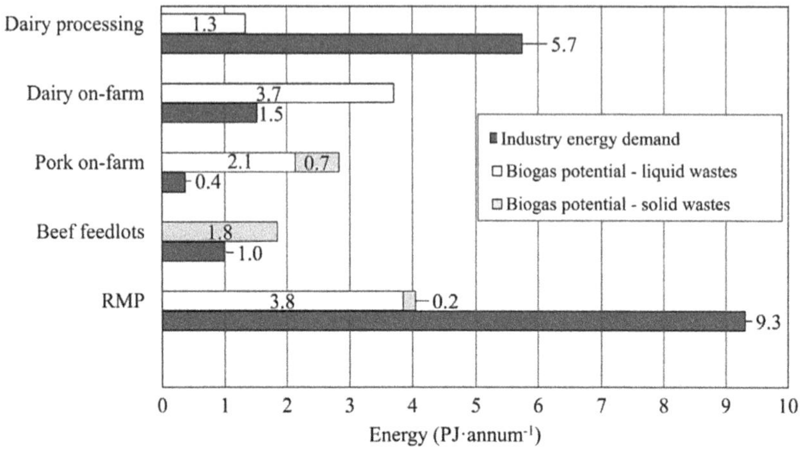

Figure 8 Energy requirements and biogas potential of animal industries in Australia. Note: RMP, red meat processing. Source: Reproduced with permission from: Tait et al. (2021).

as an alternative energy source (Patel et al., 2016). However, there are technical challenges to ensure the necessary biological parameters for anaerobic (co-) digestion are met and remain cost-effective (Tait et al., 2021). Application of whey to the land has been a common practice for whey. However, the correct dilution is required to prevent negative environmental and health outcomes (Ghaly et al., 2007). The soil type and landscape (e.g. the slope and proximity to rivers) can restrict the ability to spray whey onto the land (Watson et al., 1977).

3.3 Retail and wholesale

At the retail and wholesale stage of the chain, options for managing FLW include redistribution of edible food for human consumption, diverting food to animal feed, recycling non-food components and recovery of nutrients and energy.

Redistributing edible food is an important activity towards the goal of reducing food waste while tackling food insecurity (Reynolds et al., 2015). This includes redistributing soon-to-expire food via food rescue organisations such as food banks or discount wholesalers. For instance, in Australia, a company called 'Yume' (https://yumefood.com.au/) connects producers and processors with excess stock to retailers and food service businesses at a discounted rate. These options ensure food remains part of the human food chain and, in the case of food rescue, help address food insecurity. Further, with the consideration of tax incentives for businesses that donate excess food (KPMG, 2020), this will likely become a more viable option for businesses along the value chain.

Much like the options available to processors, feeding expired dairy products can also be diverted to animal feed (e.g. via BSF larvae production) (Khyade and Tamhane, 2021) or recovery through industrial composting facilities or AD. Extracting biopolymers (i.e. recycling) for use is technically feasible from expired milk products but is not a viable commercial solution currently (Jefferson et al., 2020).

3.4 Consumption and service

The consumption stage of the chain includes households, food service (e.g. restaurants) and institutions (e.g. hospitals and schools). Management options here are relatively similar to other stages of the value chain in that FLW can be redistributed to humans, used as animal feed, non-food components recycled and, lastly, soil nutrients, energy and water are recovered.

For food service and institutions, due to their relative scale, management options available are similar to retailers; i.e. redistribution of edible excess food to consumers; diversion to animal feed (e.g. BSF); and recovery through industrial composting facilities or AD.

For households, it is likely that significant reductions in food waste can be achieved, especially in developed countries like Australia (Ananda et al., 2021). Options include purchasing extended-shelf life or shelf-stable milk products, availability of smaller containers and improving planning behaviours in relation to purchasing and consumption so less is wasted (Williams et al., 2020). However, for waste that cannot be completely prevented and in-keeping with circular-economy-thinking, options include applying small or diluted amounts of milk to home gardens, composting and using small biogas digesters (Alexander et al., 2019). However, these options will probably not be available to all households as access to gardens, available space (e.g. for biogas digesters) and income may be limiting factors. Therefore, municipal waste management (solid waste) and wastewater treatment (for milk poured down the sink) would be needed to allow for the recovery of nutrients, energy and water.

3.5 Recovery of soil nutrients, energy and water

In this step, all of the nutrients not reused in human and animal feed, or processed to extract valuable materials, need to go through aerobic or anaerobic digestion to recover the remaining nutrients, energy and water. Again, in-keeping with circular economy thinking, these will be directed back as fertilisers for crops or as biofuel which, in the case of dairy production and processing, can be very energy intensive (Tait et al., 2021).

Management options like AD can be combined with effluent from dairy cattle. While not being classified as 'recovery' in the Food Waste Hierarchy, effluent from dairy cattle can also be used as feedstock for BSF (Myers et al.,

2014). This represents a significant opportunity for businesses that produce and process milk and also have access to other animal industries.

4 Improving decision-making in managing food loss and waste in dairy value chains

As highlighted, there are a number of management options for addressing dairy FLW. As per the Food Waste Hierarchy, a prescribed order of priority has been established for considering these management options. Where possible, the management options higher in the hierarchy should be exhausted before considering any in the lower tiers. For instance, the Food Waste Hierarchy suggests that AD should not be considered if feeding to pigs or other animals is an available option. This approach has been used by proponents of the Food Waste Hierarchy, as it provides guidance to decision-makers (e.g. governments, industry and consumers) about what would reduce the impacts of their waste, including from an environmental and social perspective (Papargyropoulou et al., 2014).

There is conflicting evidence with regard to which management options are the most effective, especially for management options in the middle tiers of the Food Waste Hierarchy (Redlingshöfer et al., 2020). Moult et al. (2018) found that, in order to minimise the GHG emissions from cheese products wasted at the UK retail level of the value chain, it was preferable to incinerate cheeses rather than to compost them. This conflicts with the generalised preferences outlined in the Food Waste Hierarchy. Many studies that examine the environmental impacts of FLW interventions do not consider multiple environmental outcomes (Redlingshöfer et al., 2020). This is important because interventions can have not only positive outcomes in some areas but also perverse outcomes in others. For example, Ridoutt et al. (2021) found that changes in diets to meet recommended dietary guidelines would reduce GHG emissions but would increase the water footprint. It is therefore important to consider multiple outcomes for each management option.

Businesses generating dairy FLW will need to identify the best management option factoring in the food product and quality, location of the business, the physical proximity of the next stage of the supply chain (e.g. piggeries, commercial composters, etc.), availability and access to technology (e.g. processing equipment), quantities of waste being generated and, most importantly, achieving minimum thresholds for a management practice that is financially viable (e.g. as shown in Fig. 7). Therefore, businesses need tools to support their FLW decision-making to identify the optimal choice for their operation, while considering the complexities and trade-offs between multiple financial and non-financial outcomes (Somlai, 2022). It is for this reason decision support tools (DSTs) have been described as a major priority for helping increase the sustainability of food systems (Béné et al., 2020).

4.1 Decision support tools used to manage dairy food loss and waste

DSTs are defined by Power (2002) as 'interactive computer-based systems that help people use computer communications, data, documents, knowledge, and models to solve problems and make decisions'. This definition covers a wide variety of technologies, including collaborative workspaces, financial models and record systems. DSTs have been referred to by many names including decision support systems, management information systems, executive information systems, business intelligence, collaborative systems and big data and cloud computing (Annosi et al., 2021; Power, 2002; Sprague, 1980).

Several DSTs have been used by decision-makers in the dairy sector to consider management options for their FLW. Eight DSTs relevant to dairy food products and waste management are summarised in Table 1. These DSTs utilise a variety of different analytical approaches including life cycle assessment (LCA)[1], life cycle costing (LCC)[2], material flow cost accounting (MCFA)[3] and benefit-cost analysis (BCA)[4].

There are a number of other DSTs that are publicly available to dairy and food business, including Simapro, Gabi, Umberto and OpenLCA (Silva et al., 2017). These all use analytical approaches such as LCA, LCC and MCFA to support decision-makers. Indeed, EPA's WARM utilises OpenLCA as the underlying programming to analyse the changes from FLW management. While these tools can be used to support more sustainable food systems, their primary aim is not to identify management options or reduce FLW issues; they are therefore not described in detail in this chapter. ERP tools like Systems Applications Products (SAP) and DEAR systems have inventory management capabilities, which can be used to identify how much food product stock needs to be managed. To make the best use of this information requires additional analysis, which can be customised or integrated with other systems to support FLW management. ERPs like SAP are not explored in detail but the authors recognise they can support businesses, especially through mass collection of inventory data. Lastly, some 'tools' are available to assist with quantifying and reporting FLW, e.g. WRAP's FLW Data Capture

[1] LCA is an evaluation of the 'inputs, outputs and the potential environmental impacts of a product system throughout its life cycle' (ISO 14040:2006).
[2] LCC is a 'methodology for systematic economic evaluation of life-cycle costs over a period of analysis, as defined in the agreed scope' (ISO 15686-5.2:2008).
[3] MCFA is a 'tool for quantifying the flows and stocks of materials in processes or production lines in both physical and monetary units' (ISO 14051:2011).
[4] BCA is a widely used economic technique to evaluate of the benefits and costs associated with an intervention, usually in monetary terms.

Table 1 Summary of decision support tools (DSTs) in dairy product and waste management

Name	Description	Reference	URL
Waste reduction model (WARM) tool (v15)	Uses life cycle assessment (LCA) and life cycle costing (LCC) to compare the environmental and economic impacts of managing waste material. This includes non-dairy and non-food materials. It is available in an Excel spreadsheet and an open-source openLCA program.	(EPA, 2019)	https://www.epa.gov/warm/versions-waste-reduction-model-warm#15 (accessed 30 March 2022)
Food loss and waste value calculator (Beta v1.1)	Uses LCA to compare impacts on environmental outcomes (GHG emissions). The Excel-based calculator is designed to complement the Food Loss Waste Protocol to assist decision-makers across the food chain to consider different management options.	(WRI, 2019)	https://flwprotocol.org/why-measure/food-loss-and-waste-value-calculator/ (accessed 30 March 2022)
Insights Engine	An interactive platform to access data and results about food loss and waste in the USA. The web-based components in the Insights Engine include a. Waste monitor - database of FLW in the USA. b. Solutions Database and Directory - quantified benefits of different management options. c. Impact Calculator - an interactive tool to compare environmental and social outcomes in the changes in FLW end destinations. d. Policy Tracker - an interactive map with updated information about relevant state and federal FLW policies.There is a combination of LCA and BCA in the Insights Engine.	(Powell and Curtis, 2020a, b)	https://insights.refed.org/ (accessed 30 March 2022)
Whey distribution chain optimisation model	Allows users to optimise the selection of whey processing equipment and allocation between different cheese manufacturers in a region and candidate whey collection and concentration centres. The model selects the most adequate location for the drying facilities (if total concentration volumes are sufficient). This is a linear programming optimisation model, which uses an Excel-based platform and generates a Google Earth output. The model also provides the return of investment to the region for capital equipment involved. It includes options for using secondhand equipment or drying facilities with spare capacities.	(García-Flores et al., 2015)	NA

Benders decomposition for a reverse logistics network	Builds on the whey distribution chain optimisation model by García-Flores et al. (2015) by accounting for whey supply uncertainty. The former model relies on two sequential stages of (1) identifying facility locations and (2) allocating whey between sites depending on the amount available day-to-day. This approach does not account for uncertainty in whey production. The model using classical HFLP with stochastic demands Benders Decomposition accounts for this uncertainty and indicates a reduction in costs by 28%.	(Esmaeilbeigi et al., 2021)	NA
Whey utilisation tool (whey tool)	The Whey Tool was developed for the Australian dairy processing sector and uses BCA to estimate the capital and operating expenditure required to process whey into new products. The web-based DST allows users to evaluate the financial viability of selling a range of whey-derived products.	(Hauser, 2013)	https://wheytool.dairyaustralia.com.au/Demonstration (accessed 30 March 2022)
FOod side flow Recovery LIFe cycle Tool (FORKLIFT)	FORKLIFT is an Excel-based DST that helps businesses think through the costs and emissions associated with FLW streams from their operation. It is tailored to priority food waste streams in Europe, including whey permeate. Others include apple pomace, brewers spent grain, rape seed press cake, tomatoes pomace and abattoir blood.	(Davis et al., 2019)	https://eu-refresh.org/forklift.html (accessed 30 March 2022)
GISWASTE	GISWASTE uses AHP, as well as LCA, BCA and GIS, to evaluate the technical requirements, and economic and environmental impacts of different waste management options for dairy, meat and vegetable waste producers. Users of this tool are waste management organisations looking to source from local producers/processors. GISWASTE uses Microsoft Sharepoint, Access and ArcGIS.	(San Martin et al., 2017)	NA
Dynamic Industry Resource Efficiency Calculation Tool (DIRECT)	DIRECT was co-developed, via a researcher partnership with food manufacturers, and later commercialised (as a Fight Food Waste CRC project). It uses mass balance to track and visualise a food organisation's product and non-product material flows. As well as using material flow cost accounting (MFCA) principles to support businesses to calculate the 'true' cost of waste, by accounting for the input and operational costs attributable to mass that is lost in a process or along the supply chain. It is a cloud-based DST and aligns with the FLW Standard (developed by the FLW Protocol) as well as the MFCA ISO 14051 (and its derivatives 14052 and 14053). The user defines the scope of the assessment and can compare multiple scenarios.	(Empauer, 2022; Verghese et al., 2018)	https://empauer.com/solutions/direct/ (accessed 30 March 2022)

(*Continued*)

Table 1 (*Continued*)

Name	Description	Reference	URL
Ecodex	Ecodex is a cloud-based lifecycle product management DST co-developed by Nestle and Selerant. The tool utilises LCA databases to estimate the environmental footprint of products in a range of sectors including packaging. For food manufacturers, it can account for food losses (e.g. spillage and the generation of by-/co-products) and the relative contribution of these to environmental outcomes such as GHG emissions and water consumption. Ecodex helps product managers and businesses during the product design stage and, therefore, can be used to help prevent or minimise FLW.	(Schenker et al., 2014)	https://empauer.com/solutions/ecodex/ (accessed 17 June 2022)
Integral Dairy Valorisation Model (IDVM)	IDVM uses a linear programming model to identify the optimum allocation of raw dairy materials available to a dairy processing company. The tool uses a specialised data analysis program and was designed in partnership with Friesland Campina in the Netherlands.	(Banaszewska et al., 2014; Banaszewska et al., 2013)	NA

LCA, life cycle assessment; LCC, life cycle cost; GHG, greenhouse gas; BCA, benefit–cost analysis; HFLP, hierarchical facility location problems; AHP, analytical hierarchy process; GIS, geographic information system; MFCA, material flow cost accounting; NA, not applicable.

Sheet (WRAP, 2022). These tools were not considered in this chapter, though we acknowledge the importance of standardised reporting frameworks to assist with managing FLW.

The following paragraphs provide an overview of each of these DSTs. It takes into consideration the type/s of dairy products and product categories; location specificity (e.g. state, country or regional data); sustainability decision criteria (i.e. social, economic and environmental); stage/s of the value chain; and FLW management options. These characteristics are outlined in Table 2. As highlighted in Table 2, the eight DSTs provide different levels of support and specificity to dairy products and waste streams. Table 2 also illustrates that no single DST adequately covers all stages of the value chain or all management options.

For instance, on the one hand, ReFED's Insights Engine provides guidance on economic, social and environmental outcomes in relation to many food products, from primary production through to consumption. It covers management practices at all stages of the Food Waste Hierarchy (i.e. prevention through to disposal). However, users are not able to separate information specific to the dairy sector, either as a category or as individual products. Users must rely on the broader sub-category of 'dairy and eggs'. Therefore, this DST will provide general guidance across a number of areas, but it is not able to provide information specific to dairy products.

On the other hand, the Integral Dairy Valorisation Model (IDVM) developed by Banaszewska et al. (2014) provides highly targeted information about 54 individual dairy by-/co-products and optimises raw materials based on milk supply, market demand, production and transportation. The model - which is specifically targeted to milk processing companies - optimises profit based on the further processing of waste streams into new value-added products. Therefore, it does not consider other FLW management options such as feeding to livestock, anaerobic digestions and so on. Also, IDVM only accounts for the economic dimension of sustainability, compared to other DSTs which also incorporate social and environmental outcomes in their analysis.

Based on the information outlined in Tables 1 and 2, the options to manage and valorise dairy FLW are captured to different degrees in the currently available DSTs, and relevance to different stages of value chain is summarised in Table 3. Further research is needed to develop tools that support dairy value chain actors to valorise dairy FLW.

4.2 Factors that need to be considered

It is becoming increasingly necessary for food businesses at any stage of the value chain, including in the dairy sector, to move towards close-looped chains. Therefore, dairy businesses need to consider the full suite of options available

Table 2 Characteristics of existing DSTs

Name	Dairy product specificity	Location specificity	Sustainability decision criteria	Stage of the chain	Management options considered
WARM tool	• 'Dairy' proxy • Mixed food waste	• USA (national and state data)	**Economic:** • Labour hours • Wages • Taxes **Environmental:** • GHG emissions (CO_2-e) • Energy	• Waste management	• Prevention • Recycling • Anaerobic digestion • Composting • Combustion • Landfill
Food loss and waste value calculator	• Cheese • Milk • Yoghurt • 'Dairy' proxy (average of cheese, milk and dairy)	• Global and regional data	**Environmental:** • Climate change (CO_2-e) • Water scarcity footprint (m^3-e) • Soil quality index (points) • Eutrophication (phosphorous and nitrogen) **Social:** • Energy (kcal) • Protein (g) • Carbohydrates (g) • Fibre (g) • Micronutrients (mg)	• Production • Processing • Distribution • Retail • Households • Waste management	• Animal feed • Bio-based materials/ biochemical processing • Co-digestion/ anaerobic digestion • Composting/ aerobic processes • Controlled combustion • Land application • Not harvested/ ploughed in • Landfill • Refuse/discards/ litter • Sewer/wastewater treatment

Insights Engine	• 'Dairy and eggs' proxy	• USA (national and state data)	**Economic:** • Cost & benefit **Environmental:** • Climate change (CO_2-e) • Water footprint (m^3e) **Social:** • Meals wasted/saved (equivalents) • Jobs	• Production • Processing • Distribution • Retail • Food service • Households	• Prevention • Donation • Industrial uses • Animal feed • Anaerobically digested • Composted • Not harvested • Incineration • Land application • Landfill • Dumping • Sewer
Whey distribution chain optimisation model	• Whey	• Non-specific (case study conducted in Brazil (published) as well as Argentina, Colombia and Uruguay (unpublished))	**Economic:** • Cost • Return on investment **Environmental:** • Regional map for whey waste avoidance	• Processing	• Further processing into powders • Option for further processing into dairy consumer products such as beverages and dairy snacks
Benders decomposition for a reverse logistics network	• Whey	• Non-specific [case study conducted in Brazil (published)]	**Economic:** • Cost	• Processing	• Further processing into powders

(Continued)

Table 2 (*Continued*)

Name	Dairy product specificity	Location specificity	Sustainability decision criteria	Stage of the chain	Management options considered
Whey tool	Whey-derived products, specifically: • Liquid raw whey • Liquid pasteurised Whey • Demineralised 40% whey (liquid and powder) • WPC 35 (liquid and powder) • Whey Powder • Ricotta Cheese • Whey Cream • Whey Permeate • Ricotta Whey	• South-east Australia • New Zealand • Europe/North America	**Economic:** • Years-to-payback	• Processing	• Further processing into food products
FORKLIFT	• Whey permeate	• EU (regional and national data)	**Economic:** • Cost **Environmental:** • GHG emissions (CO_2-e)	• Processing	• Further processing into whey permeate powder • Animal feed • Anaerobic digestion • Bioethanol production

GISWASTE	• Dairy	• Basque country, Spain	**Economic:** • Cost • Revenue/income **Environmental:** • Distance travelled • GHG emissions (CO_2-e) • Water footprint (m^3-e) • Wastewater generated • Eutrophication potential (kg P-e) • Biogas/animal feed produced	• Waste management	• Animal feed • Anaerobic digestion
DIRECT	• >50 dairy-related product groups, classes and subclasses outlined in the UN Central Product Classification (CPC), Version 2.1	• Non-specific	**Economic:** • 'True' cost of waste **Environmental:** • Tonnes of FLW • GHG emissions (CO_2-e) • Non-renewal energy and minerals • Water footprint (m^3-e) • Land use • Ecosystem quality	• Processing • Distribution • Retail • Food service	• Prevention • Donation • Industrial uses • Animal feed • Anaerobically digested • Composted • Not harvested • Incineration • Land application • Landfill • Dumping

(Continued)

Table 2 (*Continued*)

Name	Dairy product specificity	Location specificity	Sustainability decision criteria	Stage of the chain	Management options considered
Ecodex	• 33 dairy products based on CPC product types are available in the life cycle inventory database	• Geographic specificity can be at the global, regional, country and provincial level	**Environmental:** • Tonnes of FLW • GHG emissions (CO_2-e) • Non-renewal energy and minerals • Water footprint (m3-e) • Land use • Ecosystem quality	• Processing (Ecodex is primarily used for decision-makers at the processing stage but represents impacts across the product lifecycle; from production to retail)	• Prevention
IDVM	• 54 individual dairy (by-/co-) products	• The Netherlands (FrielandCampina)	**Economic:** • Profit	• Processing	• Prevention • Further processing of by-products into powders and individual food components

Table 3 FLW DST that are available to decision-makers at different stages of the dairy value chain

	Primary production	Processing	Retail and wholesale	Consumption and service	Recovery/ waste management
WARM Tool					•
Food loss and waste value calculator	•	•	•	•	•
Insights Engine	•	•	•	•	
Whey distribution chain optimisation model		•			
Benders decomposition for a reverse logistics network		•			
Whey tool		•			
FORKLIFT		•			
GISWASTE					•
DIRECT	•	•	•	•	
Ecodex		•			
IDVM		•			

(i.e. reduce, reuse, recycle and recover) to manage waste streams. However, new DSTs need to account for the specific problems facing the dairy sector and consider their stage of the value chain, location and proximity to other industries, and the cost and access to technology. A list of example research questions that would need to be addressed during the decision-making process for managing whey, and therefore need to be integrated into a DST, is as follows:

- How much whey is currently produced, including seasonal variations?
- What are the current practices in place to utilise whey and the associated costs, revenue, environmental and social impacts?
- Is there a demand for alternative products (e.g. food products, animal feed, biogas and feedstock)?
- What management pathways are being considered (e.g. processing in-house with existing infrastructure, investing in new infrastructure, outsourcing, selling raw materials to other businesses or combining with other businesses' waste products)?

- What are the minimum volumes/capacity required for each management option being considered?
- What are the associated transportation requirements?
- What investment would be needed to self-process and what would be the return on investment?
- What are the benefits and costs to transform/utilise the whey?
- What are the environmental benefits and trade-offs for each management option (e.g. GHG emissions, water footprint, etc.)?
- What are the social benefits and trade-offs for each management option (e.g. meals saved, nutrients, etc.)?
- What type of whey, and what is the quality and composition of the whey?

5 Conclusion

FLW in dairy chains is a significant issue for society but there are many opportunities to re-purpose and valorise this resource by considering circular economy principles. These include feeding more humans, using the waste as animal feed (including livestock, fish and insects), recycling non-food components, or recovering soil nutrients, energy and water. This chapter outlines several management practices that are available to decision-makers in the dairy sector. The proposed framework for utilising dairy food waste streams in a close-looped paradigm (Fig. 5) is an important contribution to the broader discussion about transitioning to circular food systems. However, knowing which practice or technology may be best suited to individual businesses and economically, environmentally and socially sustainable can be a challenge for the decision-makers involved.

DSTs can help business decision-makers to navigate the Food Waste Hierarchy and complex broader issues related to FLW. Current DSTs address some of the needs of decision-makers but do not provide full coverage of decision support for dairy businesses. Therefore, further work is required to develop DSTs that integrate various aspects of the existing tools into one tool. The ideal DST would enable companies or company groups (e.g. cooperatives or joint ventures).

Additionally, as discussed at the beginning of this chapter, the challenges facing different regions vary. This chapter highlights the unique issues and opportunities facing regions with developed dairy chains. Further research is needed to better understand the unique challenges and potential solutions facing the regions with underdeveloped dairy chains.

By providing the tools and information to decision-makers through the dairy value chain, adopting practices that align with the circular economy will be more achievable and contribute to the goal of halving global FLW.

6 Acknowledgements

This work has been supported by the following organisations: (1) Fight Food Waste Cooperative Research Centre whose activities are funded by the Australian Government's Cooperative Research Centre Program; (2) the Commonwealth Scientific and Industry Research Organisation (CSIRO); and (3) the University of Adelaide. Thanks to Paul-Antoinne Bontinck (Lifecycles) for providing access to the National Food Waste Baseline data and to Victor Barichello (Empauer) and Dejan Popovic (Selerant) and Alister Hall and Simon Lockrey (RMIT) for their feedback to this chapter.

7 Where to look for further information

There has been a considerable amount of information published on the topic of whey value addition, including:

- Smithers, G. W. (2008). Whey and whey proteins–From 'gutter-to-gold', *International Dairy Journal* 18(7), 695–704. https://doi.org/10.1016/j.idairyj.2008.03.008.
- Oliveira, D., Fox, P. and O'Mahony, J. A. (2020). Chapter 4: Byproducts from Dairy Processing, In: Simpson, B. K., Aryee, A. N. A. and Toldrá, F. (Eds.), *Byproducts from Agriculture and Fisheries*. John Wiley & Sons Ltd, West Sussex, UK. https://doi.org/10.1002/9781119383956.ch4.
- Fenwick, R., Harper, J., Hobman, P., Huffman, L., Kirkpatrick, K., MacGibbon, J., Marshall, K., Matthews, M., Wilson, A. and Woodhams, D. (2014). Whey to go: whey protein concentrate: a New Zealand success story. Ngaio Press, Martinborough, New Zealand, https://nzifst.org.nz/page-18143#:~:text=Whey%20to%20Go%20is%20the,%2C%20Robin%20Fenwick%2C%20Arthur%20Wilson.&text=A%20fascinating%20story%20about%20how%20industrial%20innovation%20really%20works%20in%20practice.
- Tetra Pak (2015). Chapter 15 - Whey Processing, In: Bylund, G. (Ed.) Dairy Processing Handbook. Tetra Pak, Pully, Switzerland.

Australia's peak body for the dairy industry recently published a food waste sector action plan, which is available here: https://www.dairyaustralia.com.au/manufacturing-support/manufacturing-sustainability/dairy-sector-food-waste-action-plan#:~:text=The%20Dairy%20Food%20Waste%20Action,across%20the%20dairy%20supply%20chain.

For further information about decision support tools refer to Table 1.

8 References

Alexander, S., Harris, P. and McCabe, B. K. (2019). Biogas in the suburbs: an untapped source of clean energy?, *Journal of Cleaner Production* 215, 1025–1035. https://doi.org/10.1016/j.jclepro.2019.01.118.

Ananda, J., Karunasena, G. G. and Pearson, D. (2021). *Priority Behaviours for Interventions to Reduce Household Food Waste in Australia*. Fight Food Waste Cooperative Research Centre (FFWCRC), Adelaide, Australia.

Annosi, M. C., Brunetta, F., Bimbo, F. and Kostoula, M. (2021). Digitalization within food supply chains to prevent food waste. Drivers, barriers and collaboration practices, *Industrial Marketing Management* 93, 208–220. https://doi.org/10.1016/j.indmarman.2021.01.005.

Asunis, F., De Gioannis, G., Dessì, P., Isipato, M., Lens, P. N. L., Muntoni, A., Polettini, A., Pomi, R., Rossi, A. and Spiga, D. (2020). The dairy biorefinery: integrating treatment processes for cheese whey valorisation, *Journal of Environmental Management* 276, 111240. https://doi.org/10.1016/j.jenvman.2020.111240.

Aydiner, C., Sen, U., Topcu, S., Sesli, D., Ekinci, D., Altınay, A. D., Ozbey, B., Koseoglu-Imer, D. Y. and Keskinler, B. (2014). Techno-economic investigation of water recovery and whey powder production from whey using UF/RO and FO/RO integrated membrane systems, *Desalination and Water Treatment* 52(1–3), 123–133. https://doi.org/10.1080/19443994.2013.786655.

Banaszewska, A., Cruijssen, F., Claassen, G. D. H. and van der Vorst, J. G. A. J. (2014). Effect and key factors of byproducts valorization: the case of dairy industry, *Journal of Dairy Science* 97(4), 1893–1908. https://doi.org/10.3168/jds.2013-7283.

Banaszewska, A., Cruijssen, F., van der Vorst, J. G. A. J., Claassen, G. D. H. and Kampman, J. L. (2013). A comprehensive dairy valorization model, *Journal of Dairy Science* 96(2), 761–779. https://doi.org/10.3168/jds.2012-5641.

Barragan-Fonseca, K. B., Dicke, M. and van Loon, J. J. A. (2017). Nutritional value of the black soldier fly (Hermetia illucens L.) and its suitability as animal feed–a review, *Journal of Insects as Food and Feed* 3(2), 105–120. https://doi.org/10.3920/JIFF2016.0055.

Béné, C., Fanzo, J., Haddad, L., Hawkes, C., Caron, P., Vermeulen, S., Herrero, M. and Oosterveer, P. (2020). Five priorities to operationalize the EAT-Lancet Commission report, *Nature Food* 1(8), 457–459. https://doi.org/10.1038/s43016-020-0136-4.

Chalermthai, B., Ashraf, M. T., Bastidas-Oyanedel, J. R., Olsen, B. D., Schmidt, J. E. and Taher, H. (2020). Techno-economic assessment of whey protein-based plastic production from a co-polymerization process, *Polymers (Basel)* 12(4), 847. https://doi.org/10.3390/polym12040847.

Chalermthai, B., Giwa, A., Schmidt, J. E. and Taher, H. (2021). Life cycle assessment of bioplastic production from whey protein obtained from dairy residues, *Bioresource Technology Reports* 15, 100695. https://doi.org/10.1016/j.biteb.2021.100695.

Chang, R., Ng, T. B. and Sun, W. Z. (2020). Lactoferrin as potential preventative and adjunct treatment for COVID-19, *International Journal of Antimicrobial Agents* 56(3), 106118. https://doi.org/10.1016/j.ijantimicag.2020.106118.

Chen, C., Chaudhary, A. and Mathys, A. (2020). Nutritional and environmental losses embedded in global food waste, *Resources, Conservation and Recycling* 160, 104912. https://doi.org/10.1016/j.resconrec.2020.104912.

Codex (1991). Guidelines for the preservation of raw milk by use of the lactoperoxidase system. Available at: http://siweb1.dss.go.th/standard/Fulltext/codex/CXG_013E.pdf. CAC/GL 13-1991. Codex Alimentarius Commission, Rome, Italy.

Davis, J., Holtz, E., Metcalf, P., De Menna, F., Scherhaufer, S., Gollnow, S., Colin, F., García Herrero, L., Vittuari, M. and Östergren, K. (2019). Valorisation models - Food side flow recovery lifecycle tool (FORKLIFT), *REFRESH*. Available at: https://eu-refresh.org/forklift.html.

de Castro, T. R., de Macedo, D. C., de Genaro Chiroli, D. M., da Silva, R. C. and Tebcherani, S. M. (2022). The potential of cleaner fermentation processes for bioplastic production: A narrative review of polyhydroxyalkanoates (PHA) and polylactic acid (PLA), *Journal of Polymers and the Environment* 30(3), 810–832. https://doi.org/10.1007/s10924-021-02241-z.

de Wit, J. N. (2001). Lecturer's handbook on whey and whey products. Available at: http://ewpa.euromilk.org/fileadmin/user_upload/Public_Documents/EWPA_Publications/Lecturer_s_Handbook_on_Whey.pdf. European Whey Products Association (EWPA), Brussels, Belgium.

Empauer (2022). DIRECT. Available at: https://empauer.com/solutions/direct/ (Accessed 16 March 2022).

EPA (2019). Waste reduction model (WARM) tool. Available at: https://www.epa.gov/warm/versions-waste-reduction-model-warm#15. Environmental Protection Agency (EPA), Washington DC.

Esmaeilbeigi, R., Middleton, R., García-Flores, R. and Heydar, M. (2021). Benders decomposition for a reverse logistics network design problem in the dairy industry, *Annals of Operations Research*. https://doi.org/10.1007/s10479-021-04309-4.

FAO (2013). Food wastage footprint. Impact on natural resources. Available at: http://www.fao.org/publications/card/en/c/000d4a32-7304-5785-a2f1-f64c6de8e7a2/. Food and Agriculture Organization, Rome, Italy.

FIAL (2021). The National Food Waste Strategy Feasibility Study–Final Report. Available at: https://www.fial.com.au/sharing-knowledge/food-waste. Food Innovation Australia Ltd (FIAL), Werribee, Australia.

Gac, A., Béline, F., Bioteau, T. and Maguet, K. (2007). A French inventory of gaseous emissions (CH4, N2O, NH3) from livestock manure management using a mass-flow approach, *Livestock Science* 112(3), 252–260. https://doi.org/10.1016/j.livsci.2007.09.006.

García-Flores, R., Martins, R., de Souza Filho, O. V., González, M., Mattos, C., Rosenthal, A. and Juliano, P. (2015). A novel facility and equipment selection model for whey utilisation: A Brazilian case study, *Computers and Electronics in Agriculture* 117, 127–140. https://doi.org/10.1016/j.compag.2015.07.016.

Ghaly, A. E., Mahmoud, N. S., Rushton, D. G. and Arab, F. (2007). Potential environmental and health impacts of high land application of cheese whey, *American Journal of Agricultural and Biological Sciences* 2(2), 106–117. https://doi.org/10.3844/ajabssp.2007.106.117.

GHD (2017). Anaerobic digestion as a treatment and energy recovery technology for dairy processing waste streams. Available at: https://www.dairyaustralia.com.au/resource-repository/2020/12/03/anaerobic-digestion-as-a-treatment-and-energy-recovery-technology-for-dairy-processing-waste-streams#.YlkmzuhByUk. Dairy Australia, Melbourne, Australia.

Gustavsson, J., Cederberg, C., Sonesson, U. and Meybeck, A. (2011). Global food losses and food waste: Extent, causes and prevention. Available at: https://www.fao.org/3/i2697e/i2697e.pdf. Food and Agriculture Organization, Rome, Italy.

Hadj Saadoun, J., Montevecchi, G., Zanasi, L., Bortolini, S., Macavei, L. I., Masino, F., Maistrello, L. and Antonelli, A. (2020). Lipid profile and growth of black soldier flies (Hermetia illucens, Stratiomyidae) reared on by-products from different food chains (Hermetia illucens, Stratiomyidae), *Journal of the Science of Food and Agriculture* 100(9), 3648–3657. https://doi.org/10.1002/jsfa.10397.

Hanson, C., Lipinski, B., Robertson, K., Dias, D., Gavilan, I., Gréverath, P., Ritter, S., Fonseca, J., VanOtterdijk, R. and Timmermans, T. (2016). *Food Loss and Waste Accounting and Reporting Standard*.

Hanson, C. and Mitchell, P. (2017). The business case for reducing food loss and waste, *Champions of 12.3*. Available at: https://champions123.org/sites/default/files/2020-08/business-case-for-reducing-food-loss-and-waste.pdf. Washington, DC.

Hauser, J. (2013). Dairy Australia whey model documentation. Available at: https://wheytool.dairyaustralia.com.au/Content/pdf/Whey%20model%20Documentation%20July%202018%20v1.pdf. Dairy Australia, Melbourne, Australia.

Hauser, J. (2017). *Analysis of South Australia Whey Processing Options*. Xcheque, Australia.

Jefferson, M. T., Rutter, C., Fraine, K., Borges, G. V. B., de Souza Santos, G. M., Schoene, F. A. P. and Hurst, G. A. (2020). Valorization of sour milk to form Bioplastics: Friend or foe?, *Journal of Chemical Education* 97(4), 1073–1076. https://doi.org/10.1021/acs.jchemed.9b00754.

Jones, R. E., Speight, R. E., Blinco, J. L. and O'Hara, I. M. (2022). Biorefining within food loss and waste frameworks: A review, *Renewable and Sustainable Energy Reviews* 154, 111781. https://doi.org/10.1016/j.rser.2021.111781.

Juliano, P. (2021). Reinforcing South American communities through whey value addition, IDF Global Dairy Conference 13–15 October 2021. Copenhagen, Denmark.

Juliano, P., González, M. A., Castells, M. L., Di Risio, J., García-Flores, R., Rosenthal, A., Silva, C. M., Krolow, A. C. R., Zoccal, R., Walter, E., Dutra, A., Caicedo, I. B., Castañeda, C. G. G., Peñaranda, R. Q., Trujillo, R. S., Valencia, J. U. S., Álvarez, L. F. G., López, T., Jorcin, S., Miraballes, M., Mattos, C., Avellaneda, Y. A. and Salazar, P. V. (2017). *Whey Value Addition* (1st edn.). Instituto Nacional de Tecnología Industrial, San Martin, Argentina - INTI.

Khyade, V. B. and Tamhane, A. (2021). Utilization of the methanol maceratives of pre-pupal stages of the black soldier fly, Hermetia illucens L. (Diptera: Stratiomyidae) for inhibition of bacterial growth', *Uttar Pradesh Journal of Zoology*, 109–118.

Kirchherr, J., Reike, D. and Hekkert, M. (2017). Conceptualizing the circular economy: An analysis of 114 definitions, *Resources, Conservation and Recycling* 127, 221–232. https://doi.org/10.1016/j.resconrec.2017.09.005.

Knapp, J. R., Laur, G. L., Vadas, P. A., Weiss, W. P. and Tricarico, J. M. (2014). Invited review: Enteric methane in dairy cattle production: Quantifying the opportunities and impact of reducing emissions, *Journal of Dairy Science* 97(6), 3231–3261. https://doi.org/10.3168/jds.2013-7234.

KPMG (2020). A national food waste tax incentive. Available at: https://assets.kpmg/content/dam/kpmg/au/pdf/2020/national-food-waste-tax-incentive.pdf. KPMG, Sydney, Australia.

Kummu, M., de Moel, H., Porkka, M., Siebert, S., Varis, O. and Ward, P. J. (2012). Lost food, wasted resources: Global food supply chain losses and their impacts on freshwater, cropland, and fertiliser use, *Science of the Total Environment* 438, 477–489. https://doi.org/10.1016/j.scitotenv.2012.08.092.

Lade, S. J., Steffen, W., de Vries, W., Carpenter, S. R., Donges, J. F., Gerten, D., Hoff, H., Newbold, T., Richardson, K. and Rockström, J. (2020). Human impacts on planetary boundaries amplified by Earth system interactions, *Nature Sustainability* 3(2), 119–128. https://doi.org/10.1038/s41893-019-0454-4.

Ling, K. C. (2008). *Whey to Ethanol: A Biofuel Role for Dairy Cooperatives?* U.S. Department of Agriculture (USDA), Washington D.C.

Lore, T. A., Omore, A. O. and Staal, S. J. (2005). Types, Levels and Causes of Post-harvest Milk and Dairy Losses in Sub-Saharan Africa and the Near East: Phase Two Synthesis Report.

Mainardis, M., Flaibani, S., Trigatti, M. and Goi, D. (2019). Techno-economic feasibility of anaerobic digestion of cheese whey in small Italian dairies and effect of ultrasound pre-treatment on methane yield, *Journal of Environmental Management* 246, 557–563. https://doi.org/10.1016/j.jenvman.2019.06.014.

March, M. D., Toma, L., Thompson, B. and Haskell, M. J. (2019). Food waste in primary production: Milk loss with mitigation potentials, *Frontiers in Nutrition* 6, 173. https://doi.org/10.3389/fnut.2019.00173.

Meneses, Y. E. and Flores, R. A. (2016). Feasibility, safety, and economic implications of whey-recovered water in cleaning-in-place systems: A case study on water conservation for the dairy industry, *Journal of Dairy Science* 99(5), 3396–3407. https://doi.org/10.3168/jds.2015-10306.

Mostafa Imeni, S., Pelaz, L., Corchado-Lopo, C., Maria Busquets, A., Ponsá, S. and Colón, J. (2019). Techno-economic assessment of anaerobic co-digestion of livestock manure and cheese whey (Cow, Goat & Sheep) at small to medium dairy farms, *Bioresource Technology* 291, 121872. https://doi.org/10.1016/j.biortech.2019.121872.

Moult, J. A., Allan, S. R., Hewitt, C. N. and Berners-Lee, M. (2018). Greenhouse gas emissions of food waste disposal options for UK retailers, *Food Policy* 77, 50–58. https://doi.org/10.1016/j.foodpol.2018.04.003.

Murari, C. S., da Silva, D. C. M. N., Schuina, G. L., Mosinahti, E. F. and Del Bianchi, V. L. (2019). Bioethanol production from dairy industrial coproducts, *Bioenergy Research* 12(1), 112–122. https://doi.org/10.1007/s12155-018-9949-5.

Myers, H. M., Tomberlin, J. K., Lambert, B. D. and Kattes, D. (2014). Development of black soldier fly (Diptera: Stratiomyidae) larvae fed dairy manure, *Environmental Entomology* 37(1), 11–15. https://doi.org/10.1093/ee/37.1.11.

Oliveira, D., Fox, P. and O'Mahony, J. A. (2020). Chapter 4: Byproducts from dairy processing. In: Simpson, B. K., Aryee, A. N. A. and Toldrá, F. (Eds). *Byproducts from Agriculture and Fisheries*. John Wiley & Sons Ltd, West Sussex, UK.

Papargyropoulou, E., Lozano, R., Steinberger, K., Wright, N. and Ujang, Zb. (2014). The food waste hierarchy as a framework for the management of food surplus and food waste, *Journal of Cleaner Production* 76, 106–115. https://doi.org/10.1016/j.jclepro.2014.04.020.

Parashar, A., Jin, Y., Mason, B., Chae, M. and Bressler, D. C. (2016). Incorporation of whey permeate, a dairy effluent, in ethanol fermentation to provide a zero waste solution for the dairy industry, *Journal of Dairy Science* 99(3), 1859–1867. https://doi.org/10.3168/jds.2015-10059.

Patel, A. K., Vaisnav, N., Mathur, A., Gupta, R. and Tuli, D. K. (2016). Whey waste as potential feedstock for biohydrogen production, *Renewable Energy* 98, 221–225. https://doi.org/10.1016/j.renene.2016.02.039.

Powell, C. and Curtis, P. (2020a). Insights engine food waste monitor - 2020 methodology. Available at: https://insights.refed.org/uploads/documents/refed_insights_engine food_waste_monitormethodology_vfinal_2021.04.12.pdf?_cchid=4f1d745d5f50363e83869ef5f3b2e962. ReFED, New York.

Powell, C. and Curtis, P. (2020b). Insights engine solutions database - 2020 methodology. Available at: https://insights.refed.org/uploads/documents/refed-insights-engine-solution-database-methodology-vfinal2021-05-27.pdf?_cchid=ccf71f4eacfac581ad228da51c320fd1. ReFED, New York.

Power, D. J. (2002). *Decision Support Systems: Concepts and Resources for Managers*. Greenwood Publishing Group, CT.

Redlingshöfer, B., Barles, S. and Weisz, H. (2020). Are waste hierarchies effective in reducing environmental impacts from food waste? A systematic review for OECD countries, *Resources, Conservation and Recycling* 156, 104723. https://doi.org/10.1016/j.resconrec.2020.104723.

Reynolds, C. J., Piantadosi, J. and Boland, J. (2015). Rescuing food from the organics waste stream to feed the food insecure: An economic and environmental assessment of Australian food rescue operations using environmentally extended waste input-output analysis, *Sustainability* 7(4), 4707–4726. https://doi.org/10.3390/su7044707.

Ridoutt, B. G., Baird, D. and Hendrie, G. A. (2021). Diets within planetary boundaries: What is the potential of dietary change alone?, *Sustainable Production and Consumption* 28, 802–810. https://doi.org/https://https://doi.org/10.1016/j.spc.2021.07.009.

Sallam, A. E., El-feky, M. M. M., Ahmed, M. S. and Mansour, A. (2022) Potential use of whey protein as a partial substitute of fishmeal on growth performance, non-specific immunity and gut histological status of juvenile European seabass, Dicentrarchus labrax, *Aquaculture Research* 53(4), 1527–1541. https://doi.org/10.1111/are.15688.

San Martin, D., Orive, M., Martínez, E., Iñarra, B., Ramos, S., González, N., de Salas, A. G., Vázquez, L. and Zufía, J. (2017) Decision Making Supporting Tool Combining AHP Method with GIS for Implementing Food Waste Valorisation Strategies, *Waste and Biomass Valorization* 8(5), 1555–1567. https://doi.org/10.1007/s12649-017-9976-z.

Saravani, M., Ehsani, A., Aliakbarlu, J. and Ghasempour, Z. (2019). Gouda cheese spoilage prevention: Biodegradable coating induced by Bunium persicum essential oil and lactoperoxidase system, *Food Science and Nutrition* 7(3), 959–968. https://doi.org/10.1002/fsn3.888.

Schenker, U., Espinoza-Orias, N. and Popovic, D. (2014). EcodEX: A simplified ecodesign tool to improve the environmental performance of product development in the food industry. In: Proceedings of the 9th International Conference on Life Cycle Assessment in the Agri-Food Sector (LCA Food 2014), San Francisco, California, USA, 8-10 October, 2014. American Center for Life Cycle Assessment. pp. 1220–1223.

Schingoethe, D. J. (1976). Whey utilization in animal feeding: A summary and evaluation, *Journal of Dairy Science* 59(3), 556–570. https://doi.org/jds.S0022-0302(76)84240-3.

Silva, D., Nunes, A. O., da Silva Moris, A., Moro, C. and Piekarski, T. O. R. (2017). How important is the LCA software tool you choose Comparative results from GaBi, openLCA, SimaPro and Umberto. In: Proceedings of the VII Conferencia Internacional de Análisis de Ciclo de Vida en Latinoamérica, Medellin, Colombia. pp. 10–15.

Somlai, R. (2022). Integrating decision support tools into businesses for sustainable development: A paradoxical approach to address the food waste challenge, *Business Strategy and the Environment* 31(4), 1607–1622. https://doi.org/10.1002/bse.2972.

Smithers, G. W. (2008). Whey and whey proteins—from 'gutter-to-gold', *International Dairy Journal* 18(7), 695–704. https://doi.org/10.1016/j.idairyj.2008.03.008.

Sprague, R. H. (1980). A framework for the development of decision support systems, *MIS Quarterly* 4(4), 1–26. https://doi.org/10.2307/248957.

Tait, S., Harris, P. W. and McCabe, B. K. (2021). Biogas recovery by anaerobic digestion of Australian agro-industry waste: A review, *Journal of Cleaner Production* 299, 126876. https://doi.org/10.1016/j.jclepro.2021.126876.

Teigiserova, D. A., Hamelin, L. and Thomsen, M. (2020). Towards transparent valorization of food surplus, waste and loss: Clarifying definitions, food waste hierarchy, and role in the circular economy, *Science of the Total Environment* 706, 136033. https://doi.org/10.1016/j.scitotenv.2019.136033.

Tsermoula, P., Khakimov, B., Nielsen, J. H. and Engelsen, S. B. (2021). Whey - the waste-stream that became more valuable than the food product, *Trends in Food Science and Technology* 118, 230–241. https://doi.org/10.1016/j.tifs.2021.08.025.

Verghese, K., Lockrey, S., Rio, M. and Dwyer, M. (2018). DIRECT, a tool for change: Co-designing resource efficiency in the food supply chain, *Journal of Cleaner Production* 172, 3299–3310. https://doi.org/10.1016/j.jclepro.2017.10.271.

Watson, K. S., Peterson, A. E. and Powell, R. D. (1977). Benefits of spreading whey on agricultural land, *Journal (Water Pollution Control Federation)* 49(1), 24–34. https://www.jstor.org/stable/25039215.

Williams, H., Lindström, A., Trischler, J., Wikström, F. and Rowe, Z. (2020). Avoiding food becoming waste in households – the role of packaging in consumers' practices across different food categories, *Journal of Cleaner Production* 265, 121775. https://doi.org/10.1016/j.jclepro.2020.121775.

WRAP (2022). Food surplus and waste data capture sheet. Available at: https://wrap.org.uk/resources/tool/food-loss-and-waste-data-capture-sheet. WRAP, Banbury, UK.

WRI (2019). Food loss and waste (FLW) value calculator. Available at: http://www.flwprotocol.org/why-measure/food-loss-and-waste-value-calculator/ (Accessed 30 March 2022).

Wu, X., Dong, C., Yao, W. and Zhu, J. (2011). Anaerobic digestion of dairy manure influenced by the waste milk from milking operations, *Journal of Dairy Science* 94(8), 3778–3786. https://doi.org/10.3168/jds.2010-4129.

Zhang, W. and Rhim, J.-W. (2022). Functional edible films/coatings integrated with lactoperoxidase and lysozyme and their application in food preservation, *Food Control* 133, 108670. https://doi.org/10.1016/j.foodcont.2021.108670.

Zotta, T., Solieri, L., Iacumin, L., Picozzi, C. and Gullo, M. (2020). Valorization of cheese whey using microbial fermentations, *Applied Microbiology and Biotechnology* 104(7), 2749–2764. https://doi.org/10.1007/s00253-020-10408-2.

Chapter 14

Aquaponics: challenges and opportunities for commercial application

Lorenzo Rossi, Martina Puccinelli, Ilaria Marchioni, Luca Incrocci, Baldassare Fronte, Carlo Bibbiani and Alberto Pardossi, University of Pisa, Italy*

1 Introduction

Aquaculture is the production, under controlled conditions, of freshwater or saltwater aquatic organisms for human consumption and non-food use: finfish (e.g. salmon, trout, seabass, seabream, carp, and tuna), molluscs (e.g. mussels, oysters, and clams), crustaceans (e.g. shrimps, crabs, and lobsters), aquatic plants, algae, and others organisms (e.g. frogs, pearls, and aquatic mammals). The growth of the aquaculture industry has been astonishing in the last 50-60 years and is known as 'blue revolution' (Carballeira Braña et al., 2021). Aquaculture production started during the 1960s in Asia and between 1970s and 1980s in Western countries and has grown at 7.5% per year since 1970 because of the increasing demand for fish proteins and the stagnation or even the decrease in capture fisheries (FAO, 2020).

The expansion of fish farming has several direct and indirect environmental impacts due to the production of feed ingredients (e.g. fishmeal and fish oil from forage fisheries), the disposal of farm effluents (organic matters, nitrogen, phosphorus, etc.), disease transmission, biological pollution (dispersal of invasive species), and destruction of habitats (Carballeira Braña et al., 2021).

* Correspondence: lorenzo.rossi@phd.unipi.it.

http://dx.doi.org/10.19103/AS.2023.0120.24

Among the aquaculture systems, conventional marine net-pens have the lowest impact while land-based recirculating aquaculture systems (RASs) have a much greater impact due to the massive use of materials and energy (Bohnes et al., 2019). Aquaponic systems are considered capable of producing more food with a much greater use efficiency of land and water and with a minor environmental impact compared to conventional food production systems such as greenhouse horticulture and in-land aquafarming (Goddek et al., 2019b; Somerville et al., 2014).

Aquaponics is an agricultural system that couples fish production in RAS with soilless plant cultivation (Baganz et al., 2022). Currently, aquaponics is moving from a production system applied on a small-scale for domestic, educational, or recreational purposes (Junge et al., 2017) to a commercial system capable of attracting investors, since it is considered as one of the most efficient and sustainable systems to produce animal proteins with high water use efficiency (Kotzen et al., 2019; Palm et al., 2018; Turnsek et al., 2020). On the other hand, several concerns on its economic viability and management complexity represent the main obstacles towards the commercial application of aquaponics (Greenfeld et al., 2020; Palm et al., 2018). The research has contributed enormously to the development of aquaponic systems, especially from a technological point of view (Greenfeld et al., 2019; Hao et al., 2020). However, several aspects need further studies such as the environmental and and socio-economic implications of aquaponic production (Greenfeld et al., 2019; Hao et al., 2020; Junge et al., 2017).

In this chapter, along with a brief description of hydroponic technologies and RASs, several issues concerning the application of aquaponics on a commercial scale are discussed, such as (1) the selection of fish and crop species; (2) the development of salt-water aquaponics systems for the diversification of water sources, fish, and crops; (3) the system monitoring; (4) fish and crop protection; (5) food safety; (6) animal welfare; and (7) how to increase economic and environmental sustainability.

2 Hydroponics and recirculating aquaculture systems

According to Baganz et al. (2022), aquaponics can be defined as the 'technology that couples tank-based animal aquaculture with hydroponics – involving microbiological processes – using water from aquaculture for plant nutrition and irrigation'. Thus, aquaponics systems are based on two production technologies: hydroponic systems for crop production and RASs for rearing fish and/or other aquatic animal species.

2.1 Hydroponics

The word 'hydroponics' includes all the techniques for cultivating plants without soil in an artificial substrate (aggregate culture) or in aerated nutrient solution (water culture) (Raviv et al., 2019).

Since the first application of hydroponics for commercial crop production in the USA more than one century ago, many hydroponic systems have been designed and tested (Pardossi et al., 2006). These systems differ in the presence and the type of substrate and container, the method used to supply the nutrient solution to the plants (drip irrigation or subirrigation; flowing, stagnant, or mist water culture), and the fate of the drainage nutrient solution (open system and closed systems). In open systems, the nutrient solution drained out of the substrate at each irrigation event is not reused to irrigate plants. Conversely, in closed systems, the drainage nutrient solution is collected and reused into the system.

Almost all plant species can be grown in soilless systems, which however are commonly applied to the production of vegetables, herbs, and cut flowers under a greenhouse. Substrate growing systems are used for fruit vegetables (e.g. tomato, pepper, eggplant, and cucumber), strawberries, and cut flowers (e.g. rose and gerbera), while leafy vegetables (e.g. lettuce, spinach, leafy beet, etc.) and herbs (e.g. basil) are generally grown in a water culture (in particular, in floating raft system; Nicola et al., 2015). The main advantages and disadvantages of hydroponic systems are summarized in Table 1 (Pardossi et al., 2006; Raviv et al., 2019).

The main components of a closed-loop hydroponic system are (Pardossi et al., 2006) as follows: (1) fertigation unit for automatic preparation of the nutrient solution; (2) electronic controller interfaced to weather station for irrigation scheduling; (3) water filtration and disinfection unit; and (4) monitoring system of the nutrient solution, in particular of pH, electrical conductivity (EC), dissolved oxygen (DO), and ion composition.

2.2 Recirculating aquaculture system

RASs are used for land-based production of fish and other aquatic species in recirculating water with a minimum discharge rate. These systems can be

Table 1 Main advantages and disadvantages of hydroponics

Advantages	Disadvantages
• Higher yield • Better product quality • High use efficiency of water and nutrients • Easier management of soil-borne disease	• High installation costs • Skilled personnel required • Quick spread of root diseases and nematodes in closed systems

Table 2 Advantages and disadvantages of RASs compared to other aquaculture systems

Advantages	Disadvantages
• Lower water consumption • Higher fish stocking density • Shorter production cycle • No/reduced risk of biological pollution • Easier marketing (production site close to the market) • Better quality of water environment • Easier waste management and nutrient recycling	• Higher capital costs • Higher energy requirements • Worse produce quality (off-flavours)

installed both indoor and outdoor and their commercial application is quite recent (Espinal and Matulić, 2019; Xiao et al., 2019). Several species are currently produced in RAS using both freshwater (e.g. African catfish, eel, and trout) and marine species (e.g. salmon, turbot, seabass, and sole) (Dalsgaard et al., 2013; Martins et al., 2010). The use of RASs is widely adopted in hatcheries (Martins et al., 2010; Mozes et al., 2011; Parisi et al., 2014), while their application for post-hatchery stages is very limited and restricted to a few countries, such as The Netherlands and Denmark (Martins et al., 2010). RASs are also temporarily used for the adaptation of fish to open-sea conditions to increase resistance to parasites, e.g. sea lice in salmon (Ytrestøyl et al., 2020).

Table 2 summarizes the main advantages and disadvantages of RASs as reported by several authors (Aadland, 2018; Kawaguchi et al., 2019; Lindholm-Lehto and Vielma, 2019; Martins et al., 2010; Tal et al., 2009; van Rijn, 2013).

To allow the recirculation of high-quality water, RASs rely on complex and advanced equipments and involve some mechanical and microbial processes to increase the average daily water recirculation rate above 95% (Xiao et al., 2019). Xiao et al. (2019) recently reviewed the most common technologies applied for water treatment in RASs, which include (1) solid separators such as solid–liquid separators, sand filters, and microscreen drum filters; (2) ammonia removal units such as aerobic nitrifying bioreactors (often called biofilter), anaerobic ammonia oxidation (ANAMMOX) (Tal et al., 2009), or ion-exchange and electrochemical oxidation (Espinal and Matulić, 2019; Gendel and Lahav, 2013); (3) aeration and oxygenation devices; (4) de-gasing unit to remove dissolved carbon dioxide (CO_2); (5) disinfection systems such as ultraviolet (UV) irradiation or ozonation; (6) anaerobic biofilter to reduce the concentration of nitrate, phosphate, sulphide, and organic carbon from the water (Tal et al., 2009; van Rijn et al., 2006); (7) protein skimming for fine solids removal; (8) heating/cooling unit; and (9) waste treatment (van Rijn, 2013).

3 Aquaponic systems

In an aquaponic system, the ammonium excreted by the fish is converted into nitrate, which is much less toxic to fish, through microbial nitrification. Nitrate tends to accumulate in the water naturally but its concentration remains relatively low due to the plant uptake. Compared to a hydroponic system, the largest part of the nutrients (>50 %) in aquaponics is provided by fish through microbial metabolism (Palm et al., 2018). Therefore, microorganisms play a key role in aquaponics, increasing nutrients through mineralization and nitrification and creating a more competitive and resilient environment against pathogens (Pantanella, 2018). Aquaponic systems differ from traditional RASs for a simpler management of solid waste (Pantanella, 2018) in addition to the presence of crop plants. Nevertheless, the presence of fish and plants in the same water loop requires some compromise as regards the characteristics of the water environment and the strategies adopted for fish and crop protection (Pantanella, 2018).

Several classifications of aquaponic systems have been proposed (Maucieri et al., 2018). Most of the classifications are based on the 'size' of the system, which is often associated with the objective of the system (e.g. educational, scientific, or commercial) (Maucieri et al., 2018; Palm et al., 2018); otherwise aquaponic systems have been classified in intensive or extensive (Junge et al., 2017). However, the most important classification refers to the water loop and divides aquaponic installations into two groups: permanently coupled or on-demand-coupled systems (Baganz et al., 2022). This recent definition overcomes the former one that divides aquaponics in coupled or decoupled systems (Goddek et al., 2019b; Palm et al., 2019).

3.1 Permanently coupled aquaponics

Permanently coupled aquaponics define a system in which fish and plants are raised in the same water loop. This type of aquaponics is the most studied since many years. A schematic representation of a coupled aquaponic system is given in Fig. 1.

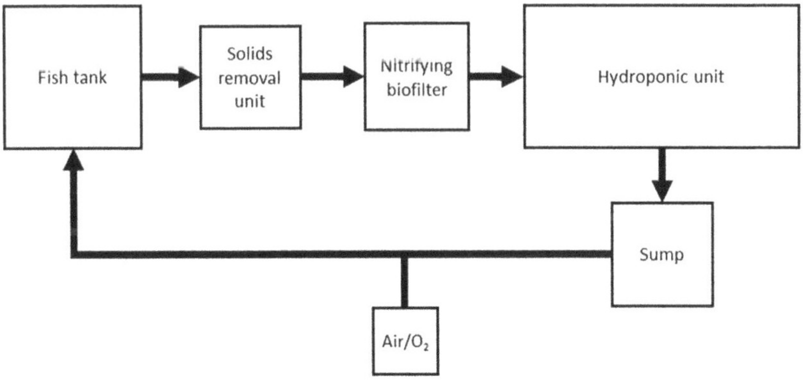

Figure 1 Schematic representation of a permanently coupled aquaponic system.

The essential design of a permanently coupled aquaponic system (Palm et al., 2019) consists of the following: (1) fish tanks; (2) solid removal unit; (3) biofilter; (4) hydroponic unit; (5) pump unit for water circulation (so-called 'sump'); and (6) blowers or liquid oxygen for water oxygenation can be also necessary in larger systems.

Fish tanks should be designed to allow a rapid removal of solids through the bottom (Davidson and Summerfelt, 2004; Palm et al., 2019). In a closed system, solids are generated from fish excreta and uneaten feed and must be removed quickly to prevent (1) the release of ammonia; (2) the development of anaerobic spots that bring to the production of toxic hydrogen sulphide (H_2S); (3) the reduction of DO level; and (4) pipeline clogging. However, in aquaponics the presence of solids contributes to the nutrient load for crops through mineralization (Pantanella, 2018). For this reason, simpler solids removal devices are used in aquaponics compared to RAS (Palm et al., 2019; Pantanella, 2018).

The aerobic nitrifying microbial community needs a high surface to establish proper water circulation and oxygenation; the biofilter provides the optimal environment for these microorganisms. In hobby installations (up to 10 m^2 of total area) the biofilter is facultative and is often replaced by the substrate used for plant growth (Palm et al., 2019; Pantanella, 2018). The effectiveness of a biofilter depends on the nitrification rate (i.e. the amount of ammonium oxidized per unit of biofilm surface area and time), which can be affected by several factors: temperature, pH, salinity, DO, and the concentration of inorganic nitrogen and organic substances (Lekang, 2007). The proper dimensioning of a biofilter depends on the production rate of total ammonia nitrogen (TAN) and the specific surface area of the medium for microbial growth (Ebeling et al., 2006; Somerville et al., 2014).

In the crop compartment of aquaponic systems, plants are normally cultivated using the following methods: substrate culture with drip irrigation or subirrigation (ebb-flow system or media bed), deep water culture, floating raft system, and nutrient film technique (NFT) (Sharma et al., 2018).

3.2 Nutrient cycling

In aquaponic systems, nutrients mainly come from the fish feed and the water added into the system (Eck et al., 2019). Fish feed formulations are generally composed of six to eight ingredients, with the aim of providing fish with a well-balanced diet in terms of proteins, carbohydrates, fats, vitamins, and minerals (Somerville et al., 2014; Timmons et al., 2018). Protein content ranges from approximately 25% for herbivorous or omnivorous fish to up to 55% for carnivorous species in early life stages (Boyd, 2015). Organic carbon ranges between 40% and 45% of fish food while organic phosphorous is around

1.2% (Eck et al., 2019; Timmons et al., 2018). Approximately 95% of feed is digested by fish and only 30–40% of this share is converted into new biomass; the remaining 60–70% is released by fish (e.g. gill excretion, urine, and faeces), of which ammonia is one of the major constituents (Khakyzadeh et al., 2015; Robaina et al., 2019). The last 5% is represented by uneaten feed. Ammonia is produced as end-product of protein catabolism and is consequently excreted by the fish through the gills as unionized ammonia (Timmons et al., 2018). Ammonia exists in two forms in water: ionized (NH_4^+) and unionized (NH_3). The latter is more toxic for fish (Timmons et al., 2018). The sum of both ammonia forms is called TAN (Wongkiew et al., 2017).

A simple model can be applied to estimate the TAN generated by the fish catabolism of proteins (Wongkiew et al., 2017):

$$TAN = FR \times PC \times 0.092 \qquad (1)$$

where TAN is expressed in g day^{-1}, FR is the feed rate (g day^{-1}), and PC is the protein content of feed (decimal value). This simple equation assumes that uneaten feed, fish faeces, and sludge are quickly eliminated from the system; proteins contain 16% N; 80% N is assimilated by fish; 80% of assimilated N is excreted as TAN (90%) and urea (10%) (Ebeling et al., 2006).

Ammonium (NH_4^+) is converted into nitrate by a two-step process called aerobic nitrification. In biofilters, ammonia-oxidizer bacteria oxidize ammonia to nitrite (NO_2^-), which in turn is transformed to nitrate (NO_3^-) by nitrite oxidizer bacteria (Ruiz et al., 2020). Nitrification is pivotal to ensure fish health since ammonia and nitrite are toxic at very low concentrations (Somerville et al., 2014). Safe values for TAN are below 1 mg L^{-1} and 3 mg L^{-1} for warm- and cold-water fish, respectively, while the desirable range for nitrite is 0–1 mg L^{-1} (Yanes et al., 2020). Nitrate is less harmful and represents the major source of nitrogen for plant growth (Wongkiew et al., 2017).

Solid wastes are the source of phosphorous, potassium, calcium, magnesium, and micronutrients (i.e. iron, zinc, copper, manganese, and molybdenum) as well (Eck et al., 2019; Goddek et al., 2015), which availability is strictly related to fish feed composition and to microbial solubilization (Eck et al., 2019; Goddek et al., 2015). Solid waste (faeces and feed leftovers) requires microbial solubilization (mineralization) to be converted from organic material to the respective ionic mineral forms, which are assimilated by plants (Goddek et al., 2015). Several other organisms are also involved in the mineralization process, such as worms, nematodes, protozoa, and fungi (Pantanella, 2018). Fish and plants nutritional requirements are satisfied when the ratio of fish feed to plant nutrient uptake is balanced and hence no nutrient accumulation or depletion occurs (Goddek et al., 2015; Somerville et al., 2014). Few studies on nutrient cycling in aquaponic systems have been published until now. Ammonia generation is linked to fish feeding rate and feed protein content, while root nitrate uptake depends on its concentration, environmental conditions, and crop developmental stage (Wongkiew et al., 2017). Crop requirement of phosphorous is also strongly related to the plant growth stage (Cerozi and Fitzsimmons, 2017). The content of iron and other micronutrients is usually low in nutrient water, since they are usually scarce in fish excreta (Doncato and Costa, 2021; Goddek et al., 2015).

The dimensioning of a permanently coupled aquaponic system should avoid nitrate accumulation in water and water discharge balancing the input and output of nitrogen in the system. One of the most common criteria considered is the feed-to-plant ratio, reported as a mass value of feed per unit of plant growing area. This ratio represents the real link between fish (that excrete nutrients through feed consumption) and plants (that use the newly generated nutrients to grow) in an aquaponic system. Thus, the amount of fish feed administrated daily is directly related to the number of plants in the crop compartment (Lennard, 2012a). Based on projects' experiences, FAO delivered rule-of-thumb feed-to-plant ratios: 40–50 g m^{-2} day^{-1} for leaf vegetables and 50–80 g m^{-2} day^{-1} for fruiting vegetables (Somerville et al., 2014). Table 3 summarizes the feed-to-plant ratio reported in the literature. Online open-access calculators exist to determine the optimal area of crops in aquaponic systems. For instance, the spreadsheet Aquaponic Media Bed Sizing Model (Ver. 2.0) is based on several important inputs (e.g. fish tank volume, fish density, and protein content) and it calculates the surface area of the media bed that should operate as a biofilter, as a solids mineralization filter, as a plant growing component, or a combination of all three (Lennard, 2012b). The

Table 3 Feed-to-plant ratio in aquaponic systems reported in the literature

Crop	Feed-to-plant ratio (g m^{-2})	Crude protein content (%)	Fish species	Reference
Leafy vegetables	40–50	32	Not specified	Somerville et al. (2014)
	60	32	Tilapia	Lennard (2012a)
Fruiting vegetables	50–80	32	Not specified	Somerville et al. (2014)
	100	32	Tilapia	Lennard (2012a)
Lettuce	56	32	Tilapia	Rakocy et al. (1997)
	16–13	32	Tilapia	Lennard (2012a)
Sweet basil	81–100	32	Tilapia	Rakocy et al. (2004)
Water spinach	15–42	32	African catfish	Endut et al. (2010)

Source: Adapted from: Pantanella (2018).

calculator proposed by Lennard (2012b) is based on a dynamic feed-to-plant ratio adapted to the protein content of the fish diet, starting from a standard value of 60 g m^{-2}. However, this ratio is strongly affected by several parameters, such as feed quantity, feed protein content, fish and crop species combination, environmental conditions, and system design. Thus, trying to generalize the feed- or fish-to-plant ratio fails to acknowledge the variability that inevitably occurs in aquaponic systems (Chu and Brown, 2021; Pantanella, 2018).

3.3 On-demand-coupled aquaponics

In a permanently coupled aquaponic system, upscaling is realized through multiunit systems, i.e. staggered fish production, diversified crops, and application of several hydroponic subsystems (Palm et al., 2019). Moreover, permanently coupled systems require trade-offs in the rearing conditions of both subsystems in terms of pH, temperature, and nutrient concentrations (Goddek et al., 2019b).

In on-demand-coupled aquaponic systems, aquaculture and hydroponic units are separate, thus providing a better control of the water environment in both systems (Goddek and Keesman, 2020). These systems can overcome the problems originated from an unbalance between fish production of nutrients and their uptake by the plants, as fish effluents can be supplemented with fertilizers to satisfy plants' demand (Delaide et al., 2016; Doncato and Costa, 2021). Moreover, it is possible to adjust the pH to make an optimal environment for plants. The separation between fish and plants can be temporarily voided to fill the hydroponic unit with RAS water and/or to use the water from hydroponic unit (two-way connected or hybrid systems) (Baganz et al., 2022; Pantanella, 2018); otherwise, the water only goes to plant from RAS (one-way connection) (Baganz et al., 2022; Monsees et al., 2017). Before being introduced in the hydroponic , water must be filtered to remove solids and avoid clogging of pipelines and drippers.

A scheme of a decoupled aquaponic system is given in Fig. 2.

Nutrients can be recovered from RAS water or solids (multi-loops systems) using bioreactors for the mineralization of organic nutrients or a distillation unit to keep the nutrient concentration of both sub-units at their respective desired levels (Goddek et al., 2019b; Goddek and Keesman, 2020). These systems keep optimal growing conditions for both animals and plants and would allow the use of the traditional animal and crop protection products without harmful effects on fish and plants (Pantanella, 2018). The main advantages of on-demand-coupled aquaponics systems are higher use efficiency of both water and nutrients and higher yields compared to stand-alone hydroponics systems (Goddek and Keesman, 2020).

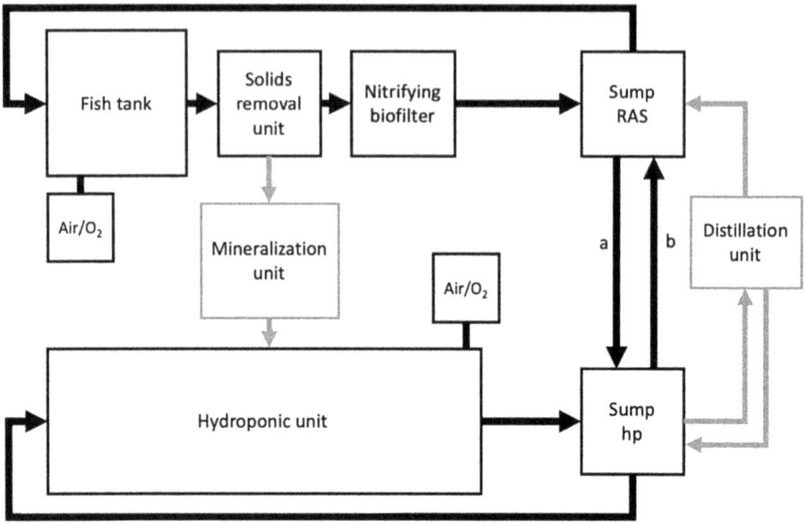

Figure 2 Schematic view of on-demand-coupled aquaponic system. Grey lines and boxes represent additional loops. Possible coupling configurations are one-way system in which the effluent from the hydroponic unit is not pumped back to the RAS unit (a) and two-way systems or hybrid systems in which water is pumped back to the RAS unit (b).

On-demand-coupled systems are more complex than permanently coupled ones; thus, simple rules of thumb may not be enough accurate for a proper dimensioning. An integrated multi-loop aquaponics and greenhouse model has been developed based on existing models for both systems (Goddek and Körner, 2019). The model was tested in three different climatic conditions: Faroe Islands, The Netherlands, and Namibia. Results indicate the importance of crop transpiration dynamics on system and sub-system sizing: optimal size of hydroponic production was 11 250 m^2, 10 250 m^2, and 5250 m^2 for tomato or 15 750 m^2, 14 000 m^2, and 9250 m^2 for lettuce in Faroe Islands, The Netherlands, and Namibia, respectively (Goddek and Körner, 2019).

3.4 Species selection

In aquaponic systems, the combination of fish and plants should be compatible with the characteristics of each production unit to balance the fish emission of nutrients that are absorbed by the plants. The choice of the proper fish species to be reared in aquaponics is crucial for the success of aquafarms. According to Barrington et al. (2009), the selection of fish species for an integrated multi-trophic aquaculture (IMTA) system should consider some criteria, in particular: (1) the geographic origin of the species (local species should be preferred); (2) complementarity with the other species in terms of growing conditions requirements; (3) growth performances; (4) market value; and (5) absence of regulatory and policy limitations. These criteria could be easily extended to aquaponics. In general, less polluting fish species should be selected together with crop species with great nutrient extractive capacity (Pantanella, 2018).

Several fish species have been tested in aquaponics, thus suggesting the large versatility of this system (Table 4). Tilapia and catfish are the most studied fish species in aquaponics; however, they are not well-accepted by all consumers (Pinho et al., 2021) and different species should be selected according to market demand. Pinho et al. (2021) have identified five fish species suitable for the south American context according to their growth performances, tolerance of polluted water, and nutritional composition: jundia or silver catfish (*Rhamdia quelen*), yellowtail lambari (*Astyanax lacustris*), pacu (*Piaractus mesopotamicus*), tambaqui (*Colossoma macropomum*), and snook (*Centropomus* spp.).

A DEXi multi-criteria decision-aid (MCDA) was recently used to select fish species for marine aquaponic systems in the Mediterranean area (Rossi et al., 2021). Different criteria were combined and evaluated using qualitative scores (e.g. high and low): natural geo-distribution, domestication degree, environmental requirements, feeding regime, growth performances, and market value. The candidate fish species were *Sparus aurata*, *Dicentrarchus labrax*, *Mugil cephalus*, *Diplodus puntazzo*, *Seriola dumerili*, *Umbrina cirrosa*, *Argyrosomus regius*, *Psetta maxima*, *Acipenser* spp., *Solea* spp., *Octopus vulgaris*. *Sparus aurata*, *Dicentrarchus labrax*, and *M. cephalus* resulted in the

Table 4 Common fish species raised in aquaponic systems in research or commercial facilities

Scope	Species	Author
Research	Tilapia (*Sarotherodon aurea*, *Oreochromis niloticus*, *Tilapia rendalli*), African catfish (*Clarias gariepinus*), Murray cod (*Macculochella peelii*), Rainbow trout (*Oncorhynchus mykiss*), common carp (*Cyprinus carpio*), Asian barramundi (*Lates calcarifer*), mullet (*Mugil cephalus*), Eurasian perch (*Perca fluviatilis*), largemouth bass (*Micropterus salmoides*), baster sturgeon (*Acipenser ruthenus* × *Huso huso*)	Pantanella (2018)
Commercial production (US)	Tilapia (*Tilapia* spp.), ornamental fish,[a] catfish (*Siluriformes* order), perch (*Perca* spp.), bluegill (*Lepomis macrochirus*), trout (*Oncorhynchus* spp., *Salmo* spp., *Salvenlius* spp.), bass (*Micropterus* spp., *Morone* spp.), shrimp and prawns (*Dendrobranchiata* suborder), crayfish (*Astacoidea* and *Parastacoidea* families), minnows,[a] carp (Cyprinidae family), pacu (*Colossoma* spp., *Piaractus* spp., etc.), barramundi (*Lates calcarifer*), pangasius (*Pangasius* spp.)	Love et al. (2015a)
Commercial (EU)	Tilapia,[a] catfish,[a] ornamental fish,[a] trout,[a] bass,[a] perch[a]	Villarroel et al. (2016)

[a]Scientific names of the species are not specified by the authors.

most suitable species. Since DEXi increases the objectivity of fish selection, its use of this MCDA analysis could be extended to the other organisms raised up in the aquaponic system, including plants (Rossi et al., 2021).

Regarding crops, many plant species can be grown in an aquaponic system (Carrasco and Shamshak, 2020). However, culinary, medicinal, and aromatic herbaceous plants are the most recommended species due to their fast growth and great versatility (Makri and Kintzios, 2008; Rains, 2007; Sánchez et al., 2011). Indeed, the most common plants grown in aquaponics are basil, kale, lettuce, mint, and watercress. In aquaponic, leafy vegetables tend to grow better; if there are enough fishes in the system, also fruit vegetables can be grown, such as beans, cucumbers, squash, tomatoes, peas, peppers, and strawberries (Bernstein, 2011).

Plant species should be also selected in consideration of the hydroponic system used (Tyson and Simonne, 2021). Plants with a little root system grow well in deep-water culture while media beds are more suitable for plants with bigger root systems, e.g. tomato and muskmelon (Bernstein, 2011). The physiological requirements of the plants are also important. Plants will grow better if their needs are like those of the companion fish species. Good results were obtained in freshwater aquaponics with the following combination of crop and fish species: tomato and pak choi with tilapia (Hu et al., 2015); basil with tilapia (Rakocy et al., 2004); basil with prawn (Ronzón-Ortega et al., 2012); mint with tilapia (Wahap et al., 2010); and oregano or spearmint with tilapia (Campos-Pulido et al., 2013; Sánchez et al., 2011).

3.5 Saltwater aquaponic systems

Although freshwater aquaponics is the most widely described and practiced, the availability of freshwater for agriculture is rapidly decreasing in many regions of the world (Fronte et al., 2016; Kotzen et al., 2019). This has led to an increased interest towards alternative water sources (e.g. brackish to highly saline water or seawater) and the production of euryhaline or saltwater fish and halophytes or salt-tolerant glycophytes (Joesting et al., 2016; Kotzen et al., 2019). Saltwater can be used also for aquaponic production.

Aquaponic systems that use saline water are relatively new and need to be consolidated. According to the salinity level of the water, aquaponic systems can be classified as follows (Gunning et al., 2016; Kotzen et al., 2019; Palm et al., 2018): haloponics (1-3 g L^{-1}), moderately saline (3-10 g L^{-1}), high salinity (10–35 g L^{-1}), and maraponics (> 35 g L^{-1}).

Saltwater aquaponics does not differ much from the freshwater aquaponics in terms of system design; however, one must consider that nitrifying bacteria are less efficient and plant mineral uptake is reduced at high salinity (Pantanella, 2018). Indeed, the rate of ammonia removal rate is markedly reduced (by approximately 37%) in saltwater RAS compared to freshwater systems (Timmons et al., 2018).

Euryhaline fish could adapt to large salinity ranges (Edwards and Marshall, 2012), even if the adaptation to different salinities imposes greater energetic costs for active ion transport (Laiz-Carrión et al., 2005) that could negatively affect fish growth and survival.

The type of farmed aquatic animal strictly affects the choice of plants or seaweeds that can be cultivated in aquaponics, based on the salinity ranges. In the case of saltwater aquaponics, a preliminary evaluation of the economic viability of the system is needed and crop selection is more important than fish selection since many fish species with high commercial value can be raised up in these systems (Gunning et al., 2016).

Table 5 summarizes the fish and plant combinations in saltwater aquaponic systems and their respective productions, as reported by some authors. Several commercially valuable fish species have been tested in saltwater aquaponics, such as sea bass (*Dicentrarchus labrax*) and seabream (*Sparus aurata*) and witheleg shrimp (*Litopenaeus vannamei*). Several halophytes, such as *Tripolium pannonicum*, *Plantago coronopus*, *Salicornia dolichostachya*, and *Crithmum maritimum*, well adapt to high salinity conditions. All these plants are cultivated for their leaves rich in nutraceutical compounds (Lombardi et al., 2022); some of them (e.g. *Salicornia* spp.) are gourmet vegetables with a high retail price.

3.6 Analysis of aquaponic solution

The species reared in an aquaponic system (fish, microorganisms, and plants) have different requirements in terms of water quality. Permanently coupled aquaponics require a trade-off between the requirements of different organisms while on-demand-coupled aquaponics allow a better management of those parameters. Fish feeding density, growth rate, food intake, and quality are among the main causes of the deterioration of water quality (Diver, 2006; Wei et al., 2019).

The concentration of macro- and micro-elements in aquaponic systems is often below the recommended level for hydroponic (Goddek et al., 2015), thus causing plant nutrient deficiency.

Dissolved oxygen is one of the most important water parameters in intensive RAS since it determines the successful rearing of fish and the optimal activity of nitrifying biofilter. In general, the oxygen requirement of a RAS is determined by stocking densities, feed addition, temperature, and fish tolerance to hypoxia (Espinal and Matulić, 2019). However, DO levels should be kept close to saturation, as fish are very sensitive to oxygen deficiency, in particular, at high temperatures. In land-based aquaculture facilities (including

Table 5 Summary of experiments conducted in saltwater aquaponics

Animal species	Initial fish density (kg m^{-3})	FR (%)	CP (%)	FCR	SGR	Plant species	Hydroponic system	Density (# m^{-2})	Crop yield (kg m^{-2})	Timespan (days)	Salinity (ppt)	Author
Sparus aurata	1.2	5	54	1.8	3.17	*Crithmum maritimum*	MB	200	ns	ns	8–20	Vlahos et al. (2019)
Dichentrarcus labrax	ns	1.5	ns	ns	ns	*Beta vulgaris* (cicla)	MB	50	ns	65	0–20	Nozzi et al. (2016)
Dichentrarcus labrax	1.2	ns	58	0.93	1.5	*Tripolium pannonicum*	DWC	39	1.0	35	15–16	Waller et al. (2015)
						Plantago coronopus			0.6			
						Salicornia dolichostachya			2.2			
Mugil cephalus	8.07	ns	54	ns	ns	*Salsola soda*	DWC	144	2.2	35	5	Pantanella (2012)
									2.4		10	
	7.37								3.2	32	10	
									2.2		20	
	7.4								5.1	30	10	
									1.0		30	
Sciaenops ocellatus	2.82–4.23	2.6	45	ns	ns	*Sesuvium portulacastrum*	DWC	47	32.1[a]	261	15	Boxman et al. (2018)
						Batis maritima			3.6[a]			
Xiphophorus	0.8	7	45	ns	ns	*Sesuvium portulacastrum*	MB	ns	0.5–0.7	30	15	Boxman et al. (2017a)
						Batis maritima			0.3			

Litopenaeus vannamei	200	3.5	35	1.4–1.7	2.1–2.4	*Atriplex hortensis*	DWC	100	0.3–0.4	28	10–20	Chu and Brown (2020)
						Salsola komarovii			0.4–0.6			
						Plantago coronopus			0.6–2.0			
Litopenaeus vannamei	200–500	3	35	1.50–2.14	1.45 1.97	*Atriplex hortensis*	DWC	100	0.6–1.2	28	15	Chu and Brown (2021)
						Salsola komarovii			0.2–0.4			
						Plantago coronopus			2.0–2.5			
Litopenaeus vannamei	250	ns	*35.00*	1.8–1.7	2.085515802	*Sarcocornia ambigua*	NFT	100	1.1–1.9	73	33.8–36.5	da Silva et al. (2021)
Litopenaeus vannamei	ns	ns	35	ns	5.0	*Ocimum basilicum*	DWC	16	6.7	81	1.7	Fierro-Sañudo et al. (2018)

FR, feeding rate; CP, crude protein; FCR, feed conversion rate; SGR, specific growth rate; MB, media bed; DWC, deep water culture; NFT, nutrient film technique; ns, not specified by the author. [a]Results are expressed on a dry matter basis.

RAS), liquid oxygen is used to keep proper DO concentration in water (Timmons et al., 2018).

The solubility of oxygen depends on temperature, salinity, and atmospheric pressure, as shown in Table 6; it decreases with increasing temperature and salinity. Solids originated from uneaten feed, fish faeces, and microbial biomass could represent an important issue for aquaculture systems due to their accumulation and consequent oxygen consumption by the microbial community (Wongkiew et al., 2017).

Plant roots also are susceptible to oxygen deficiencies, although they are more tolerant than fish. In hydroponics, DO should not drop below 5 mg L^{-1} or 65% of saturation (Holtman et al., 2014). Hypoxia markedly inhibits plant growth and increases the incidence of root diseases (Gorbe and Calatayud, 2010). The risk of the occurrence of hypoxic conditions in the root zone is greater in deep water culture and bed systems than in NFT, where the roots are directly exposed to atmospheric oxygen (Colt et al., 2022).

Therefore, in aquaponics, an automated monitoring and alarm system connected to a backup electric generator is necessary to avoid the interruption of aeration or oxygenation, which will rapidly cause fish death (Colt et al., 2022). The luminescent DO probes are more appropriate for monitoring DO in aquaponics systems since they do not require stirring and are more accurate than the traditional membrane sensors (Colt et al., 2022).

In addition to DO, the water environment must be regularly monitored to ensure an optimal water environment for fish, plants, and bacteria in aquaponics (Somerville et al., 2014). Water parameters can be monitored manually

Table 6 The solubility of oxygen as a function of temperature (°C) and salinity (g L^{-1}) at atmospheric pressure of 1013 mbar

	Salinity (g L^{-1})								
Temperature (°C)	0	5	10	15	20	25	30	35	40
15	10.1	9.8	9.5	9.2	8.9	8.6	8.4	8.1	7.9
16	9.9	9.6	9.3	9.0	8.7	8.5	8.2	8.0	7.7
17	9.7	9.4	9.1	8.8	8.6	8.3	8.1	7.8	7.6
18	9.5	9.2	8.9	8.7	8.4	8.1	7.9	7.7	7.4
19	9.3	9.0	8.7	8.5	8.2	8.0	7.8	7.5	7.3
20	9.1	8.8	8.6	8.3	8.1	7.8	7.6	7.4	7.2
21	8.9	8.7	8.4	8.2	7.9	7.7	7.5	7.3	7.1
22	8.7	8.5	8.2	8.0	7.8	7.6	7.3	7.1	6.9
23	8.6	8.3	8.1	7.9	7.6	7.4	7.2	7.0	6.8
24	8.4	8.2	7.9	7.7	7.5	7.3	7.1	6.9	6.7
25	8.3	8.0	7.8	7.6	7.4	7.2	7.0	6.8	6.6

Source: Adapted from: USGS (2022).

or automatically with a probe or colorimetric kit (Somerville et al., 2014; Thorarinsdottir, 2015). Table 7 shows the reference values and the frequency of measurements of several water parameters in freshwater aquaponics.

Optimal pH ranges between 5.0 and 6.5 in hydroponics and between 7.0 and 8.0 in RAS; in aquaponics systems, pH values of 7.0–8.5 are optimal for nitrifying bacteria (Lennard and Goddek, 2019). The pH can be monitored directly on-site by using manual or automatic electronic probes (Yanes et al., 2020).

Several properties of the water are affected by pH, such as the ammonia form (i.e. NH_3 and NH_4^+, the latter less toxic for fish) (Timmons et al., 2018) or the availability of nutrients for plants (Lennard and Goddek, 2019). The molar ratio of NH_3 and NH_4^+ in water is related to pH, temperature, and salinity (Timmons et al., 2018): it increases with pH and temperature, as shown in Fig. 3, and decreases with salinity. For instance, at 20°C and pH of 8, the NH_3:NH_4^+ molar ratio is 0.040 in freshwater (0 g L^{-1}) and 0.031 in seawater (35 g L^{-1}).

The pH tends to decrease in aquaponic systems and RAS as a consequence of nitrification (Lekang, 2007), thus favouring the presence of ionized ammonia.

Table 7 Optimal range of several water parameters and the frequency and method for their monitoring in aquaponic system according to several authors

Parameter	Optimal value	Frequency of monitoring	Monitoring method
pH	6–8.5	Daily	Automatic/manual[a]
Dissolved oxygen (DO; mg L^{-1})	>5	Daily	Automatic/manual[a]
T (°C)	18–30	Daily	Automatic/manual[a]
Total ammonia nitrogen (TAN; mg L^{-1})	<1	Weekly	Manual[b]
Nitrite (mg L^{-1})	<1	Weekly	Manual[b]
Nitrate (mg L^{-1})	5-150	Weekly	Manual[b]
Alkalinity (mg L^{-1})	>100	Weekly	Manual[c]
Biological oxygen demand (BOD; mg L^{-1})	<20	Monthly	Manual[c]
Carbon dioxide (mg L^{-1})	<20	Weekly	Automatic/manual[a,b]
Electrical conductivity (EC, dS m^{-1})	2–4	Daily	Automatic/manual[a]
Total dissolved solids (TDS, g L^{-1})	< 1	Monthly	Manual[c]

[a] By means of hand-held probe.
[b] By colorimetric test kit on-site.
[c] Laboratory analysis. Source: Lennard and Goddek (2019); Pantanella (2018); Somerville et al. (2014); Thorarinsdottir (2015); Yanes et al. (2020).

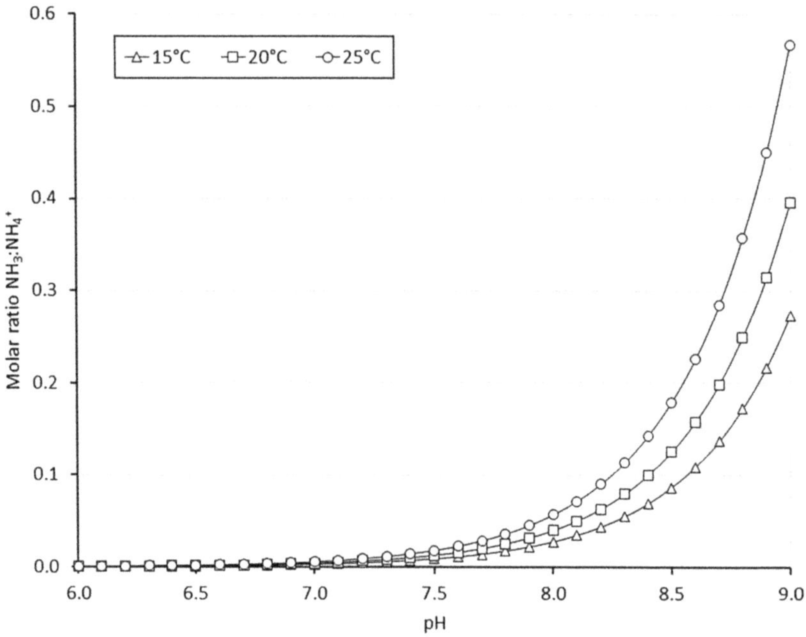

Figure 3 Molar ratio between un-ionized (NH_3) and ionized (NH_4^+) forms of ammonia in freshwater as a function of pH and temperature.

It is possible to adjust pH through acidification or alkalinization by adding inorganic acids or buffers (salts of calcium, potassium, or magnesium) (Goddek et al., 2015; Somerville et al., 2014).

The EC is a measurement of the ability of an aqueous solution to conduct electric current, and therefore it is highly correlated to salinity (Nagayo et al., 2017). The EC must be regularly monitored and adjusted to the desired values. Portable potentiometric probes are generally used for EC measurements (Yanes et al., 2020). In saltwater aquaponics, the salinity can be easily measured with an optical refractometer or a densimeter.

Ammonium, nitrite, nitrate, and phosphorous should be also regularly monitored to avoid toxicity phenomena in fish and nutrient deficiency in plants. These parameters can be quantified in the laboratory with standard analytical methods, which are time consuming and require trained staff or directly on-site using quick colorimetric tests or probes (Yanes et al., 2020). Specific test kits exist for analysing saltwater in which the presence of chlorine represents the main source of interference for some analytical methods.

Other water parameters can be measured less frequently (every 1–2 weeks) such as the total dissolved solids (TDS) and the concentration of micronutrients. The TDS level, which depends on both inorganic salts and organic matter, should be less than 1 g L^{-1} to avoid toxicity for most fish species (Yanes et al., 2020).

Smart monitoring systems are under development for an aquaponic system with the aim to reduce manual labour, increase the accessibility and connectivity of the parameters, and use computer capabilities to make data-driven decisions (see Yanes et al., 2020, for review).

3.7 Crop protection

Root-borne pathogens such as *Fusarium* spp., *Phytophthora* spp., and *Pythium* spp. are the most problematic plant diseases in aquaponics. *Phytophthora* spp. and *Pythium* spp. can spread easily in aquaponics systems because their zoospores easily move in the recirculating water.

In permanently coupled aquaponic systems, the application of pesticides and fungicides is limited by their possible toxicity for fish and beneficial bacteria and the lack of registered products (Somerville et al., 2014). Thus, in aquaponics crop protection is primarily based on preventive actions, including water disinfection (Goddek et al., 2019a).

An appropriate identification of the pathogen that causes a specific disease is pivotal to choose the best curative measures (Lepoivre, 2003). Filtration, heat, or UV treatments are physical water treatments that can be applied to contrast the spread of potential pathogens in water, without harmful effects on fish and plant health (Ebeling and Timmons, 2010; Postma et al., 2008). *Bacillus* and/or *Pseudomonas* spp. are responsible for the microbial suppressive activity in the filter (Renault et al., 2012) due to the competition for nutrients, antibiosis, and the production of antifungal and/or antibacterial substances (Beneduzi et al., 2012; Déniel et al., 2004). Heat treatment is a very effective method against plant pathogens; however, it needs a lot of energy and also kills beneficial microorganisms (Postma et al., 2008). Disinfection by UV radiation (200–280 nm) directly damages the DNA of microorganisms (Villarroel et al., 2016). Filtration also removes plant debris, algae, small particles, and some soil-borne pathogens, and it is the most used technique because of its effectiveness and lower cost compared with the other systems (Postma et al., 2008; Vallance et al., 2011).

The application of chemicals to disinfect the recirculating water is limited. Ozonation (or ozonization) removes all pathogens, including viruses in certain conditions, and ozone is rapidly transformed into oxygen, thus avoiding toxic effects on both fish and plants (Gonçalves and Gagnon, 2011). However, ozonation can produce by-products and oxidants that must be removed from the water and oxidizes the mineral elements dissolved in the water, making them unavailable for plants (Gonçalves and Gagnon, 2011).

Other preventive measures can be adopted for the control of crop pests and diseases, such as the use of healthy propagation materials; the use of traps and nets; timely removal of infested plants; access restricted to operators with disinfected shoes and clothes; and climate control in greenhouse aquaponic systems.

Among the methods to control plant diseases in aquaponics, there are the addition of some microbial metabolites such as biosurfactants (Nielsen et al., 2006) and the use of a complex of antagonistic agents (Vallance et al., 2011). In aquaponic systems, there are beneficial microorganisms that interact with plants and plant pathogens (Sirakov et al., 2016). These microorganisms elicit plant defence response, such as the production of plant metabolites with antibiotic effect (Beneduzi et al., 2012; Narayanasamy, 2013) and can rapidly develop and spread in aquaponic systems due to the recirculation of water (Vallance et al., 2011).

In addition, the suppressiveness of aquaponic systems depends on the content of organic matter that is contained in uneaten feeds, fish faeces, organic substrate, root exudates, and plant residues and can enhance the presence of antagonistic agents, inducing a microbial ecosystem less favourable for plant pathogens (Waechter-Kristensen et al., 1999; Vallance et al., 2011). On the other hand, a higher C/N ratio increases the abundance of heterotrophic bacteria and reduces that of autotrophic nitrifying bacteria with a negative impact on the system.

3.8 Animal welfare

Fish welfare represents the physiological, behavioural, and cognitive responses, as a consequence of stressful stimuli (Yavuzcan Yildiz et al., 2017). In a confined environment, such as aquaponic systems or RAS, the appearance of stressful stimuli results from the interaction of several factors: stocking density (i.e. the biomass of fish reared in a certain volume of water), water quality, diet, management procedures, and system design. These factors can increase the level of stress, negatively affect animal health, and induce the development of impaired or aggressive behaviour (Espinal and Matulić, 2019; Yavuzcan Yildiz et al., 2017). These conditions have a negative impact on the whole aquaponic system. Moreover, assuring a proper fish welfare can be an important driver for the consumers' acceptance of aquaponic products (Miličić et al., 2017).

Stocking density is a pivotal factor affecting fish welfare in aquaculture, especially in RASs where high densities are generally used to increase fish productivity (Espinal and Matulić, 2019). In general, the recommended fish stocking density is much lower in aquaponics than in RAS; for instance, it is 20 kg m^{-3} in small-scale aquaponics (Somerville et al., 2014). In an aquaponic system rearing European Carp (*Cyprinus carpio*), an initial stocking density of 2.5 kg m^{-3} increased the rearing water quality and the yield of catalogna chicory, lettuce, and Swiss Chard, compared with a density of 4.6 kg m^{-3} (Maucieri et al., 2019). A reduced growth of fish raised at high density was also associated with higher

TAN levels and reduced DO concentration (Maucieri et al., 2019). The effect of different stocking densities (3.81 kg m^{-3} and 7.26 kg m^{-3}) were also evaluated in rainbow trout (*Oncorhynchus mykiss*) (Birolo et al., 2020). Water parameters (DO, pH, temperature, TAN, and nitrite) and fish growth were not affected by stocking density while the presence of Pseudomonas and H2S-producing bacteria on fish skin increased with fish density (Birolo et al., 2020). The effects of sub-optimal water parameters on fish welfare are summarized in Table 8.

Fish raised in RAS and aquaponic systems require good water quality conditions, in particular, regarding DO, CO_2, TAN, nitrite, nitrate, and pH. Carrying capacity represents the highest fish biomass with acceptable water quality (Yavuzcan Yildiz et al., 2017). Basically, the carrying capacity of a system is determined by the oxygen consumption by the fish and the accumulation of TAN, CO_2, and other potentially toxic metabolic wastes (Yavuzcan Yildiz et al., 2017). In an aquaponic system, the carrying capacity of the water is one of the major concerns for maintaining the balance between plant and fish without negative effects on fish welfare.

Both plants and algae release chemical compounds (allelochemicals) to make a more favourable environment for themselves (allelophathy). So far, negative effects of potential allelochemicals excreted by aquaponically grown plants on fish have not been reported (Yavuzcan Yildiz et al., 2017). However, algae (blue-green algae) can release cyanotoxins, which can directly affect fish health, and the algal spikes of diatoms (e.g. *Chaetoceros* spp.) can damage the gills (Chalmers, 2004; Yavuzcan Yildiz et al., 2017).

Another critical issue regarding water quality is the accumulation of solids. As already mentioned, solids should be removed as soon as possible to avoid

Table 8 The effect of non-optimal values of some water parameters on fish welfare in aquaponics

Parameter	Negative effects
Low DO content	Increased rate of opercular ventilation; gasping mouths; greater nitrite toxicity; reduced swimming activity; reduced growth; low feed efficiency; death
High pH	Increase un-ionized form of ammonia
High temperature	Infectiousness of many pathogens; toxicity of dissolved contaminants; reduction of DO level
High TAN content	Neurological disorder (acute); decrease survival; reduced growth; appearance of physiological dysfunctions
High nitrite content	Conversion of haemoglobin into methaemoglobin; increased rate of opercular ventilation; morphological and physiological alterations of gills; hypoxia
High nitrite content	Reduced swimming activity; decreased survival; reduced growth
High CO_2 content	Hypercarbia; moribund fish; gaping mouths; bright red gill lamellae

Source: Yavuzcan Yildiz et al. (2017).

severe consequences on other components of the system. The presence of an excess of suspended solids in the water is a stressful factor for the fish and can affect growth performance and increase the risk of diseases. Suspended solids can damage fish directly (e.g. damage to gills epithelium or smothering) (Yavuzcan Yildiz et al., 2017).

Fish feed has a high impact on the carrying capacity of aquaponic water from a quantitative (feeding rate) and qualitative (feed composition and digestibility) viewpoint. Fish feed needs to fulfil fish and plant nutrient demand in aquaponics (Robaina et al., 2019). Most aquaponic systems use specific feeds formulated for RASs (Yavuzcan Yildiz et al., 2017) since little is known on specific formulations for fish reared in aquaponic systems. Specific supplementation is required for some micronutrients in aquaponics since the amount excreted by fish is generally inadequate to meet plants' demand (Yavuzcan Yildiz et al., 2017). Tailored feed formulation for aquaponics needs higher content of plant nutrients or making them more bio-available after excretion and biotransformation in fish (Robaina et al., 2019). Fish welfare can be directly affected by this practice. Since fish can take up micronutrients both from water and feeds, understanding the effect of excess feed and waterborne micronutrients on fish and plants is crucial (Yavuzcan Yildiz et al., 2017).

Another factor affecting fish welfare is noise, which is generated by water pumps, blowers, and filtration systems. Noise can determine a stressful condition for the fish (Espinal and Matulić, 2019). Several studies have shown that fish behaviour and physiology can be negatively impacted by background noise and intense sound (Espinal and Matulić, 2019). However, these findings should not be generalized across farmed fish species, because many of them (e.g. catfish and cyprinids) have much greater hearing sensitivity than others (e.g. rainbow trout) (Espinal and Matulić, 2019). No significant differences were observed in the growth rate, feed conversion, and survival of rainbow trout after five months of exposure to different noise levels (Davidson et al., 2009).

3.9 Animal health issues

The occurrence of fish diseases usually results from the interaction among various factors (e.g. climate, nutrition, immune response, and presence of pathogens) and are species- and system-specific. So far, specific aquaponic diseases have not been described but they may be similar to those in RASs (Yavuzcan Yildiz et al., 2019). In aquaculture, non-infectious diseases are related to environmental, nutrition, and genetic factors; bacteria cause the largest part of infectious diseases followed by viruses, parasites, and fungi (Yavuzcan Yildiz et al., 2019). Fluctuations in water quality facilitate the development of fish pathogen. For instance, ubiquitous bacteria, which are present in any water containing organic matter, can become opportunistic pathogens under certain environmental conditions (Joyce et al., 2019; Yavuzcan Yildiz et al., 2019). Nevertheless, the capability of a microorganism to cause a clinical disease depends on the interaction of several factors such as physiological status, host, husbandry, environment, and nutrition (Yavuzcan Yildiz et al., 2019). The characteristics of an aquaponic system often reduce the risk of pathogen introduction and disease outbreaks due to better control of inputs and water management (Somerville et al., 2014).

Since animals and vegetables grow together, the use of chemicals (e.g. disinfectants and antibiotics) should be limited to avoid microbial community dysfunction and the accumulation in plant edible tissues. Moreover, the use of antibiotics is not allowed in plant cultivation (European Commission, 2001). Therefore, in aquaponics, more than in aquaculture, preventive measures should be applied to reduce the risk of infectious disease outbreaks (Chalmers, 2004; Somerville et al., 2014; Yavuzcan Yildiz et al., 2019).

Along with the use of pathogen-free water and stock, an important biosecurity measure is the quarantine for newly introduced larvae, fry, or adult animals (Somerville et al., 2014; Yavuzcan Yildiz et al., 2019). Somerville et al. (2014) recommend a quarantine period of 45 days before the introduction of fish in the definitive rearing tanks. During quarantine, fish are checked and monitored for apparent disease signs and possibly exposed to prophylactic treatments, for instance, with vaccines for viruses and bacteria, salt baths for ectoparasites, or chemicals that do not leave residues in water (Sirakov et al., 2016; Somerville et al., 2014; Yavuzcan Yildiz et al., 2019).

Another way to prevent the spread of infectious and non-infectious diseases is the use of functional feeds or water enrichments. Several functional ingredients, such as probiotics and immunostimulants, can be added to formulated feed or directly in the rearing water (Yavuzcan Yildiz et al., 2019). Among immunostimulants, glucans are the most studied and promising for aquaculture industry (Chalmers, 2004; Fronte et al., 2019). Also, the addition of marine-derived polysaccharides in feed formulations is promising for their immunostimulant effect (Mohan et al., 2019). Moreover, the extraction of these compounds (e.g. ulvan) could be integrated with the phytoremediation of aquaculture wastewater with seaweeds (e.g. *Ulva* spp.) (Gennaro et al., 2006).

Mori and Smith (2019) review the most common methods adopted for water treatment. In soilless cultivation and RASs, UV irradiation is one of the most common disinfection methods (Mori and Smith, 2019). Continuous activity of the UV lamps and reduced load of suspended solids increase the efficacy of water treatment (Mori and Smith, 2019).

A quick individuation of symptomatic fish is essential to prevent the transmission of infective agents and reduce fish mortality. Fish should be checked almost daily, looking for any unusual behaviour (e.g. feeding and swimming activity) or lesions, then immediately removed from the rearing water and thoroughly inspected and transferred to a separate (quarantine) tank for any eventual treatment (Somerville et al., 2014; Yavuzcan Yildiz et al., 2019).

Comparing the two aquaponic design, on-demand-coupled systems allow a better monitoring and the use of conventional treatments for fish disease that are not possible in permanently coupled systems (Yavuzcan Yildiz et al., 2019).

3.10 Food safety

Foodborne illnesses are significant hazards to human health. Human pathogens and chemical contaminants must be maintained within safe limits or completely prevented (Jennings et al., 2016). Food safety risks exist in integrated food production systems and aquaponics is not an exception.

In aquaponics, the main concern is due to the copresence of aquatic animals and edible plants grown in recirculating water, which could provide suitable conditions for the development and spread of pathogens (Kasozi et al., 2021; Ljubojević et al., 2017; Mori and Smith, 2019). Special attention should be paid for vegetables that are usually consumed raw, in particular, for leafy vegetables and herbs that grow closer to the recirculating water than fruity plants and then can be easily contaminated (Elumalai et al., 2017).

Aquatic food and plant products are highly perishable and deteriorate quickly if not handled and stored properly. Good storage and food hygiene practices are essential to prevent spoilage of aquaponic products (Jennings et al., 2016).

Aquatic animals are not considered high-probability vectors of zoonotic diseases to humans, due to significant physiological differences between them (i.e. fish are cold-blooded animals, while humans are warm-blooded mammals) (Hollyer et al., 2009). However, several microbiological agents (e.g. *Streptococcus* spp., *Aeromonas* spp., *Listeria monocytogenes*, *Escherichia coli*, *Salmonella* spp., *Vibrio* spp., and others) were found to trig diseases in humans and hence to endanger the safety of aquaponic products if a contamination occurred (Ljubojevic and Djordjevic, 2016). Shiga toxin-producing *Escherichia coli* was recently isolated from fish faeces, water, and on the surface of the roots of lettuce, basil, and tomato without internalization in edible portions (Wang et al., 2020). Possible sources of bacterial pathogens are water, fish feed, fish juveniles introduced into the system, plant propagation materials, and humans themselves (staff and visitors) (Kasozi et al., 2021). Therefore, strict hygiene measures should be adopted in aquaponics. The young plants introduced into the system must be free of coliforms and other human pathogenic bacteria (Pantanella et al., 2015). Staff is required to wear gloves when handling fish and pay attention to avoid spines and to work with open wounds (Hollyer et al., 2009). Working with bare hands could introduce new pathogens into the system. Harvesting equipment and other tools have to be kept clean and sanitized.

Plant products should be harvested to avoid contact with water; nevertheless, it is advisable to wash any aquaponic products before consumption (Somerville et al., 2014).

In addition, proper practices for solids management are required, and fish faeces should be continuously removed from the system to limit the release of coliform bacteria (Somerville et al., 2014). Even diseased fish or plants should be promptly removed to hamper the subsequent shedding of potential pathogens in the water (Mori and Smith, 2019).

The microbiological quality of the water in aquaculture and hydroponic systems can be ensured using physical disinfection and filtration methods, such as UV, ozone, or hydrogen peroxide treatment, blue light-emitting diodes (LED), heat, sonication, media, or membrane filtration (see El-Kazzaz (2017) and Mori and Smith (2019), for reviews).

3.11 Environmental sustainability

Increasing concerns about the environmental sustainability of aquaculture production have been raised, and the standards and requirements imposed on the aquaculture sector are becoming stricter. A common tool to assess the environmental sustainability of products or production systems is the life cycle assessment (LCA), which has been also applied to aquaculture systems (Bibbiani et al., 2018). So far, however, few LCA studies have been conducted on aquaponics (Junge et al., 2017). LCA was first applied to a small-scale aquaponic system (Love et al., 2015b) or focused on technical and economic analyses (Xie and Rosentrater, 2015). Then, LCA was applied to analyse specific case studies (Boxman, 2015; Boxman et al., 2017b; Ghamkhar et al., 2020; Maucieri et al., 2018) to compare different aquaponic systems (Forchino et al., 2017) or locations (Körner et al., 2021) and separate hydroponic systems and RAS with aquaponics (Chen et al., 2020; Greenfeld et al., 2021; Jaeger et al., 2019; Körner et al., 2021). These studies were conducted from the cradle-to-farm gate, thus excluding upstream (e.g. hatchery and seedlings production) and downstream processes (i.e. distribution, market, and waste disposal) and considered freshwater aquaponic systems, apart from the one conducted by Boxman (2015) on salt water aquaponic system. All the systems analysed were permanently coupled, except the one studied by Körner et al. (2021). The functional unit was always mass based, except one case in which the economic value was chosen (Chen et al., 2020). The typical multifunctionality (i.e. production of fish and plants within the same system) was solved through system expansion (Boxman et al., 2017b; Jaeger et al., 2019) or mass allocation (Forchino et al., 2017).

The main hotspots for the environmental sustainability of aquaponic are energy, feed, and infrastructures, including equipment and devices (Bohnes et al., 2019). Water saving is, on the contrary, one of the major strengths of aquaponic systems (Maucieri et al., 2018).

Aquaponics require a large amount of electric energy for water aeration, pumping, heating, and lighting, if applied (Boxman et al., 2017b; Chen et al., 2020; Forchino et al., 2017; Ghamkhar et al., 2020). Energy use can be reduced through the optimization of aeration (Boxman et al., 2017b) and pumping (Forchino et al., 2017) and using renewable energies (Chen et al., 2020; Forchino et al., 2017).

Likewise, in aquaculture (Aubin et al., 2009; d'Orbcastel et al., 2009; Jerbi et al., 2012), fish feed is one of the main sources of environmental impact of aquaponics. The use of crop products and by-products contributes to typical impact categories associated with land and resource use (Boxman et al., 2017b; Jaeger et al., 2019). The environmental impact of feed production is greater than the one of fertilizers (Chen et al., 2020). Therefore, lowering the feed conversion rate (FCR, i.e. the ratio between feed administrated and fish biomass gain) is crucial to increase the sustainability of aquaponic systems (Boxman et al., 2017b; Forchino et al., 2017).

When hydroponic subunit is considered, media beds have a greater impact than deep water culture mainly because a larger amount of inert material is used (Forchino et al., 2017) while NFT could reduce energy consumption since this system does not require aeration (Boxman et al., 2017b). The environmental impact of aquaponics is lower than hydroponics and aquaculture (Chen et al., 2020; Greenfeld et al., 2021; Körner et al., 2021). Aquaponic systems in Northern Europe strongly influence the energy-related impact categories, while Southern Europe systems have a higher impact on water-related categories (Körner et al., 2021).

Jaeger et al. (2019) have calculated the budgets of nitrogen and phosphorus in an aquaponic system. Feed represents the largest part of nitrogen and phosphorus input (93% and 57%, respectively). Nitrogen retention in fish and plants account for 48% (47.4% fish and 0.6% plants), while a total of 22% of the total nitrogen input was released from the aquaponic system with water discharge (16.9%), sediments (4.6%), and leachate (0.5%), while 5.4 % is dissolved in water. The unaccounted amount of nitrogen (not retained by fish and plants), approx. 25%, is probably lost with gas emissions of N_2, NH_3, and NO (Jaeger et al., 2019; Wongkiew et al., 2017). Similar findings have been reported by Maucieri et al. (2020). As regards phosphorous, 37% of the total phosphorus input was released from the aquaponic system, mainly in sediments (23%), discharged water (13%), and leachate (2%) (Jaeger et al., 2019). The higher amount of phosphorus removed from the system compared to other studies (Cerozi and Fitzsimmons, 2017) can mostly attributed to the high efficiency of the drum filter adopted in the reference system (Jaeger et al., 2019). According to Jaeger et al. (2019), nutrients recovery (especially phosphorus) can be optimized by mineralizing the fish sludge and maximizing the crop area.

The environmental sustainability of aquaponics should be considered holistically, therefore including economic and social assessment. le Féon et al. (2021) propose a multi-criteria analysis, implemented in DEXi software, to assess the environmental (i.e. LCA and Emergy accounting), economic (i.e. life cycle cost), and social implications (i.e. Social-LCA, SLCA) of IMTA systems.

3.12 Economic issues

Small-scale aquaponics has been popular worldwide for many decades, especially for recreational or research purposes, while commercial aquaponics is still in its infancy (Araújo et al., 2021). However, the global aquaponics market is expected to reach a value of US$1.34 billion before 2024, especially in North America where the systems are technically advanced compared with other geographic areas (MRF, 2022).

When referring to commercial aquaponic, not only the production of edible organisms (fish and vegetables) is considered but also the sale of aquaponic materials, equipments and/or services (e.g. consulting) (Turnšek et al., 2019). Love et al. (2015a) surveyed commercial aquaponic producers mostly in the United States: only 37% of these participants could be considered commercial producers *sensu stricto* (i.e. with revenue from selling only fish or plants) while 36% combined the production of fish and plants with the sale of material or services for aquaponics. Only 31% of aquaponic producers reported profits in the previous year (Love et al., 2015a). Villarroel et al. (2016) estimated approximately 20 aquaponic companies settled in Europe. More recently, Turnsek et al. (2020) reported 30 European companies belonging to the Association of Commercial Aquaponics Companies (ACAC), of which only a third produce only fish and/or plants.

Several methodologies can be applied for the evaluation of the profitability of aquaponic systems (Greenfeld et al. (2019), which are as follows: (1) the calculation of the internal return rate (IRR) of the initial investment; (2) the comparison of profits obtained from aquaponics with those from separate hydroponics and aquaculture, for the same fish and crop species; and (3) survey about profitability of commercial aquaponic farms. Nevertheless, overall economic evaluation of an aquaponic system is difficult due to the great heterogeneity of aquaponic systems and geographic contexts and the difficulties in achieving financial data from commercial operations (Cammies et al., 2021; Turnšek et al., 2019).

Challenges that affect the economic viability of aquaponics investment costs are location, energy consumption and sources, workload, wages, lack of skilled workers, legal constraints fish and plant combinations and prices, and public acceptance (Araújo et al., 2021). At a farm scale, the first barrier is the high investment costs for the system start-up (Greenfeld et al., 2019). The realistic return of investment is roughly 7 years (Turnšek et al., 2019). Then, systems should be optimized in terms of fish-plant combination, nutrient cycling, and technological solutions to increase system efficiency and reduce production costs (Greenfeld et al., 2019). On-demand coupled aquaponic systems offer the possibility to increase the water use efficiency and total production compared to permanently coupled systems (Goddek and Keesman, 2020); however, information on commercial production of the former systems is scarce for accurate economic assessment (Greenfeld et al., 2019).

Labour and energy are the main cost drivers of aquaponics, besides fish feed, juveniles, and plant propagation materials (Turnšek et al., 2019). Moreover, farmers tend to specialize in specific crops and animals while aquaponics requires a complex set of skills that may represent a notable knowledge barrier (Greenfeld et al., 2020).

While energy depends mostly on the location and greenhouse management (e.g. temperature control and lighting), the workload may represent a bottleneck due to the scarce automatization of these systems. On average, for commercial aquaponic production 5–10 workers per hectare are required, likewise the workload of hydroponics (Turnšek et al., 2019). On the other hand, aquaponics could represent a great opportunity for job creation in marginal areas suitable for aquaponic production.

Since aquaponics is a relatively new technology, several barriers exist at a larger scale, such as the absence of a specific legislative framework (Reinhardt et al., 2019), consumer awareness (Turnšek et al., 2019), and the possibility of obtaining organic certification (Fruscella et al., 2021).

Consumers normally accept to pay more for healthier, organic or pesticide-free, local food products (Greenfeld et al., 2019). The challenge is to increase consumers' awareness of aquaponics through a proper communication and create a stable market niche (Araújo et al., 2021; Short et al., 2017; Turnsek et al., 2020). In a recent survey, nearly half of the interviewed consumers had never heard of aquaponics before, but their acceptance of aquaponics was in general positive although only 17% of respondents were willing to pay more (up to 40%) for aquaponic products (Miličić et al., 2017).

The sale of fish and plants as organic is an opportunity for aquaponic farmers due to the higher price and easier product recognition by consumers (Cammies et al., 2021; Greenfeld et al., 2019). However, in European Union (EU) the Regulation No 848/2018 (European Pariament and Council of the Euopean Union, 2018) on organic agriculture does not allow hydroponics and RASs, except during the hatchery stage. Therefore, aquaponic products cannot be certified as organic in EU in contrast to other countries such as the United States and Canada (Fruscella et al., 2021). To overcome this issue, several strategies can be adopted, such as a soil-based plant cultivation (organically certified), the use of environmental enrichments in RAS, or adjusting the fish stocking density (Cammies et al., 2021; Fruscella et al., 2021). The recommended density for aquaponic production fulfils the density limitation imposed by Regulation (EU) No 2009/710 (European Commission, 2009), which allows a maximum density 30 kg m^{-3} for sturgeons raised in freshwater.

3.13 Multi-trophic aquaponic system: the self-sufficient integrated multi-trophic aquaponics project

In aquaponic systems, solids originating from fish rearing unit (i.e. faeces and uneaten feed) should be quickly removed from the recirculating water to prevent system collapse (Thorarinsdottir, 2015). In doing so, however, most of the nutrients introduced in the system are lost since they accumulate in the solids (Delaide et al., 2019). Delaide et al. (2019) reviewed the aerobic and anaerobic processes that could be used in aquaponic systems to reduce the presence of solids and, at the same time, mineralize the nutrients to be re-used as fertilizers for crop plants. Some of these processes produce also other valuable compounds, such as methane (from anaerobic digestion) that can be used as an energy source (Delaide et al., 2019). Besides these technologies, other methods can be applied for solids reduction in aquaponics. Kotzen et al. (2019) named 'vermiponics' the combination of red wriggler worms (*Eisinia fetida*) within an aquaponic system. Worms can be directly introduced in the flood and drain beds for plants, where they can break down any solid waste from fish and detritus from plants, thus providing nutrients and avoiding clogging phenomena (Kotzen et al., 2019). The production of earthworms on aquaculture sewage has been proposed but remains poorly understood (Kouba et al., 2018). The integration of low-trophic organisms (e.g. terrestrial or aquatic worms) can both bio-remediate aquaculture wastes and produce a valuable biomass that could be used as feed ingredients for farmed fish.

Marine polychaetes have been studied for their bio-remediation potential of aquaculture wastes. Among these, *Hediste diversicolor* is supposed to be one of the most interesting species for IMTA systems (Pombo et al., 2020). Polychaetes worms, including Nereididae., are harvested on the intertidal flats or are cultured in intensive or semi-intensive (often integrated with other species) systems (Pombo et al., 2020). The trade of *H. diversicolor* has expanded in the last years not only thanks to their use as baits by anglers but also as feed ingredient for fish and shrimps, mainly for broodstock and hatcheries (Pombo et al., 2020). There are many studies in which *H. diversicolor* are used as bio-mitigating species of effluents from both freshwater and saltwater aquaponics thanks to their euryhaline attitude (Bischoff et al., 2009; Jerónimo et al., 2021a,b, 2020; Marques et al., 2018, 2017; Pajand et al., 2017; Santos et al., 2016; Wang et al., 2019a,b; Yousefi-Garakouei et al., 2019).

The SIMTAP (acronym of Self-sufficient Integrated Multi-Trophic AquaPonics; www.simtap.eu) project aims to evaluate the interactions of species that belong to different trophic levels in order to produce, in a permanently coupled aquaponic system, an alternative feed for fish. SIMTAP takes inspiration from the IMTA system, where two groups of organisms live synergistically in the same water body: the generating (fed) species (i.e. fish) produce waste through their catabolism (e.g. faeces) while the extractive organisms use this waste for their own anabolic processes (Rossi et al., 2021). The SIMTAP system is designed for the production of euryhaline fish species in diluted seawater or brackish water, and this drastically affects the selection of crop species. Various extractive organisms can coexist in the SIMTAP system, such as halophyte and salt-tolerant glycophyte plants (e.g. *Atriplex hortensis*; *Beta vulgaris* var. *cicla*; *B. vulgaris* spp. *maritima*; *Portulaca oleracea*; *Salicornia europaea*); filter-feeding and detritivorous organisms (i.e. shellfish and polychaetes); seaweeds (e.g. *Ulva* spp.); and micro-algae (e.g. *Chlorella* spp.).

There is a growing interest in the greenhouse cultivation of halophytic plants to produce either raw or minimally processed (i.e. ready-to-eat) vegetables, in particular for gourmet cuisine, due to their salty taste and the high content of antioxidants and essential nutrients (Lombardi et al., 2022).

Macroalgae (e.g. *Ulva* spp.) can be cultured on aquaculture effluents (Gennaro et al., 2006). The biomass obtained can be used as a fish feed ingredient (Shpigel et al., 2017) or processed to extract bioactive compounds (e.g. ulvan as immunostimulants) (Kidgell et al., 2019).

The production of microalgae is a fundamental step within the process since it represents the main external contribution to the system. The role of microalgae is to integrate the nourishment of filter-feeders and detritivorous organisms to improve their growth performance and nutritional composition. The production of microalgae can be integrated with the bio-remediation of greenhouse hydroponic effluents.

Thus, SIMTAP adds a further step to IMTA: the possibility to reuse the extractive organisms as feed for the generative species. A schematic representation of SIMTAP process is given in Fig. 4.

The SIMTAP project is still in progress, and several experiments are currently being conducted using four different prototypes installed in different coastal zones: the Atlantic coast in France and the Mediterranean coast in Italy, Turkey, and Malta. The fish species tested are Gilthead sea bream (*Sparus aurata*) and European sea bass (*Dicentrarchus labrax*), while several experiments have been conducted on several halophytic species, especially on *Salicornia europaea* and on *Ulva rigida*. Several detritivores and filter-feeder organism species have been also tested in the SIMTAP prototypes: marine anellids (*H. diversicolor*), kuruma shrimps (*Penaeus japonicus*), oysters (*Crassostrea gigas*), and clams (*Ruditapes decussatus*). Finally, different strains of *Chlorella sorokiniana* and a

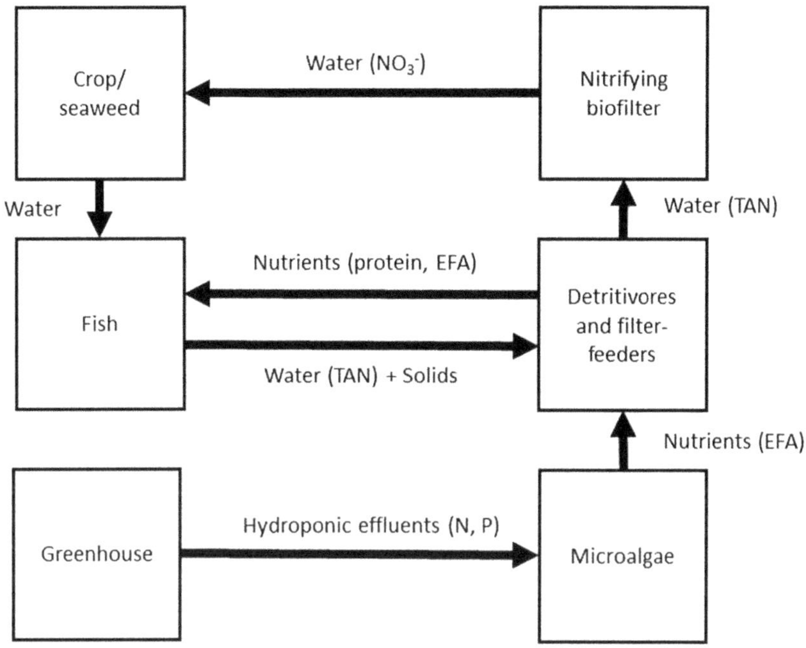

Figure 4 Schematic representation of SIMTAP process. DFFO, detritivores and filter-feeder organisms; TAN, total ammonium nitrogen; EFA, essential fatty acids.

Chlorella-like strain (Ciurli et al., 2021) are currently cultured on greenhouse effluents in low-cost photobioreactors.

4 Conclusion and future trends

Aquaponics is an emerging technique that combines hydroponic cultivation with fish production. Currently, aquaponics is popular at hobby scale, while its commercial exploitation is limited. Aquaponics offers a wide variety of configurations from a technological point of view, even if on-demand-coupled aquaponic systems seem the best option for commercial production. Several bottlenecks for the development of aquaponic systems have been identified and discussed here.

Research is making a crucial contribution to the development of aquaponics for food production. Currently, research mainly focuses on the system components, wastewater treatment, nutrient management, and system production, while other issues have not been deeply addressed so far, for instance, how to prevent disease and pests outbreaks, to recycle macro- and micro-nutrients for better crop performance, and to increase the social awareness of aquaponics. Specific feed formulations for aquaponics systems are also necessary. Since feed is the most important input of aquaponic systems, it could also be the carrier of micronutrients that are often scarce; in this case, possible interactions with fish welfare and health must be studied. A specific feed for aquaponics should also consider alternative protein sources, such as insect meal, to reduce the use of fish meal and fish oil. Research is also needed to study the development of zoonotic etiological agents within aquaponic systems and identify the methods to contrast their spread. In order to increase the efficiency of aquaponic systems, strategies to reduce direct emissions (e.g. solid wastes) should be also studied.

The possibility of integrating low-trophic organisms within aquaponics is one of the principal objectives of the SIMTAP project; these organisms play the double role of bioremediation agents of the solid waste produced in the aquaculture and ingredients of fish feed. The application of multitrophic aquaponic systems, however, requires further studies on many issues including the regulatory implications, the best combination of organism, and the strategies to improve their productivity.

5 Where to look for further information

5.1 Key research articles

- Goddek, S., Joyce, A., Kotzen, B. and Burnell, G. M. (Eds.). (2019). *Aquaponics food production systems: combined aquaculture and hydroponic production technologies for the future*. Springer Nature.
- Hai, F. I., Visvanathan, C. and Boopathy, R. (Eds.). (2018). *Sustainable aquaculture*. Springer.
- Somerville, C., Cohen, M., Pantanella, E., Stankus, A. and Lovatelli, A. (Eds.) (2014). Small-scale aquaponic food production: integrated fish and plant farming. *FAO Fisheries and Aquaculture Technical Paper* 589.
- Tórarinsdóttir, R. I. (Ed.) (2015). *Aquaponics guidelines*. University of Iceland. DOI: 10.13140/RG.2.1.4975.6880.

5.2 Major international research projects

- ACQUACOMBINE - Integrated on-farm Aquaponics systems for co-production of fish, halophyte vegetable, bioactive compounds. Horizon 2020, 2019–2023, https://www.aquacombine.eu.
- HORTIMED - Towards circular horticulture: closing the loop on Mediterranean greenhouses. PRIMA 2019, 2020-onwards, https://www.hortimed-prima.eu/7752-2/.

- IMPAQT - Intelligent management system for integrated multi-trophic aquaculture; 2018-2021. Horizon 2020, 2018-2021, https://impaqtproject.eu/.
- MEDAID - Mediterranean Aquaculture Integrated Development. Horizon 2020, 2017-2021, http://www.medaid-h2020.eu/.
- SIMTAP - Self-sufficient Integrated Multi-Trophic AquaPonics, PRIMA 2018, 2018-2022, www.simtap.eu.

5.3 Key journals/conferences

- Aquaculture, https://www.journals.elsevier.com/aquaculture.
- European Aquaculture Society (EAS), https://www.aquaeas.eu/.

6 Acknowledgements

This study was conducted within the framework of the EU-PRIMA (2, 2018) project entitled 'SIMTAP Self-sufficient Integrated MultiTrophic AquaPonic' and financially supported by the Ministry of Education, University and Research (MIUR) of Italy.

7 References

Aadland, C. 2018. New analysis finds greater costs for land-based aquaculture [WWW Document]. Available at: https://www.intrafish.com/aquaculture/new-analysis-finds-greater-costs-for-land-based-aquaculture/2-1-444543 (accessed 12.16.21). Intrafish.

Araújo, L. S., Keesman, K. J. and Goddek, S. 2021. Making aquaponics a business: A framework. *Water* 13(21), 2978. https://doi.org/10.3390/W13212978.

Aubin, J., Papatryphon, E., van der Werf, H. M. G. and Chatzifotis, S. 2009. Assessment of the environmental impact of carnivorous finfish production systems using life cycle assessment. *Journal of Cleaner Production* 17(3), 354–361. https://doi.org/10.1016/j.jclepro.2008.08.008.

Baganz, G. F. M., Junge, R., Portella, M. C., Goddek, S., Keesman, K. J., Baganz, D., Staaks, G., Shaw, C., Lohrberg, F. and Kloas, W. 2022. The aquaponic principle—it is all about coupling. *Reviews in Aquaculture* 14, 252–264. https://doi.org/10.1111/RAQ.12596.

Barrington, K., Chopin, T. and Robinson, S. 2009. Integrated multi-trophic aquaculture (IMTA) in marine temperate waters. In: Soto, D. (Ed.). *Integrated Mariculture: A Global Review*. Food and Agriculture Organization, Rome, pp. 7-46.

Beneduzi, A., Ambrosini, A. and Passaglia, L. M. P. 2012. Plant growth-promoting rhizobacteria (PGPR): their potential as antagonists and biocontrol agents. *Genetics and Molecular Biology* 35(4) ((Suppl.)), 1044–1051.

Bernstein, S. 2011. *Aquaponic Gardening: a Step-by-Step Guide to Raising Vegetables and Fish Together*. New Society Publishers.

Bibbiani, C., Fronte, B., Incrocci, L. and Campiotti, C. A. 2018. Life cycle impact of industrial aquaculture systems: a review. *Quality - Access to Success* 19, 67–71.

Birolo, M., Bordignon, F., Trocino, A., Fasolato, L., Pascual, A., Godoy, S., Nicoletto, C., Maucieri, C. and Xiccato, G. 2020. Effects of stocking density on the growth and flesh quality of rainbow trout (*Oncorhynchus mykiss*) reared in a low-tech aquaponic system. *Aquaculture* 529, 735653. https://doi.org/10.1016/J.AQUACULTURE.2020.735653.

Bischoff, A. A., Fink, P. and Waller, U. 2009. The fatty acid composition of *Nereis diversicolor* cultured in an integrated recirculated system: possible implications for aquaculture. *Aquaculture* 296, 271–276. https://doi.org/10.1016/j.aquaculture.2009.09.002.

Bohnes, F. A., Hauschild, M. Z., Schlundt, J. and Laurent, A. 2019. Life cycle assessments of aquaculture systems: a critical review of reported findings with recommendations for policy and system development. *Reviews in Aquaculture* 11(4), 1061–1079. https://doi.org/10.1111/raq.12280.

Boxman, S. E. 2015. *Resource Recovery through Halophyte Production in Marine Aquaponics: an Evaluation of the Nutrient Cycling and the Environmental Sustainability of Aquaponics.*

Boxman, S. E., Nystrom, M., Capodice, J. C., Ergas, S. J., Main, K. L. and Trotz, M. A. 2017a. Effect of support medium, hydraulic loading rate and plant density on water quality and growth of halophytes in marine aquaponic systems. *Aquaculture Research* 48(5), 2463–2477. https://doi.org/10.1111/are.13083.

Boxman, S. E., Zhang, Q., Bailey, D. and Trotz, M. A. 2017b. Life cycle assessment of a commercial-scale freshwater aquaponic system. *Environmental Engineering Science* 34(5), 299–311. https://doi.org/10.1089/ees.2015.0510.

Boxman, S. E., Nystrom, M., Ergas, S. J., Main, K. L. and Trotz, M. A. 2018. Evaluation of water treatment capacity, nutrient cycling, and biomass production in a marine aquaponic system. *Ecological Engineering* 120, 299–310. https://doi.org/10.1016/j.ecoleng.2018.06.003.

Boyd, C. E. 2015. Overview of aquaculture feeds: global impacts of ingredient use. In: Davis, A. D. (Ed.). *Feed and Feeding Practices in Aquaculture*. Woodhead Publishing, pp. 3-25.

Cammies, C., Mytton, D. and Crichton, R. 2021. Exploring economic and legal barriers to commercial aquaponics in the EU through the lens of the UK and policy proposals to address them. *Aquaculture International* 29(3), 1245–1263. https://doi.org/10.1007/s10499-021-00690-w.

Campos-Pulido, R., Alonso-López, A., Avalos-de la Cruz, D. A., Asiain-Hoyos, A. and Reta-Mendiola, J. L. 2013. Caracterización fisicoquímica de un efluente salobre de tilapia en acuaponía. *Revista Mexicana de Ciencias Agrícolas* 4, 939–950.

Carballeira Braña, C. B., Cerbule, K., Senff, P. and Stolz, I. K. 2021. Towards environmental sustainability in marine finfish aquaculture. *Frontiers in Marine Science* 8, 343. https://doi.org/10.3389/fmars.2021.666662.

Carrasco, G. and Shamshak, G. 2020. *A Practical Introduction to Aquaponics EC 400: Independent Study in Aquaponics*.

Cerozi, B. S. and Fitzsimmons, K. 2017. Phosphorus dynamics modeling and mass balance in an aquaponics system. *Agricultural Systems* 153, 94–100. https://doi.org/10.1016/j.agsy.2017.01.020.

Chalmers, G. A. 2004. *Aquaponics and Food Safety 1–74*.

Chen, P., Zhu, G., Kim, H. J., Brown, P. B. and Huang, J. Y. 2020. Comparative life cycle assessment of aquaponics and hydroponics in the Midwestern United States.

Journal of Cleaner Production 275, 122888. https://doi.org/10.1016/j.jclepro.2020.122888.

Chu, Y. T. and Brown, P. B. 2021. Sustainable marine aquaponics: effects of shrimp to plant ratios and C/N ratios. *Frontiers in Marine Science* 8, 1825. https://doi.org/10.3389/fmars.2021.771630.

Chu, Y.-T. and Brown, P. B. 2020. Evaluation of Pacific whiteleg shrimp and three halophytic plants in marine aquaponic systems under three salinities. *Sustainability* 13(1), 269. https://doi.org/10.3390/su13010269.

Ciurli, A., Modeo, L., Pardossi, A. and Chiellini, C. 2021. Multidisciplinary integrated characterization of a native Chlorella-like microalgal strain isolated from a municipal landfill leachate. *Algal Research* 54. https://doi.org/10.1016/J.ALGAL.2021.102202.

Colt, J., Schuur, A. M., Weaver, D. and Semmens, K. 2022. Engineering design of aquaponics systems. *Reviews in Fisheries Science and Aquaculture* 30(1), 33–80. https://doi.org/10.1080/23308249.2021.1886240.

da Silva, H. V., Martins, M. A., do Espírito Santo, C. M., do Nascimento Vieira, F., Rezende, P. C., Gonzaga, L. V., Fett, R. and Seiffert, W. Q. 2021. Aquaponic production of sea asparagus and Pacific white shrimp using biofloc technology: different irrigation regimes affect plant production of bioactive compounds and antioxidant capacity. *Aquaculture Research* 00, 1–10. https://doi.org/10.1111/ARE.15641.

Dalsgaard, J., Lund, I., Thorarinsdottir, R., Drengstig, A., Arvonen, K. and Pedersen, P. B. 2013. Farming different species in RAS in Nordic countries: current status and future perspectives. *Aquacultural Engineering* 53, 2–13. https://doi.org/10.1016/j.aquaeng.2012.11.008.

Davidson, J., Bebak, J. and Mazik, P. 2009. The effects of aquaculture production noise on the growth, condition factor, feed conversion, and survival of rainbow trout, Oncorhynchus mykiss. *Aquaculture* 288, 337–343. https://doi.org/10.1016/J.AQUACULTURE.2008.11.037.

Davidson, J. and Summerfelt, S. 2004. Solids flushing, mixing, and water velocity profiles within large (10 and 150 m3) circular 'Cornell-type' dual-drain tanks. *Aquacultural Engineering* 32(1), 245–271. https://doi.org/10.1016/J.AQUAENG.2004.03.009.

Delaide, B., Goddek, S., Gott, J., Soyeurt, H. and Jijakli, M. H. 2016. Lettuce (*Lactuca sativa* L. var. Sucrine) growth performance in complemented aquaponic solution outperforms hydroponics. *Water* 8(10), 467. https://doi.org/10.3390/W8100467.

Delaide, B., Monsees, H., Gross, A. and Goddek, S. 2019. Aerobic and anaerobic treatments for aquaponic sludge reduction and mineralisation. In: Goddek, S., Joyce, A., Kotzen, B. and Burnell, G. M. (Eds). *Aquaponics Food Production Systems*. Springer International Publishing, Cham, pp. 247–266.

Déniel, F., Rey, P., Chérif, M., Guillou, A. and Tirilly, Y. 2004. Indigenous bacteria with antagonistic and plant-growth-promoting activities improve slow-filtration efficiency in soilless cultivation. *Canadian Journal of Microbiology* 50(7), 499–508. https://doi.org/10.1139/w04-034.

Diver, S. 2006. *Aquaponics – Integration of Hydroponics with Aquaculture. ATTRA*.

Doncato, K. B. and Costa, C. S. B. 2021. Micronutrient supplementation needs for halophytes in saline aquaponics with BFT system water. *Aquaculture* 531, 735815. https://doi.org/10.1016/J.AQUACULTURE.2020.735815.

d'Orbcastel, E. R., Blancheton, J. P. and Aubin, J. 2009. Towards environmentally sustainable aquaculture: comparison between two trout farming systems using

Life Cycle Assessment. *Aquacultural Engineering* 40(3), 113–119. https://doi.org/10.1016/J.AQUAENG.2008.12.002.

Ebeling, J. M. and Timmons, M. B. 2010. *Recirculating Aquaculture. Cayuga Aqua Ventures*.

Ebeling, J. M., Timmons, M. B. and Bisogni, J. J. J. 2006. Engineering analysis of the stoichiometry of photoautotrophic, autotrophic, and heterotrophic removal of ammonia-nitrogen in aquaculture systems. *Aquaculture* 257(1-4), 346–358. https://doi.org/10.1016/j.aquaculture.2006.03.019.

Eck, M., Körner, O. and Jijakli, M. H. 2019. Nutrient cycling in aquaponics systems. In: Goddek, S., Joyce, A., Kotzen, B. and Burnell, G. M. (Eds). *Aquaponics Food Production Systems*. Springer International Publishing, Cham, pp. 231–246. https://doi.org/10.1007/978-3-030-15943-6_9.

Edwards, S. L. and Marshall, W. S. 2012. Principles and patterns of osmoregulation and euryhalinity in fishes. In: *Fish Physiology*. Academic Press, pp. 1–44.

El-Kazzaz, A. 2017. Soilless agriculture a new and advanced method for agriculture development: an introduction. *Agricultural Research and Technology:Open Access Journal* 3(2). https://doi.org/10.19080/ARTOAJ.2017.03.555610.

Elumalai, S., Shaw, A., Pattillo, D., Currey, C., Rosentrater, K. and Xie, K. 2017. Influence of UV treatment on the food safety status of a model aquaponic system. *Water* 9(1), 27. https://doi.org/10.3390/w9010027.

Endut, A., Jusoh, A., Ali, N., Wan Nik, W. B. and Hassan, A. 2010. A study on the optimal hydraulic loading rate and plant ratios in recirculation aquaponic system. *Bioresource Technology* 101(5), 1511–1517. https://doi.org/10.1016/j.biortech.2009.09.040.

Espinal, C. A. and Matulić, D. 2019. Recirculating aquaculture technologies. In: Goddek, S., Joyce, A., Kotzen, B. and Burnell, G. M. (Eds). *Aquaponics Food Production Systems*. Springer International Publishing, Cham, pp. 35–76.

European Commission 2009. Commission Regulation (EC) no 710/2009 of 5 August 2009 amending Regulation (EC). Available at: https://eur-lex.europa.eu/legal-content/EN/ALL/?uri=celex%3A32009R0710 (accessed 3.22.22). As Regards Laying down Detailed Rules on Organic Aquaculture Animal and Seaweed Production [WWW Document] (vol. 889/2008) laying down detailed rules for the implementation of Council Regulation (EC) No 834/2007.

European Commission 2001. *Communication From the Commission on a Community Strategy Against Antimicrobial Resistanc*. Brussels.

European Parliament and Council of the European Union 2018. *REGULATION (EU) 2018/848 OF The European Parliament AND OF THE COUNCIL of 30 May 2018*.

FAO 2020. The State of World Fisheries and Aquaculture 2020. Food and Agriculture Organization.

Fierro-Sañudo, J. F., Rodríguez-Montes De Oca, G. A., León-Cañedo, J. A., Alarcón-Silvas, S. G., Martin Mariscal-Lagarda, M., Díaz-Valdés, T. and Páez-Osuna, F., & 2018. Production and management of shrimp-basil co-culture 63 Production and management of shrimp (*Penaeus vannamei*) in co-culture with basil (*Ocimum basilicum*) using two sources of low-salinity water. *Latin American Journal of Aquatic Research* 46, 63–71. https://doi.org/10.3856/vol46-issue1-fulltext-8.

Forchino, A. A., Lourguioui, H., Brigolin, D. and Pastres, R. 2017. Aquaponics and sustainability: the comparison of two different aquaponic techniques using the Life Cycle Assessment (LCA). *Aquacultural Engineering* 77, 80–88. https://doi.org/10.1016/j.aquaeng.2017.03.002.

Fronte, B., Galliano, G. and Bibbiani, C. 2016. From freshwater to marine aquaponic: new opportunities for marine fish species production. *Biotechnical Centre Naklo* 21, 20.

Fronte, B., Kim, C. H., Bagliacca, M., Casini, L. and de Zoysa, M. 2019. 1,3-1-6 ß-glucans enhance tissue regeneration in zebrafish (*Danio rerio*): potential advantages for aquaculture applications. *Aquaculture Research* 50(11), 3163–3170. https://doi.org/10.1111/ARE.14270.

Fruscella, L., Kotzen, B. and Milliken, S. 2021. Organic aquaponics in the European Union: towards sustainable farming practices in the framework of the new EU regulation. *Reviews in Aquaculture* 13, 1661–1682. https://doi.org/10.1111/raq.12539.

Gendel, Y. and Lahav, O. 2013. A novel approach for ammonia removal from fresh-water recirculated aquaculture systems, comprising ion exchange and electrochemical regeneration. *Aquacultural Engineering* 52, 27–38. https://doi.org/10.1016/J.AQUAENG.2012.07.005.

Gennaro, P., Guidotti, M., Funari, E., Porrello, S. and Lenzi, M. 2006. Reduction of land based fish-farming impact by phytotreatment pond system in a marginal lagoon area. *Aquaculture* 256, 246–254. https://doi.org/10.1016/j.aquaculture.2006.02.028.

Ghamkhar, R., Hartleb, C., Wu, F. and Hicks, A. 2020. Life cycle assessment of a cold weather aquaponic food production system. *Journal of Cleaner Production* 244, 118767. https://doi.org/10.1016/j.jclepro.2019.118767.

Goddek, S., Delaide, B., Mankasingh, U., Ragnarsdottir, K., Jijakli, H. and Thorarinsdottir, R. 2015. Challenges of sustainable and commercial aquaponics. *Sustainability* 7(4), 4199–4224. https://doi.org/10.3390/su7044199.

Goddek, S., Joyce, A., Kotzen, B. and Burnell, G. M. 2019a. *Aquaponics Food Production Systems - Combined Aquaculture and Hydroponic Production Technologies for the Future, Aquaponics Food Production Systems*. https://doi.org/10.1007/978-3-030-15943-6.

Goddek, S., Joyce, A., Wuertz, S., Körner, O., Bläser, I., Reuter, M. and Keesman, K. J. 2019b. Decoupled aquaponics systems. In: Goddek, S., Joyce, A., Kotzen, B. and Burnell, G. M. (Eds). *Aquaponics Food Production Systems*. Springer International Publishing, Cham, pp. 201–229.

Goddek, S. and Keesman, K. J. 2020. Improving nutrient and water use efficiencies in multi-loop aquaponics systems. *Aquaculture International* 28, 2481–2490. https://doi.org/10.1007/S10499-020-00600-6/FIGURES/3.

Goddek, S. and Körner, O. 2019. A fully integrated simulation model of multi-loop aquaponics: A case study for system sizing in different environments. *Agricultural Systems* 171, 143–154. https://doi.org/10.1016/j.agsy.2019.01.010.

Gonçalves, A. A. and Gagnon, G. A. 2011. Ozone application in recirculating aquaculture system: an overview. *Ozone: Science and Engineering* 33(5), 345–367.

Gorbe, E. and Calatayud, Á. 2010. Optimization of nutrition in soilless systems: a review. *Advances in Botanical Research* 53, 193–245.

Greenfeld, A., Becker, N., Bornman, J. F. and Angel, D. L. 2020. Identifying knowledge levels of aquaponics adopters. *Environmental Science and Pollution Research International* 27(4), 4536–4540. https://doi.org/10.1007/s11356-019-06758-8.

Greenfeld, A., Becker, N., Bornman, J. F., Spatari, S. and Angel, D. L. 2021. Monetizing environmental impact of integrated aquaponic farming compared to separate systems. *Science of the Total Environment* 792, 148459. https://doi.org/10.1016/J.SCITOTENV.2021.148459.

Greenfeld, A., Becker, N., McIlwain, J., Fotedar, R. and Bornman, J. F. 2019. Economically viable aquaponics? Identifying the gap between potential and current uncertainties. *Reviews in Aquaculture* 11(3), 848–862. https://doi.org/10.1111/raq.12269.

Gunning, D., Maguire, J. and Burnell, G. 2016. The development of sustainable saltwater-based food production systems: a review of established and novel concepts. *Water* 8(12), 598. https://doi.org/10.3390/w8120598.

Hao, Y., Ding, K., Xu, Y., Tang, Y., Liu, D. and Li, G. 2020. States, trends, and future of aquaponics research. *Sustainability* 12(18), 7783. https://doi.org/10.3390/su12187783.

Hollyer, J., Tamaru, C., Riggs, A., Klinger-Bowen, R., Howerton, R., Okimoto, D., Castro, L., Troegner, V. and Martinez, G. 2009. *On-Farm Food Safety: Aquaponics. Food Safety and Technology*.

Holtman, W. L., Oppedijk, B. J., Vennik, M. and van Duijn, B. 2014. Low oxygen stress in horticultural practice. In: *Plant Cell Monographs*. Springer, Vienna, 371–380.

Hu, Z., Lee, J. W., Chandran, K., Kim, S., Brotto, A. C. and Khanal, S. K. 2015. Effect of plant species on nitrogen recovery in aquaponics. *Bioresource Technology* 188, 92–98. https://doi.org/10.1016/J.BIORTECH.2015.01.013.

Jaeger, C., Foucard, P., Tocqueville, A., Nahon, S. and Aubin, J. 2019. Mass balanced based LCA of a common carp-lettuce aquaponics system. *Aquacultural Engineering* 84, 29–41. https://doi.org/10.1016/j.aquaeng.2018.11.003.

Jennings, S., Stentiford, G. D., Leocadio, A. M., Jeffery, K. R., Metcalfe, J. D., Katsiadaki, I., Auchterlonie, N. A., Mangi, S. C., Pinnegar, J. K., Ellis, T., Peeler, E. J., Luisetti, T., Baker-Austin, C., Brown, M., Catchpole, T. L., Clyne, F. J., Dye, S. R., Edmonds, N. J., Hyder, K., Lee, J., Lees, D. N., Morgan, O. C., O'Brien, C. M., Oidtmann, B., Posen, P. E., Santos, A. R., Taylor, N. G. H., Turner, A. D., Townhill, B. L. and Verner-Jeffreys, D. W. 2016. Aquatic food security: insights into challenges and solutions from an analysis of interactions between fisheries, aquaculture, food safety, human health, fish and human welfare, economy and environment. *Fish and Fisheries* 17(4), 893–938. https://doi.org/10.1111/FAF.12152.

Jerbi, M. A., Aubin, J., Garnaoui, K., Achour, L. and Kacem, A. 2012. Life cycle assessment (LCA) of two rearing techniques of sea bass (*Dicentrarchus labrax*). *Aquacultural Engineering* 46, 1–9. https://doi.org/10.1016/j.aquaeng.2011.10.001.

Jerónimo, D., Lillebø, A. I., Cremades, J., Cartaxana, P. and Calado, R. 2021a. Recovering wasted nutrients from shrimp farming through the combined culture of polychaetes and halophytes. *Scientific Reports* 11(1), 6587. https://doi.org/10.1038/s41598-021-85922-y.

Jerónimo, D., Lillebø, A. I., Maciel, E., Domingues, M. R. M., Cremades, J. and Calado, R. 2021b. Unravelling the fatty acid profiles of different polychaete species cultured under integrated multi-trophic aquaculture (IMTA). *Scientific Reports* 11(1), 10812. https://doi.org/10.1038/s41598-021-90185-8.

Jerónimo, D., Lillebø, A. I., Santos, A., Cremades, J. and Calado, R. 2020. Performance of polychaete assisted sand filters under contrasting nutrient loads in an integrated multi-trophic aquaculture (IMTA) system. *Scientific Reports* 10(1), 20871. https://doi.org/10.1038/s41598-020-77764-x.

Joesting, H. M., Blaylock, R., Biber, P. and Ray, A. 2016. The use of marine aquaculture solid waste for nursery production of the salt marsh plants *Spartina alterniflora* and *Juncus roemerianus*. *Aquaculture Reports* 3, 108–114. https://doi.org/10.1016/J.AQREP.2016.01.004.

Joyce, A., Timmons, M., Goddek, S. and Pentz, T. 2019. Bacterial relationships in aquaponics: new research directions. In: Goddek, S., Joyce, S., Kotzen, B. and Burnell, G. M. (Eds). *Aquaponics Food Production Systems*. Springer International Publishing, Cham, pp. 145–161.

Junge, R., König, B., Villarroel, M., Komives, T. and Jijakli, M. 2017. Strategic points in aquaponics. *Water* 9(3), 182. https://doi.org/10.3390/w9030182.

Kasozi, N., Abraham, B., Kaiser, H. and Wilhelmi, B. 2021. The complex microbiome in aquaponics: significance of the bacterial ecosystem. *Annals of Microbiology* 71, 1-13. https://doi.org/10.1186/S13213-020-01613-5/TABLES/2.

Kawaguchi, O., Tanaka, M., Yoshii, M., Iwamoto, Y., Midooka, A., Toutani, F., Nagao, N., Matsumoto, T., Mabuchi, R. and Tanimoto, S. 2019. Off-flavor of red sea bream Pagrus major reared in recirculating aquaculture systems with low salinity is caused by 2-methylisoborneol. *Fisheries Science* 85, 553–560. https://doi.org/10.1007/S12562-019-01307-1/TABLES/3.

Khakyzadeh, V., Luque, R., Zolfigol, M. A., Vahidian, H. R., Salehzadeh, H., Moradi, V., Soleymani, A. R., Moosavi-Zare, A. R. and Xu, K. 2015. Waste to wealth: a sustainable aquaponic system based on residual nitrogen photoconversion. *RSC Advances* 5(5), 3917–3921. https://doi.org/10.1039/C4RA15242E.

Kidgell, J. T., Magnusson, M., de Nys, R. and Glasson, C. R. K. 2019. Ulvan: A systematic review of extraction, composition and function. *Algal Research* 39, 101422. https://doi.org/10.1016/j.algal.2019.101422.

Körner, O., Bisbis, M. B., Baganz, G. F. M., Baganz, D., Staaks, G. B. O., Monsees, H., Goddek, S. and Keesman, K. J. 2021. Environmental impact assessment of local decoupled multi-loop aquaponics in an urban context. *Journal of Cleaner Production* 313, 127735. https://doi.org/10.1016/J.JCLEPRO.2021.127735.

Kotzen, Benz, Emerenciano, M. G. C., Moheimani, N., Burnell and Gavin, M. 2019. Aquaponics: alternative types and approaches. In: Goddek, S., Joyce, A., Kotzen, B. and Burnell, G. M. (Eds). *Aquaponics Food Production Systems*. Springer International Publishing, Cham, pp. 301–330.

Kouba, A., Lunda, R., Hlaváč, D., Kuklina, I., Hamáčková, J., Randák, T., Kozák, P., Koubová, A. and Buřič, M. 2018. Vermicomposting of sludge from recirculating aquaculture system using *Eisenia andrei*: technological feasibility and quality assessment of end-products. *Journal of Cleaner Production* 177, 665–673. https://doi.org/10.1016/J.JCLEPRO.2017.12.216.

Laiz-Carrión, R., Sangiao-Alvarellos, S., Guzmán, J. M., Martín Del Río, M. P., Soengas, J. L. and Mancera, J. M. 2005. Growth performance of gilthead sea bream *Sparus aurata* in different osmotic conditions: implications for osmoregulation and energy metabolism. *Aquaculture* 250(3–4), 849–861. https://doi.org/10.1016/j.aquaculture.2005.05.021.

le Féon, S., Dubois, T., Jaeger, C., Wilfart, A., Akkal-Corfini, N., Bacenetti, J., Costantini, M. and Aubin, J. 2021. DEXiAqua, a model to assess the sustainability of aquaculture systems: methodological development and application to a French salmon farm. *Sustainability* 13(14). https://doi.org/10.3390/SU13147779.

Lekang, O. I. 2007. Aquaculture engineering, second. ed. Aquaculture Engineering. Wiley-Blackwell.

Lennard, W. 2012a. *Aquaponic System Design Parameters: Fish to Plant Ratios (Feeding Rate Ratios)*. Aquaponic Fact Sheet Series.

Lennard, W. 2012b. *Aquaponic Media Bed Sizing Calculator-Metric*.

Lennard, W. and Goddek, S. 2019. Aquaponics: the basics. In: Goddek, S., Joyce, A., Kotzen, B. and Burnell, G. M. (Eds). *Aquaponics Food Production Systems: Combined Aquaculture and Hydroponic Production Technologies for the Future*. Springer International Publishing, Cham, pp. 113–143.

Lepoivre, P. 2003. Phytopathologie, 1st edition. In: de Gembloux Bruxelles, L. P. A. (Ed.). De Boeck.

Lindholm-Lehto, P. C. and Vielma, J. 2019. Controlling of geosmin and 2-methylisoborneol induced off-flavours in recirculating aquaculture system farmed fish–a review. *Aquaculture Research* 50, 9–28. https://doi.org/10.1111/ARE.13881.

Ljubojevic, D. and Djordjevic, V. 2016. Bacterial hazards in fish meat: the aetiologic agents of foodborne diseases. Scientific journal. *Meat Technology* 57, 31–42.

Ljubojević, D., Pelić, M., Radosavljević, V. and Ćirković, V. 2017. Food safety hazards related to fish produced in aquaponics. In: *Aquaculture Europe* (vol. 2017). Dubrovnik.

Lombardi, T., Bertacchi, A., Pistelli, L., Pardossi, A., Pecchia, S., Toffanin, A. and Sanmartin, C. 2022. Biological and agronomic traits of the main halophytes widespread in the Mediterranean region as potential new vegetable crops. *Horticulturae* 8(3). https://doi.org/10.3390/horticulturae8030195.

Love, D. C., Fry, J. P., Li, X., Hill, E. S., Genello, L., Semmens, K. and Thompson, R. E. 2015a. Commercial aquaponics production and profitability: findings from an international survey. *Aquaculture* 435, 67–74. https://doi.org/10.1016/j.aquaculture.2014.09.023.

Love, D. C., Uhl, M. S. and Genello, L. 2015b. Energy and water use of a small-scale raft aquaponics system in Baltimore, Maryland, United States. *Aquacultural Engineering* 68, 19–27. https://doi.org/10.1016/j.aquaeng.2015.07.003.

Makri, O. and Kintzios, S. 2008. *Ocimum* sp. (Basil): Botany, Cultivation, Pharmaceutical Properties, and Biotechnology. *Journal of Herbs, Spices, and Medicinal Plants* 13, 123–150. https://doi.org/10.1300/J044v13n03_10.

Marques, B., Calado, R. and Lillebø, A. I. 2017. New species for the biomitigation of a super-intensive marine fish farm effluent: combined use of polychaete-assisted sand filters and halophyte aquaponics. *Science of the Total Environment* 599–600, 1922–1928. https://doi.org/10.1016/j.scitotenv.2017.05.121.

Marques, B., Lillebø, A. I., Ricardo, F., Nunes, C., Coimbra, M. A. and Calado, R. 2018. Adding value to ragworms (*Hediste diversicolor*) through the bioremediation of a super-intensive marine fish farm. *Aquaculture Environment Interactions* 10, 79–88. https://doi.org/10.3354/aei00255.

Martins, C. I. M., Eding, E. H., Verdegem, M. C. J., Heinsbroek, L. T. N., Schneider, O., Blancheton, J. P., d'Orbcastel, E. R. and Verreth, J. A. J. 2010. New developments in recirculating aquaculture systems in Europe: A perspective on environmental sustainability. *Aquacultural Engineering* 43(3), 83–93. https://doi.org/10.1016/j.aquaeng.2010.09.002.

Maucieri, C., Forchino, A. A., Nicoletto, C., Junge, R., Pastres, R., Sambo, P. and Borin, M. 2018. Life cycle assessment of a micro aquaponic system for educational purposes built using recovered material. *Journal of Cleaner Production* 172, 3119–3127. https://doi.org/10.1016/j.jclepro.2017.11.097.

Maucieri, C., Nicoletto, C., Zanin, G., Birolo, M., Trocino, A., Sambo, P., Borin, M. and Xiccato, G. 2019. Effect of stocking density of fish on water quality and growth performance of European Carp and leafy vegetables in a low-tech aquaponic system. *PLoS ONE* 14(5), e0217561. https://doi.org/10.1371/journal.pone.0217561.

Maucieri, C., Nicoletto, C., Zanin, G., Birolo, M., Xiccato, G., Sambo, P. and Borin, M. 2020. Nitrogen budget in recirculating aquaponic systems with different fish stocking density. *Italian Journal of Agronomy* 15(3), 239–245. https://doi.org/10.4081/ija.2020.1639.

Miličić, V., Thorarinsdottir, R., Santos, M. and Hančič, M. 2017. Commercial aquaponics approaching the European Market: to consumers' perceptions of aquaponics products in Europe. *Water* 9(2), 80. https://doi.org/10.3390/w9020080.

Mohan, K., Ravichandran, S., Muralisankar, T., Uthayakumar, V., Chandirasekar, R., Seedevi, P., Abirami, R. G. and Rajan, D. K. 2019. Application of marine-derived polysaccharides as immunostimulants in aquaculture: a review of current knowledge and further perspectives. *Fish and Shellfish Immunology* 86, 1177–1193. https://doi.org/10.1016/j.fsi.2018.12.072.

Monsees, H., Kloas, W. and Wuertz, S. 2017. Decoupled systems on trial: eliminating bottlenecks to improve aquaponic processes. *PLoS ONE* 12(9), e0183056. https://doi.org/10.1371/JOURNAL.PONE.0183056.

Mori, J. and Smith, R. 2019. Transmission of waterborne fish and plant pathogens in aquaponics and their control with physical disinfection and filtration: A systematized review. *Aquaculture* 504, 380–395. https://doi.org/10.1016/j.aquaculture.2019.02.009.

Mozes, N., Papandroulakis, N., Vergara, J. M., Biswas, A., Takii, K. and Ntatsopoulos, A. 2011. Production systems. In: *Sparidae*. John Wiley & Sons, Ltd, pp. 169–198.

MRFR 2022. Aquaponics Market SWOT analysis and Key Business Strategies, Demand & Forecast 2027. Digital Journal.

Nagayo, A. M., Mendoza, C., Vega, E., al Izki, R. K. S. and Jamisola, R. S. 2017. An automated solar-powered aquaponics system towards agricultural sustainability in the Sultanate of Oman IEEE International Conference on Smart Grid and Smart Cities, ICSGSC 2017 (vol. 2017), pp. 42–49.

Narayanasamy, P. 2013. Biological management of diseases of crops. In *Volume 1. Characteristics of Biological Control Agents*. Springer, Netherlands.

Nicola, S., Egea-Gilabert, C., Niñirola, D., Conesa, E., Pignata, G., Fontana, E. and Fernández, J. A. 2015. Nitrogen and aeration levels of the nutrient solution in soilless cultivation systems as important growing conditions affecting inherent quality of baby leaf vegetables: a review. *Acta Horticulturae* (1099), 167–177. https://doi.org/10.17660/ActaHortic.2015.1099.17.

Nielsen, C. J., Ferrin, D. M. and Stanghellini, M. E. 2006. Efficacy of biosurfactants in the management of *Phytophthora capsici* on pepper in recirculating hydroponic systems. *Canadian Journal of Plant Pathology* 28(3), 450–460. https://doi.org/10.1080/07060660609507319.

Nozzi, V., Parisi, G., di Crescenzo, D., Giordano, M. and Carnevali, O. 2016. Evaluation of Dicentrarchus labrax meats and the vegetable quality of *Beta vulgaris* var. *Cicla* farmed in freshwater and saltwater aquaponic systems. *Water* 8(10), 1–14. https://doi.org/10.3390/w8100423.

Pajand, Z. O., Soltani, M., Bahmani, M. and Kamali, A. 2017. The role of polychaete *Nereis diversicolor* in bioremediation of wastewater and its growth performance and fatty acid composition in an integrated culture system with *Huso huso* (Linnaeus, 1758). *Aquaculture Research* 48(10), 5271–5279. https://doi.org/10.1111/are.13340.

Palm, H. W., Knaus, U., Appelbaum, S., Goddek, S., Strauch, S. M., Vermeulen, T., Haïssam Jijakli, M. and Kotzen, B. 2018. Towards commercial aquaponics: a review of systems, designs, scales and nomenclature. *Aquaculture International* 26(3), 813–842. https://doi.org/10.1007/s10499-018-0249-z.

Palm, H. W., Knaus, U., Appelbaum, S., Strauch, S. M. and Kotzen, B. 2019. Coupled aquaponics systems. In: Goddek, S., Joyce, A., Kotzen, B. and Burnell, G. M. (Eds). *Aquaponics Food Production Systems: Combined Aquaculture and Hydroponic Production Technologies for the Future*. Springer International Publishing, Cham, pp. 163–199.

Pantanella, E. 2018. Aquaponics production, practices and opportunities. In: Hai, F. I., Visvanathan, C. and Boopathy, R. (Eds). *Sustainable Aquaculture*. Springer International Publishing, Cham, pp. 191–248.

Pantanella, E. 2012. *Nutrition and Quality of Aquaponic Systems*.

Pantanella, E., Cardarelli, M., di Mattia, E. and Colla, G. 2015. Aquaponics and food safety: effects of UV sterilization on total coliforms and lettuce production. *Acta Horticulturae* (1062), 71–76. https://doi.org/10.17660/ActaHortic.2015.1062.8.

Pardossi, A., Malorgio, F., Incrocci, L. and Tognoni, F. 2006. Hydroponic technologies for greenhouse crops. In: Ramdame, D. (Ed.). *Crops: Growth, Quality and Biotechnology*. WFL Publisher Science & Tecnology, Helsinki, pp. 360–378.

Parisi, G., Terova, G., Gasco, L., Piccolo, G., Roncarati, A., Moretti, V. M., Centoducati, G., Gatta, P. P. and Pais, A. 2014. Current status and future perspectives of Italian finfish aquaculture. *Reviews in Fish Biology and Fisheries* 24(1), 15–73. https://doi.org/10.1007/s11160-013-9317-7.

Pinho, S. M., David, L. H., Garcia, F., Keesman, K. J., Portella, M. C. and Goddek, S. 2021. South American fish species suitable for aquaponics: a review. *Aquaculture International* 29(4), 1427–1449. https://doi.org/10.1007/s10499-021-00674-w.

Pombo, A., Baptista, T., Granada, L., Ferreira, S. M. F., Gonçalves, S. C., Anjos, C., Sá, E., Chainho, P., Cancela da Fonseca, L., Fidalgo e Costa, P. and Costa, J. L. 2020. Insight into aquaculture's potential of marine annelid worms and ecological concerns: a review. *Reviews in Aquaculture* 12(1), 107–121. https://doi.org/10.1111/raq.12307.

Postma, J. Os, E., Bonants, P 2008. *Pathogen Detection and Management Strategies in Soilless Plant Growing Systems*.

Rains, J. 2007. Plants that grow well in aquaponics. Backyard Aquaponics.

Rakocy, J. E., Bailey, D. S., Shultz, K. A. and Cole, W. M. 1997. Evaluation of a commercial-scale aquaponic unit for the production of tilapia and lettuce. In: *Tilapia Aquaculture*. Proceedings of the Fourth International Symposium on Tilapia in Aquaculture, Orlando, Florida, pp. 357–372.

Rakocy, J. E., Shultz, R. C., Bailey, D. S. and Thoman, E. S. 2004. Aquaponic production of tilapia and basil: comparing a batch and staggered cropping system. *Acta Horticulturae* (648), 63–69. https://doi.org/10.17660/ActaHortic.2004.648.8.

Raviv, M., Lieth, J. H. and Bar-Tal, A. (Eds.) 2019. *Soilless Culture: Theory and Practice*, Second ed. Academic Press.

Reinhardt, T., Hoevenaars, K. and Joyce, A. 2019. Regulatory frameworks for aquaponics in the European Union. In: Goddek, S., Joyce, A., Kotzen, B. and Burnell, G. M. (Eds). *Aquaponics Food Production Systems*. Springer International Publishing, Cham, pp. 501–522.

Renault, D., Vallance, J., Déniel, F., Wery, N., Godon, J. J., Barbier, G. and Rey, P. 2012. Diversity of bacterial communities that colonize the filter units used for controlling plant pathogens in soilless cultures. *Microbial Ecology* 63(1), 170–187. https://doi.org/10.1007/s00248-011-9961-1.

Robaina, L., Pirhonen, J., Mente, E., Sánchez, J. and Goosen, N. 2019. Fish diets in aquaponics. In: Goddek, S., Joyce, A., Kotzen, B. and Burnell, G. M. (Eds). *Aquaponics Food Production Systems*. Springer International Publishing, Cham, pp. 333–352.

Ronzón-Ortega, M., Patricia Hernández-Vergara, M. and Iván Pérez-Rostro, C. 2012. Produccion hidropónica y acuapónica de albahaca (*Ocimum basilicum*) y langostino malayo (*Macrobrachium rosenbergii*). *Tropical and Subtropical Agroecosystems* 15, 63-71.

Rossi, L., Bibbiani, C., Fierro-Sañudo, J. F., Maibam, C., Incrocci, L., Pardossi, A. and Fronte, B. 2021. Selection of marine fish for integrated multi-trophic aquaponic production in the Mediterranean area using DEXi multi-criteria analysis. *Aquaculture* 535. https://doi.org/10.1016/j.aquaculture.2021.736402.

Ruiz, P., Vidal, J. M., Sepúlveda, D., Torres, C., Villouta, G., Carrasco, C., Aguilera, F., Ruiz-Tagle, N. and Urrutia, H. 2020. Overview and future perspectives of nitrifying bacteria on biofilters for recirculating aquaculture systems. *Reviews in Aquaculture* 12, 1478–1494. https://doi.org/10.1111/RAQ.12392.

Sánchez, L. M. R., Trujillo, M. M. P., Jiménez, P., Giraldo, H. H. and Ramírez, E. G. 2011. Evaluación Preliminar de Sistemas Acuapónicos e Hidropónicos en Cama Flotante para el Cultivo de Orégano (*Origanum vulgare*: Lamiaceae). *Revista Facultad de Ciencias Básicas* 7, 242–259. https://doi.org/10.18359/RFCB.2057.

Santos, A., Granada, L., Baptista, T., Anjos, C., Simões, T., Tecelão, C., Fidalgo e Costa, P., Costa, J. L. and Pombo, A. 2016. Effect of three diets on the growth and fatty acid profile of the common ragworm *Hediste diversicolor* (O. F. Müller, 1776). *Aquaculture* 465, 37–42. https://doi.org/10.1016/j.aquaculture.2016.08.022.

Sharma, N., Acharya, S., Kumar, K., Singh, N. and Chaurasia, O. P. 2018. Hydroponics as an advanced technique for vegetable production: an overview. *Journal of Soil and Water Conservation* 17(4), 364. https://doi.org/10.5958/2455-7145.2018.00056.5.

Short, G., Yue, C., Anderson, N., Russell, C. and Phelps, N. 2017. Consumer perceptions of aquaponic systems. *Horttechnology* 27, 358–366. https://doi.org/10.21273/HORTTECH03606-16.

Shpigel, M., Guttman, L., Shauli, L., Odintsov, V., Ben-Ezra, D. and Harpaz, S. 2017. Ulva Lactuca from an integrated multi-trophic aquaculture (IMTA) biofilter system as a protein supplement in gilthead seabream (*Sparus aurata*) diet. *Aquaculture* 481, 112–118. https://doi.org/10.1016/j.aquaculture.2017.08.006.

Sirakov, I., Lutz, M., Graber, A., Mathis, A., Staykov, Y., Smits, T. H. M. and Junge, R. 2016. Potential for combined biocontrol activity against fungal fish and plant pathogens by bacterial isolates from a model aquaponic system. *Water* 8(11), 518. https://doi.org/10.3390/W8110518.

Somerville, C., Cohen, M., Pantanella, E., Stankus, A. and Lovatelli, A. 2014. *Small-Scale Aquaponic Food Production, FAO Fisheries and Aquaculture*.

Tal, Y., Schreier, H. J., Sowers, K. R., Stubblefield, J. D., Place, A. R. and Zohar, Y. 2009. Environmentally sustainable land-based marine aquaculture. *Aquaculture* 286(1-2), 28–35. https://doi.org/10.1016/j.aquaculture.2008.08.043.

Thorarinsdottir, R. I. 2015. *Aquaponics Guidelines*. Reykjavik.

Timmons, M. ben, Guerdat, T. and Vinci, B. J. 2018. *Recirculating Aquaculture* (4th edn.). Ithaca Publishing Company LLC.

Turnsek, M., Joly, A., Thorarinsdottir, R. and Junge, R. 2020. Challenges of commercial aquaponics in Europe: beyond the hype. *Water* 12(1), 306. https://doi.org/10.3390/W12010306.

Turnšek, M., Morgenstern, R., Schröter, I., Mergenthaler, M., Hüttel, S. and Leyer, M. 2019. Commercial aquaponics: A long road ahead. In: Goddek, S., Joyce, A., Kotzen, B. and Burnell, G. M. (Eds). *Aquaponics Food Production Systems*. Springer International Publishing, Cham, pp. 453–485.

Tyson, R. v and Simonne, E. H. 2021. *A Practical Guide for Aquaponics as an Alternative Enterprise 1. Horticultural Sciences Department, UF/IFAS Extension*.

USGS 2022. DOTABLES [WWW Document]. Available at: https://water.usgs.gov/water-resources/software/DOTABLES/ (accessed 3.28.22).

Vallance, J., Déniel, F., Floch, G. le, Guérin-Dubrana, L., Blancard, D. and Rey, P. 2011. Pathogenic and beneficial microorganisms in soilless cultures. *Sustainable Agriculture* 2, 711–726. https://doi.org/10.1007/978-94-007-0394-0_31.

van Rijn, J. 2013. Waste treatment in recirculating aquaculture systems. *Aquacultural Engineering* 53, 49–56. https://doi.org/10.1016/j.aquaeng.2012.11.010.

van Rijn, J., Tal, Y. and Schreier, H. J. 2006. Denitrification in recirculating systems: theory and applications. *Aquacultural Engineering* 34, 364–376. https://doi.org/10.1016/j.aquaeng.2005.04.004.

Villarroel, M., Junge, R., Komives, T., König, B., Plaza, I., Bittsánszky, A. and Joly, A. 2016. Survey of aquaponics in Europe. *Water* 8, 468. https://doi.org/10.3390/w8100468.

Vlahos, N., Levizou, E., Stathopoulou, P., Berillis, P., Antonopoulou, E., Bekiari, V., Krigas, N., Kormas, K. and Mente, E. 2019. An experimental brackish aquaponic system using juvenile gilthead sea bream (*Sparus aurata*) and rock samphire (*Crithmum maritimum*). *Sustainability* 11(18), 4820. https://doi.org/10.3390/su11184820.

Waechter-Kristensen, B., Caspersen, S., Adalsteinsson, S., Sundin, P. and Jensén, P. 1999. Organic compounds and micro-organisms in closed, hydroponic culture: occurrence and effects on plant growth and mineral nutrition. *Acta Horticulturae* (481), 197–204. https://doi.org/10.17660/ActaHortic.1999.481.20.

Wahap, N., Estim, A., Kian, A. Y. S., Senoo, S. and Mustafa, S. 2010. *Producing Organic Fish and Mint in an Aquaponic System. Borneo Marine Research Institue, Sabah, Malaysia*, pp. 29–33.

Waller, U., Buhmann, A. K., Ernst, A., Hanke, V., Kulakowski, A., Wecker, B., Orellana, J. and Papenbrock, J. 2015. Integrated multi-trophic aquaculture in a zero-exchange recirculation aquaculture system for marine fish and hydroponic halophyte production. *Aquaculture International* 23(6), 1473–1489. https://doi.org/10.1007/s10499-015-9898-3.

Wang, H., Hagemann, A., Reitan, K. I., Ejlertsson, J., Wollan, H., Handå, A. and Malzahn, A. M. 2019a. Potential of the polychaete *Hediste diversicolor* fed on aquaculture and biogas side streams as an aquaculture food source. *Aquaculture Environment Interactions* 11, 551-562. https://doi.org/10.3354/AEI00331.

Wang, H., Seekamp, I., Malzahn, A., Hagemann, A., Carvajal, A. K., Slizyte, R., Standal, I. B., Handå, A. and Reitan, K. I. 2019b. Growth and nutritional composition of the polychaete *Hediste diversicolor* (OF Müller, 1776) cultivated on waste from land-based salmon smolt aquaculture. *Aquaculture* 502, 232–241. https://doi.org/10.1016/j.aquaculture.2018.12.047.

Wang, Y. J., Deering, A. J. and Kim, H. J. 2020. The occurrence of Shiga toxin-producing *E. coli* in aquaponic and hydroponic systems. *Horticulturae* 6(1), 1 6. https://doi.org/10.3390/HORTICULTURAE6010001.

Wei, Y., Li, W., An, D., Li, D., Jiao, Y. and Wei, Q. 2019. Equipment and intelligent control system in aquaponics: a review. *IEEE Access* 7, 169306–169326. https://doi.org/10.1109/ACCESS.2019.2953491.

Wongkiew, S., Hu, Z., Chandran, K., Lee, J. W. and Khanal, S. K. 2017. Nitrogen transformations in aquaponic systems: a review. *Aquacultural Engineering* 76, 9–19. https://doi.org/10.1016/j.aquaeng.2017.01.004.

Xiao, R., Wei, Y., An, D., Li, D., Ta, X., Wu, Y. and Ren, Q. 2019. A review on the research status and development trend of equipment in water treatment processes of recirculating aquaculture systems. *Reviews in Aquaculture* 11(3), 863–895. https://doi.org/10.1111/RAQ.12270.

Xie, K. and Rosentrater, K. 2015. Life cycle assessment (LCA) and Techno-economic analysis (TEA) of tilapia-basil aquaponics. ASABE Annual International Meeting. New Orleans.

Yanes, A. R., Martinez, P. and Ahmad, R. 2020. Towards automated aquaponics: a review on monitoring, IoT, and smart systems. *Journal of Cleaner Production* 263, 121571. https://doi.org/10.1016/J.JCLEPRO.2020.121571.

Yavuzcan Yildiz, H., Radosavljevic, V., Parisi, G. and Cvetkovikj, A. 2019. Insight into risks in aquatic animal health in aquaponics. In: Goddek, S., Joyce, A., Kotzen, B. and Burnell, G. M. (Eds). *Aquaponics Food Production Systems*. Springer International Publishing, Cham, pp. 435–452.

Yavuzcan Yildiz, H., Robaina, L., Pirhonen, J., Mente, E., Domínguez, D. and Parisi, G. 2017. Fish welfare in aquaponic systems: its relation to water quality with an emphasis on feed and faeces–a review. *Water* 9(1), 13. https://doi.org/10.3390/w9010013.

Yousefi-Garakouei, M., Kamali, A. and Soltani, M. 2019. Effects of rearing density on growth, fatty acid profile and bioremediation ability of polychaete *Nereis diversicolor* in an integrated aquaculture system with rainbow trout (*Oncorhynchus mykiss*). *Aquaculture Research* 50, 725–735. https://doi.org/10.1111/are.13918.

Ytrestøyl, T., Takle, H., Kolarevic, J., Calabrese, S., Timmerhaus, G., Rosseland, B. O., Teien, H. C., Nilsen, T. O., Handeland, S. O., Stefansson, S. O., Ebbesson, L. O. E. and Terjesen, B. F. 2020. Performance and welfare of Atlantic salmon, *Salmo salar* L. post-smolts in recirculating aquaculture systems: importance of salinity and water velocity. *Journal of the World Aquaculture Society* 51(2), 373–392. https://doi.org/10.1111/JWAS.12682.

Index

www.ingramcontent.com/pod-product-compliance
Ingram Content Group UK Ltd.
Pitfield, Milton Keynes, MK11 3LW, UK
UKHW021824100926
8556IPUK00003B/150

9 781801 462563